CAMBRIDGE STUDIES IN ADVANCED MATHEMATICS 177

*Editorial Board*
B. BOLLOBÁS, W. FULTON, F. KIRWAN,
P. SARNAK, B. SIMON, B. TOTARO

# FORMAL GEOMETRY AND BORDISM OPERATIONS

This text organizes a range of results in chromatic homotopy theory, running a single thread through theorems in bordism and a detailed understanding of the moduli of formal groups. It emphasizes the naturally occurring algebro-geometric models that presage the topological results, taking the reader through a pedagogical development of the field. In addition to forming the backbone of the stable homotopy category, these ideas have found application in other fields: the daughter subject "elliptic cohomology" abuts mathematical physics, manifold geometry, topological analysis, and the representation theory of loop groups. The common language employed when discussing these subjects showcases their unity and guides the reader breezily from one domain to the next, ultimately culminating in the construction of Witten's genus for String manifolds.

This text is an expansion of a set of lecture notes for a topics course delivered at Harvard University during the spring term of 2016.

**Eric Peterson** works in quantum compilation for near-term supremacy hardware at Rigetti Computing in Berkeley, California. He was previously a Benjamin Peirce Fellow at Harvard University.

CAMBRIDGE STUDIES IN ADVANCED MATHEMATICS

*Editorial Board:*
B. Bollobás, W. Fulton, F. Kirwan, P. Sarnak, B. Simon, B. Totaro

All the titles listed below can be obtained from good booksellers or from Cambridge University Press. For a complete series listing, visit www.cambridge.org/mathematics.

*Already published*
140   R. Pemantle & M. C. Wilson *Analytic Combinatorics in Several Variables*
141   B. Branner & N. Fagella *Quasiconformal Surgery in Holomorphic Dynamics*
142   R. M. Dudley *Uniform Central Limit Theorems* (2nd Edition)
143   T. Leinster *Basic Category Theory*
144   I. Arzhantsev, U. Derenthal, J. Hausen & A. Laface *Cox Rings*
145   M. Viana *Lectures on Lyapunov Exponents*
146   J.-H. Evertse & K. Győry *Unit Equations in Diophantine Number Theory*
147   A. Prasad *Representation Theory*
148   S. R. Garcia, J. Mashreghi & W. T. Ross *Introduction to Model Spaces and Their Operators*
149   C. Godsil & K. Meagher *Erdős–Ko–Rado Theorems: Algebraic Approaches*
150   P. Mattila *Fourier Analysis and Hausdorff Dimension*
151   M. Viana & K. Oliveira *Foundations of Ergodic Theory*
152   V. I. Paulsen & M. Raghupathi *An Introduction to the Theory of Reproducing Kernel Hilbert Spaces*
153   R. Beals & R. Wong *Special Functions and Orthogonal Polynomials*
154   V. Jurdjevic *Optimal Control and Geometry: Integrable Systems*
155   G. Pisier *Martingales in Banach Spaces*
156   C. T. C. Wall *Differential Topology*
157   J. C. Robinson, J. L. Rodrigo & W. Sadowski *The Three-Dimensional Navier–Stokes Equations*
158   D. Huybrechts *Lectures on K3 Surfaces*
159   H. Matsumoto & S. Taniguchi *Stochastic Analysis*
160   A. Borodin & G. Olshanski *Representations of the Infinite Symmetric Group*
161   P. Webb *Finite Group Representations for the Pure Mathematician*
162   C. J. Bishop & Y. Peres *Fractals in Probability and Analysis*
163   A. Bovier *Gaussian Processes on Trees*
164   P. Schneider *Galois Representations and ($\varphi$, $\Gamma$)-Modules*
165   P. Gille & T. Szamuely *Central Simple Algebras and Galois Cohomology* (2nd Edition)
166   D. Li & H. Queffelec *Introduction to Banach Spaces, I*
167   D. Li & H. Queffelec *Introduction to Banach Spaces, II*
168   J. Carlson, S. Müller-Stach & C. Peters *Period Mappings and Period Domains* (2nd Edition)
169   J. M. Landsberg *Geometry and Complexity Theory*
170   J. S. Milne *Algebraic Groups*
171   J. Gough & J. Kupsch *Quantum Fields and Processes*
172   T. Ceccherini-Silberstein, F. Scarabotti & F. Tolli *Discrete Harmonic Analysis*
173   P. Garrett *Modern Analysis of Automorphic Forms by Example, I*
174   P. Garrett *Modern Analysis of Automorphic Forms by Example, II*
175   G. Navarro *Character Theory and the McKay Conjecture*
176   P. Fleig, H. P. A. Gustafsson, A. Kleinschmidt & D. Persson *Eisenstein Series and Automorphic Representations*
177   E. Peterson *Formal Geometry and Bordism Operators*
178   A. Ogus *Lectures on Logarithmic Algebraic Geometry*
179   N. Nikolski *Hardy Spaces*

# Formal Geometry and Bordism Operations

ERIC PETERSON
*Harvard University, Massachusetts*

CAMBRIDGE
UNIVERSITY PRESS

# CAMBRIDGE
UNIVERSITY PRESS

University Printing House, Cambridge CB2 8BS, United Kingdom

One Liberty Plaza, 20th Floor, New York, NY 10006, USA

477 Williamstown Road, Port Melbourne, VIC 3207, Australia

314–321, 3rd Floor, Plot 3, Splendor Forum, Jasola District Centre,
New Delhi – 110025, India

79 Anson Road, #06–04/06, Singapore 079906

Cambridge University Press is part of the University of Cambridge.

It furthers the University's mission by disseminating knowledge in the pursuit of
education, learning, and research at the highest international levels of excellence.

www.cambridge.org
Information on this title: www.cambridge.org/9781108428033
DOI: 10.1017/9781108552165

© Eric Peterson 2019

This publication is in copyright. Subject to statutory exception
and to the provisions of relevant collective licensing agreements,
no reproduction of any part may take place without the written
permission of Cambridge University Press.

First published 2019

Printed in the United Kingdom by TJ International Ltd. Padstow Cornwall

*A catalogue record for this publication is available from the British Library*

Library of Congress Cataloging-in-Publication data
Names: Peterson, Eric, 1987- author.
Title: Formal geometry and bordism operations / Eric Peterson (Rigetti Computing,
Berkeley, California).
Description: Cambridge ; New York, NY : Cambridge University Press, [2019] |
Series: Cambridge studies in advanced mathematics ; 177
Identifiers: LCCN 2018034169 | ISBN 9781108428033 (hardback : alk. paper)
Subjects: LCSH: Topological manifolds. | Manifolds (Mathematics) | Geometry,
Algebraic. | Boundary value problems.
Classification: LCC QA613.2 .P48 2019 | DDC 514/.2–dc23
LC record available at https://lccn.loc.gov/2018034169

ISBN 978-1-108-42803-3 Hardback

Cambridge University Press has no responsibility for the persistence or accuracy of
URLs for external or third-party internet websites referred to in this publication
and does not guarantee that any content on such websites is, or will remain,
accurate or appropriate.

Let us be glad we don't work in algebraic geometry.
  J. F. Adams [Ada78b, Section 2.1]

# Contents

| | | |
|---|---|---|
| *Foreword by Matthew Ando* | | *page* ix |
| *Preface* | | xi |
| **Introduction** | | **1** |
| | Conventions | 7 |
| **1** | **Unoriented Bordism** | **9** |
| | 1.1 Thom Spectra and the Thom Isomorphism | 10 |
| | 1.2 Cohomology Rings and Affine Schemes | 16 |
| | 1.3 The Steenrod Algebra | 22 |
| | 1.4 Hopf Algebra Cohomology | 31 |
| | 1.5 The Unoriented Bordism Ring | 39 |
| **2** | **Complex Bordism** | **49** |
| | 2.1 Calculus on Formal Varieties | 51 |
| | 2.2 Divisors on Formal Curves | 60 |
| | 2.3 Line Bundles Associated to Thom Spectra | 65 |
| | 2.4 Power Operations for Complex Bordism | 73 |
| | 2.5 Explicitly Stabilizing Cyclic $MU$–Power Operations | 83 |
| | 2.6 The Complex Bordism Ring | 91 |
| **3** | **Finite Spectra** | **97** |
| | 3.1 Descent and the Context of a Spectrum | 99 |
| | 3.2 The Structure of $\mathcal{M}_{\mathbf{fg}}$ I: The Affine Cover | 111 |
| | 3.3 The Structure of $\mathcal{M}_{\mathbf{fg}}$ II: Large Scales | 121 |
| | 3.4 The Structure of $\mathcal{M}_{\mathbf{fg}}$ III: Small Scales | 132 |
| | 3.5 Nilpotence and Periodicity in Finite Spectra | 140 |
| | 3.6 Chromatic Fracture and Convergence | 154 |

| | | |
|---|---|---|
| **4** | **Unstable Cooperations** | **165** |
| | 4.1 Unstable Contexts and the Steenrod Algebra | 167 |
| | 4.2 Algebraic Mixed Unstable Cooperations | 178 |
| | 4.3 Unstable Cooperations for Complex Bordism | 188 |
| | 4.4 Dieudonné Modules | 194 |
| | 4.5 Ordinary Cooperations for Landweber Flat Theories | 203 |
| | 4.6 Cooperations among Geometric Points on $\mathcal{M}_{\mathbf{fg}}$ | 211 |
| **5** | **The $\sigma$-Orientation** | **219** |
| | 5.1 Coalgebraic Formal Schemes | 220 |
| | 5.2 Special Divisors and the Special Splitting Principle | 226 |
| | 5.3 Chromatic Analysis of $BU\langle 6, \infty\rangle$ | 238 |
| | 5.4 Analysis of $BU\langle 6, \infty\rangle$ at Infinite Height | 247 |
| | 5.5 Modular Forms and $MU\langle 6, \infty\rangle$-Manifolds | 256 |
| | 5.6 Chromatic *Spin* and *String* Orientations | 270 |
| *Appendix A* | **Power Operations** | **283** |
| | A.1 Rational Chromatic Phenomena (Nathaniel Stapleton) | 285 |
| | A.2 Orientations and Power Operations | 307 |
| | A.3 The Spectrum of Modular Forms | 322 |
| | A.4 Orientations by $E_\infty$ Maps | 333 |
| *Appendix B* | **Loose Ends** | **349** |
| | B.1 Historical Retrospective (Michael Hopkins) | 349 |
| | B.2 The Road Ahead | 352 |
| | *References* | 379 |
| | *Index* | 401 |

# Foreword

This book does a remarkable job of introducing some of the interactions between algebraic topology and algebraic geometry, which these days – thanks to Doug Ravenel – goes by the name "chromatic stable homotopy theory."

Chromatic homotopy theory had its origins in the work of Novikov and Quillen, who first investigated the relationship between complex cobordism and formal groups and perceived its potential for investigating the stable homotopy groups of spheres. That this works as well as it does still boggles my mind, and there is still research to be done to understand why this is so (for example, the recent work of Beardsley, mentioned in Appendix B).

We are fortunate that Jack Morava perceived that the work of Novikov and Quillen hinted at a deep relationship between the structure of the stable category and the structure of the stack of formal groups, and that he was persuasive enough to get others, including Landweber, Miller, Ravenel, and Wilson, excited about this approach. The remarkable activity that followed culminated in Ravenel's periodicity conjectures and their resolution by Devinatz, Hopkins, and Smith.

The chromatic homotopy theorists of the 1970s took Morava's vision as inspiration and proved amazing results, but in their published work they usually did not make use of modern algebro-geometric methods, such as the theory of stacks, which was more or less simultaneously under development. (Although in "Forms of $K$-theory" [Mor89], which dates as far back as 1973, Morava sketches a stacky proof of Landweber's exact functor theorem.)

Around 1990, the study of chromatic stable homotopy underwent a qualitative change. Mike Hopkins took the lead in showing that algebraic geometry and the theory of stacks provide powerful tools for proving theorems in chromatic homotopy theory; at the same time, the conceptual picture of the subject became much simpler. The simplifications that resulted made it possible for many more people, including me, to enter the subject.

I was fortunate to have Haynes Miller and Mike Hopkins as teachers. I was also fortunate to have Adams's "blue book" [Ada95] and Ravenel's "green book" [Rav86]. Until recently, students entering the subject since the 1990s have not had access to comparable sources that introduce them to the mix of algebraic topology and algebraic geometry that form the context for modern chromatic stable homotopy theory (although Strickland's lovely "Formal Schemes and Formal Groups" [Str99b] is a notable exception). This has begun to change, and there are several expositions of aspects of the subject – the list in the Preface is a good starting point.

This brings me to this book. I had the good fortune to meet Eric as an undergraduate and convince him to work on some problems I was interested in. The things that make Eric fun to work with are well reflected in this book. It has a down-to-earth and inviting style (no small achievement in a book about functorial algebraic geometry). It is elegant, precise, and incisive, and it is strong on both theory and calculation. An important feature of the book is that it takes the time to give elegant proofs of some "theory-external" results: theorems you might care about even if chromatic stable homotopy theory isn't your subject.

There is a huge amount yet to be discovered: Appendix B indicates some possible directions for future research. It is great to see this material assembled here to help the next generation of researchers get started on an exciting subject.

> Matthew Ando
> October, 2017

# Preface

This book owes an incredible amount to a number of incredible people. Understanding the research program summarized here has been one of the primary pursuits – if not *the* primary pursuit – of my young academic career. It has been an unmistakable and enormous privilege not only to be granted the time and freedom to learn about these beautiful ideas, but also to learn directly from their progenitors. I owe very large debts to Matthew Ando, Michael Hopkins, and Neil Strickland, for each having shown me such individual attention and care, as well as for having worked out the tail of this long story. Matt, in particular, is the person who got me into higher mathematics, and I feel that for a decade now I have been paying forward the good will and deep friendliness he showed me during my time at Urbana–Champaign. Mike and Neil are not far behind. Mike has been my supervisor in one sense or another for years running, in which role he has been continually encouraging and giving. Among other things, Neil shared with me a note of his that eventually blossomed into my thesis problem, which is an awfully nice gift to have given.

Less directly in ideas but no less directly in stewardship, I also owe a very large debt to my Ph.D. adviser, Constantin Teleman. By the time I arrived at UC Berkeley, I was already too soaked through with homotopy theory to develop a flavor for his sort of mathematical physics, and he nonetheless endeavored to meet me where I was. It was Constantin who encouraged me to focus special attention on making these ideas accessible, speaking understandably, and highlighting the connections with nearby fields. He emphasized that mathematics done in isolation, rather than in maximal connection to other people, is mathematics wasted. He has very much contributed to my passion for communication and clarity, which – in addition to the literal mathematics presented here – is the main goal of this text.[1,2]

More broadly, the topology community has been very supportive of me as I have learned about, digested, and sometimes erroneously recapitulated the ideas of chromatic homotopy theory, tolerating me as a very loud and public learner. Haynes Miller, Doug Ravenel, and Steve Wilson (the *BP* Mafia [Hop08]) have all been invaluable resources: They have answered my questions tirelessly, they are each charming and friendly, they are prolific and meticulous authors, and they literally invented this subfield of mathematics. Jack Morava has played no smaller a role in both the discovery of chromatic homotopy theory and my own personal education. It has been an incredible treat to know him and to have received pushes from him at critical moments. Nat Stapleton and Charles Rezk also deserve special mention: power operations were among the last things that I managed to understand while writing this book, and it is an enormous credit first to their intelligence that they are so comfortable with something so bottomlessly complicated, and second to their inexhaustible patience that they walked me through understanding this material time and time again, in the hopes that I would someday get it.

The bulk of this book began as a set of lecture notes for a topics course that I was invited to teach at Harvard University in the spring term of 2016[3], and I would like to thank the department for the opportunity and for the very enriching time that I spent there. The *germ* of these notes, however, took root at the workshop *Flavors of Cohomology*, organized and hosted by Hisham Sati in June 2015. In particular, this was the first time that I tried to push the idea of a "context" on someone else, which – for better or worse – has grown into the backbone of this book. The book also draws on feedback from lecturing in the *In-Formal Groups Seminar*, which took place during the MSRI semester program in homotopy theory in the spring term of 2014, attended primarily by David Carchedi, Achim Krause, Matthew Pancia, and Sean Tilson. Finally, my thoughts about the material in this book and its presentation were further refined by many, many, *many* conversations with other students at UC Berkeley, primarily: my undergraduate readers Hood Chatham and Geoffrey Lee, the

---

[1] I have a clear memory of delivering a grad student seminar talk during my first year, where my mathematical sibling, George Melvin, asked me why $\widehat{\mathbb{G}}_m$ was called the multiplicative formal group. I looked at him, looked at the board, and cautiously offered, "because of the mixed term in the group law?" Silly as it seems, this book has been significantly shaped by striving to correct for this single inarticulate incident, where it was revealed that I did not understand the original context of these tools that topologists were borrowing. Although everyone starts learning from somewhere and not knowing something is no cause for lasting embarrassment, it is certainly helpful to receive pushes in the direction of intellectual responsibility. Thanks, George.

[2] It is, of course, up to the reader to determine whether I have actually succeeded at this, and, of course, any failure of mine in this regard can't possibly be visited upon Constantin.

[3] MATH 278 (159627), spring 2016.

visiting student Catherine Ray, and my officemates Aaron Mazel-Gee and Kevin Wray – who, poor guys, put up with listening to me for years on end.

This book also draws on a lot of unpublished material. The topology community gets some flak for this reluctance to publish certain documents, but I think it is to our credit that they are made available anyway, essentially without redaction. Reference materials of this sort that have influenced this book include: Matthew Ando's *Dieudonné Crystals Associated to Lubin–Tate Formal Groups*; Michael Hopkins's *Complex Oriented Cohomology Theories and the Language of Stacks*; Jacob Lurie's *Chromatic Homotopy (252x)*; Haynes Miller's *Notes on Cobordism*; Charles Rezk's *Elliptic Cohomology and Elliptic Curves*, as well as his *Notes on the Hopkins–Miller Theorem*, and his *Supplementary Notes for Math 512*; Neil Strickland's *Formal Schemes for K-Theory Spaces* as well as his *Functorial Philosophy for Formal Phenomena*; the Hopf archive preprint version of the Ando–Hopkins–Strickland article *Elliptic Spectra, the Witten Genus, and the Theorem of the Cube*;[4] and the unpublished Ando–Hopkins–Strickland article *Elliptic Cohomology of $BO\langle 8\rangle$ in Positive Characteristic*, recovered from the mists of time by Gerd Laures.[5] I would not have understood the material presented here without access to these resources, never mind the supplementary guidance.

In addition to their inquisitive presences in the lecture hall, the students who took the Harvard topics course under me often contributed directly to the notes. These contributors are: Eva Belmont; Hood Chatham (especially his marvelous spectral sequence package, `spectralsequences`); Dexter Chua (an outside consultant who translated the picture in Figure 3.3 from a scribble on a scrap of paper into something of professional caliber, who provided a mountain of valuable feedback, and who came up with the book's clever epigraph); Arun Debray (a student at UT Austin); Jun Hou Fung; Jeremy Hahn (especially the material in Case Study 2 and Appendix A.4); Mauro Porta; Krishanu Sankar; Danny Shi; and Allen Yuan (especially, again, Appendix A.4, which I may never have tried to understand without his insistence that I speak about it and his further help in preparing that talk).

More broadly, the following people contributed to the course just by attending, where I have highlighted those who additionally survived to the end of the semester: Colin Aitken, Adam Al-Natsheh, **Eva Belmont**, Jason Bland, Dorin Boger, **Lukas Brantner**, **Christian Carrick**, **Hood Chatham**, David Corwin,

---

[4] This earlier version contains a lot of information that didn't make it to publication, as the referee (perhaps rightly) found it too dense to make head or tail of. Once the reader already has the sketch of the argument established, however, the original version is a truly invaluable resource to go back and re-read.

[5] There is a bit of a funny story here: None of the authors of this article could find their own preprint, but a graduate student had held on to a paper copy. In their defense, two decades had passed – but that in turn only makes Gerd's organizational skills more heroic.

**Jun Hou Fung**, **Meng Guo**, **Jeremy Hahn**, Changho Han, Chi-Yun Hsu, **Erick Knight**, Benjamin Landon, Gabriel Long, Yusheng Luo, Jake Marcinek, **Jake McNamara**, **Akhil Mathew**, Max Menzies, Morgan Opie, Alexander Perry, Mauro Porta, **Krishanu Sankar**, **Jay Shah**, Ananth Shankar, **Danny Shi**, Koji Shimizu, Geoffrey Smith, Hunter Spink, Philip Tynan, Yi Xie, David Yang, Zijian Yao, Lynnelle Ye, Chenglong Yu, **Allen Yuan**, Adrian Zahariuc, Yifei Zhao, Rong Zhou, and Yihang Zhu. Their energy and enthusiasm were overwhelming – I felt duty bound to keep telling them things they didn't already know, and despite my best efforts to keep out ahead I also felt like they were constantly nipping at my heels. As I've gone through my notes during the editing process, it has been astonishing to see how reliably they asked exactly the right question at exactly the right time, often despite my own confusion. They're a *very* bright group. Of the highlighted names, Erick Knight deserves special mention: He was an arithmetic geometer living among the rest of us topologists, and he attentively listened to me butcher his native field without once making me feel self-conscious.[6]

On top of the students, various others have contributed in this way or that during the long production of this book, from repairing typos to long conversations and between. Such helpful people include: Tobias Barthel, Tilman Bauer, Jonathan Beardsley, Martin Bendersky, Sanath Devalapurkar, Ben Gadoua, Joe Harris, Mike Hill, Nitu Kitchloo, Johan Konter, Achim Krause, Akhil Mathew, Pedro Mendes de Araujo, Denis Nardin, Justin Noel, Sune Precht Reeh, Emily Riehl, Andrew Senger, Robert Smith, Reuben Stern, Sean Tilson, Dylan Wilson, and Steve Wilson. Of course, Matt Ando, Mike Hopkins, and Nat Stapleton deserve further special mention as direct contributors of their respective sections. Additionally, the referees and the publishers themselves have been enormously helpful in editing this into a passable document.

Lastly, but by far most importantly, this book – and, frankly, *I* – would not have made it out of the gates without Samrita Dhindsa's love, support, and patience. She made living in Boston a completely different experience: a balanced life instead of being quickly and totally overwhelmed by work, a lively circle of friends instead of what would have been a much smaller world, new experiences instead of worn-through ones, and love throughout. Without her compassion, tenderness, and understanding I would not be half the open and vibrant person that I am today, and I would know so much less of myself. It is awe-inspiring to think about, and it is a pleasure and an honor to acknowledge her like this and to dedicate this book to her.

Thanks to my many friends here, and thanks also to Thomas Dunning especially. Thanks, everyone. Theveryone.

---

[6] Between him and Dorin, I do wonder if there's a Fossey-esque novel in the works: *Topologists in the Mists*. I hope we came across as gentle creatures, uncannily similar to proper mathematicians once they got to know us.

# Introduction

The goal of this book is to communicate a certain *Weltanschauung* uncovered in pieces by many different people working in bordism theory, and the goal just for this introduction is to tell a story about one theorem where it is especially apparent.

To begin, we will define a homology theory called *bordism homology*. Recall that the singular homology of a space $X$ comes about by probing $X$ with simplices: Beginning with the collection of continuous maps $\sigma \colon \Delta^n \to X$, we take the free $\mathbb{Z}$-module on each of these sets and construct a chain complex

$$\cdots \xrightarrow{\partial} \mathbb{Z}\{\Delta^n \to X\} \xrightarrow{\partial} \mathbb{Z}\{\Delta^{n-1} \to X\} \xrightarrow{\partial} \cdots .$$

Bordism homology is constructed analogously, but using manifolds $Z$ as the probes instead of simplices:[1]

$$\cdots \xrightarrow{\partial} \{Z^n \to X \mid Z^n \text{ a compact } n\text{-manifold}\}$$
$$\xrightarrow{\partial} \{Z^{n-1} \to X \mid Z^{n-1} \text{ a compact } (n-1)\text{-manifold}\}$$
$$\xrightarrow{\partial} \cdots .$$

**Lemma 1** ([Koc78, section 4]) *This forms a chain complex of monoids under disjoint union of manifolds, and its homology is written $MO_*(X)$. These are naturally abelian groups,[2] and moreover they satisfy the axioms of a generalized homology theory.* □

In fact, we can define a bordism theory $MG$ for any suitable family of structure

---

[1] One does not need to take the free abelian group on anything, since the disjoint union of two manifolds is already a (disconnected) manifold, whereas the disjoint union of two simplices is not a simplex.

[2] For instance, the inverse map comes from the cylinder construction: For a manifold $M$, the two components of $\partial(I \times M)$ witness the existence of an inverse to $M$ in the bordism groups.

groups $G(n) \to O(n)$. The coefficient ring of $MG$, or its value $MG_*(*)$ on a point, gives the ring of $G$-bordism classes, and generally $MG_*(Y)$ gives a kind of "bordism in families over the space $Y$." There are comparison morphisms for the most ordinary kinds of bordism, given by replacing a chain of manifolds with an equivalent simplicial chain:

$$MO \to H\mathbb{Z}/2, \qquad MSO \to H\mathbb{Z}.$$

In both cases, we can evaluate on a point to get ring maps, called *genera*:

$$MO_*(*) \to \mathbb{Z}/2, \qquad MSO_*(*) \to \mathbb{Z},$$

neither of which is very interesting, since they are both zero in positive degrees.

However, having maps of homology theories (or, really, of spectra) is considerably more data than just the genus. For instance, we can use such a map to extract a theory of integration by considering the following special case of oriented bordism, where we evaluate $MSO_*$ on an infinite loopspace:

$$MSO_n K(\mathbb{Z}, n) = \{\text{oriented } n\text{-manifolds mapping to } K(\mathbb{Z}, n)\} / \sim$$
$$= \left\{ \begin{array}{c} \text{oriented } n\text{-manifolds } Z \\ \text{with a specified class } \omega \in H^n(Z; \mathbb{Z}) \end{array} \right\} \Big/ \sim .$$

Associated to such a representative $(Z, \omega)$, the yoga of stable homotopy theory then allows us to build a composite

$$\mathbb{S} \xrightarrow{(Z,\omega)} MSO \wedge (\mathbb{S}^{-n} \wedge \Sigma^\infty_+ K(\mathbb{Z}, n))$$
$$\xrightarrow{\text{colim}} MSO \wedge H\mathbb{Z}$$
$$\xrightarrow{\varphi \wedge 1} H\mathbb{Z} \wedge H\mathbb{Z}$$
$$\xrightarrow{\mu} H\mathbb{Z},$$

where $\varphi$ is the orientation map. Altogether, this composite gives us an element of $\pi_0 H\mathbb{Z}$, i.e., an integer.

**Lemma 2** *The integer obtained by the above process is $\int_Z \omega$.* □

Many theorems accompany this definition of $\int_Z \omega$ for free, entailed by the general machinery of stable homotopy theory. The definition is also very general: Given a ring map off of any bordism spectrum, a similar sequence of steps will furnish us with an integral tailored to that situation.

In the case of the trivial structure group $G = e$, this construction gives the bordism theory of stably framed manifolds, and the Pontryagin–Thom theorem amounts to an equivalence $\mathbb{S} \xrightarrow{\simeq} Me$. Through this observation, these techniques gain widespread application in stable homotopy theory. For a ring spectrum $E$, we can reconsider the unit map as a ring map

$$Me \xrightarrow{\simeq} \mathbb{S} \to E,$$

and by following the same path of ideas we learn that $E$ is therefore equipped with a theory of integration for framed manifolds:

$$\int : \left\{ \begin{array}{c} \text{stably framed } n\text{-manifolds } Z \\ \text{with a specified class } \omega \in E^{n-m}(Z) \end{array} \right\}\bigg/_\sim \to \pi_m E.$$

Sometimes, as in the examples above, this unit map factors:

$$\mathbb{S} \simeq Me \to MO \to H\mathbb{Z}/2.$$

This is a witness to the overdeterminacy of $H\mathbb{Z}/2$'s integral for framed bordism: If the framed manifold is pushed all the way down to an unoriented manifold, there is still enough residual data to define the integral.[3] Given any ring spectrum $E$, we can ask the analogous question: If we filter $O$ by a decreasing system of structure groups, through what stage does the unit map $Me \to E$ factor? For instance, the map

$$\mathbb{S} = Me \to MSO \to H\mathbb{Z}$$

considered above does *not* factor further through $MO$ – an orientation is *required* to define the integral of an integer-valued cohomology class. Recognizing $SO \to O$ as the zeroth Postnikov–Whitehead truncation of $O$, we are inspired to use the rest of the Postnikov filtration as our filtration of structure groups. Here is a diagram of this filtration and some interesting minimally factored integration theories related to it, circa 1970:

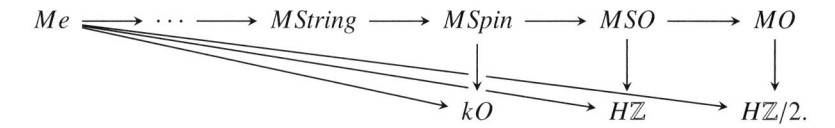

$$\begin{array}{ccccccccc} Me & \Longrightarrow & \cdots & \longrightarrow & M\text{String} & \longrightarrow & M\text{Spin} & \longrightarrow & MSO & \longrightarrow & MO \\ & & & & & & \downarrow & & \downarrow & & \downarrow \\ & & & & & & kO & \longrightarrow & H\mathbb{Z} & \longrightarrow & H\mathbb{Z}/2. \end{array}$$

This is the situation homotopy theorists found themselves in some decades ago, when Ochanine and Witten proved the following mysterious theorem using analytical and physical methods:

**Theorem 3** (Ochanine [Och87b, Och91], Witten [Wit87, Wit88]) *There is a map of rings*

$$\sigma : M\text{Spin}_* \to \mathbb{C}((q)).$$

---

[3] It is literally more information than this – even unframeable unoriented manifolds acquire a compatible integral.

*Moreover, if Z is a Spin manifold such that half[4] its first Pontryagin class vanishes – that is, if Z lifts to a String manifold – then $\sigma(Z)$ lands in the subring $MF \subseteq \mathbb{Z}[\![q]\!]$ of q-expansions of modular forms with integral coefficients.* □

However, neither party gave indication that their result should be valid "in families," and no attendant theory of integration was formally produced. From the perspective of a homotopy theorist, it was not clear what such a claim would mean: To give a topological enrichment of these theorems would mean finding a ring spectrum $E$ such that $E_*(*)$ had something to do with modular forms.

Around the same time, Landweber, Ravenel, and Stong began studying *elliptic cohomology* for independent reasons [LRS95]; some time much earlier, Morava had constructed an object "$K^{\text{Tate}}$" associated to the Tate elliptic curve [Mor89, section 5]; and a decade later Ando, Hopkins, and Strickland [AHS01] put all these together in the following theorem:

**Theorem 4** ([AHS01, Theorem 2.59])  *If $E$ is an "elliptic cohomology theory," then there is a canonical map of homotopy ring spectra $M String \to E$ called the $\sigma$-orientation (for $E$). Additionally, there is an elliptic spectrum $K^{\text{Tate}}$ whose $\sigma$-orientation gives Witten's genus $M String_* \to K^{\text{Tate}}_*$ .* □

We now come to the motivation for this text: The homotopical $\sigma$-orientation was actually first constructed using formal geometry. The original proof of Ando, Hopkins, and Strickland begins with a reduction to maps of the form

$$MU[6, \infty) \to E.$$

They then work to show that in especially good cases they can complete the missing arrow in the diagram

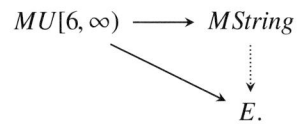

$$MU[6, \infty) \longrightarrow MString$$
$$\searrow \quad \downarrow$$
$$E.$$

Leaving aside the extension problem for the moment, their main theorem is the following description of the cohomology ring $E^*MU[6, \infty)$:

**Theorem 5** (Ando–Hopkins–Strickland [AHS01], see Singer [Sin68] and Stong [Sto63])  *For $E$ an even-periodic cohomology theory, there is an isomorphism*

$$\operatorname{Spec} E_*MU[6, \infty) \cong C^3(\widehat{\mathbb{G}}_E; \mathcal{I}(0)),$$

*where "$C^3(\widehat{\mathbb{G}}_E; \mathcal{I}(0))$" is the affine scheme parametrizing cubical structures on*

---
[4] It is a special property of *Spin* manifolds that this class is always divisible by 2.

*the line bundle* $I(0)$ *over* $\widehat{\mathbb{G}}_E$. *When $E$ is taken to be elliptic, so that there is a specified elliptic curve $C$ and a specified isomorphism* $\widehat{\mathbb{G}}_E \cong C_0^\wedge$, *the theory of elliptic curves gives a canonical such cubical structure and hence a preferred class* $MU[6, \infty) \to E$. *This assignment is natural in the choice of elliptic $E$.* □

Our real goal is to understand theorems like this last one, where algebraic geometry asserts some real control over something squarely in the domain of homotopy theory. In the course of this text, we will work through a sequence of case studies where this perspective shines through most brightly. In particular, rather than taking an optimal route to the Ando–Hopkins–Strickland result, we will use it as a gravitational slingshot to cover many ancillary topics which are also governed by the technology of formal geometry. We will begin by working through Thom's calculation of the homotopy of $MO$, which holds the simultaneous attractive features of being approachable while revealing essentially all of the structural complexity of the general situation, so that we know what to expect later on. Having seen that through, we will then venture on to other examples: the complex bordism ring, structure theorems for finite spectra, unstable cooperations, and, finally, the $\sigma$-orientation and its extensions. Again, the overriding theme of the text will be that algebraic geometry is a good organizing principle that gives us one avenue of insight into how homotopy theory functions: It allows us to organize "operations" of various sorts between spectra derived from bordism theories.

We should also mention that we will specifically *not* discuss the following aspects of this story:

- Analytic techniques will be completely omitted. Much of modern research stemming from the above problem is an attempt to extend index theory across Witten's genus, or to find a "geometric cochains" model of certain elliptic cohomology theories. These often mean heavy analytic work, and we will strictly confine ourselves to the domain of homotopy theory.
- As sort of a subpoint (and despite the motivation provided in this introduction), we will also mostly avoid manifold geometry. Again, much of the contemporary research about *tmf* is an attempt to find a geometric model, so that geometric techniques can be imported – including equivariance and the geometry of quantum field theories, to name two.
- In a different direction, our focus will not linger on actually computing bordism rings $MG_*$, nor will we consider geometric constructions on manifolds and their behavior after imaging into the bordism ring. This is also the source of active research: the structure of the symplectic bordism ring remains, to a large extent, mysterious, and what we do understand of it comes through a mix of formal geometry and raw manifold geometry. This could be a topic

that fits logically into this document, were it not for time limitations and the author's inexpertise.
- The geometry of $E_\infty$-rings will also be avoided, at least to the extent possible. Such objects become inescapable by the conclusion of our story, but there are better resources from which to learn about $E_\infty$-rings, and the pre-$E_\infty$ story is not told so often these days. So, we will focus on the unstructured part, relegate the $E_\infty$-rings to Appendix A, and leave their details to other authors.
- As a related note, much of the contents of this book could be thought of as computational foundations for the derived algebraic geometry of even-periodic ring spectra. We will make absolutely no attempt to set up such a theory here, but we will endeavor to phrase our results in a way that will, hopefully, be forward-compatible with any such theory arising in the future.
- There will be plenty of places where we will avoid making statements in maximum generality or with maximum thoroughness. The story we are interested in telling draws from a blend of many others from different subfields of mathematics, many of which have their own dedicated textbooks. Sometimes this will mean avoiding stating the most beautiful theorem in a subfield in favor of a theorem we will find more useful. Other times this will mean abbreviating someone else's general definition to one more specialized to the task at hand. In any case, we will give references to other sources where you can find these cast in starring roles.

Finally, we must mention that there are several good companions to these notes. Essentially none of the material here is original – it is almost all cribbed either from published or unpublished sources – but the source documents are quite scattered and individually dense. We will make a point to cite useful references as we go. One document stands out above all others, though: Neil Strickland's *Functorial Philosophy for Formal Phenomena* [Strb]. These lecture notes can basically be viewed as an attempt to make it through this paper in the span of a semester.

## Conventions

Throughout this book, we use the following conventions:

- Categories will be consistently typeset as in the examples

    Spaces, FormalGroups, GradedHopfAlgebras.

- C($X, Y$) will denote the mapping set of arrows $X \to Y$ in a category C. If C is an $\infty$-category, this will be interpreted instead as a mapping *space*. If C has a self-enrichment, we will often write $\underline{C}(X, Y)$ (or, e.g., $\underline{Aut}(X)$) to distinguish the internal mapping object from the classical mapping set C($X, Y$). As a first exception to this uniform notation, we will sometimes abbreviate $\underline{Spaces}(X, Y)$ to $F(X, Y)$, and similarly we will sometimes abbreviate $\underline{Spectra}(X, Y)$ to $F(X, Y)$, with "$F$" short for "function." As a second exception, for two formal groups $\widehat{\mathbb{G}}$ and $\widehat{\mathbb{H}}$, we denote the function scheme by

    $$\underline{FormalGroups}(\widehat{\mathbb{G}}, \widehat{\mathbb{H}}),$$

    even though this is a *scheme* rather than a *formal scheme*.

- Following Lurie, for an object $X \in C$ we will write $C_{/X}$ for the slice category of objects *over* $X$ and $C_{X/}$ for the slice category of objects *under* $X$.

- For a spectrum $E$, we will write $E^*(X)$ for the unreduced $E$-cohomology of a space $X$ and $E_*(X)$ for the unreduced $E$-homology of $X$. We denote the reduced $E$-cohomology of a pointed space $X$ by $\widetilde{E}^*(X)$ and the reduced $E$-homology by $\widetilde{E}_*(X)$. Finally, for $F$ another spectrum, we write $E^*(F)$ and $E_*(F)$ for the $E$-cohomology and $E$-homology respectively of $F$. Altogether, these satisfy the relations

    $$E_*(X) = E_*(\Sigma_+^\infty X) = E_*(\Sigma^\infty X) \oplus E_* = \widetilde{E}_*(X) \oplus E_*,$$

    and similarly for cohomology.

- For a spectrum $E$, we will write $\underline{E}_n$ for the $n$th space in the $\Omega$-spectrum representing $E$. The homotopy type of this space is determined by the formula

    $$h\text{Spaces}(X, \underline{E}_n) = h\text{Spaces}(X, \Omega^\infty \Sigma^n E) = h\text{Spaces}(\Sigma^\infty X, \Sigma^n E) = \widetilde{E}^n(X).$$

- For a ring spectrum $E$, we will write $E_* = \pi_* E$ for its coefficient ring, $E^* = \pi_{-*} E$ for its coefficient ring with the opposite grading, and $E_0 = E^0 = \pi_0 E$ for the zeroth degree component of its coefficient ring. This allows us to make sense of expressions like $E^* [\![ x ]\!]$, which we interpret as

    $$E^* [\![ x ]\!] = (E^*)[\![ x ]\!] = (\pi_{-*} E)[\![ x ]\!] = \left\{ \sum_{j=0}^\infty a_j x^j \,\middle|\, \begin{array}{l} a_j \text{ is of degree } * - j|x| \\ \text{for some fixed degree } * \end{array} \right\}.$$

- For a space or spectrum, we will write $X[n, \infty) \to X$ for the upward $n$th Postnikov truncation over $X$ and $X \to X(-\infty, n)$ for the downward $n$th Postnikov truncation under $X$. There is thus a natural fiber sequence

$$X[n, \infty) \to X \to X(-\infty, n).$$

This notation extends naturally to objects like $X(a, b)$ or $X[a, b]$, where $-\infty \le a \le b \le \infty$ denote the (closed or open) endpoints of any interval.
- We will write $S^n$ for the $n$th sphere when considered as a space and $\mathbb{S}^n$ for its suspension spectrum. We will often abbreviate $\mathbb{S}^0$ to simply $\mathbb{S}$.
- We prefer the notation $O_X$ for the ring of functions on a scheme $X$ and $\mathcal{I}_D$ for ideal sheaf determined by a subscheme $D$, but we will also denote these by the synonyms $O(X)$ and $\mathcal{I}(D)$ when the subscripts reach sufficient complexity.
- We write $KO$ and $KU$ for periodic real and complex $K$-theory, and we write $kO$ and $kU$ for their respective connective variants. (Other authors write $ko$ and $ku$, or $bo$ and $bu$, or even the ill-advised $BO$ and $BU$ for these spectra.)
- We primarily treat 2-periodic spectra, though "in the wild" many of the spectra we consider here are taken to have lower periodicity (e.g., $E(d)$ is typically taken to have periodicity $2(p^d - 1)$) or no periodicity at all (e.g., the ordinary homology spectrum $H\mathbb{F}_2$). Where confusion might otherwise arise, we have done our best to insert a "$P$" into the names of our standard spectra as a clear indication that we are speaking about the 2-periodic version.

In all these cases, I have done my best to be absolutely consistent in these regards, and I apologize profusely for any erratic typesetting that might have slipped through.

# Case Study 1

## Unoriented Bordism

A simple observation about the bordism ring $MO_*(*)$ (or $MO_*(X)$ more generally, for any space $X$) is that it consists entirely of 2-torsion: Any chain $Z \to X$ can be bulked out to a constant cylinder $Z \times I \to X$, which has as its boundary the chain $2 \cdot (Z \to X)$. Accordingly, $MO_*(X)$ is always an $\mathbb{F}_2$-vector space. Our goal in this case study is to arrive at two remarkable calculations: First, in Corollary 1.5.7 we will make an explicit calculation of this $\mathbb{F}_2$-vector space in the case of the bordism homology of a point; and second, in Lemma 1.5.8 we will show that there is a natural isomorphism

$$MO_*(X) = H\mathbb{F}_{2*}(X) \otimes_{\mathbb{F}_2} MO_*(*).$$

Our goal in discussing these results in the first case study of the book is to take the opportunity to introduce several key concepts that will serve us throughout. First and foremost, we will require a definition of bordism spectrum that we can manipulate computationally, using just the tools of abstract homotopy theory. Once that is established, we immediately begin to bring algebraic geometry into the mix: The main idea is that the cohomology ring of a space is better viewed as a scheme (with plenty of extra structure), and the homology groups of a spectrum are better viewed as a representation for a certain elaborate algebraic group. This data actually finds familiar expression in homotopy theory: We show that a form of group cohomology for this representation forms the input to the classical Adams spectral sequence, which classically takes the form

$$\mathrm{Cotor}^{*,*}_{\mathcal{A}_*}(\mathbb{F}_2, H\mathbb{F}_{2*}(Y)) \Rightarrow \pi_*(Y),$$

converging for certain very nice spectra $Y$ – including, for instance, $Y = MO$. In particular, we can bring the tools from the preceding discussion to bear on the homology and cohomology of $MO$, where we make an explicit calculation of its representation structure. Finding that it is suitably free, we thereby gain control of the Adams spectral sequence, finish the computation, and prove the desired result.

Our *real* goal in this case study, however, is to introduce one of the main phenomena guiding this text: There is some governing algebro-geometric object, the formal group $\mathbb{RP}^\infty_{H\mathbb{F}_2}$, which exerts an extraordinary amount of control over everything in sight. We will endeavor to rephrase as much of this classical computation as possible so as to highlight its connection to this central object, and we will use this as motivation in future case studies to pursue similar objects, which will lead us down much deeper and more rewarding rabbit holes. The counterbalance to this is that, at least for now, we will not introduce concepts or theorems in their maximum generality.[1] Essentially everything mentioned in this case study will be re-examined more thoroughly in future case studies, so the reader is advised to look to those for the more expansive set of results.

## 1.1 Thom Spectra and the Thom Isomorphism

Our goal is a sequence of theorems about the unoriented bordism spectrum $MO$. We will begin by recalling a definition of the spectrum $MO$ using just abstract homotopy theory, because it involves ideas that will be useful to us throughout the text and because we cannot compute effectively with the chain-level definition given in the Introduction.

**Definition 1.1.1** For a spherical bundle $S^{n-1} \to \xi \to X$, its *Thom space* is given by the cofiber

$$\xi \to X \xrightarrow{\text{cofiber}} T_n(\xi).$$

*"Proof" of definition* There is a more classical construction of the Thom space: Take the associated disk bundle by gluing an $n$-disk fiberwise, and add a point at infinity by collapsing $\xi$:

$$T_n(\xi) = (\xi \cup_{X \times S^{n-1}} (X \times D^n))^+.$$

To compare this with the cofiber definition, recall that the thickening of $\xi$ to an $n$-disk bundle is the same as taking the mapping cylinder on $\xi \to X$. Since the inclusion into the mapping cylinder is now a cofibration, the quotient by this subspace agrees with both the cofiber of the map and the introduction of a point at infinity. □

Before proceeding, here are two important examples:

---

[1] For an obvious example, everything in this case study will be done relative to $\operatorname{Spec} \mathbb{F}_2$.

## 1.1 Thom Spectra and the Thom Isomorphism

*Example 1.1.2* If $\xi = S^{n-1} \times X$ is the trivial bundle, then $T_n(\xi) = S^n \wedge (X_+)$. This is supposed to indicate what Thom spaces are "doing": if you feed in the trivial bundle then you get the suspension out, so if you feed in a twisted bundle you should think of it as a *twisted suspension*.

*Example 1.1.3* Let $\xi$ be the tautological $S^0$-bundle over the unpointed space $\mathbb{RP}^\infty = BO(1)$. Because $\xi$ has contractible total space, $EO(1)$, the cofiber degenerates and it follows that $T_1(\xi) = \mathbb{RP}^\infty$, a pointed space.[2] More generally, arguing by cells shows that the Thom space for the tautological bundle over $\mathbb{RP}^n$ is $\mathbb{RP}^{n+1}$.

Now we catalog a bunch of useful properties of the Thom space construction. First, recall that a spherical bundle over $X$ is the same data [May75] as a map $X \to BGL_1 S^{n-1}$, where $GL_1 S^{n-1}$ is the subspace of $F(S^{n-1}, S^{n-1})$ expressed by the pullback of spaces

$$\begin{array}{ccc}
GL_1 S^{n-1} & \longrightarrow & \text{Spaces}(S^{n-1}, S^{n-1}) \\
\downarrow & & \downarrow \\
h\text{Spaces}(S^{n-1}, S^{n-1})^\times & \to & h\text{Spaces}(S^{n-1}, S^{n-1}) = \pi_0 \text{Spaces}(S^{n-1}, S^{n-1}).
\end{array}$$

**Lemma 1.1.4** *The construction $T_n$ can be viewed as a functor from the slice category over $BGL_1 S^{n-1}$ to* Spaces. *Maps of slices*

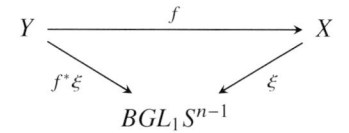

$$\begin{array}{ccc}
Y & \xrightarrow{f} & X \\
& \searrow{f^*\xi} \quad \swarrow{\xi} & \\
& BGL_1 S^{n-1} &
\end{array}$$

*induce maps $T_n(f^*\xi) \to T_n(\xi)$, and $T_n$ is suitably homotopy-invariant.* □

Next, the spherical subbundle of a vector bundle gives a common source of spherical bundles. The action of $O(n)$ on $\mathbb{R}^n$ preserves the unit sphere, and hence gives a map $O(n) \to GL_1 S^{n-1}$. These are maps of topological groups, and the block-inclusion maps $i^n: O(n) \to O(n+1)$ commute with the suspension map $GL_1 S^{n-1} \to GL_1 S^n$. In fact, much more can be said:

**Lemma 1.1.5** *The block-sum maps $O(n) \times O(m) \to O(n+m)$ are compatible with the join maps $GL_1 S^{n-1} \times GL_1 S^{m-1} \to GL_1 S^{n+m-1}$.* □

Taking a cue from $K$-theory, we pass to the colimit as $n$ grows large, using the maps

---

[2] If you already know what's coming, this should comport with the Thom isomorphism, which asserts $x \cdot H\mathbb{F}_2^* \mathbb{RP}^\infty \cong \widetilde{H\mathbb{F}_2}^{*+1} \mathbb{RP}^\infty$.

## 12    Unoriented Bordism

$$\begin{array}{ccccccc}
BO(n) & = & BO(n) \times * & \xrightarrow{\text{id} \times \text{triv}} & BO(n) \times BO(1) & \xrightarrow{\oplus} & BO(n+1) \\
\downarrow & & \downarrow & & \downarrow & & \downarrow \\
BGL_1 S^{n-1} & = & BGL_1 S^{n-1} \times * & \xrightarrow{\text{id} \times \text{triv}} & BGL_1 S^{n-1} \times BGL_1 S^0 & \xrightarrow{*} & BGL_1 S^n.
\end{array}$$

**Corollary 1.1.6** *The operations of block-sum and topological join imbue the colimiting spaces $BO$ and $BGL_1 \mathbb{S}$ with the structure of $H$-groups. Moreover, the colimiting map*

$$J_{\mathbb{R}} \colon BO \to BGL_1 \mathbb{S},$$

*called the J-homomorphism, is a morphism of $H$-groups.* □

Finally, we can ask about the compatibility of Thom constructions with all of this. In order to properly phrase the question, we need a version of the construction which operates on stable spherical bundles, i.e., whose source is the slice category over $BGL_1 \mathbb{S}$. By calculating

$$T_{n+1}(\xi * \text{triv}) \simeq \Sigma T_n(\xi),$$

we are inspired to make the following definition:

**Definition 1.1.7**   For $\xi$ an $S^{n-1}$-bundle, we define the *Thom spectrum* of $\xi$ to be

$$T(\xi) := \Sigma^{-n} \Sigma^{\infty} T_n(\xi).$$

By filtering the base space by compact subspaces, this begets a functor

$$T \colon \text{Spaces}_{/BGL_1 \mathbb{S}} \to \text{Spectra}.$$

**Lemma 1.1.8**   *$T$ is monoidal: It carries external fiberwise joins to smash products of Thom spectra. Correspondingly, $T \circ J_{\mathbb{R}}$ carries external direct sums of stable vector bundles to smash products of Thom spectra.* □

**Definition 1.1.9**   The spectrum $MO$ arises as the universal example of all these constructions, strung together:

$$MO := T(J_{\mathbb{R}}) = \underset{n}{\text{colim}}\, T(J_{\mathbb{R}}^n) = \underset{n}{\text{colim}}\, \Sigma^{-n} T_n J_{\mathbb{R}}^n.$$

The spectrum $MO$ has several remarkable properties. First, it can be shown that the generalized homology theory that this spectrum encodes matches the one described in the Introduction [Swi02, theorem 12.30], [Rud98, theorem 7.27]. The most basic homotopical property is that this spectrum is naturally a ring spectrum, and this follows immediately from $J_{\mathbb{R}}$ being a homomorphism of $H$-spaces. Much more excitingly, we can also deduce the presence of Thom isomorphisms just from the properties stated thus far. That $J_{\mathbb{R}}$ is a homomorphism means that the square in the following diagram commutes:

## 1.1 Thom Spectra and the Thom Isomorphism

$$\begin{array}{ccccc} BO \times BO & \xrightarrow{\sigma} & BO \times BO & \xrightarrow{\mu} & BO \\ & \searrow & \downarrow{\scriptstyle J_{\mathbb{R}} \times J_{\mathbb{R}}} & & \downarrow{\scriptstyle J_{\mathbb{R}}} \\ & & BGL_1\mathbb{S} \times BGL_1\mathbb{S} & \xrightarrow{\mu} & BGL_1\mathbb{S}. \end{array}$$

We have extended this square very slightly by a certain shearing map $\sigma$ defined by $\sigma(x, y) = (xy^{-1}, y)$. It is evident that $\sigma$ is a homotopy equivalence, since just as we can de-scale the first coordinate by $y$ we can re-scale by it – indeed, this is the observation that $BO$ is a torsor for itself. We can calculate directly the behavior of the long composite:

$$J_{\mathbb{R}} \circ \mu \circ \sigma(x, y) = J_{\mathbb{R}} \circ \mu(xy^{-1}, y) = J_{\mathbb{R}}(xy^{-1}y) = J_{\mathbb{R}}(x).$$

It follows that the second coordinate plays no role, and that the bundle classified by the long composite can be written as $J_{\mathbb{R}} \times 0$.[3] We are now in a position to see the Thom isomorphism:

**Lemma 1.1.10** (Thom isomorphism, universal example, see [Mah79]) *As $MO$-modules,*

$$MO \wedge MO \simeq MO \wedge \Sigma_+^\infty BO.$$

*Proof* Stringing together the naturality properties of the Thom functor outlined above, we can thus make the following calculation:

$$\begin{aligned} T(\mu \circ (J_{\mathbb{R}} \times J_{\mathbb{R}})) &\simeq T(\mu \circ (J_{\mathbb{R}} \times J_{\mathbb{R}}) \circ \sigma) & \text{(homotopy invariance)} \\ &\simeq T(\mu \circ (J_{\mathbb{R}} \times 0)) & \text{(constructed lift)} \\ &\simeq T(J_{\mathbb{R}}) \wedge T(0) & \text{(monoidality)} \\ &\simeq T(J_{\mathbb{R}}) \wedge \Sigma_+^\infty BO & \text{(Example 1.1.2)} \\ T(J_{\mathbb{R}}) \wedge T(J_{\mathbb{R}}) &\simeq T(J_{\mathbb{R}}) \wedge \Sigma_+^\infty BO & \text{(monoidality)} \\ MO \wedge MO &\simeq MO \wedge \Sigma_+^\infty BO. & \text{(definition of } MO\text{)} \end{aligned}$$

In order to verify that this equivalence is one of $MO$-modules, one performs an analogous computation with $J_{\mathbb{R}}^{\times 3}$. □

From here, the general version of Thom's theorem follows quickly:

**Definition 1.1.11** A map $\varphi \colon MO \to E$ of homotopy ring spectra is called a *(real) orientation* of $E$ (by $MO$).

---

[3] This factorization does *not* commute with the rest of the diagram, just with the little lifting triangle it forms.

14     Unoriented Bordism

**Theorem 1.1.12** (Thom isomorphism)  *Let $\xi \colon X \to BO$ classify a vector bundle and let $\varphi \colon MO \to E$ be a map of ring spectra. Then there is an equivalence of $E$-modules*

$$E \wedge T(\xi) \simeq E \wedge \Sigma_+^\infty X.$$

*Modifications to above proof*  To accommodate $X$ rather than $BO$ as the base, we redefine $\sigma \colon BO \times X \to BO \times X$ by

$$\sigma(x, y) = \sigma(x\xi(y)^{-1}, y).$$

Follow the same proof as before with the diagram

$$\begin{array}{ccccccc}
BO \times X & \xrightarrow{\sigma}_{\simeq} & BO \times X & \xrightarrow{1 \times \xi} & BO \times BO & \xrightarrow{\mu} & BO \\
 & & & & \downarrow{J_\mathbb{R} \times J_\mathbb{R}} & & \downarrow{J_\mathbb{R}} \\
 & & & & BGL_1\mathbb{S} \times BGL_1\mathbb{S} & \xrightarrow{\mu} & BGL_1\mathbb{S}.
\end{array}$$

This gives an equivalence $\theta_{MO} \colon MO \wedge T(\xi) \to MO \wedge \Sigma_+^\infty X$ of $MO$-modules. To introduce $E$, note that there is a diagram of $MO$-module spectra

$$\begin{array}{ccc}
E \wedge T(\xi) & \xrightarrow{\theta_E} & E \wedge \Sigma_+^\infty X \\
{\scriptstyle \eta_{MO} \wedge 1 \wedge 1}\Big\downarrow & & \Big\uparrow{\scriptstyle \eta_{MO} \wedge 1 \wedge 1} \\
MO \wedge E \wedge T(\xi) & \xrightarrow{\theta_{MO} \wedge 1} & MO \wedge E \wedge \Sigma_+^\infty X \\
{\scriptstyle (\mu \circ (\varphi \wedge 1)) \wedge 1}\Big\downarrow & & \Big\uparrow{\scriptstyle (\mu \circ (\varphi \wedge 1)) \wedge 1} \\
E \wedge T(\xi) & \xrightarrow{\theta_E} & E \wedge \Sigma_+^\infty X,
\end{array}$$

whose columns are retractions. The dotted arrows $\theta_E$ are defined by following the solid path from corner to corner. Similarly, the inverse $\alpha_{MO}$ to $\theta_{MO}$ induces a map $\alpha_E$. We claim that $\theta_E$ and $\alpha_E$ are themselves inverses, which is shown by staggering and rearranging where the collapse of the $MO$ factor into $E$ happens. The composite $\alpha_E \circ \theta_E$ is equivalent to the following:

$$E \wedge \Sigma_+^\infty X \xrightarrow{1 \wedge \eta \wedge 1} E \wedge MO \wedge \Sigma_+^\infty X$$

$$\xrightarrow{1 \wedge 1 \wedge \eta \wedge 1} E \wedge MO \wedge MO \wedge \Sigma_+^\infty X$$

$$\xrightarrow{1 \wedge 1 \wedge \theta_{MO}} E \wedge MO \wedge MO \wedge T(\xi)$$

$$\xrightarrow{1 \wedge \tau \wedge 1} E \wedge MO \wedge MO \wedge T(\xi)$$

## 1.1 Thom Spectra and the Thom Isomorphism

$$\xrightarrow{1\wedge 1\wedge \alpha_{MO}} E\wedge MO\wedge MO\wedge \Sigma^\infty_+ X$$

$$\xrightarrow{\mu\wedge 1\wedge 1} E\wedge MO\wedge \Sigma^\infty_+ X$$

$$\xrightarrow{\mu\wedge 1} E\wedge \Sigma^\infty_+ X,$$

where $\tau$ is the twist map. Using the commuting square

$$\begin{array}{ccc} E\wedge MO\wedge MO & \xrightarrow{1\wedge \mu} & E\wedge MO \\ \downarrow{\mu\wedge 1} & & \downarrow{\mu} \\ E\wedge MO & \xrightarrow{\mu} & E, \end{array}$$

we have that the middle composite is equivalent to

$$E\wedge \Sigma^\infty_+ X \xrightarrow{1\wedge \eta\wedge 1} E\wedge MO\wedge \Sigma^\infty_+ X$$

$$\xrightarrow{1\wedge \theta_{MO}} E\wedge MO\wedge T(\xi)$$

$$\xrightarrow{1\wedge \alpha_{MO}} E\wedge MO\wedge \Sigma^\infty_+ X$$

$$\xrightarrow{\mu\wedge 1} E\wedge \Sigma^\infty_+ X,$$

which is the identity on $E\wedge \Sigma^\infty_+ X$. The proof that $\alpha_E$ is also the right-inverse of $\theta_E$ is similar. □

*Remark 1.1.13* One of the tentpoles of the theory of Thom spectra is that Theorem 1.1.12 has a kind of converse: If a ring spectrum $E$ has suitably natural and multiplicative Thom isomorphisms for Thom spectra formed from real vector bundles, then one can define an essentially unique ring map $MO \to E$ that realizes these isomorphisms via the machinery of Theorem 1.1.12.

*Remark 1.1.14* There is also a cohomological version of the Thom isomorphism. Suppose that $E$ is a ring spectrum under $MO$ and let $\xi$ be the spherical bundle of a vector bundle on a space $X$. The spectrum $F(\Sigma^\infty_+ X, E)$ is a ring spectrum under $E$ (hence under $MO$), so there is a Thom isomorphism as well as an evaluation map

$$F(\Sigma^\infty_+ X, E)\wedge T(\xi) \xrightarrow{\simeq} F(\Sigma^\infty_+ X, E)\wedge \Sigma^\infty_+ X \xrightarrow{\text{eval}} E.$$

Passing through the exponential adjunction, the map

$$F(\Sigma^\infty_+ X, E) \xrightarrow{\simeq} F(T(\xi), E)$$

can be seen to give the cohomological Thom isomorphism

$$E^* X \cong E^* T(\xi).$$

*Example 1.1.15* We will close out this section by using this to actually make a calculation. Recall from Example 1.1.3 that $T_1 \circ J_\mathbb{R}(\mathcal{L}$ over $\mathbb{RP}^n)$ is given by $\mathbb{RP}^{n+1}$. Because $MO$ is a connective spectrum, the truncation map

$$MO \to MO(-\infty, 0] = H\pi_0 MO = H\mathbb{F}_2$$

is a map of ring spectra [May77, lemma II.2.12]. Hence, we can apply the Thom isomorphism theorem to the mod-2 homology of Thom complexes coming from real vector bundles:

$$\pi_*(H\mathbb{F}_2 \wedge T(\mathcal{L} - 1)) \cong \pi_*(H\mathbb{F}_2 \wedge T(0)) \quad \text{(Thom isomorphism)}$$
$$\pi_*(H\mathbb{F}_2 \wedge \Sigma^{-1}\Sigma^\infty \mathbb{RP}^{n+1}) \cong \pi_*(H\mathbb{F}_2 \wedge \Sigma^\infty_+ \mathbb{RP}^n) \quad \text{(Example 1.1.3)}$$
$$\widetilde{H\mathbb{F}_2}_{*+1} \mathbb{RP}^{n+1} \cong H\mathbb{F}_{2*} \mathbb{RP}^n. \quad \text{(generalized homology)}$$

This powers an induction that shows that $H\mathbb{F}_{2*}\mathbb{RP}^\infty$ has a single class in every degree. The cohomological version of the Thom isomorphism in Remark 1.1.14, together with the $H\mathbb{F}_2^* \mathbb{RP}^n$-module structure of $H\mathbb{F}_2^* T(\mathcal{L} - 1)$, also gives the ring structure:

$$H\mathbb{F}_2^* \mathbb{RP}^n = \mathbb{F}_2[x]/x^{n+1}.$$

## 1.2 Cohomology Rings and Affine Schemes

An abbreviated summary of this book is that we are going to put "Spec" in front of rings appearing in algebraic topology and see what happens. Before actually doing any algebraic topology, we should recall what this means on the level of algebra. The core idea is to replace an $\mathbb{F}_2$-algebra $A$ by the functor it corepresents, which we will denote by Spec $A$. For any other "test $\mathbb{F}_2$-algebra" $T$, we set

$$(\text{Spec } A)(T) := \text{Algebras}_{\mathbb{F}_2/}(A, T) \cong \text{Schemes}_{/\mathbb{F}_2}(\text{Spec } T, \text{Spec } A).$$

More generally, we have the following definition:

**Definition 1.2.1** An *affine $\mathbb{F}_2$-scheme* is a functor $X \colon \text{Algebras}_{\mathbb{F}_2/} \to \text{Sets}$ which is (noncanonically) isomorphic to Spec $A$ for some $\mathbb{F}_2$-algebra $A$. Given such an isomorphism, we will refer to Spec $A \to X$ as a *parameter* for $X$ and its inverse $X \to \text{Spec } A$ as a *coordinate* for $X$.

**Lemma 1.2.2** *There is an equivalence of categories*

$$\text{Spec} \colon \text{Algebras}^{\text{op}}_{\mathbb{F}_2/} \to \text{AffineSchemes}_{/\mathbb{F}_2}. \qquad \square$$

## 1.2 Cohomology Rings and Affine Schemes

The centerpiece of thinking about rings in this way, for us and for now, is to translate between a presentation of $A$ as a quotient of a free algebra and a presentation of $(\operatorname{Spec} A)(T)$ as selecting tuples of elements in $T$ subject to certain conditions. Consider the following example:

*Example 1.2.3* Set $A_n = \mathbb{F}_2[x_1, \ldots, x_n]$. Elements of $(\operatorname{Spec} A_n)(T)$ can be identified with $n$-tuples of elements of $T$, since a function in

$$(\operatorname{Spec} A_n)(T) = \operatorname{ALGEBRAS}_{\mathbb{F}_2/}(\mathbb{F}_2[x_1, \ldots, x_n], T)$$

is entirely determined by where the $x_j$ are sent. Consider also what happens when we impose a relation by passing to $A_n^J = \mathbb{F}_2[x_1, \ldots, x_n]/(x_k^{j_k+1})$: a function in

$$(\operatorname{Spec} A_n^J)(T) = \operatorname{ALGEBRAS}_{\mathbb{F}_2/}(\mathbb{F}_2[x_1, \ldots, x_n]/(x_k^{j_k+1}), T)$$

is again determined by where the $x_j$ are sent, but now $x_j$ can only be sent to an element which is nilpotent of order $j_k + 1$. These schemes are both important enough that we give them special names:

$$\mathbb{A}^n := \operatorname{Spec} \mathbb{F}_2[x_1, \ldots, x_n], \qquad \mathbb{A}^{n,J} := \operatorname{Spec} \mathbb{F}_2[x_1, \ldots, x_n]/(x_k^{j_k+1}).$$

The functor $\mathbb{A}^n$ is called *affine n-space* – reasonable, since the value $\mathbb{A}^n(T)$ is isomorphic to $T^n$. We refer to $\mathbb{A}^1$ as the *affine line*. Note that the quotient map $A_1 \to A_1^{(j)}$ induces an inclusion $\mathbb{A}^{1,(j)} \to \mathbb{A}^1$ and that $\mathbb{A}^{1,(0)}$ is a constant functor:

$$\mathbb{A}^{1,(0)}(T) = \{f \colon \mathbb{F}_2[x] \to T \mid f(x) = 0\}.$$

Accordingly, we declare "$\mathbb{A}^{1,(0)}$" to be the *origin on the affine line* and "$\mathbb{A}^{1,(j)}$" to be the $(n+1)$*st order (nilpotent) neighborhood of the origin* in the affine line.

We can also use this language to re-express another common object arising in algebraic topology: the Hopf algebra, which appears when taking the mod-2 cohomology of an $H$-group. In addition to the usual ring structure on cohomology groups, the $H$-group multiplication, unit, and inversion maps additionally induce a diagonal map $\Delta$, an augmentation map $\varepsilon$, and an antipode $\chi$, respectively. Running through the axioms, one quickly checks the following:

**Lemma 1.2.4** *For a Hopf $\mathbb{F}_2$-algebra $A$, the functor* $\operatorname{Spec} A$ *is naturally valued in groups. Such functors are called group schemes. Conversely, a choice of group structure on* $\operatorname{Spec} A$ *endows $A$ with the structure of a Hopf algebra.*

*Proof sketch* This is a matter of recognizing the product in $\operatorname{ALGEBRAS}_{\mathbb{F}_2/}^{\operatorname{op}}$ as the tensor product, then using the Yoneda lemma to transfer the structure around. □

*Example 1.2.5* The functor $\mathbb{A}^1$ previously introduced is naturally valued in groups: Since $\mathbb{A}^1(T) \cong T$, we can use the addition on $T$ to make it into an abelian group. When considering $\mathbb{A}^1$ with this group scheme structure, we denote it as $\mathbb{G}_a$. Applying the Yoneda lemma, one deduces the following formulas for the Hopf algebra structure maps:[4]

$$\mathbb{G}_a \times \mathbb{G}_a \xrightarrow{\mu} \mathbb{G}_a \qquad\qquad x_1 + x_2 \hookleftarrow x,$$

$$\mathbb{G}_a \xrightarrow{\chi} \mathbb{G}_a \qquad\qquad -x \hookleftarrow x,$$

$$\operatorname{Spec} \mathbb{F}_2 \xrightarrow{\eta} \mathbb{G}_a \qquad\qquad 0 \hookleftarrow x,$$

where we have written $x_1 = x \otimes 1$ and $x_2 = 1 \otimes x$ for the elements of

$$O_{\mathbb{G}_a \times \mathbb{G}_a} \cong O_{\mathbb{G}_a} \otimes O_{\mathbb{G}_a} \cong \mathbb{F}_2[x] \otimes \mathbb{F}_2[x].$$

As an example of how to reason this out, consider the following diagram:

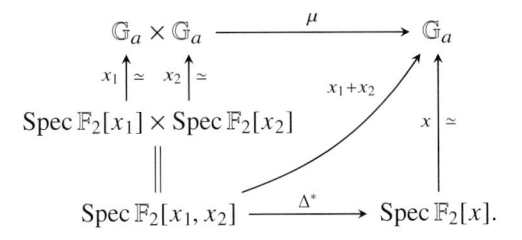

$$\begin{array}{ccc}
\mathbb{G}_a \times \mathbb{G}_a & \xrightarrow{\mu} & \mathbb{G}_a \\
{\scriptstyle x_1}\uparrow\simeq \ \ {\scriptstyle x_2}\uparrow\simeq & {\scriptstyle x_1+x_2}\nearrow & \uparrow {\scriptstyle x}\simeq \\
\operatorname{Spec} \mathbb{F}_2[x_1] \times \operatorname{Spec} \mathbb{F}_2[x_2] & & \\
\| & & \\
\operatorname{Spec} \mathbb{F}_2[x_1, x_2] & \xrightarrow{\Delta^*} & \operatorname{Spec} \mathbb{F}_2[x].
\end{array}$$

It follows that the bottom map of affine schemes is induced by the algebra map

$$\mathbb{F}_2[x] \xrightarrow{\Delta} \mathbb{F}_2[x_1, x_2], \qquad\qquad x \mapsto x_1 + x_2.$$

*Remark 1.2.6* In fact, $\mathbb{A}^1$ is naturally valued in *rings*. It models the inverse functor to Spec in the equivalence of these categories, i.e., the elements of a ring $A$ always form a complete collection of $\mathbb{A}^1$-valued functions on some affine scheme Spec $A$.

*Example 1.2.7* We define the *multiplicative group scheme* by

$$\mathbb{G}_m = \operatorname{Spec} \mathbb{F}_2[x, y]/(xy - 1).$$

Its value $\mathbb{G}_m(T)$ on a test algebra $T$ is the set of pairs $(x, y)$ such that $y$ is a multiplicative inverse to $x$, and hence $\mathbb{G}_m$ is valued in groups. Applying the Yoneda lemma, we deduce the following formulas for the Hopf algebra structure maps:

---

[4] Of course, since we are working over $\mathbb{F}_2$, we could just as well write $\chi(x) = x$.

## 1.2 Cohomology Rings and Affine Schemes

$$\mathbb{G}_m \times \mathbb{G}_m \xrightarrow{\mu} \mathbb{G}_m \qquad x_1 \otimes x_2 \mapsfrom x$$
$$\qquad\qquad\qquad\qquad y_1 \otimes y_2 \mapsfrom y,$$
$$\mathbb{G}_m \xrightarrow{\chi} \mathbb{G}_m \qquad (y, x) \mapsfrom (x, y),$$
$$\operatorname{Spec} \mathbb{F}_2 \xrightarrow{\eta} \mathbb{G}_m \qquad 1 \mapsfrom x, y.$$

*Remark 1.2.8* As presented earlier the multiplicative group comes with a natural inclusion $\mathbb{G}_m \to \mathbb{A}^2$. Specifically, the subset $\mathbb{G}_m \subseteq \mathbb{A}^2$ consists of pairs $(x, y)$ in the graph of the hyperbola $y = 1/x$. However, the element $x$ also gives an $\mathbb{A}^1$-valued function $x \colon \mathbb{G}_m \to \mathbb{A}^1$, and because multiplicative inverses in a ring are unique, we see that this map too is an inclusion. These two inclusions have rather different properties relative to their ambient spaces, and we will think harder about these essential differences later on.

*Example 1.2.9* (see Example 4.4.12) This example showcases the complications that algebraic geometry introduces to this situation, and is meant as discouragement from thinking of the theory of affine group schemes as a strong analog of the theory of linear complex Lie groups. It jumps ahead of the present narrative by a fair amount – the reader should feel completely comfortable skipping this example for now. We set $\alpha_2 = \operatorname{Spec} \mathbb{F}_2[x]/(x^2)$, with group scheme structure given by

$$\alpha_2 \times \alpha_2 \xrightarrow{\mu} \alpha_2 \qquad x_1 + x_2 \mapsfrom x,$$
$$\alpha_2 \xrightarrow{\chi} \alpha_2 \qquad -x \mapsfrom x,$$
$$\operatorname{Spec} \mathbb{F}_2 \xrightarrow{\eta} \alpha_2 \qquad 0 \mapsfrom x.$$

This group scheme has several interesting properties which we will merely state, reserving their proofs for Example 4.4.12.

1. $\alpha_2$ has the same underlying structure ring as $\mu_2 := \mathbb{G}_m[2]$, the 2-torsion points of $\mathbb{G}_m$, but is not isomorphic to it.
2. There is no commutative group scheme $G$ of rank four such that $\alpha_2 = G[2]$.
3. If $E/\mathbb{F}_2$ is the supersingular elliptic curve, then there is a short exact sequence
$$0 \to \alpha_2 \to E[2] \to \alpha_2 \to 0.$$
However, this short exact sequence does not split (even after base change).
4. The subgroups of $\alpha_2 \times \alpha_2$ of order 2 are parameterized by the scheme $\mathbb{P}^1$, i.e., for $A$ an $\mathbb{F}_2$-algebra the subgroup schemes of $\alpha_2 \times \alpha_2$ of order 2 *which are defined over $A$* are parameterized by the set $\mathbb{P}^1(A)$.

We now turn to a different class of examples, which will wind up being the key players in our upcoming topological story. To begin, consider the colimit of the sets $\operatorname{colim}_{j \to \infty} \mathbb{A}^{1,(j)}(T)$, which is of use in algebra: it is the collection of nilpotent elements in $T$. These kinds of conditions that are "unbounded in $j$" appear frequently enough that we are moved to give these functors a name too:

**Definition 1.2.10** An *affine formal scheme* is an ind-system of finite affine schemes. The morphisms between two formal schemes are computed[5,6] by

$$\text{FORMALSCHEMES}(\{X_\alpha\}, \{Y_\beta\}) = \lim_\alpha \text{colim}_\beta \text{SCHEMES}(X_\alpha, Y_\beta).$$

Given affine charts $X_\alpha = \text{Spec } A_\alpha$, we will glibly suppress the system from the notation and write

$$\text{Spf } A := \{\text{Spec } A_\alpha\}.$$

*Example 1.2.11* The individual schemes $\mathbb{A}^{1,(j)}$ do not support group structures. After all, the sum of two elements which are nilpotent of order $j + 1$ can only be guaranteed to be nilpotent of order $2j + 1$. It follows that the entire ind-system $\{\mathbb{A}^{1,(j)}\} =: \widehat{\mathbb{A}}^1$ supports a group structure, even though none of its constituent pieces do. We call such an object a *formal group scheme*, and this particular formal group scheme we denote by $\widehat{\mathbb{G}}_a$.

*Example 1.2.12* Similarly, one can define the scheme $\mathbb{G}_m[j]$ of elements of unipotent order $j$:

$$\mathbb{G}_m[j] = \text{Spec } \frac{\mathbb{F}_2[x, y]}{(xy - 1, x^j - 1)} \subseteq \mathbb{G}_m.$$

These *are* all group schemes, and they nest together in a complicated way: There is an inclusion of $\mathbb{G}_m[j]$ into $\mathbb{G}_m[jk]$. There is also a second filtration along the lines of the one considered in Example 1.2.11:

$$\mathbb{G}_m^{(j)} = \text{Spec } \frac{\mathbb{F}_2[x, y]}{(xy - 1, (x - 1)^j)}.$$

These schemes form a sequential system, but they are only occasionally group

---

[5] For the categorical reader, we include a significant categorical aside: Passing to ind-systems has the effect of formally adjoining colimits of filtered diagrams to a category. The formula for the mapping set comes from asserting that the assignment $C \to \text{Ind}(C)$ is fully faithful, that its image consists of compact objects, and that each diagram $D \to C$ which can be interpreted as a member of $\text{Ind}(C)$ is its own colimit. To control the difference between this category and the original category, it is often useful to restrict attention to diagrams of objects *which are already compact in* $C$, as in our definition of a formal scheme.

[6] Along the same lines, one can show the following recognition principle [Str99b, proposition 4.6]: a functor $X: \text{ALGEBRAS} \to \text{SETS}$ which preserves finite limits is a formal scheme exactly when there exists a family of maps $X_i \to X$ from a set of affine schemes $X_i$ such that for all test algebras $T$ the following map is onto:

$$\coprod_i X_i(T) \to X(T).$$

## 1.2 Cohomology Rings and Affine Schemes

schemes. Specifically, $\mathbb{G}_m^{(2^k)}$ is a group scheme, in which case $\mathbb{G}_m^{(2^k)} \cong \mathbb{G}_m[2^k]$.[7] We define $\widehat{\mathbb{G}}_m$ using this common subsystem:

$$\widehat{\mathbb{G}}_m := \{\mathbb{G}_m^{(2^k)}\}_{k=0}^{\infty}.$$

Let us now consider the example that we closed with in the previous Lecture, where we calculated $H\mathbb{F}_2^*(\mathbb{R}P^n) = \mathbb{F}_2[x]/(x^{n+1})$. Putting "Spec" in front of this, we could reinterpret this calculation as

$$\mathrm{Spec}\, H\mathbb{F}_2^*(\mathbb{R}P^n) \cong \mathbb{A}^{1,(n)}.$$

This is so useful that we will give it a notation all of its own:

**Definition 1.2.13** Let $X$ be a finite cell complex, so that $H\mathbb{F}_2^*(X)$ is a ring which is finite dimensional as an $\mathbb{F}_2$-vector space. We will write

$$X_{H\mathbb{F}_2} = \mathrm{Spec}\, H\mathbb{F}_2^* X$$

for the corresponding finite affine scheme.

*Example 1.2.14* Putting together the discussions from this Lecture and the previous one, in the new notation we have calculated

$$\mathbb{R}P^n_{H\mathbb{F}_2} \cong \mathbb{A}^{1,(n)}.$$

So far, this example just restates what we already knew in a mildly different language. Our driving goal for the next section is to incorporate as much information as we have about these cohomology rings $H\mathbb{F}_2^*(\mathbb{R}P^n)$ into this description, which will result in us giving a more "precise" name for this object. Along the way, we will discover why $X$ had to be a *finite* complex and how to think about more general $X$. For now, though, we will content ourselves with investigating the Hopf algebra structure on $H\mathbb{F}_2^* \mathbb{R}P^\infty$, the cohomology of an infinite complex.

*Example 1.2.15* Recall that $\mathbb{R}P^\infty$ is an $H$-space in two equivalent ways:

1. There is an identification $\mathbb{R}P^\infty \simeq K(\mathbb{F}_2, 1)$, and the $H$-space structure is induced by the sum on cohomology.
2. There is an identification $\mathbb{R}P^\infty \simeq BO(1)$, and the $H$-space structure is induced by the tensor product of real line bundles.

In either case, this induces a Hopf algebra diagonal

$$H\mathbb{F}_2^* \mathbb{R}P^\infty \otimes H\mathbb{F}_2^* \mathbb{R}P^\infty \xleftarrow{\Delta} H\mathbb{F}_2^* \mathbb{R}P^\infty.$$

---

[7] Additionally, the *only* values of $j$ for which $\mathbb{G}_m[j]$ is an infinitesimal thickening of $\mathbb{G}_m[1]$ are those of the form $j = 2^k$.

which we would like to analyze. This map is determined by where it sends the class $x$, and because it must respect gradings it must be of the form $\Delta x = ax_1 + bx_2$ for some constants $a, b \in \mathbb{F}_2$. Furthermore, because it belongs to a Hopf algebra structure, it must satisfy the unitality axiom

$$H\mathbb{F}_2^*\mathbb{R}P^\infty \xleftarrow{\binom{\varepsilon \otimes 1}{1 \otimes \varepsilon}} H\mathbb{F}_2^*\mathbb{R}P^\infty \otimes H\mathbb{F}_2^*\mathbb{R}P^\infty \xleftarrow{\Delta} H\mathbb{F}_2^*\mathbb{R}P^\infty,$$

$$\underbrace{\phantom{H\mathbb{F}_2^*\mathbb{R}P^\infty \otimes H\mathbb{F}_2^*\mathbb{R}P^\infty}}_{1}$$

and hence it takes the form

$$\Delta(x) = x_1 + x_2.$$

Noticing that this is exactly the diagonal map in Example 1.2.5, we tentatively identify "$\mathbb{R}P^\infty_{H\mathbb{F}_2}$" with the additive group. This is extremely suggestive but does not take into account the fact that $\mathbb{R}P^\infty$ is an infinite complex, so we have not yet allowed ourselves to write "$\mathbb{R}P^\infty_{H\mathbb{F}_2}$." In light of the rest of the material discussed in this section, we have left open a very particular point: It is not clear if we should use the name "$\mathbb{G}_a$" or "$\widehat{\mathbb{G}}_a$." We will straighten this out in the subsequent Lecture.

## 1.3 The Steenrod Algebra

We left off in the previous Lecture with an ominous finiteness condition in Definition 1.2.13, and we produced a pair of reasonable guesses as to what "$\mathbb{R}P^\infty_{H\mathbb{F}_2}$" could mean in Example 1.2.15. We will decide which of the two guesses is correct by rigidifying the target category of Definition 1.2.13 so as to incorporate the following extra structures:

1. Cohomology rings are *graded*, and maps of spaces respect this grading.
2. Cohomology rings receive an action of the Steenrod algebra, and maps of spaces respect this action.
3. Both of these are made somewhat more complicated when taking the cohomology of an infinite complex.
4. (Cohomology rings for more elaborate cohomology theories are only skew-commutative, but "Spec" requires a commutative input.)

In this Lecture, we will address all these deficiencies of $X_{H\mathbb{F}_2}$ except for #4, which does not matter with mod-2 coefficients but which will be something of a bugbear throughout the rest of the book.

## 1.3 The Steenrod Algebra

We will begin by considering the grading on $H\mathbb{F}_2^* X$, where $X$ is a finite complex. In algebraic geometry, the following standard construction is used to track gradings:[8]

**Definition 1.3.1** ([Str99b, definition 2.95]) A $\mathbb{Z}$-grading on a ring $A$ is a system of additive subgroups $A_k$ of $A$ satisfying $A = \bigoplus_k A_k$, $1 \in A_0$, and $A_j A_k \subseteq A_{j+k}$. Additionally, a map $f \colon A \to S$ of graded rings is said to *respect the grading* if $f(A_k) \subseteq S_k$.[9]

**Lemma 1.3.2** ([Str99b, proposition 2.96]) *A graded ring $A$ is equivalent data to an affine scheme* $\operatorname{Spec} A$ *with an action by* $\mathbb{G}_m$. *Additionally, a map* $A \to B$ *is homogeneous exactly when the induced map* $\operatorname{Spec} B \to \operatorname{Spec} B$ *is* $\mathbb{G}_m$-*equivariant.*

*Proof* A $\mathbb{G}_m$-action on $\operatorname{Spec} A$ is equivalent data to a coaction map

$$\alpha^* \colon A \to A \otimes \mathbb{F}_2[x^{\pm}].$$

Define $A_k$ to be those points in $a$ satisfying $\alpha^*(a) = a \otimes x^k$. It is clear that we have $1 \in A_0$ and that $A_j A_k \subseteq A_{j+k}$. To see that $A = \bigoplus_k A_k$, note that every tensor can be written as a sum of pure tensors. Conversely, given a graded ring $A$, define the coaction map on $A_k$ by

$$(a_k \in A_k) \mapsto x^k a_k$$

and extend linearly. □

This notion from algebraic geometry is somewhat different from what we are used to in algebraic topology, essentially because the algebraic topologist's "cohomology ring" is not *really* a ring at all – one is only allowed to consider sums of homogeneous degree elements. This restriction stems directly from the provenance of cohomology rings: recall that

$$H\mathbb{F}_2^n X := \pi_{-n} F(\Sigma_+^\infty X, H\mathbb{F}_2).$$

One can only form sums internal to a *particular* homotopy group, using the cogroup structure on $\mathbb{S}^{-n}$. On the other hand, the most basic ring of algebraic geometry is the polynomial ring, and hence their definition is adapted to handle, for instance, the potential degree drop when taking the difference of two (nonhomogeneous) polynomials of the same degree.

---

[8] Strickland gives an alternative formalism for tracking gradings [Strb, sections 11 and 14] called a *polarization*, which amounts to choosing a trivialization $\pi_2 E \cong \pi_0 E$ and considering the isomorphisms $\pi_{2n} E \cong (\pi_2 E)^{\otimes \pi_0 E(n)}$.

[9] The terminology "$\mathbb{Z}$-filtering" might be more appropriate, but this is the language commonly used.

We can modify our perspective very slightly to arrive at that of the algebraic geometers' by replacing $H\mathbb{F}_2$ with the periodified spectrum

$$H\mathbb{F}_2 P = \bigvee_{j=-\infty}^{\infty} \Sigma^j H\mathbb{F}_2.$$

This spectrum forms a ring in the homotopy, where the ring multiplication is given by the following map for each summand of the smash product:

$$\Sigma^j H\mathbb{F}_2 \wedge \Sigma^k H\mathbb{F}_2 \simeq \Sigma^{j+k}(H\mathbb{F}_2 \wedge H\mathbb{F}_2) \xrightarrow{\Sigma^{j+k}\mu} \Sigma^{j+k} H\mathbb{F}_2.$$

This has the property that $H\mathbb{F}_2 P^0(X)$ is isomorphic to $\bigoplus_n H\mathbb{F}_2^n(X)$ as ungraded rings,[10] but now we can make topological sense of the sum of two classes that used to live in different $H\mathbb{F}_2$-degrees. At this point we can manually craft the desired coaction map $\alpha^*$ from Lemma 1.3.2, but we will shortly find that algebraic topology gifts us with it on its own.

Our route to finding this internally occurring $\alpha^*$ is by turning to the next supplementary structure: the action of the Steenrod algebra. Naively approached, this does not fit into the framework we have been sketching so far: The Steenrod algebra arises as the homotopy endomorphisms of $H\mathbb{F}_2$ and so is a *noncommutative* algebra. In turn, the action map

$$\begin{array}{ccc}
\mathcal{A}^* \otimes H\mathbb{F}_2^* X & \longrightarrow & H\mathbb{F}_2^* X \\
\| & & \| \\
[H\mathbb{F}_2, H\mathbb{F}_2]_* \otimes [X, H\mathbb{F}_2]_* & \xrightarrow{\circ} & [X, H\mathbb{F}_2]
\end{array}$$

will be difficult to squeeze into any kind of algebro-geometric framework. Milnor was the first person to see a way around this, with two crucial observations. First, the Steenrod algebra is a Hopf algebra,[11] using the map

$$[H\mathbb{F}_2, H\mathbb{F}_2]_* \xrightarrow{\mu^*} [H\mathbb{F}_2 \wedge H\mathbb{F}_2, H\mathbb{F}_2]_* \cong [H\mathbb{F}_2, H\mathbb{F}_2]_* \otimes [H\mathbb{F}_2, H\mathbb{F}_2]_*$$

as the diagonal. This Hopf algebra structure is actually cocommutative – this is a rephrasing of the symmetry of the Cartan formula:

$$\mathrm{Sq}^n(xy) = \sum_{i+j=n} \mathrm{Sq}^i(x)\, \mathrm{Sq}^j(y).$$

It follows that the linear-algebraic dual of the Steenrod algebra $\mathcal{A}_*$ is a commutative ring, and hence $\operatorname{Spec} \mathcal{A}_*$ would make a reasonable algebro-geometric object.

---

[10] This follows from the equivalence $\bigvee_{j=-\infty}^{\infty} \Sigma^j H\mathbb{F}_2 \to \prod_{j=-\infty}^{\infty} \Sigma^j H\mathbb{F}_2$ and the finiteness of $X$.

[11] The construction of both the Hopf algebra diagonal here and the coaction map that follows is somewhat ad hoc. We will give a more robust presentation in Lecture 3.1.

## 1.3 The Steenrod Algebra

Second, we want to identify the relationship between $\mathcal{A}_*$ and $H\mathbb{F}_2^* X$. Under the assumption that $X$ is a finite complex, the action map

$$H\mathbb{F}_2^* X \leftarrow \mathcal{A}^* \otimes H\mathbb{F}_2^* X$$

transposes under $\mathbb{F}_2$-linear duality to give a coaction map:

$$\mathcal{A}_* \otimes H\mathbb{F}_2^* X \leftarrow H\mathbb{F}_2^* X.$$

Finally, we pass to schemes to interpret this as an action map:

$$\operatorname{Spec} \mathcal{A}_* \times X_{H\mathbb{F}_2} \xrightarrow{\alpha} X_{H\mathbb{F}_2}.$$

Having produced the action map $\alpha$, we are now moved to study $\alpha$ as well as the structure group $\operatorname{Spec} \mathcal{A}_*$ itself. Milnor works out the Hopf algebra structure of $\mathcal{A}_*$ by defining elements $\xi_j \in \mathcal{A}_*$ as follows. Taking $X = \mathbb{R}P^n$ and $x \in H\mathbb{F}_2^1(\mathbb{R}P^n)$ the generator, he observes two things: first, $\operatorname{Sq}^{2^{j-1}} \cdots \operatorname{Sq}^{2^0} x = x^{2^j}$ is nonzero; and second, any other admissible sequence of squares applied to $x$ vanishes for degree reasons. It follows that the coaction applied to $x$ is supported exactly in cohomological degrees of the form $2^j$, which determines elements $\xi_j$ as in

$$\lambda^*(x) = \sum_{j=0}^{\lfloor \log_2 n \rfloor} x^{2^j} \otimes \xi_j \quad (\text{in } H\mathbb{F}_2^* \mathbb{R}P^n).$$

Noticing that taking the limit $n \to \infty$ gives a well-defined infinite sum, he then makes the following calculation, stable in $n$:

$$(\lambda^* \otimes 1) \circ \lambda^*(x) = (1 \otimes \Delta) \circ \lambda^*(x), \qquad \text{(coassociativity)}$$

$$(\lambda^* \otimes 1)\left(\sum_{j=0}^{\infty} x^{2^j} \otimes \xi_j\right)$$

$$= \sum_{j=0}^{\infty} \left(\sum_{i=0}^{\infty} x^{2^i} \otimes \xi_i\right)^{2^j} \otimes \xi_j \qquad \text{(ring homomorphism)}$$

$$= \sum_{j=0}^{\infty} \left(\sum_{i=0}^{\infty} x^{2^{i+j}} \otimes \xi_i^{2^j}\right) \otimes \xi_j = (1 \otimes \Delta) \circ \lambda^*(x).$$

$$\text{(characteristic 2)}$$

Then, turning to the right-hand side:

$$\sum_{j=0}^{\infty} \left(\sum_{i=0}^{\infty} x^{2^{i+j}} \otimes \xi_i^{2^j}\right) \otimes \xi_j = (1 \otimes \Delta)\left(\sum_{m=0}^{\infty} x^{2^m} \otimes \xi_m\right),$$

$$\sum_{j=0}^{\infty} \left(\sum_{i=0}^{\infty} x^{2^{i+j}} \otimes \xi_i^{2^j}\right) \otimes \xi_j = \sum_{m=0}^{\infty} x^{2^m} \otimes \Delta(\xi_m),$$

from which it follows that

$$\Delta \xi_m = \sum_{i+j=m} \xi_i^{2^j} \otimes \xi_j.$$

Finally, Milnor shows that this is the complete story:

**Theorem 1.3.3** (Milnor [Mil58, theorem 2], [MT68, chapter 6], [Lura, proposition 13.1]) *There is an isomorphism*

$$\mathcal{A}_* \cong \mathbb{F}_2[\xi_1, \xi_2, \ldots, \xi_j, \ldots].$$

*Flippant proof* The definition of the elements $\xi_j$ determines a Hopf algebra map $\mathbb{F}_2[\xi_1, \xi_2, \ldots] \to \mathcal{A}_*$ and hence a dual map

$$\mathcal{A}^* \to \mathbb{F}_2[\xi_j \mid j \geq 1]^\vee.$$

This second map is injective: If it had a kernel, then this would produce a subalgebra of $\mathcal{A}^*$ that acts trivially on $H\mathbb{F}_2^*\mathbb{RP}^\infty$, but any nonconstant member of $\mathcal{A}^*$ acts nontrivially on $H\mathbb{F}_2^*(\mathbb{RP}^\infty)^{\times k}$ for $k \gg 0$, and hence nontrivially on $H\mathbb{F}_2^*\mathbb{RP}^\infty$ itself. Then, since $\mathcal{A}_*$ and the polynomial algebra are both of graded finite type, Milnor can conclude his argument by counting how many elements he has produced, comparing against how many Adem and Cartan found (which we will do ourselves in Lecture 4.1), and noting that he has exactly enough.[12] □

We are now in a position to uncover the desired map $\alpha^*$ from earlier. In order to retell Milnor's story with $H\mathbb{F}_2P$ in place of $H\mathbb{F}_2$, note that there is a topological construction involving $H\mathbb{F}_2$ from which $\mathcal{A}_*$ emerges:

$$\mathcal{A}_* := \pi_*(H\mathbb{F}_2 \wedge H\mathbb{F}_2).$$

Performing substitution on this formula gives the periodified dual Steenrod algebra:

$$\mathcal{A}P_0 := \pi_0(H\mathbb{F}_2P \wedge H\mathbb{F}_2P) = H\mathbb{F}_2P_0(H\mathbb{F}_2P) \cong \mathcal{A}_*[\xi_0^\pm],$$

and taking the *continuous* linear dual of the action map

$$H\mathbb{F}_2P^0X \leftarrow \mathcal{A}P^0 \otimes H\mathbb{F}_2P^0X$$

gives a coaction map as before, and hence an action map

$$\operatorname{Spec} \mathcal{A}P_0 \times X_{H\mathbb{F}_2P} \to X_{H\mathbb{F}_2P}.$$

---

[12] The elements dual to $\xi_j$ can be explicitly identified as the *Milnor primitives* $Q_j$, defined by $Q_1 = \operatorname{Sq}^1$ and by $Q_j = [Q_{j-1}, \operatorname{Sq}^{2^{j-1}}]$ for $j > 1$ [Mil58, corollary 2].

## 1.3 The Steenrod Algebra

**Lemma 1.3.4** ([Goe08, formula 3.4, remark 3.14]) *Projecting to the quotient Hopf algebra $\mathcal{A}P_0 \to \mathbb{F}_2[\xi_0^{\pm}]$ gives exactly the $\mathbb{G}_m$-action map provided by Lemma 1.3.2.*

*Calculation* The main topological observation is that there is an inclusion

$$(H\mathbb{F}_2 \wedge H\mathbb{F}_2)P =$$
$$\bigvee_j \Sigma^j(H\mathbb{F}_2 \wedge H\mathbb{F}_2) \subseteq \bigvee_{r,s} \Sigma^r H\mathbb{F}_2 \wedge \Sigma^s H\mathbb{F}_2$$
$$= H\mathbb{F}_2 P \wedge H\mathbb{F}_2 P$$

which is compatible with the unit map as in

$$H\mathbb{F}_2 P \wedge H\mathbb{F}_2 P \wedge X$$
$$H\mathbb{F}_2 P \wedge X \longrightarrow (H\mathbb{F}_2 \wedge H\mathbb{F}_2)P \wedge X.$$

We now study how these inclusions interact with homogenization. Starting with an auxiliary cohomology class $x \in H\mathbb{F}_2^n(X)$, we produce a homogenized cohomology class $x \cdot u^n \in H\mathbb{F}_2 P^0(X)$, and a class $\kappa \in \mathcal{A}_n$ similarly gives a homogenized class $\kappa u^n \in \mathcal{A}P_0$. The Steenrod coaction on $x$ admits expression as a sum

$$\psi(x) = \sum_I x_I \prod_j \xi_j^{I_j}$$

for classes $x_I$ with $|x_I| = |x| - \sum_j (2^j - 1)I_j$. Under the periodified coaction map, the homogenized element $xu^n$ is sent to

$$\psi(xu^n) = \sum_I \left( x_I u^{n - \sum_j (2^j - 1)I_j} \right) \otimes \xi_0^{n - \sum_j (2^j - 1)I_j} \prod_j \left( u^{2^j - 1} \xi_j \right)^{I_j},$$

where the invertible element $\xi_0 = u^{-1} \otimes u \in \mathcal{A}P_0$ of degree 0 appears when shearing the powers of $u$ into the desired places in the formula. Projecting to the tensor factor of $\mathcal{A}P_0$ spanned by $\xi_0$, we thus compute

$$H\mathbb{F}_2 P^0(X) \xrightarrow{\alpha^*} H\mathbb{F}_2 P^0(X) \otimes \mathcal{A}P_0 \longrightarrow H\mathbb{F}_2 P^0(X) \otimes \mathbb{F}_2[\xi_0^{\pm}]$$
$$x \cdot u^n \longmapsto \qquad\qquad\qquad\qquad\qquad\qquad x \cdot u^n \otimes \xi_0^n.$$

Applying Lemma 1.3.2 thus selects the original degree $n$ classes. □

Early on in this discussion, trading the language "graded map" for "$\mathbb{G}_m$-equivariant map" did not seem to have much of an effect on our mathematics – so little, in fact, that we will freely move between the graded homotopy groups of a spectrum and the zeroth homotopy group of a periodic spectrum throughout the rest of this book.[13] The thrust of this lemma, however, is that "Steenrod-equivariant map" already includes "$\mathbb{G}_m$-equivariant map," which is a visible gain in brevity. To study the rest of the content of Steenrod equivariance algebro-geometrically, we need only identify what the series $\lambda^*(x)$ embodies. Note that this necessarily involves some creativity, and the only justification we can supply will be moral, borne out over time, as our narrative encompasses more and more phenomena. With that caveat in mind, here is one such description. Recall the map induced by the $H$-space multiplication:

$$H\mathbb{F}_2 P^0 \mathbb{R}P^\infty \otimes H\mathbb{F}_2 P^0 \mathbb{R}P^\infty \leftarrow H\mathbb{F}_2 P^0 \mathbb{R}P^\infty.$$

Since this map comes from a map of spaces, it is equivariant for the Steenrod coaction, and since the action on the left is furthermore diagonal, we deduce the formula $\lambda^*(x_1 + x_2) = \lambda^*(x_1) + \lambda^*(x_2)$.

**Lemma 1.3.5**  *The series $\lambda(x) = \sum_{j=0}^{\infty} x^{2^j} \otimes \xi_j$ is the universal example of a series satisfying $\lambda(x_1 + x_2) = \lambda(x_1) + \lambda(x_2)$. Points $f \in (\operatorname{Spec} \mathcal{A}P_0)(T)$ biject with those power series $\lambda_f$ with coefficients in the $\mathbb{F}_2$-algebra $T$ satisfying*

$$\lambda_f(x_1 + x_2) = \lambda_f(x_1) + \lambda_f(x_2).$$

*Proof*  Given a point $f \in (\operatorname{Spec} \mathcal{A}P_0)(T)$, we extract such a series by setting

$$\lambda_f(x) := f^*\lambda(x) = \sum_{j=0}^{\infty} f(\xi_j) x^{2^j} \in T[\![x]\!].$$

Conversely, any series $\kappa(x)$ satisfying this homomorphism property must have nonzero terms appearing only in integer powers of 2, and hence we can construct a point $f$ by declaring that $f$ sends $\xi_j$ to the $(2^j)$th coefficient of $\kappa$. □

We close our discussion by codifying what Milnor did when he stabilized against $n$. Each $\mathbb{R}P^n_{H\mathbb{F}_2 P}$ is a finite affine scheme, and to make sense of the object $\mathbb{R}P^\infty_{H\mathbb{F}_2 P}$ Milnor's technique was to consider the ind-system $\{\mathbb{R}P^n_{H\mathbb{F}_2 P}\}_{n=0}^{\infty}$ of finite affine schemes. We will record this as our technique to handle general infinite complexes:

**Definition 1.3.6** (see Definition 2.1.13)  When $X$ is an infinite complex, filter it by its subskeleta $X^{(n)}$ and define $X_{H\mathbb{F}_2 P}$ to be the ind-system $\{X^{(n)}_{H\mathbb{F}_2 P}\}_{n=0}^{\infty}$ of finite schemes.

---

[13] Generally speaking, certain *computations* are expressed most comfortably for an algebraic topologist in terms of graded objects, but we will endeavor to express the surrounding *theory* in the language of equivariance.

## 1.3 The Steenrod Algebra

This choice to follow Milnor resolves our uncertainty about the topological example from the previous lecture:

*Example 1.3.7* (see Examples 1.2.11 and 1.2.15) Write $\widehat{\mathbb{G}}_a$ for the ind-system $\mathbb{A}^{1,(n)}$ with the group scheme structure given in Example 1.2.15. That this group scheme structure filters in this way is a simultaneous reflection of two facts:

1. Algebraic: The set $\widehat{\mathbb{G}}_a(T)$ consists of all nilpotent elements in $T$. The sum of two nilpotent elements of orders $n$ and $m$ is guaranteed to itself be nilpotent with order at most $n + m$.
2. Topological: There is a factorization of the multiplication map on $\mathbb{RP}^\infty$ as $\mathbb{RP}^n \times \mathbb{RP}^m \to \mathbb{RP}^{n+m}$ purely for dimensional reasons.

As group schemes, we have thus calculated

$$\mathbb{RP}^\infty_{H\mathbb{F}_2 P} \cong \widehat{\mathbb{G}}_a.$$

*Example 1.3.8* Noting the homomorphism condition appearing in Lemma 1.3.5, we would like to connect $\operatorname{Spec} \mathcal{A}P_0$ with $\widehat{\mathbb{G}}_a$ more directly. Toward this end, we define a "hom functor"[14] for two formal schemes:

$$\underline{\operatorname{FormalSchemes}}(X, Y)(\mathbb{F}_2 \xrightarrow{u} T) = \{f \colon u^*X \to u^*Y\}.$$

Restricting attention to homomorphisms, we see that a name for $\operatorname{Spec} \mathcal{A}P_0$ is

$$\operatorname{Spec} \mathcal{A}P_0 \cong \underline{\operatorname{Aut}} \widehat{\mathbb{G}}_a.$$

To check this, consider a point $g \in (\operatorname{Spec} \mathcal{A}P_0)(T)$ for an $\mathbb{F}_2$-algebra $T$. The $\mathbb{F}_2$-algebra structure of $T$ (which is uniquely determined by a property of $T$) give rise to a map $u \colon \operatorname{Spec} T \to \operatorname{Spec} \mathbb{F}_2$. The rest of the data of $g$ give rise to a power series in $T[\![x]\!]$ as in the proof of Lemma 1.3.5, which can be reinterpreted as an automorphism $g \colon u^*\widehat{\mathbb{G}}_a \to u^*\widehat{\mathbb{G}}_a$ of formal group schemes.[15]

*Remark 1.3.9* The projection $\mathcal{A}P_0 \to \mathbb{F}_2[\xi_0^\pm]$ is split as Hopf algebras, and hence there is a decomposition

$$\underline{\operatorname{Aut}} \widehat{\mathbb{G}}_a \cong \mathbb{G}_m \times \underline{\operatorname{Aut}}_1 \widehat{\mathbb{G}}_a,$$

where $\underline{\operatorname{Aut}}_1 \widehat{\mathbb{G}}_a$ consists of those automorphisms with leading coefficient $\xi_0$ exactly equal to 1. This can be read to mean that the "interesting" part of the Steenrod algebra, $\underline{\operatorname{Aut}}_1 \widehat{\mathbb{G}}_a$, consists of stable operations, in the sense that their action is independent of the degree-tracking mechanism.

---

[14] We are careful to say "functor" here because it is *not* generally another scheme.
[15] This description, too, is sensitive to the difference between $\widehat{\mathbb{G}}_a$ and $\mathbb{G}_a$. The scheme $\underline{\operatorname{End}} \mathbb{G}_a$ is populated by *polynomials* satisfying a homomorphism condition, and essentially none of them have inverses.

*Example 1.3.10* Remembering the slogan

$$\operatorname{Spec} \mathcal{A} P_0 \cong \underline{\operatorname{Aut}} \, \widehat{\mathbb{G}}_a$$

also makes it easy to recall the structure formulas for the dual Steenrod algebra. For instance, consider the antipode map, which has the effect on $\underline{\operatorname{Aut}} \, \widehat{\mathbb{G}}_a$ of sending a power series to its compositional inverse. That is:

$$\sum_{j=0}^{\infty} \chi(\xi_j) \left( \sum_{k=0}^{\infty} \xi_k x^{2^k} \right)^{2^j} = \sum_{j=0}^{\infty} \sum_{k=0}^{\infty} \chi(\xi_j) \xi_k^{2^j} x^{2^{j+k}}$$

$$= \sum_{n=0}^{\infty} \left( \sum_{j+k=n} \chi(\xi_j) \xi_k^{2^j} \right) x^{2^n} = 1,$$

from which we can extract formulas like

$$\chi(\xi_0) = \xi_0^{-1}, \quad \chi(\xi_1) = \xi_0^{-3} \xi_1, \quad \chi(\xi_2) = \xi_0^{-7} \xi_1^3 + \xi_0^{-5} \xi_2, \quad \ldots.$$

*Remark 1.3.11* Our interest in $\operatorname{colim}_n H\mathbb{F}_2 P^0 \mathbb{R}P^n$ is the culmination of quite a lot of definitions in basic algebraic topology, quietly lurking in the background. For instance, infinite CW-complexes are *defined* to have the weak topology coming from their skeleta, so that a continuous map off of such an infinite complex is continuous if and only if it is the colimit of a compatible system of maps. The cohomology of an infinite complex comes with a Milnor sequence, and if we arrange the situation so as not to encounter a $\lim^1$ term, it is exactly equal to the limit of the cohomology groups of the finite stages – but because all maps between complexes filter through the finite stages, all the induced maps on cohomology are necessarily continuous for the adic topology. This is exactly the phenomenon we are capturing (or, indeed, enforcing in algebra) when we track the system of finite schemes $\{\mathbb{R}P^n_{H\mathbb{F}_2 P}\}_{n=0}^{\infty}$.

In summary, the formula $\mathbb{R}P^{\infty}_{H\mathbb{F}_2 P} \cong \widehat{\mathbb{G}}_a$ embodies the extremely good compression ratio of this approach: It simultaneously encodes the cohomology ring of $\mathbb{R}P^{\infty}$ as the formal scheme, its diagonal as the group scheme structure, and the coaction of the dual Steenrod algebra by the identification with $\underline{\operatorname{Aut}} \, \widehat{\mathbb{G}}_a$. As a separate wonder, it is also remarkable that the single cohomological calculation $\mathbb{R}P^{\infty}_{H\mathbb{F}_2 P}$ exerts such enormous control over mod-2 cohomology itself (e.g., the entire structure of the dual Steenrod algebra). We will eventually give a concrete reason for this in Lecture 4.1, but it will also turn out to be a surprisingly common occurrence even in many situations in which such a direct link is not available. This is, already, one of the mysteries of the subject.

## 1.4 Hopf Algebra Cohomology

In this section, we will focus on an important classical tool: the Adams spectral sequence. We are going to study this in greater earnest later on, so we will avoid giving a satisfying construction in this lecture. Still, even without a construction, it is instructive to see how this comes about from a moral perspective. Because we intend to use the Adams spectral sequence to perform computations, in this lecture we prefer to work with graded homology and cohomology.

Begin by considering the following three self-maps of the stable sphere:

$$\mathbb{S}^0 \xrightarrow{0} \mathbb{S}^0, \qquad \mathbb{S}^0 \xrightarrow{1} \mathbb{S}^0, \qquad \mathbb{S}^0 \xrightarrow{2} \mathbb{S}^0.$$

If we apply mod-2 homology to each line, the induced maps are

$$\mathbb{F}_2 \xrightarrow{0} \mathbb{F}_2, \qquad \mathbb{F}_2 \xrightarrow{1} \mathbb{F}_2, \qquad \mathbb{F}_2 \xrightarrow{0} \mathbb{F}_2.$$

We see that mod-2 homology can immediately distinguish between the null map and the identity map just by its behavior on morphisms, but it cannot distinguish between the null map and the multiplication-by-2 map. To try to distinguish between these two, we use the only other tool available to us: Homology theories send cofiber sequences to long exact sequences, and moreover the data of a map $f$ and the data of the inclusion map $\mathbb{S}^0 \to C(f)$ into its cone are equivalent in the stable category. So, we trade our maps 0 and 2 for the following cofiber sequences:

$$\mathbb{S}^0 \longrightarrow C(0) \longrightarrow \mathbb{S}^1, \qquad \mathbb{S}^0 \longrightarrow C(2) \longrightarrow \mathbb{S}^1.$$

The homology groups of these spectra $C(0)$ and $C(2)$ are more complicated than just that of $\mathbb{S}^0$, and we will draw them according to the following conventions: each "•" in the row labeled "[$j$]" indicates an $\mathbb{F}_2$-summand in the $j$th $H\mathbb{F}_2$-homology of the spectrum. Applying homology to these cofiber sequences and drawing the results, these again appear to be identical:

[1]          • ⟶ •                • ⟶ •

[0]    • ⟶ •                • ⟶ •

$$H\mathbb{F}_{2*}\mathbb{S}^0 \to H\mathbb{F}_{2*}C(0) \to H\mathbb{F}_{2*}\mathbb{S}^1, \qquad H\mathbb{F}_{2*}\mathbb{S}^0 \to H\mathbb{F}_{2*}C(2) \to H\mathbb{F}_{2*}\mathbb{S}^1.$$

However, if we enrich our picture with the data we discussed in Lecture 1.3, we can finally see the difference. Recall the topological equivalences

$$C(0) \simeq \mathbb{S}^0 \vee \mathbb{S}^1, \quad C(2) \simeq \Sigma^{-1}\Sigma^{\infty}\mathbb{R}\mathrm{P}^2.$$

In the two cases, the coaction map $\lambda_*$ is given by

$$\lambda_* : H\mathbb{F}_{2*}C(0) \to H\mathbb{F}_{2*}C(0) \otimes \mathcal{A}_* \qquad \lambda_* : H\mathbb{F}_{2*}C(2) \to H\mathbb{F}_{2*}C(2) \otimes \mathcal{A}_*$$
$$\lambda^* : e_0 \mapsto e_0 \otimes 1 \qquad\qquad\qquad \lambda^* : e_0 \mapsto e_0 \otimes 1 + e_1 \otimes \xi_1$$
$$\lambda^* : e_1 \mapsto e_1 \otimes 1, \qquad\qquad\qquad \lambda^* : e_1 \mapsto e_1 \otimes 1.$$

We use a vertical line to indicate the nontrivial coaction involving $\xi_1$:

[1]     •⟶•                •⟶•
                                   |
                                   $\xi_1$
                                   |
[0]     •⟶•                •⟶•

$$H\mathbb{F}_{2*}\mathbb{S}^0 \to H\mathbb{F}_{2*}C(0) \to H\mathbb{F}_{2*}\mathbb{S}^1, \quad H\mathbb{F}_{2*}\mathbb{S}^0 \to H\mathbb{F}_{2*}C(2) \to H\mathbb{F}_{2*}\mathbb{S}^1.$$

We can now see what trading maps for cofiber sequences has brought us: Mod-2 homology can distinguish the defining sequences for $C(0)$ and $C(2)$ by considering their induced extensions of comodules over $\mathcal{A}_*$.[16] The Adams spectral sequence bundles this thought process into a single machine:

**Theorem 1.4.1** ([Rav86, definition 2.1.8, lemma 2.1.16], [MT68, chapter 18]) *The Adams spectral sequence is convergent and has signature*

$$\mathrm{Ext}^{*,*}_{\mathcal{A}_*}(\mathbb{F}_2, \mathbb{F}_2) \Rightarrow (\pi_*\mathbb{S}^0)^{\wedge}_2. \qquad \square$$

In effect, this asserts that the above process is *exhaustive*: Every element of $(\pi_*\mathbb{S}^0)^{\wedge}_2$ can be detected and distinguished by some representative class of extensions of comodules for the dual Steenrod algebra. Mildly more generally, if $X$ is a bounded-below spectrum of finite type, then there is even an Adams spectral sequence of signature

$$\mathrm{Ext}^{*,*}_{\mathcal{A}_*}(\mathbb{F}_2, H\mathbb{F}_{2*}X) \Rightarrow \pi_*X^{\wedge}_2.$$

We could now work through the construction of the Adams spectral sequence, but it will fit more nicely into a story later on in Lecture 3.1. Before moving on to other pursuits, however, we will record the following utility lemma. It is

---

[16] More complicated homotopy elements may require more complicated extensions: For instance, the element $4 = 2 \cdot 2$ has to be written as a length-2 extension to distinguish it from other elements.

## 1.4 Hopf Algebra Cohomology

believable based on the above discussion, and we will need to use it before we get around to examining the guts of the construction.

**Lemma 1.4.2** (see Remark 3.1.16) *The 0-line of the Adams spectral sequence consists of exactly those elements visible to the Hurewicz homomorphism.* □

For the rest of this section, we will focus on an interpretation of the algebraic input "$\mathrm{Ext}_{\mathcal{A}_*}^{*,*}(\mathbb{F}_2, H\mathbb{F}_{2*}X)$," which will require us to grapple with the homological algebra of comodules for a Hopf algebra. To start that discussion, it's both reassuring and instructive to see that homological algebra can, in fact, even be done with comodules. In the usual development of homological algebra for *modules*, the key observations are the existence of projective and injective modules, and there is something similar at work here.

*Remark 1.4.3* ([Rav86, appendix A1]) Much of the results that follow do not rely on working with a Hopf algebra over the field $k = \mathbb{F}_2$. In fact, $k$ can usually be taken to be a ring rather than a field. More generally, the theory goes through in the context of comodules over flat Hopf algebroids, see also Lemma 3.1.15.

**Lemma 1.4.4** ([Rav86, definition A1.2.1]) *Let $A$ be a Hopf $k$-algebra, let $M$ be an $A$-comodule, and let $N$ be a $k$-module. There is a cofree adjunction:*

$$\mathrm{Comodules}_A(M, N \otimes_k A) \cong \mathrm{Modules}_k(M, N),$$

*where $N \otimes_k A$ is given the structure of an $A$-comodule by the coaction map*

$$N \otimes_k A \xrightarrow{1 \otimes \Delta} N \otimes_k (A \otimes_k A) = (N \otimes_k A) \otimes_k A.$$

*Proof* Given a map $f \colon M \to N$ of $k$-modules, we can build the composite

$$M \xrightarrow{\psi_M} M \otimes_k A \xrightarrow{f \otimes 1} N \otimes_k A.$$

Alternatively, given a map $g \colon M \to N \otimes_k A$ of $A$-comodules, we build the composite

$$M \xrightarrow{g} N \otimes_k A \xrightarrow{1 \otimes \varepsilon} N \otimes_k k = N. \qquad \square$$

**Corollary 1.4.5** ([Rav86, lemma A1.2.2]) *The category $\mathrm{Comodules}_A$ has enough injectives. Namely, if $M$ is an $A$-comodule and $M \to I$ is an inclusion of $k$-modules into an injective $k$-module $I$, then $M \to I \otimes_k A$ is an injective $A$-comodule under $M$.* □

*Remark 1.4.6* In our case, $M$ itself is always $k$-injective, so there's already an injective map $\psi_M \colon M \to M \otimes A$: the coaction map. The assertion that this map is coassociative is identical to saying that it is a map of comodules.

Satisfied that "Ext" at least makes sense, we're free to pursue more conceptual ends. Recall from algebraic geometry that a module $M$ over a ring $A$ is equivalent data to quasicoherent sheaf $\widetilde{M}$ over $\mathrm{Spec}\, A$. We now give a definition of "quasicoherent sheaf" that fits with our functorial perspective:

**Definition 1.4.7** ([Hov02, definition 1.1], [Str99b, definition 2.42]) An assignment $\mathcal{F}\colon X(T) \to \mathrm{Modules}_T$ is said to be a *presheaf (of modules) over* $X$ when it satisfies lax functoriality in $T$: There is a natural transformation $\tau$ such that for each map $f\colon T \to T'$, $\tau$ describes a map (between the *not necessarily equal composites*)

$$\begin{array}{ccc} X(T) & \xrightarrow{\mathcal{F}(T)} & \mathrm{Modules}_T \\ \downarrow{X(f)} \;\; {}^{\tau(f)}\!\!\nearrow & & \downarrow{-\otimes_T T'} \\ X(T') & \xrightarrow{\mathcal{F}(T')} & \mathrm{Modules}_{T'}. \end{array}$$

We think of the image of a particular point $t\colon \mathrm{Spec}\, T \to X$ in $\mathrm{Modules}_T$ as the module of "sections over $t$." Such a presheaf is said to be a *quasicoherent sheaf* when these natural transformations are all natural isomorphisms.

**Lemma 1.4.8** ([Str99b, proposition 2.47]) *An $A$-module $M$ gives rise to a quasicoherent sheaf $\widetilde{M}$ on $\mathrm{Spec}\, A$ by the rule*

$$(\mathrm{Spec}\, T \to \mathrm{Spec}\, A) \mapsto M \otimes_A T.$$

*Conversely, every quasicoherent sheaf over an affine scheme arises in this way.* □

The tensoring operation appearing in the definition of a presheaf appears more generally as an operation on the category of sheaves.

**Definition 1.4.9** A map $f\colon \mathrm{Spec}\, B \to \mathrm{Spec}\, A$ induces maps $f^* \dashv f_*$ of categories of quasicoherent sheaves. At the level of modules, these are given by

$$\begin{array}{ccc} \mathrm{QCoh}_{\mathrm{Spec}\, A} & \underset{f_*}{\overset{f^*}{\rightleftarrows}} & \mathrm{QCoh}_{\mathrm{Spec}\, B} \\ \| & & \| \\ \mathrm{Modules}_A & \underset{N \leftarrow\!\shortmid N}{\overset{M \mapsto M \otimes_A S}{\rightleftarrows}} & \mathrm{Modules}_B. \end{array}$$

One of the main uses of these operations is to define the cohomology of a sheaf. Let $\pi\colon X \to \mathrm{Spec}\, k$ be a scheme over $\mathrm{Spec}\, k$, $k$ a field, and let $\mathcal{F}$ be a sheaf over $X$. The adjunction above induces a derived adjunction

$$\mathrm{Ext}_X(\pi^* k, \mathcal{F}) \cong \mathrm{Ext}_{\mathrm{Spec}\, k}(k, R\pi_* \mathcal{F}),$$

## 1.4 Hopf Algebra Cohomology

which is used to translate the *definition* of sheaf cohomology to that of the cohomology of the derived pushforward $R\pi_*\mathcal{F}$, itself interpretable as a mere complex of $k$-modules. This pattern is very general: The sense of "cohomology" relevant to a situation is often accessed by taking the derived pushforward to a suitably terminal object.[17] To invent a notion of cohomology for comodules over a Hopf algebra, we are thus moved to produce push and pull functors for a map of Hopf algebras, and this is best motivated by another example.

*Example 1.4.10* A common source of Hopf algebras is through group-rings: Given a group $G$, we can define the Hopf $k$-algebra $k[G]$ consisting of formal $k$-linear combinations of elements of $G$. This Hopf algebra is commutative exactly when $G$ is abelian, and $k[G]$-modules are naturally equivalent to $k$-linear $G$-representations. Dually, the ring $k^G$ of $k$-valued functions on $G$ is always commutative, using pointwise multiplication of functions, and it is *cocommutative* exactly when $G$ is abelian. If $G$ is finite, then $k^G$ and $k[G]$ are $k$-linear dual Hopf algebras, and hence finite-dimensional $k^G$-comodules are naturally equivalent to finite-dimensional $k$-linear $G$-representations.[18]

A map of groups $f\colon G \to H$ induces a map $k^f\colon k^H \to k^G$ of Hopf algebras, and it is reasonable to expect that the induced push and pull maps of comodules mimic those of $G$- and $H$-representations. Namely, given an $H$-representation $M$, we can produce a corresponding $G$-representation by precomposition with $f$. However, given a $G$-representation $N$, two features may have to be corrected to extract an $H$-representation:

1. If $f$ is not surjective, we must decide what to do with the extra elements in $H$.
2. If $f$ is not injective – say, $f(g_1) = f(g_2)$ – then we must force the behavior of the extracted $H$-representation to agree on $f(g_1)$ and $f(g_2)$, even if $g_1$ and $g_2$ act differently on $N$. In the extreme case of $f\colon G \to 1$, we expect to recover the fixed points of $N$, since this pushforward computes $H^0_{\text{gp}}(G; N)$.

These concerns, together with the definition of a tensor product as a coequalizer, motivate the following:

---

[17] This perspective often falls under the heading of "six-functor formalism."
[18] There is a variation on this equivalence that uses fewer dualities and which is instructive to expand. The Hopf algebra $k^G = \prod_{g \in G} k$ is the ring of functions on the constant group scheme $G$, and its $k$-points $(\operatorname{Spec} k^G)(k)$ biject with points in $G$. Namely, given $g \in G$ we can form a projection map $g\colon k^G \to k$ and hence a composite $M \to M \otimes_k k^G \xrightarrow{1 \otimes g} M \otimes_k k \cong M$. Collectively, this determines a map $G \times M \to M$ witnessing $M$ as a $G$-representation. In the other direction, if $G$ is finite then we can construct a map $M \to M \otimes_k \prod_{g \in G} k$ sending $m \in M$ to $g \cdot m$ in the $g$th labeled component of the target.

36                    Unoriented Bordism

**Definition 1.4.11**  Given $A$-comodules $M$ and $N$, their *cotensor product* is the $k$-module defined by the equalizer

$$M \square_A N \to M \otimes_k N \xrightarrow{\psi_M \otimes 1 - 1 \otimes \psi_N} M \otimes_k A \otimes_k N.$$

**Lemma 1.4.12**  *Given a map $f \colon A \to B$ of Hopf $k$-algebras, the induced adjunction $f^* \dashv f_*$ is given at the level of comodules by*

$$\begin{array}{c} \text{QCoh}_{\operatorname{Spec} k /\!\!/ \operatorname{Spec} A} \underset{f_*}{\overset{f^*}{\rightleftarrows}} \text{QCoh}_{\operatorname{Spec} k /\!\!/ \operatorname{Spec} B} \\ \| \qquad\qquad\qquad \| \\ \text{Comodules}_A \underset{N \square_B A \hookleftarrow N}{\overset{M \mapsto M}{\rightleftarrows}} \text{Comodules}_B. \end{array} \qquad \square$$

*Remark 1.4.13*  In Lecture 3.1 (and Definition 3.1.13 specifically), we will explain the notation "Spec $k /\!\!/$ Spec $A$" used above. For now, suffice it to say that there again exists a functor-of-points notion of "quasicoherent sheaf" associated to a Hopf $k$-algebra $A$, and such sheaves are equivalent to $A$-comodules.

As an example application, cotensoring gives rise to a concise description of what it means to be a comodule map:

**Lemma 1.4.14** ([Rav86, lemma A1.1.6b])  *Let $M$ and $N$ be $A$-comodules with $M$ projective as a $k$-module. Then there is an equivalence*

$$\text{Comodules}_A(M, N) = \text{Modules}_k(M, k) \square_A N. \qquad \square$$

This has two pleasant consequences. The first is that we can deduce a connection between the push–pull flavor of comodule cohomology described above and the input to the Adams spectral sequence:

**Corollary 1.4.15**  *There is an isomorphism*

$$\text{Comodules}_A(k, N) = \text{Modules}_k(k, k) \square_A N = k \square_A N$$

*and hence*

$$\text{Ext}_A(k, N) \cong \text{Cotor}_A(k, N) \; (= H^* R\pi_* N).$$

*Proof*  Resolve $N$ using the cofree modules described above, then apply either functor $\text{Comodules}_A(k, -)$ or $k \square_A -$. In both cases, you get the same complex. $\qquad\square$

The second is that cofree comodules are cotorsion-free:

**Corollary 1.4.16**  *Let $N = N' \otimes_k A$ be a cofree comodule. Then $N \square_A k = N'$.*

## 1.4 Hopf Algebra Cohomology

*Proof* Picking $M = k$, we have

$$\text{Modules}_k(k, N') = \text{Comodules}_A(k, N)$$
$$= \text{Modules}_k(k, k) \square_A N$$
$$= k \square_A N. \qquad \square$$

*Example 1.4.17* In Lecture 1.3, we identified $\mathcal{A}_*$ with the ring of functions on the group scheme $\underline{\text{Aut}}_1(\widehat{\mathbb{G}}_a)$ of strict automorphisms of $\widehat{\mathbb{G}}_a$, which is defined by the kernel sequence

$$0 \to \underline{\text{Aut}}_1(\widehat{\mathbb{G}}_a) \to \underline{\text{Aut}}(\widehat{\mathbb{G}}_a) \to \mathbb{G}_m \to 0.$$

The punchline is that this is analogous to Example 1.4.10: The object $\text{Cotor}_{\mathcal{A}_*}(\mathbb{F}_2, H\mathbb{F}_{2*}X)$ is thought of as the "derived fixed points" of $\underline{\text{Aut}}_1(\widehat{\mathbb{G}}_a)$ acting on $H\mathbb{F}_{2*}X$.

We now give several examples to get a sense of how the Adams spectral sequence behaves.

*Example 1.4.18* Consider the degenerate case $X = H\mathbb{F}_2$. Then $H\mathbb{F}_{2*}(H\mathbb{F}_2) = \mathcal{A}_*$ is a cofree comodule, and hence Cotor is concentrated on the 0-line:

$$\text{Cotor}^{*,*}_{\mathcal{A}_*}(\mathbb{F}_2, H\mathbb{F}_{2*}(H\mathbb{F}_2)) = \mathbb{F}_2.$$

The Adams spectral sequence collapses to show the wholly unsurprising equality $\pi_* H\mathbb{F}_2 = \mathbb{F}_2$, and indeed this is the element in the image of the Hurewicz map $\pi_* H\mathbb{F}_2 \to H\mathbb{F}_{2*}H\mathbb{F}_2$.

*Example 1.4.19* In the slightly less degenerate case of $X = H\mathbb{Z}$, one can calculate

$$\text{Cotor}^{*,*}_{\mathcal{A}_*}(\mathbb{F}_2, H\mathbb{F}_{2*}H\mathbb{Z}) \cong \text{Cotor}^{*,*}_{\Lambda[\xi_1]}(\mathbb{F}_2, \mathbb{F}_2) \cong \mathbb{F}_2[h_0].$$

This spectral sequence collapses, and the additive extensions cause it to converge to $\mathbb{Z}_2^\wedge$ in degree 0 and 0 in all other degrees. The governing element $\xi_1$ is precisely the *2-adic Bockstein*, which mediates the difference between trivial and nontrivial extensions of $\mathbb{Z}/2^j$ by $\mathbb{Z}/2$.

*Example 1.4.20* Next, we consider the more computationally serious case of $X = kO$, the connective real $K$-theory spectrum. The main input we need is the structure of $H\mathbb{F}_{2*}kO$ as an $\mathcal{A}_*$-comodule, so that we can compute

$$\text{Cotor}^{*,*}_{\mathcal{A}_*}(\mathbb{F}_2, H\mathbb{F}_{2*}kO) \Rightarrow \pi_* kO_2^\wedge.$$

There is a slick trick for doing this: By working in the category of $kO$-modules rather than in all spectra, we can construct a relative Adams spectral sequence

$$\text{Cotor}^{*,*}_{\pi_* H\mathbb{F}_2 \wedge_{kO} H\mathbb{F}_2}(\mathbb{F}_2, \pi_* H\mathbb{F}_2 \wedge_{kO} (kO \wedge H\mathbb{F}_2)) \Rightarrow \pi_*(kO \wedge H\mathbb{F}_2).$$

The second argument is easy to identify:

$$\pi_* H\mathbb{F}_2 \wedge_{kO} (kO \wedge H\mathbb{F}_2) = \pi_* H\mathbb{F}_2 \wedge H\mathbb{F}_2 = \mathcal{A}_*.$$

The Hopf algebra requires further input. Consider the following trio of cofiber sequences:[19]

$$\Sigma kO \xrightarrow{\eta} kO \to kU, \quad \Sigma^2 kU \xrightarrow{\beta} kU \to H\mathbb{Z}, \quad H\mathbb{Z} \xrightarrow{\cdot 2} H\mathbb{Z} \to H\mathbb{F}_2.$$

These combine to give a resolution of $H\mathbb{F}_2$ via an iterated cofiber of free $kO$-modules, with Poincaré series

$$((1+t^2) + t^3(1+t^2)) + t((1+t^2) + t^3(1+t^2)) = 1 + t + t^2 + 2t^3 + t^4 + t^5 + t^6.$$

Repeatedly using the identity $kO \wedge_{kO} H\mathbb{F}_2 \simeq H\mathbb{F}_2$ gives a small presentation of the Hopf algebra $\pi_* H\mathbb{F}_2 \wedge_{kO} H\mathbb{F}_2$: It is a commutative Hopf algebra over $\mathbb{F}_2$ with the above Poincaré series. The Borel–Milnor–Moore classification of commutative Hopf algebras over $\mathbb{F}_2$ [MM65, theorem 7.11] shows that the algebra structure is either

$$\frac{\mathbb{F}_2[a,b,c]}{(a^2 = 0, b^2 = 0, c^2 = 0)} \quad \text{or} \quad \frac{\mathbb{F}_2[a,b,c]}{(a^2 = b, b^2 = 0, c^2 = 0)}$$

for $|a| = 1$, $|b| = 2$, and $|c| = 3$. By knowing that the natural map $\mathcal{A} \to \pi_* H\mathbb{F}_2 \wedge_{kO} H\mathbb{F}_2$ winds up inducing an isomorphism $\pi_{*\leq 2}\mathbb{S} \to \pi_{*\leq 2}kO$, we conclude that we are in the latter case, which gives a presentation of the Hopf algebra as a whole:

$$\begin{array}{ccc}
\pi_* H\mathbb{F}_2 \wedge H\mathbb{F}_2 & \longrightarrow & \pi_* H\mathbb{F}_2 \wedge_{kO} H\mathbb{F}_2 \\
\| & & \| \\
\mathcal{A}_* & \longrightarrow & \dfrac{\mathbb{F}_2[\xi_1, \xi_2, \xi_3, \xi_4, \ldots]}{(\xi_1^4, \xi_2^2), (\xi_n \mid n \geq 3)}.
\end{array}$$

This Hopf algebra is commonly denoted $\mathcal{A}(1)_*$, and its corresponding subgroup scheme $\operatorname{Spec} \pi_* H\mathbb{F}_2 \wedge_{kO} H\mathbb{F}_2 \subseteq \underline{\operatorname{Aut}}_1 \widehat{\mathbb{G}}_a$ allows easy memorization: It is the subscheme of automorphisms of the form $x + \xi_1 x^2 + \xi_2 x^4$, with exactly the additional relations imposed on $\xi_1$ and $\xi_2$ so that this set is stable under composition and inversion.[20,21]

---

[19] This first sequence, known as the Wood cofiber sequence, is a consequence of a very simple form of Bott periodicity [Har80, section 5]: There is a fiber sequence of infinite-loopspaces $O/U \to BO \to BU$, and $\underline{kO}_1 = O/U$.

[20] A similar analysis shows that $\pi_*(H\mathbb{F}_2 \wedge_{H\mathbb{Z}} H\mathbb{F}_2)$ corepresents the subscheme of automorphisms of the form $x + \xi_1 x^2$, which are stable under composition and inversion.

[21] There is also an accidental isomorphism of this Hopf algebra with $\mathbb{F}_2^{D_4}$, where $D_4$ is the dihedral group with eight elements.

Formal geometry aside, this $kO$-based Adams spectral sequence collapses, giving an isomorphism

$$\mathrm{Cotor}^{*,*}_{\mathcal{A}(1)_*}(\mathbb{F}_2, \mathcal{A}_*) = \mathcal{A}_*/\!\!/\mathcal{A}(1)_* \cong H\mathbb{F}_{2*}kO.$$

In turn, the original Adams spectral sequence takes the form

$$\mathrm{Cotor}^{*,*}_{\mathcal{A}_*}(\mathbb{F}_2, \mathcal{A}/\!\!/\mathcal{A}(1)_*) \cong \mathrm{Cotor}^{*,*}_{\mathcal{A}(1)_*}(\mathbb{F}_2, \mathbb{F}_2) \Rightarrow \pi_* kO,$$

where we have used the cofreeness property of Corollary 1.4.16. This spectral sequence is also collapsing, and we provide a picture of it in Figure 1.1. In particular, eight-fold real Bott periodicity can be quickly read off from this picture.

*Example 1.4.21* At the other extreme, we can pick the extremely nondegenerate case $X = \mathbb{S}$, where $\underline{\mathrm{Aut}}_1 \widehat{\mathbb{G}}_a$ acts maximally nonfreely on $\widetilde{\mathbb{F}}_2$. The resulting spectral sequence is pictured through a range in Figure 1.2. It is worth remarking that some of the stable stems receive identifiable names in this language: for instance, the elements of $H^1_{\mathrm{gp}}(\underline{\mathrm{Aut}}_1(\widehat{\mathbb{G}}_a); \widetilde{\mathbb{F}}_2)$ are exactly the 1-cocycles

$$h_j : \left( f = x + \sum_{j=1}^{\infty} \xi_j x^{2^j} \right) \mapsto \xi_j.$$

The cocycle $h_j$ transforms in the $\mathbb{G}_m$-character $z \mapsto z^{2^j}$, hence $h_1$ is a name for the lone element in the spectral sequence contributing to $\pi_1 \mathbb{S}$, which we also know to be $\eta$. In general, the elements $h_j$ selecting the power series coefficients are called the *Hopf invariant 1 elements*, and their survival or demise in the Adams spectral sequence is directly related to the problem of putting $H$-space structures on spheres [Ada62].

## 1.5 The Unoriented Bordism Ring

Our goal in this section is to use our results so far to make a calculation of $\pi_* MO$, the unoriented bordism ring. Our approach is the same as in the examples at the end of the previous section: We will want to use the Adams spectral sequence of signature

$$H^*_{\mathrm{gp}}(\underline{\mathrm{Aut}}_1(\widehat{\mathbb{G}}_a); H\mathbb{F}_2 P_0(MO)\widetilde{\ }) \Rightarrow \pi_* MO,$$

where $H\mathbb{F}_2 P_0(MO)\widetilde{\ }$ denotes the associated quasicoherent sheaf over $\mathrm{Spec}\, k /\!\!/ \mathrm{Spec}\, A_*$ as in Remark 1.4.13 – and we will therefore want to understand $H\mathbb{F}_2 P_0(MO)$ as a comodule for the dual Steenrod algebra.

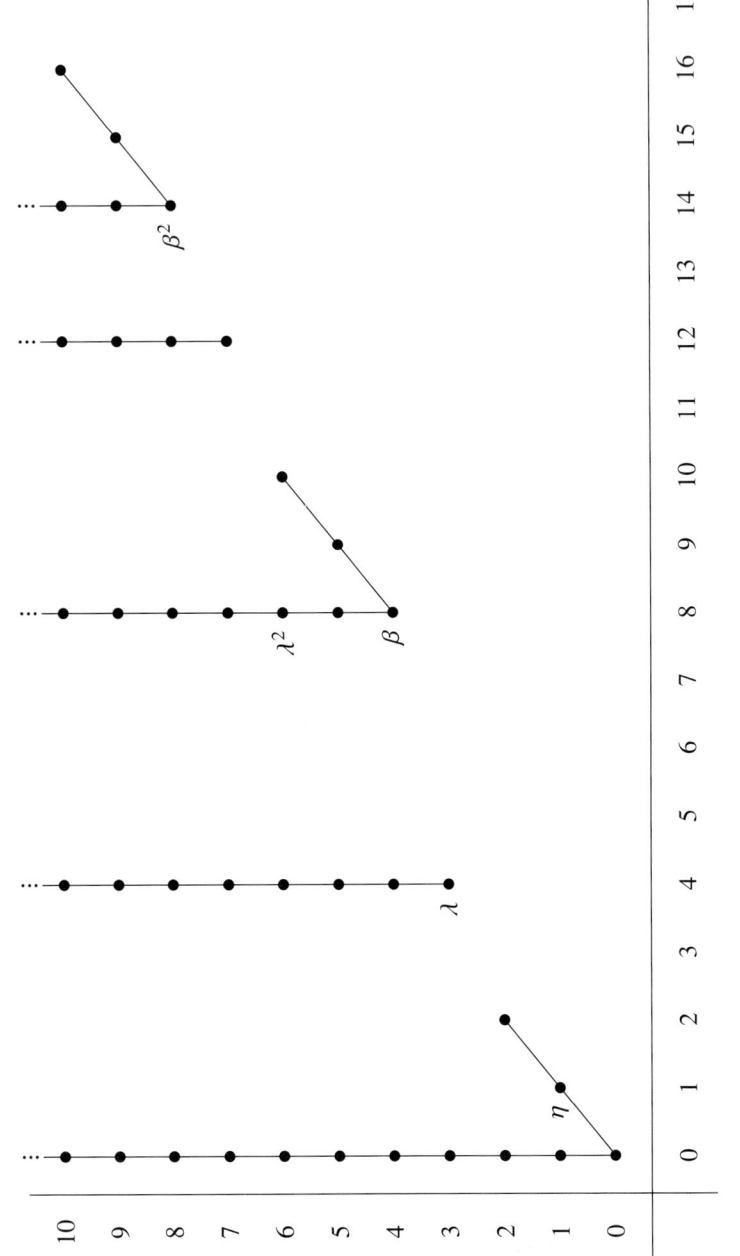

Figure 1.1 The $H\mathbb{F}_2$-Adams spectral sequence for $kO$, which collapses at the second page. North and northeast lines denote multiplication by 2 and by $\eta$.

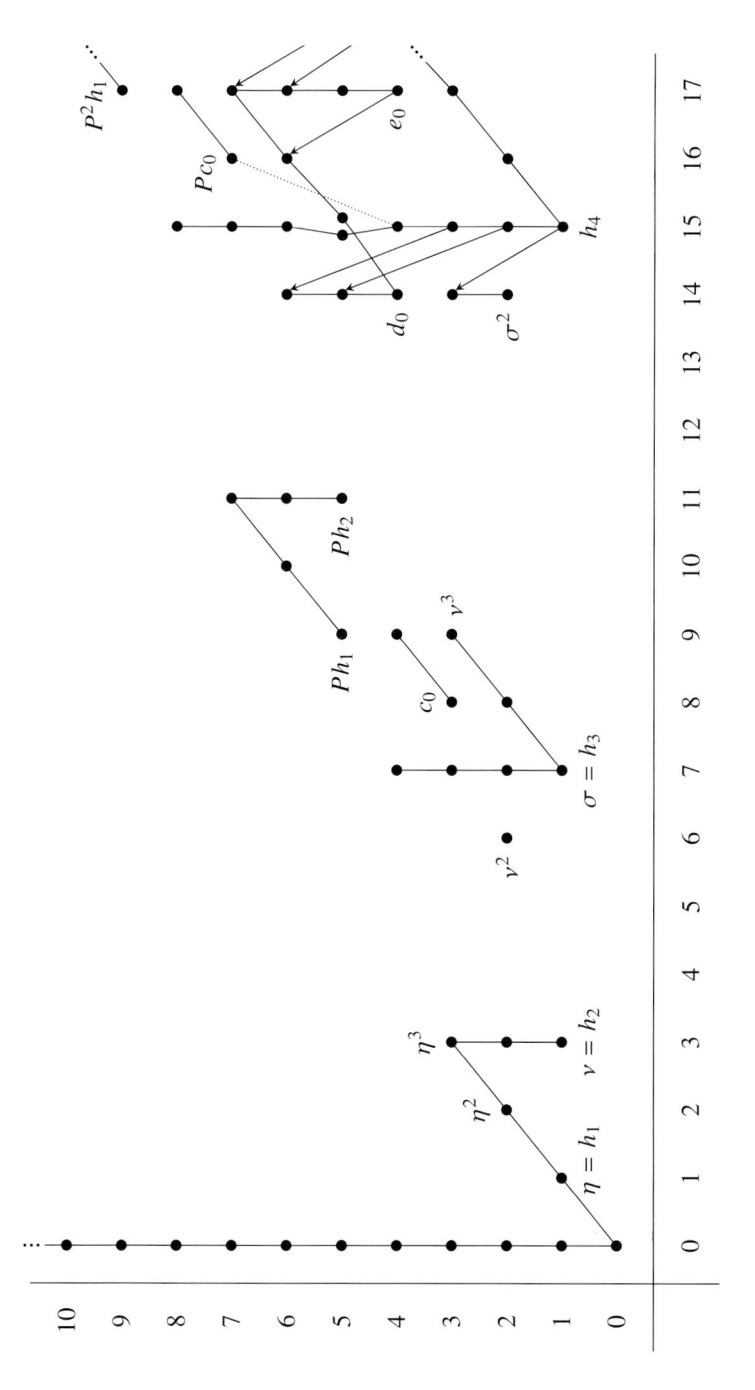

Figure 1.2 A small piece of the $H\mathbb{F}_2$-Adams spectral sequence for the sphere, beginning at the second page [Rav78b, p. 412]. North and northeast lines denote multiplication by 2 and by $\eta$, northwest lines denote $d_2$- and $d_3$-differentials.

Our first step toward this is the following calculation:

**Lemma 1.5.1** ([Swi02, theorem 16.17])  *The natural map*
$$\widetilde{H\mathbb{F}_2 P_0}(BO(1)) \to H\mathbb{F}_2 P_0(BO)$$
*induces a map*
$$\mathrm{Sym}\, \widetilde{H\mathbb{F}_2 P_0}(BO(1)) = \frac{\mathrm{Sym}\, H\mathbb{F}_2 P_0(BO(1))}{\alpha_0 = 1} \xrightarrow{\simeq} H\mathbb{F}_2 P_0(BO),$$
*which is an isomorphism of Hopf algebras and of comodules for the dual Steenrod algebra.*

*Proof* This follows from several standard facts about Stiefel–Whitney classes. First, these classes generate the cohomology ring $H\mathbb{F}_2 P^0 BO(n)$ [Swi02, theorem 16.10]:
$$H\mathbb{F}_2 P^0 BO(n) \cong \mathbb{F}_2[\![w_1, \ldots, w_n]\!].$$
Then, the total Stiefel–Whitney class $w(V) = \sum_j w_j(V) t^j$ is exponential [Swi02, theorem 16.3], in the sense that
$$w(V \oplus W) = w(V) \cdot w(W).$$
It follows that the natural map
$$H\mathbb{F}_2 P^0 BO(n) \xrightarrow{\bigoplus_{j=1}^n \mathcal{L}_j} H\mathbb{F}_2 P^0 BO(1)^{\times n} \cong (H\mathbb{F}_2 P^0 BO(1))^{\otimes n}$$
is the inclusion of the symmetric polynomials, by calculating the total Stiefel–Whitney class:
$$w\left(\bigoplus_{j=1}^n \mathcal{L}_j\right) = \prod_{j=1}^n (1 + w_1(\mathcal{L}_j)) = \sum_{j=0}^n \sigma_j(w_1(\mathcal{L}_1), \ldots, w_1(\mathcal{L}_n)) t^j.$$
Dually, the homological map
$$(H\mathbb{F}_2 P_0 BO(1))^{\otimes n} \to H\mathbb{F}_2 P_0 BO(n)$$
is surjective, modeling the quotient from the tensor product to the symmetric tensor product. Stabilizing as $n \to \infty$, we recover the statement of the lemma. □

With this in hand, we now turn to the homotopy ring $H\mathbb{F}_2 P_0 MO$. There are two equivalences that we might consider employing. We have the Thom isomorphism:
$$\begin{array}{rcl} H\mathbb{F}_2 P_0(BO(1)) & =\!=\!= & H\mathbb{F}_2 P_0(MO(1)) \\ \alpha_j, j \geq 0 & \longmapsto & \alpha'_j, j \geq 0, \end{array}$$

## 1.5 The Unoriented Bordism Ring

and we also have the equivalence induced by the topological map in Example 1.1.3:

$$\widetilde{H\mathbb{F}_2 P_0}(BO(1)) == H\mathbb{F}_2 P_0(\Sigma MO(1))$$
$$\alpha_j, j \geq 1 \longmapsto \alpha'_{j-1}, j \geq 1.$$

We will use them both in turn.

**Corollary 1.5.2** ([Ada95, section I.3], [Hop, proposition 6.2]) *There is again an isomorphism*

$$H\mathbb{F}_2 P_0(MO) \cong \frac{\operatorname{Sym} H\mathbb{F}_2 P_0 MO(1)}{b'_0 = 1}$$

*of rings and of comodules for the dual Steenrod algebra.*

*Proof* The block sum maps

$$BO(n) \times BO(m) \to BO(n+m)$$

Thomify to give compatible maps

$$MO(n) \wedge MO(m) \to MO(n+m).$$

Taking the colimit, this gives a ring structure on $MO$ compatible with that on $\Sigma_+^\infty BO$ and compatible with the Thom isomorphism. □

We now seek to understand the utility of the scheme $\operatorname{Spec} H\mathbb{F}_2 P_0(MO)$, as well as its action of $\underline{\operatorname{Aut}}(\widehat{\mathbb{G}}_a)$. The first of these tasks comes from untangling some of the topological dualities we've been using thus far.

**Lemma 1.5.3** *The following square commutes:*

$$\begin{array}{ccc} \operatorname{Modules}_{\mathbb{F}_2}(H\mathbb{F}_2 P_0(MO), \mathbb{F}_2) & === & \operatorname{Spectra}(MO, H\mathbb{F}_2 P) \\ \uparrow & & \uparrow \\ \operatorname{Algebras}_{\mathbb{F}_2/}(H\mathbb{F}_2 P_0(MO), \mathbb{F}_2) & === & \operatorname{RingSpectra}(MO, H\mathbb{F}_2 P). \end{array}$$

*Proof* The top isomorphism asserts that $\mathbb{F}_2$-cohomology and $\mathbb{F}_2$-homology are linearly dual to one another. The second follows immediately from investigating the effect of the ring homomorphism diagrams in the bottom-right corner in terms of the subset they select in the top-left. □

**Corollary 1.5.4** *There is a bijection between homotopy classes of ring maps $MO \to H\mathbb{F}_2 P$ and homotopy classes of factorizations*

$$\begin{array}{ccc} \mathbb{S}^0 & \longrightarrow & MO(1) \\ & \searrow & \vdots \\ & & H\mathbb{F}_2 P. \end{array}$$

*Proof* We extend the square in Lemma 1.5.3 using the following diagram:

$$\begin{array}{ccc}
\text{Modules}_{\mathbb{F}_2}(H\mathbb{F}_2 P_0(MO(1)), \mathbb{F}_2) & \longleftarrow & \text{Modules}_{\mathbb{F}_2}(H\mathbb{F}_2 P_0(MO), \mathbb{F}_2) \\
\uparrow & & \uparrow \\
\{f\colon H\mathbb{F}_2 P_0(MO(1)) \to \mathbb{F}_2 \,|\, f(\alpha_0') = 1\} & = & \text{Algebras}_{\mathbb{F}_2/}(H\mathbb{F}_2 P_0(MO), \mathbb{F}_2),
\end{array}$$

where the equality at bottom follows from the universal property of $H\mathbb{F}_2 P_0(MO)$ in $\mathbb{F}_2$-algebras expressed in Corollary 1.5.2. Noting that $\alpha_0'$ is induced by the topological map $\mathbb{S}^0 \to MO(1)$, the condition $f(\alpha_0') = 1$ is exactly the condition expressed in the statement of the corollary. $\square$

**Corollary 1.5.5** ([Str06, definition 12.1, theorem 12.3]) *There is an $\underline{\text{Aut}}(\widehat{\mathbb{G}}_a)$-equivariant isomorphism of schemes*

$$\text{Spec } H\mathbb{F}_2 P_0(MO) \cong \text{Coord}_1(\mathbb{RP}^\infty_{H\mathbb{F}_2 P}),$$

*where the latter is the subscheme of functions $\mathbb{RP}^\infty_{H\mathbb{F}_2 P} \to \widehat{\mathbb{A}}^1$ which are coordinates (i.e., which are* isomorphisms *of formal schemes – or, equivalently, which restrict to the canonical identification of tangent spaces $\mathbb{RP}^1_{H\mathbb{F}_2 P} = \widehat{\mathbb{A}}^{1,(1)}$).*

*Proof* The conclusion of the previous corollary is that the $\mathbb{F}_2$-points of $\text{Spec } H\mathbb{F}_2 P_0(MO)$ biject with classes in $H\mathbb{F}_2 P^0 MO(1) \cong \widetilde{H\mathbb{F}_2 P}^0 \mathbb{RP}^\infty$ satisfying the condition that they give an isomorphism $\mathbb{RP}^\infty_{H\mathbb{F}_2 P} \cong \widehat{\mathbb{G}}_a$. Because $H\mathbb{F}_2 P_0(MO)$ is a polynomial algebra, this holds in general: For $u\colon \mathbb{F}_2 \to T$ an $\mathbb{F}_2$-algebra, the $T$-points of $\text{Spec } H\mathbb{F}_2 P_0(MO)$ will biject with coordinates on $u^* \mathbb{RP}^\infty_{H\mathbb{F}_2 P}$. The isomorphism of schemes follows, though we have not yet discussed equivarience.

To compute the action of $\underline{\text{Aut}}\,\widehat{\mathbb{G}}_a$, we turn to the map in Example 1.1.3:

$$\Sigma^\infty BO(1) \xrightarrow{c,\simeq} \Sigma MO(1).$$

Writing $\alpha(t) = \sum_{j=0}^\infty \alpha_j t^j$ and $\xi(t) = \sum_{k=0}^\infty \xi_k t^{2^k}$, the dual Steenrod coaction on $H\mathbb{F}_2 P_0 BO(1)$ is encoded by the formula

$$\sum_{j=0}^\infty \psi(\alpha_j) t^j = \psi(\alpha(t)) = \alpha(\xi(t)) = \sum_{j=0}^\infty \alpha_j \left( \sum_{k=0}^\infty \xi_k t^{2^k} \right)^j.$$

Because $c_*(\alpha_j) = \alpha_{j-1}'$, this translates to the formula $\psi(\alpha'(t)) = \alpha'(\xi(t))$, where

$$\alpha'(t) = \sum_{j=0}^\infty \alpha_j' t^{j+1}.$$

## 1.5 The Unoriented Bordism Ring

Passing from $H\mathbb{F}_2 P_0(MO(1))$ to

$$H\mathbb{F}_2 P_0(MO) \cong \operatorname{Sym} H\mathbb{F}_2 P_0(MO(1))/(\alpha'_0 = 1),$$

we see that this is precisely the formula for precomposing a coordinate with a strict automorphism. This is exactly the assertion that a point in $\underline{\operatorname{Aut}}_1(\widehat{\mathbb{G}}_a)$ acts on a point in $\operatorname{Coord}(\mathbb{R}\mathrm{P}^\infty_{H\mathbb{F}_2 P})$ in the way claimed. □

We are now ready to analyze the group cohomology of $\underline{\operatorname{Aut}}(\widehat{\mathbb{G}}_a)$ with coefficients in the comodule $H\mathbb{F}_2 P_0(MO)$. This is the last piece of input we need to assess the Adams spectral sequence computing $\pi_* MO$.

**Theorem 1.5.6** ([Str06, theorem 12.2], [Mit83, proposition 2.1]) *The action of $\underline{\operatorname{Aut}}_1(\widehat{\mathbb{G}}_a)$ on $\operatorname{Coord}_1(\widehat{\mathbb{G}}_a)$ is free:*

$$\operatorname{Coord}_1(\widehat{\mathbb{G}}_a) \cong \operatorname{Spec} \mathbb{F}_2[b_j \mid j \ne 2^k - 1] \times \underline{\operatorname{Aut}}_1(\widehat{\mathbb{G}}_a).$$

*Proof* Recall, again, that $\underline{\operatorname{Aut}}_1(\widehat{\mathbb{G}}_a)$ is defined by the (split) kernel sequence

$$0 \to \underline{\operatorname{Aut}}_1(\widehat{\mathbb{G}}_a) \to \underline{\operatorname{Aut}}(\widehat{\mathbb{G}}_a) \to \mathbb{G}_m \to 0.$$

Consider a point $f \in \operatorname{Coord}_1(\widehat{\mathbb{G}}_a)(R)$, which in terms of the standard coordinate can be expressed as

$$f(x) = \sum_{j=1}^\infty b_{j-1} x^j,$$

where $b_0 = 1$. Decompose this series as $f(x) = f_{\operatorname{typ}}(x) + f_{\operatorname{rest}}(x)$, with

$$f_{\operatorname{typ}}(x) = \sum_{k=0}^\infty b_{2^k - 1} x^{2^k}, \qquad f_{\operatorname{rest}}(x) = \sum_{j \ne 2^k} b_{j-1} x^j.$$

Because we assumed $b_0 = 1$ and $f_{\operatorname{typ}}$ is concentrated in power-of-2 degrees, it follows that $f_{\operatorname{typ}}$ gives a point $f_{\operatorname{typ}} \in \underline{\operatorname{Aut}}_1(\widehat{\mathbb{G}}_a)(R)$. We can use it to de-scale and get a new coordinate

$$g(x) = f_{\operatorname{typ}}^{-1}(f(x)) = f_{\operatorname{typ}}^{-1}(f_{\operatorname{typ}}(x)) + f_{\operatorname{typ}}^{-1}(f_{\operatorname{rest}}(x)),$$

which itself has an analogous decomposition into series $g_{\operatorname{typ}}(x)$ and $g_{\operatorname{rest}}(x)$. Finally, note that $g_{\operatorname{typ}}(x) = x$, since $f_{\operatorname{typ}}^{-1}(f_{\operatorname{rest}}(x))$ cannot contribute to the power-of-2 degrees in the composite, and that $f_{\operatorname{typ}}$ is the unique point in $\underline{\operatorname{Aut}}_1(\widehat{\mathbb{G}}_a)(R)$ that has this property. Altogether, this shows that the map $f \mapsto (f_{\operatorname{typ}}, g_{\operatorname{rest}})$ is an isomorphism. □

**Corollary 1.5.7** ([Str06, remark 12.3]) $\pi_* MO = \mathbb{F}_2[a_j \mid j \ne 2^k - 1, j \ge 1]$ *with* $|a_j| = j$.

*Proof* Set $M = \mathbb{F}_2[a_j \mid j \neq 2^k - 1]$, and write $\mathcal{A}P_0'$ for the ring of functions on $\underline{\mathrm{Aut}}_1(\widehat{\mathbb{G}}_a)$. It follows from Corollary 1.4.16 applied to Theorem 1.5.6 that the $\underline{\mathrm{Aut}}_1(\widehat{\mathbb{G}}_a)$-cohomology of $H\mathbb{F}_2 P_0(MO)$ has amplitude 0:

$$\mathrm{Cotor}^{*,*}_{\mathcal{A}P_0'}(\mathbb{F}_2, H\mathbb{F}_2 P_0(MO)) = \mathrm{Cotor}^{*,*}_{\mathcal{A}P_0'}(\mathbb{F}_2, \mathcal{A}P_0' \otimes_{\mathbb{F}_2} M)$$
$$= \mathbb{F}_2 \square_{\mathcal{A}P_0'} (\mathcal{A}P_0' \otimes_{\mathbb{F}_2} M)$$
$$= \mathbb{F}_2 \otimes_{\mathbb{F}_2} M = M.$$

Since the Adams spectral sequence

$$H^*_{\mathrm{gp}}(\underline{\mathrm{Aut}}_1(\widehat{\mathbb{G}}_a); H\mathbb{F}_2 P_0(MO)) \Rightarrow \pi_* MO$$

is concentrated on the 0-line, it collapses. We can infer the grading from the residual $\mathbb{G}_m$-action:

$$f_{\mathrm{rest}}(x) \mapsto \xi_0^{-1} f_{\mathrm{rest}}(\xi_0 x) = \sum_{j \neq 2^k} \xi_0^{j-1} a_{j-1} x^j.$$

We therefore deduce

$$\pi_* MO = \mathbb{F}_2[a_j \mid j \neq 2^k - 1]. \qquad \square$$

This is pretty remarkable: Some statement about manifold geometry came down to understanding how we could reparameterize a certain formal group, itself a (fairly simple) purely algebraic problem. The connection between these two problems seems fairly miraculous: We needed a small object, $\mathbb{R}P^\infty$, which controlled the whole story; we needed to be able to compute everything about it; and we needed various other "generation" or "freeness" results to work out in our favor. It is not obvious that we will get this lucky twice, should we try to reapply these ideas to other cases. Nevertheless, trying to push our luck as far as possible is the main thrust of the rest of the book. We could close this section with this accomplishment, but there are two easy consequences of this calculation that are worth recording before we leave.

**Lemma 1.5.8** *$MO$ splits as a wedge of shifts of $H\mathbb{F}_2$.*

*Proof* Referring to Lemma 1.4.2, we find that the Hurewicz map induces a $\pi_*$-injection $MO \to H\mathbb{F}_2 \wedge MO$. Pick an $\mathbb{F}_2$-basis $\{v_\alpha\}_\alpha$ for $\pi_* MO$ and extend it to an $\mathbb{F}_2$-basis $\{v_\alpha\}_\alpha \cup \{w_\beta\}_\beta$ for $\pi_* H\mathbb{F}_2 \wedge MO$. Altogether, this larger basis can be represented as a single map,

$$\bigvee_\alpha \Sigma^{|v_\alpha|} \mathbb{S} \vee \bigvee_\beta \Sigma^{|w_\beta|} \mathbb{S} \xrightarrow{\bigvee_\alpha v_\alpha \vee \bigvee_\beta w_\beta} H\mathbb{F}_2 \wedge MO.$$

## 1.5 The Unoriented Bordism Ring

Smashing through with $H\mathbb{F}_2$ gives an equivalence

$$\bigvee_\alpha \Sigma^{|v_\alpha|} H\mathbb{F}_2 \vee \bigvee_\beta \Sigma^{|w_\beta|} H\mathbb{F}_2 \xrightarrow{\simeq} H\mathbb{F}_2 \wedge MO.$$

The composite map

$$MO \to H\mathbb{F}_2 \wedge MO \xleftarrow{\simeq} \bigvee_\alpha \Sigma^{|v_\alpha|} H\mathbb{F}_2 \vee \bigvee_\beta \Sigma^{|w_\beta|} H\mathbb{F}_2 \to \bigvee_\alpha \Sigma^{|v_\alpha|} H\mathbb{F}_2$$

is a weak equivalence. □

*Remark 1.5.9* Just using that $\pi_* MO$ is connective and $\pi_0 MO = \mathbb{F}_2$, we can produce a ring spectrum map $MO \to H\mathbb{F}_2$. What we have learned is that this map has a splitting: Lemma 1.5.8 can be used to factor the unit map of $MO$ through $H\mathbb{F}_2$ in a way that makes it an $H\mathbb{F}_2$-algebra.

*Remark 1.5.10* We are also in a position to understand the stable cooperations $MO_* MO$. We may rewrite this as

$$MO_* MO = \pi_* MO \wedge MO = \pi_* MO \wedge_{H\mathbb{F}_2} (H\mathbb{F}_2 \wedge MO)$$
$$\Leftarrow \mathrm{Tor}^{\mathbb{F}_2}_{*,*}(MO_*, H\mathbb{F}_{2*} MO) = MO_* \otimes_{\mathbb{F}_2} H\mathbb{F}_{2*} MO.$$

Hence, a point in $\mathrm{Spec}\, MO_* MO$ consists of a pair of points in $\mathrm{Spec}\, MO_*$ and $\mathrm{Spec}\, H\mathbb{F}_{2*} MO$, which we have already identified respectively as formal group laws with vanishing 2-series, a property, and formal group laws with specified logarithms, data. This description can be amplified to capture all of the structure maps: A formal group law with vanishing 2-series admits a logarithm, which indicates how the composition and conjugation maps of Definition 3.1.13 behave.

# Case Study 2

## Complex Bordism

Having totally dissected unoriented bordism, we can now turn our attention to other sorts of bordism theories, and there are many available: oriented, *Spin*, *String*, complex, ... the list continues. We would like to replicate the results from Case Study 1 for these other cases, but upon even a brief inspection we quickly see that only one of the bordism theories mentioned supports this program. Specifically, the space $\mathbb{R}P^\infty = BO(1)$ was a key player in the unoriented bordism story, and the only other similar ground object is $\mathbb{C}P^\infty = BU(1)$ in complex bordism. This informs our choice to spend this Case Study focused on it. To begin, the contents of Lecture 1.1 can be replicated essentially *mutatis mutandis*, resulting in the following theorems:

**Theorem 2.0.1** (see Lemma 1.1.5 and surrounding discussion)  *There is a map of infinite-loopspaces*

$$J_\mathbb{C} \colon BU \to BGL_1 \mathbb{S}$$

*called the complex J-homomorphism.* □

**Definition 2.0.2** (see Definition 1.1.9)  The associated Thom spectrum is written "$MU$" and called *complex bordism*. A map $MU \to E$ of ring spectra is said to be a *complex orientation of $E$*.

**Theorem 2.0.3** (see Theorem 1.1.12)  *For a complex vector bundle $\xi$ on a space $X$ and a complex-oriented ring spectrum $E$, there is a natural equivalence*

$$E \wedge T(\xi) \simeq E \wedge \Sigma^\infty_+ X.$$  □

**Corollary 2.0.4** (see Example 1.1.15)  *In particular, for a complex-oriented ring spectrum $E$ it follows that $E^* \mathbb{C}P^\infty$ is isomorphic to a one-dimensional power series ring.* □

We would like to then review the results of Lecture 1.3 and conclude (by reinterpreting Corollary 2.0.4) that $\mathbb{CP}_E^\infty$ gives a 1-dimensional formal group over Spec $E_*$. In order to make this statement honestly, however, we are first required to describe more responsibly the algebraic geometry we outlined in Lecture 1.2. Because the unoriented bordism ring is of characteristic 2, it could be studied using only $H\mathbb{F}_2$. In turn, $H\mathbb{F}_2$ has many nice properties – for example, it has a duality between homology and cohomology, and it supports a Künneth isomorphism – and these are reflected in the extremely simple algebraic geometry of Spec $\mathbb{F}_2$. By contrast, the complex bordism ring is considerably more complicated, not least because it is a characteristic 0 ring, and more generally we have essentially no control over the behavior of the coefficient ring $E_*$ of some other complex-oriented theory. Nonetheless, once the background theory and construction of "$X_E$" are taken care of in Lecture 2.1, we indeed find that $\mathbb{CP}_E^\infty$ is a one-dimensional formal group over Spec $E_*$.

However, where we could explicitly calculate $\mathbb{RP}_{H\mathbb{F}_2}^\infty$ to be $\widehat{\mathbb{G}}_a$, we again have little control over what formal group $\mathbb{CP}_E^\infty$ could possibly be. In the universal case, $\mathbb{CP}_{MU}^\infty$ comes equipped with a natural coordinate, and this induces a map

$$\text{Spec } MU_* \to \mathcal{M}_{\text{fgl}}$$

from the spectrum associated to the coefficient ring of complex bordism to the moduli of formal group laws. The conclusion of this case study in Corollary 2.6.10 (modulo an algebraic result, shown in the next case study as Theorem 3.2.2) states that this map is an isomorphism, so that $\mathbb{CP}_{MU}^\infty$ carries the universal – i.e., maximally complicated – formal group law. Our route for proving this passes through the foothills of the theory of "$p$th power operations," which simultaneously encode many possible natural transformations from $MU$-cohomology to itself glommed together in a large sum, one term of which is the literal $p$th power. Remarkably, the identity operation also appears in this family of operations, and the rest of the operations are in some sense controlled by this naturally occurring formal group law. A careful analysis of this sum begets the inductive proof in Corollary 2.6.5 that $\mathcal{O}_{\mathcal{M}_{\text{fgl}}} \to MU_*$ is surjective.

The execution of this proof requires some understanding of cohomology operations for complex-oriented cohomology theories generally. Stable such operations correspond to homotopy classes $MU \to E$, i.e., elements of $E^0 MU$, which correspond via the Thom isomorphism to elements of $E^0 BU$. This object is the repository of $E$-characteristic classes for complex vector bundles, which we describe in terms of divisors on formal curves. This amounts to a description of the formal schemes $BU(n)_E$, which underpins our understanding of the whole story and which significantly informs our study of connective orientations in Case Study 5.

## 2.1 Calculus on Formal Varieties

In light of the introduction, we see that it would be prudent to develop some of the theory of formal schemes and formal varieties outside of the context of $\mathbb{F}_2$-algebras. However, writing down a list of definitions and checking that they have good enough properties is not especially enlightening or fun. Instead, it will be informative to understand where these objects come from in algebraic geometry, so that we can carry the accompanying geometric intuition along with us as we maneuver our way back toward homotopy theory and bordism. Our overarching goal in this lecture is to develop a notion of calculus (and analytic expansions in particular) in the context of affine schemes. The place to begin is with definitions of cotangent and tangent spaces, as well as some supporting vocabulary.

**Definition 2.1.1** (see Definition 1.2.1) Fix a ring $R$. For an $R$-algebra $A$, the functor

$$\operatorname{Spec} A \colon \text{Algebras}_R \to \text{Sets},$$
$$T \mapsto \text{Algebras}_R(A, T)$$

is called the *spectrum of A*. A functor $X$ which is naturally isomorphic to some $\operatorname{Spec} A$ is called an *affine (R-)scheme*, and $A = \mathcal{O}_{\operatorname{Spec} A}$ is called its *ring of functions*. A subfunctor $Y \subseteq X$ is said to be a *closed*[1] *subscheme* when an identification[2] $X \cong \operatorname{Spec} A$ induces a further identification

$$\begin{array}{ccc} Y & \longrightarrow & X \\ \uparrow \simeq & & \uparrow \simeq \\ \operatorname{Spec}(A/I) & \longrightarrow & \operatorname{Spec} A. \end{array}$$

**Definition 2.1.2** Take $S = \operatorname{Spec} R$ to be our base scheme, let $X = \operatorname{Spec} A$ be an affine scheme over $S$, and consider an $S$-point $s \colon S \to X$ of $X$. This is automatically a closed subscheme, so that $s$ is presented as $\operatorname{Spec} A/I \to \operatorname{Spec} A$ for some ideal $I$. The *cotangent space* $T_s^* X$ is defined by the quotient $R$-module

$$T_s^* X := I/I^2,$$

consisting of functions vanishing at $s$ as considered up to first order. Examples

---

[1] The word "closed" is meant to suggest properties of these inclusions: In suitable senses, they are closed under finite unions and arbitrary intersections. The complementary concept of "open" is harder to describe: Open subschemes of affine schemes are merely "covered" by finitely many affines, which requires a discussion of coverings, which we remit to the actual algebraic geometers [Strb, definition 8.1].
[2] This property is independent of choice of chart.

52                          Complex Bordism

of these include the linear parts of curves passing through $s$, so we additionally define the *tangent space* $T_s X$ by

$$T_s X = \text{Schemes}_{\text{Spec } R}/(\text{Spec } R[\varepsilon]/\varepsilon^2, X),$$

i.e., maps $\text{Spec } R[\varepsilon]/\varepsilon^2 \to X$ which restrict to $s\colon S \to X$ upon setting $\varepsilon = 0$.

**Remark 2.1.3**  In the situation above, there is a naturally occurring map

$$T_s X \to \text{Modules}_R(T_s^* X, R).$$

Namely, a map $O_X \to R[\varepsilon]/\varepsilon^2$ induces a map $I \to (\varepsilon)$, and hence induces a further map

$$I/I^2 \to (\varepsilon)/(\varepsilon^2) \cong R,$$

which can be interpreted as a point in $(T_s^* X)^*$.

Harkening back to Example 1.3.8, the definition of the $R$-module tangent space begs promotion to an $S$-scheme.

**Lemma 2.1.4**  *There is an affine scheme $T_s X$ defined by*

$$(T_s X)(u\colon \text{Spec } T \to S) := \{f \mid f \in T_{u^* x} u^* X\}.$$

*Proof sketch*  We specialize an argument of Strickland [Str99b, proposition 2.94] to the case at hand.[3] We start by seeking an $R$-algebra $B$ that corepresents the entire *tangent bundle*, $T_* X$: we ask that $R$-algebra maps $B \to T$ biject with pairs of a map $u\colon R \to T$ and a $T$-algebra map

$$f\colon A \otimes_R T \to R[\varepsilon]/\varepsilon^2 \otimes_R T.$$

Such a map $f$ is equivalent data to an $R$-algebra map

$$A \to R[\varepsilon]/\varepsilon^2 \otimes_R T.$$

Using the sequence of inclusions

$$\text{Algebras}_{R/}(A, R[\varepsilon]/\varepsilon^2 \otimes_R T) \subseteq \text{Modules}_R(A, R[\varepsilon]/\varepsilon^2 \otimes_R T)$$
$$\cong \text{Modules}_R(A \otimes_R (R[\varepsilon]/\varepsilon^2)^*, T)$$
$$\cong \text{Algebras}_{R/}(\text{Sym}_R(A \otimes_R (R[\varepsilon]/\varepsilon^2)^*), T),$$

---

[3]  Strickland also shows that mapping schemes between formal schemes exist considerably more generally [Str99b, theorem 4.69]. The source either has to be "finite" in some sense, in which case the proof proceeds along the lines presented here, or it has to be *coalgebraic*, which is an important technical tool that we discuss much later in Definition 5.1.6.

## 2.1 Calculus on Formal Varieties

we see that we can pick out the original mapping set by passing to a quotient of the domain. After some thought, we arrive at the equation

$$T_*X = \underline{\text{Schemes}}_{/S}(\text{Spec } R[\varepsilon]/\varepsilon^2, X) = \text{Spec } \frac{A\{1, da \mid a \in A\}}{\left(\begin{array}{c} dr = 0 \text{ for } r \in R, \\ d(a_1 a_2) = da_1 \cdot a_2 + a_1 \cdot da_2 \end{array}\right)}.$$

To extract the scheme $T_s X$ from this, we construct the pullback

$$T_s X := \underline{\text{Schemes}}_S(\text{Spec } R[\varepsilon]/\varepsilon^2, X) \times_X S,$$

where the structure maps are given on the left by setting $\varepsilon = 0$, and on the right using the point $s$. Expanding the formulas again shows that the coordinate ring of this affine scheme is given by

$$\mathcal{O}_{T_s X} = A/I^2 \cong R \oplus T_s^* X. \qquad \square$$

**Definition 2.1.5** The ring of functions appearing in the previous proof fits into an exact sequence

$$0 \to \Omega_{A/R}$$

$$\to A\{1, da \mid a \in A\} \Big/ \left(\begin{array}{c} dr = 0 \text{ for } r \in R, \\ d(a_1 a_2) = da_1 \cdot a_2 + a_1 \cdot da_2 \end{array}\right)$$

$$\to A\{1\} \to 0.$$

The kernel $\Omega_{A/R}$ is called the module of *Kähler differentials* (of $A$, relative to $R$). The map $d: A \to \Omega^1_{A/R}$ is the universal $R$-linear derivation into an $A$-module, i.e.,

$$\text{Derivations}_R(A, M) = \text{Modules}_A(\Omega^1_{A/R}, M).$$

The upshot of this calculation is that $\text{Spec } A/I^2$ is a natural place to study the linear behavior of functions on $X$ near $s$. We have also set the definitions up so that we can easily generalize to higher-order approximations:

**Definition 2.1.6** More generally, the *nth jet space* of $X$ at $s$, or the *nth order neighborhood* of $s$ in $X$, is defined by

$$\underline{\text{Schemes}}_S(\text{Spec } R[\varepsilon]/\varepsilon^{n+1}, X) \times_X S \cong \text{Spec } A/I^{n+1}.$$

Each jet space has an inclusion from the one before, modeled by the closed subscheme $\text{Spec } A/I^n \to \text{Spec } A/I^{n+1}$.

In order to study analytic expansions of functions, we bundle these jet spaces together into a single object embodying formal expansions in $X$ at $s$:

**Definition 2.1.7** Fix a scheme $S$. A *formal $S$-scheme* $X = \{X_\alpha\}_\alpha$ is an ind-system of $S$-schemes $X_\alpha$.[4] Given a closed subscheme $Y$ of an affine $S$-scheme $X$, we define the *nth order neighborhood of $Y$ in $X$* to be the scheme $\operatorname{Spec} A/I^{n+1}$. The *formal neighborhood of $Y$ in $X$* is then defined to be the formal scheme

$$X_Y^\wedge := \operatorname{Spf} A_I^\wedge := \left\{ \operatorname{Spec} A/I \to \operatorname{Spec} A/I^2 \to \operatorname{Spec} A/I^3 \to \cdots \right\}.$$

In the case that $Y = S$, this specializes to the system of jet spaces as in Definition 2.1.6.

Although we will make use of these definitions generally, the following ur-example captures the most geometrically intuitive situation.

*Example 2.1.8* Picking the affine scheme $X = \operatorname{Spec} R[x_1, \ldots, x_n] = \mathbb{A}^n$ and the point $s = (x_1 = 0, \ldots, x_n = 0)$ gives a formal scheme known as *formal affine n-space*, given explicitly by

$$\widehat{\mathbb{A}}^n = \operatorname{Spf} R[\![x_1, \ldots, x_n]\!].$$

Evaluated on a test algebra $T$, $\widehat{\mathbb{A}}^1(T)$ yields the ideal of nilpotent elements in $T$ and $\widehat{\mathbb{A}}^n(T)$ its $n$-fold Cartesian power. Pointed maps $\widehat{\mathbb{A}}^n \to \widehat{\mathbb{A}}^m$ naturally biject with $m$-tuples of $n$-variate power series with no constant term.[5]

Part of the point of the geometric language is to divorce abstract rings (e.g., $E^0 \mathbb{C}P^\infty$) from concrete presentations (e.g., $E^0[\![x]\!]$), so we additionally reserve some vocabulary for the property of being isomorphic to $\widehat{\mathbb{A}}^n$:

**Definition 2.1.9** A *formal affine variety* (of dimension $n$, over a base $R$) is a formal scheme $V$ which is (noncanonically) isomorphic to $\widehat{\mathbb{A}}^n$ – or, equivalently, $\mathcal{O}_V$ is (noncanonically) isomorphic to a power series ring. The two maps in an isomorphism pair

$$V \xrightarrow{\sim} \widehat{\mathbb{A}}^n, \quad V \xleftarrow{\sim} \widehat{\mathbb{A}}^n$$

are called a *coordinate (system)* and a *parameter (system)* respectively. Finally, an $S$-point $s \colon S \to X$ is called *formally smooth* when $X_s^\wedge$ gives a formal variety.

---

[4] This definition, owing to Strickland [Str99b, definition 4.1], is somewhat idiosyncratic. Its generality gives it good categorical properties, but it is somewhat disconnected from the formal schemes familiar to algebraic geometers, which primarily arise through linearly topologized rings [Har77, p. 194]. For functor-of-points definitions that hang more tightly with the classical definition, the reader is directed toward Strickland's solid formal schemes [Str99b, definition 4.16] or to Beilinson and Drinfeld [BD, section 7.11.1].

[5] In some sense, this lemma is a full explanation for why anyone would even think to involve formal geometry in algebraic topology (never mind how useful the program has been in the long run). Calculations in algebraic topology have long been expressed in terms of power series rings, and with this lemma we are provided geometric interpretations for such statements.

## 2.1 Calculus on Formal Varieties

This definition allows local theorems from analytic differential geometry to be imported in coordinate-free language. For instance, there is the following version of the inverse function theorem:

**Theorem 2.1.10** (Inverse function theorem, [Laz75, theorem I.8.1]) *A pointed map $f: V \to W$ of finite-dimensional formal varieties is an isomorphism if and only if the induced map $T_0 f : T_0 V \to T_0 W$ is an isomorphism of R-modules.* □

Coordinate-free theorems are only really useful if we can verify their hypotheses by coordinate-free methods as well. The following two results are indispensable in this regard:

**Theorem 2.1.11** *Let $R$ be a Noetherian ring and $F:$ Algebras$_{R/} \to$ Sets$_{*/}$ be a functor such that $F(R) = *$, $F$ takes surjective maps to surjective maps, and there is a fixed finite free R-module $M$ such that $F$ carries square-zero extensions of Noetherian R-algebras $I \to B \to B'$ to product sequences*

$$* \to I \otimes_R M \to F(B) \to F(B') \to *.$$

*Then, a basis $M \cong R^n$ determines an isomorphism $F \cong \widehat{\mathbb{A}}^n$.*

*Proof sketch* In the motivating case where $F \cong \widehat{\mathbb{A}}^n$ is given, we can define $M$ to be

$$M := F(R[\varepsilon]/\varepsilon^2) = (\varepsilon)^{\times n} = R\{\varepsilon_1, \ldots, \varepsilon_n\}.$$

In fact, this is always the case: The square-zero extension

$$(\varepsilon) \to R[\varepsilon]/\varepsilon^2 \to R$$

induces a product sequence and hence an isomorphism

$$* \to (\varepsilon) \otimes_R M \xrightarrow{\cong} F(R[\varepsilon]/\varepsilon^2) \to * \to *.$$

A choice of basis $M \cong R^{\times n}$ thus induces an isomorphism

$$F(R[\varepsilon]/\varepsilon^2) = (\varepsilon) \otimes_R M = M \cong R^{\times n} = \widehat{\mathbb{A}}^n(R[\varepsilon]/\varepsilon^2).$$

Lastly, induction shows that if the maps between the outer terms of the set-theoretic product sequence exist, then so must the middle:

$$\begin{array}{ccccccccc}
* & \to & I \otimes_R M & \to & F(B) & \to & F(B') & \to & * \\
& & \| & & \simeq \uparrow & & \simeq \uparrow & & \\
* & \to & I \otimes_R M & \to & \widehat{\mathbb{A}}^n(B) & \to & \widehat{\mathbb{A}}^n(B') & \to & *.
\end{array}$$

□

**Corollary 2.1.12** ([Gro71, théorème III.2.1]) *We define a nilpotent thickening to be a closed inclusion of schemes whose associated ideal sheaf is nilpotent. An S-point $s\colon S \to X$ of a Noetherian scheme is formally smooth exactly when $T_s X$ is a projective R-module and for any nilpotent thickenings $S \to \operatorname{Spec} B \to \operatorname{Spec} B'$ and any solid diagram*

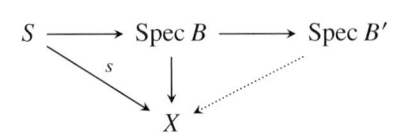

$$\begin{array}{ccc} S & \longrightarrow & \operatorname{Spec} B & \longrightarrow & \operatorname{Spec} B' \\ & \searrow_s & \downarrow & \swarrow & \\ & & X & & \end{array}$$

*there exists a dotted arrow extending the diagram.* □

With all this algebraic geometry in hand, we now return to our original motivation: extracting formal schemes from the rings appearing in algebraic topology.

**Definition 2.1.13** (see Definition 1.3.6) Let $E$ be an even-periodic ring spectrum, and let $X$ be a CW-space. Because $X$ is compactly generated, it can be written as the colimit of its compact subspaces $X^{(\alpha)}$, and we set[6,7]

$$X_E := \operatorname{Spf} E^0 X := \{\operatorname{Spec} E^0 X^{(\alpha)}\}_\alpha.$$

Consider the example of $\mathbb{C}P^\infty_E$ for $E$ a complex-oriented cohomology theory. We saw in Corollary 2.0.4 that the complex-orientation determines an isomorphism $\mathbb{C}P^\infty_E \cong \widehat{\mathbb{A}}^1$ (i.e., an isomorphism $E^0\mathbb{C}P^\infty \cong E^0[\![x]\!]$). However, the object "$E^0\mathbb{C}P^\infty$" is something that exists independent of the orientation map $MU \to E$, and the language of Definition 2.1.9 allows us to make the distinction between the property and the data:

**Lemma 2.1.14** *A cohomology theory $E$ is complex orientable (i.e., it is able to receive a ring map from $MU$) precisely when $\mathbb{C}P^\infty_E$ is a formal curve (i.e., it is a formal variety of dimension 1). A choice of orientation $MU \to E$ determines a coordinate $\mathbb{C}P^\infty_E \cong \widehat{\mathbb{A}}^1$ via the first Chern class associated to the orientation.* □

*Remark 2.1.15* An even-periodic ring spectrum is automatically complex-orientable: the Atiyah–Hirzebruch spectral sequence for the cohomology of $\mathbb{C}P^\infty$ collapses.

---

[6] The properties satisfied by this construction suffer greatly if it is not the case that the $E$-cohomology of $X$ is even-concentrated, i.e., if $E^*X \cong E^* \otimes_{E^0} E^0 X$ is violated. In fact, even the intermediate failure of $E^*X_\alpha \cong E^* \otimes_{E^0} E^0 X_\alpha$ can cause issues, but passing to a clever cofinal subsystem often alleviates them. For instance, such a subsystem exists if $H\mathbb{Z}_*X$ is free and even [Str99b, definition 8.15, proposition 8.17].

[7] In cases of "large" spaces and cohomology theories, the technical points underlying this definition are necessary: $BU_{KU}$ is an instructive example, as it is *not* the formal scheme associated to $KU^0(BU)$ by any adic topology.

As in Example 1.3.7, the formal scheme $\mathbb{CP}^\infty_E$ has additional structure: It is a group. We close this lecture with some remarks about such objects.

**Definition 2.1.16** A *formal group* is a formal variety endowed with an abelian group structure.[8] If $E$ is a complex-orientable cohomology theory, then $\mathbb{CP}^\infty_E$ naturally forms a (one-dimensional) formal group using the map classifying the tensor product of line bundles.

*Remark 2.1.17* As with formal schemes, formal groups can arise as the formal completion of an algebraic group at its identity point. It turns out that there are many more formal groups than come from this procedure, a phenomenon that is of keen interest to stable homotopy theorists – see Appendix B.2.

We give the following corollary as an example of how nice the structure theory of formal varieties is – in particular, formal groups often behave like physical groups.

**Corollary 2.1.18** *The formal group addition map on $\widehat{\mathbb{G}}$ determines the inverse law.*

*Proof* Consider the shearing map
$$\widehat{\mathbb{G}} \times \widehat{\mathbb{G}} \xrightarrow{\sigma} \widehat{\mathbb{G}} \times \widehat{\mathbb{G}},$$
$$(x, y) \mapsto (x, x + y).$$
The induced map $T_0 \sigma$ on tangent spaces is evidently invertible, so by Theorem 2.1.10 there is an inverse map $(x, y) \mapsto (x, y - x)$. Setting $y = 0$ and projecting to the second factor gives the inversion map. □

**Definition 2.1.19** Let $\widehat{\mathbb{G}}$ be a formal group. In the presence of a coordinate $\varphi \colon \widehat{\mathbb{G}} \cong \widehat{\mathbb{A}}^n$, the addition law on $\widehat{\mathbb{G}}$ begets a map

$$\begin{array}{ccc} \widehat{\mathbb{G}} \times \widehat{\mathbb{G}} & \longrightarrow & \widehat{\mathbb{G}} \\ {\scriptstyle \simeq} \downarrow & & \downarrow {\scriptstyle \simeq} \\ \widehat{\mathbb{A}}^n \times \widehat{\mathbb{A}}^n & \longrightarrow & \widehat{\mathbb{A}}^n, \end{array}$$

and hence an $n$-tuple of $(2n)$-variate power series "$+_\varphi$," satisfying

$$\underline{x} +_\varphi \underline{y} = \underline{y} +_\varphi \underline{x}, \quad \text{(commutativity)}$$
$$\underline{x} +_\varphi \underline{0} = \underline{x}, \quad \text{(unitality)}$$
$$\underline{x} +_\varphi (\underline{y} +_\varphi \underline{z}) = (\underline{x} +_\varphi \underline{y}) +_\varphi \underline{z}. \quad \text{(associativity)}$$

---

[8] Formal groups in dimension 1 are automatically commutative if and only if the ground ring has no elements which are simultaneously nilpotent and torsion [Haz12, theorem I.6.1].

Such a (tuple of) series $+_\varphi$ is called a *formal group law*, and it is the concrete data associated to a formal group.

Let us now consider two examples of complex-orientable ring spectra $E$ and describe these invariants for them.

*Example 2.1.20* There is an isomorphism $\mathbb{CP}^\infty_{H\mathbb{Z}P} \cong \widehat{\mathbb{G}}_a$. This follows from reasoning identical to that given in Example 1.3.7.

*Example 2.1.21* There is also an isomorphism $\mathbb{CP}^\infty_{KU} \cong \widehat{\mathbb{G}}_m$. The standard choice of first Chern class is given by the topological map
$$c_1 \colon \Sigma^{-2}\Sigma^\infty \mathbb{CP}^\infty \xrightarrow{1-\beta\mathcal{L}} KU,$$
and a formula for the first Chern class of the tensor product is thus
$$\begin{aligned} c_1(\mathcal{L}_1 \otimes \mathcal{L}_2) &= 1 - \beta(\mathcal{L}_1 \otimes \mathcal{L}_2) \\ &= -\beta^{-1}\left((1-\beta\mathcal{L}_1) \cdot (1-\beta\mathcal{L}_2)\right) + (1-\beta\mathcal{L}_1) + (1-\beta\mathcal{L}_2) \\ &= c_1(\mathcal{L}_1) + c_1(\mathcal{L}_2) - \beta^{-1} c_1(\mathcal{L}_1) c_1(\mathcal{L}_2). \end{aligned}$$
In this coordinate on $\mathbb{CP}^\infty_{KU}$, the group law is then $x_1 +_{\mathbb{CP}^\infty_{KU}} x_2 = x_1 + x_2 - \beta^{-1} x_1 x_2$. Using the coordinate function $1-t$, this is also the coordinate that arises on the formal completion of $\mathbb{G}_m$ at $t=1$:
$$\begin{aligned} x_1(t_1) +_{\mathbb{G}_m} x_2(t_2) &= 1 - (1-t_1)(1-t_2) \\ &= t_1 + t_2 - t_1 t_2. \end{aligned}$$

As an application of all these tools, we will show that the *rational* theory of formal groups is highly degenerate: Every rational formal group admits a *logarithm*, i.e., an isomorphism to $\widehat{\mathbb{G}}_a$. Suppose now that $R$ is a $\mathbb{Q}$-algebra and that $A = R[\![x]\!]$ is the coordinatized ring of functions on a formal line over $R$. What's special about this rational curve case is that differentiation gives an isomorphism between the Kähler differentials $\Omega^1_{A/R}$ and the ideal $(x)$ of functions vanishing at the origin (i.e., the ideal sheaf selecting the closed subscheme $0 \colon \operatorname{Spec} R \to \operatorname{Spf} A$). Its inverse is formal integration:
$$\int \colon \left( \sum_{j=0}^\infty c_j x^j \right) \mathrm{d}x \mapsto \sum_{j=0}^\infty \frac{c_j}{j+1} x^{j+1}.$$

In pursuit of the construction of a logarithm for a formal group $\widehat{\mathbb{G}}$ over $R$, we now take a cue from classical Lie theory:

**Definition 2.1.22** A 1-form $\omega \in \Omega^1_{A/R}$ is said to be *invariant (under a group law $+_\varphi$)* when $\omega = \tau_y^* \omega$ for all translations $\tau_y(x) = x +_\varphi y$. We write $\omega_{\widehat{\mathbb{G}}} \subseteq \Omega^1_{A/R}$ for the subsheaf of such invariant 1-forms.

## 2.1 Calculus on Formal Varieties

**Lemma 2.1.23** *For R a $\mathbb{Q}$-algebra, integration gives an isomorphism*
$$\int : \omega_{\widehat{\mathbb{G}}} \to \text{FormalGroups}(\widehat{\mathbb{G}}, \widehat{\mathbb{G}}_a).$$

*Proof* Suppose that $f \in \omega_{\widehat{\mathbb{G}}}$ is an invariant differential, which in terms of a coordinate $x$ indicates the equation
$$f(x)\,dx = \omega = \tau_y^* \omega = f(x +_\varphi y)\,d(x +_\varphi y).$$

Writing $F(x) = \int f(x)\,dx$ for its formal integral, we use the multivariate chain rule to extend this equation to
$$\frac{\partial F(x)}{\partial x}\,dx = f(x)\,dx = \omega = \tau_y^* \omega = f(x +_\varphi y)\,d(x +_\varphi y) = \frac{\partial F(x +_\varphi y)}{\partial x}\,dx.$$

It follows that $F(x +_\varphi y)$ and $F(x)$ differ by a constant. Checking at $x = 0$ shows that the constant is $F(y)$, hence
$$F(x +_\varphi y) = F(x) + F(y).$$

This chain of steps can be read in reverse: Starting with an $F(x)$ satisfying this last equation, differentiating against $x$ yields the long equation above, and hence an invariant differential $\omega = \frac{\partial F(x)}{\partial x}\,dx$. □

**Lemma 2.1.24** *For R any ring, restriction to the identity point yields an isomorphism $\omega_{\widehat{\mathbb{G}}} \cong T_0^* \widehat{\mathbb{G}}$.*

*Proof* We set out to analyze the space of invariant differentials in terms of a coordinate $x$. As above, a differential $\omega$ admits expression as $\omega = f(x)\,dx$, and the invariance condition above becomes
$$f(x)\,dx = f(y +_\varphi x)\,d(y +_\varphi x) = f(y +_\varphi x)\frac{\partial(y +_\varphi x)}{\partial x}\,dx.$$

Restricting this equation to the origin by setting $x = 0$, we produce the condition
$$f(0) = f(y) \cdot \left.\frac{\partial(y +_\varphi x)}{\partial x}\right|_{x=0}.$$

The partial differential is a multiplicatively invertible power series, and hence we may rewrite $f(y)$ as
$$f(y) = f(0) \cdot \left(\left.\frac{\partial(y +_\varphi x)}{\partial x}\right|_{x=0}\right)^{-1}.$$

This shows that the assignment $f(y) \mapsto f(0)$ is bijective, establishing the desired isomorphism. □

**Theorem 2.1.25** *If $R$ is a $\mathbb{Q}$-algebra, there is a natural logarithm*
$$\log_{\widehat{\mathbb{G}}} \colon \widehat{\mathbb{G}} \to \widehat{\mathbb{G}}_a \otimes T_0\widehat{\mathbb{G}}.$$

*Proof* First, Lemma 2.1.24 shows that a choice of section of the cotangent space at the identity of $\widehat{\mathbb{G}}$ uniquely specifies an invariant differential. Then, Lemma 2.1.23 shows that this uniquely specifies a logarithm function. This describes an isomorphism
$$T_0^*\widehat{\mathbb{G}} \to \text{FormalGroups}(\widehat{\mathbb{G}}, \widehat{\mathbb{G}}_a),$$
which we transpose to give the desired isomorphism
$$\log_{\widehat{\mathbb{G}}} \colon \widehat{\mathbb{G}} \to \widehat{\mathbb{G}}_a \otimes T_0\widehat{\mathbb{G}}. \qquad \square$$

*Example 2.1.26* Consider the formal group law $x_1(t_1) +_{\widehat{\mathbb{G}}_m} x_2(t_2) = t_1 + t_2 - t_1 t_2$ studied in Example 2.1.21. Its associated rational logarithm is computed as
$$\log_{\widehat{\mathbb{G}}_m}(t_2) = f(0) \cdot \int \frac{1}{1-t_2}\, dt_2 = -f(0)\log(1-t_2) = -f(0)\log(x_2),$$
where "$\log(x_2)$" refers to Napier's classical natural logarithm of $x_2$.

## 2.2 Divisors on Formal Curves

We continue to develop vocabulary and accompanying machinery used to give algebro-geometric reinterpretations of the results in the introduction to this case study. In the previous section we deployed the language of formal schemes to recast Corollary 2.0.4 in geometric terms, and we now turn toward reencoding Theorem 2.0.3. In Definition 1.4.7 and Lemma 1.4.8 we discussed a general correspondence between $R$-modules and quasicoherent sheaves over Spec $R$, and the isomorphism of one-dimensional $E^*X$-modules appearing in Theorem 2.0.3 moves us to study sheaves over $X_E$ which are one-dimensional – i.e., line bundles. In fact, for the purposes of Theorem 2.0.3, we will find that it suffices to understand the basics of the geometric theory of line bundles *just over formal curves*. This is our goal in this lecture, and we leave the applications to algebraic topology aside for later. For the rest of this section we fix the following three pieces of data: a base formal scheme $S$, a formal curve $C$ over $S$, and a distinguished point $\zeta \colon S \to C$ on $C$.

To begin, we will be interested in a very particular sort of line bundle over $C$: for any function $f$ on $C$ which is not a zero-divisor, the (ideal) subsheaf $\mathcal{I}_f = f \cdot \mathcal{O}_C$ of functions on $C$ which are divisible by $f$ form a one-dimensional $\mathcal{O}_C$-submodule of the ring of functions $\mathcal{O}_C$ itself – i.e., a line bundle on $C$. By

## 2.2 Divisors on Formal Curves

interpreting $\mathcal{I}_f$ as an ideal sheaf, this gives rise to a second interpretation of these data in terms of a closed subscheme

$$\operatorname{Spec} \mathcal{O}_C/f \subseteq C,$$

which we will refer to as the *divisor* associated to $\mathcal{I}_f$. In general these can be somewhat pathological, so we specialize further to an extremely nice situation:

**Definition 2.2.1** ([Str99b, section 5.1]) A formal subscheme $D$ of a formal scheme $X = \operatorname{colim}_\alpha X_\alpha$ is said to be *closed* when it pulls back along any of the defining affine schemes $X_\alpha \to X$ to give a closed subscheme of $X_\alpha$. An *effective Weil divisor* $D$ on a formal curve $C$ is a closed subscheme of $C$ whose structure map $D \to S$ presents $D$ as finite and free. We say that the *rank* of $D$ is $n$ when its ring of functions $\mathcal{O}_D$ is free of rank $n$ over $\mathcal{O}_S$.

**Lemma 2.2.2** ([Str99b, proposition 5.2], see also [Str99b, example 2.10]) *There is a scheme $\operatorname{Div}_n^+ C$ of effective Weil divisors of rank $n$. It is a formal variety of dimension $n$. In fact, a coordinate $x$ on $C$ determines an isomorphism $\operatorname{Div}_n^+ C \cong \widehat{\mathbb{A}}^n$ where a divisor $D$ is associated to a monic polynomial $f_D(x)$ with nilpotent lower-order coefficients.*

*Proof sketch* To pin down the functor we wish to analyze, we make the definition

$$\operatorname{Div}_n^+(C)(u\colon \operatorname{Spec} T \to S) = \{D \text{ an effective divisor on } u^*C \text{ of rank } n\}.$$

To show that this is a formal variety, we pursue the final claim and select a coordinate $x$ on $C$, as well as a point $D \in \operatorname{Div}_n^+(C)(T)$. The coordinate presents $u^*C$ as

$$u^*C = \operatorname{Spf} T[\![x]\!],$$

and the characteristic polynomial $f_D(x)$ of $x$ in $\mathcal{O}_D$ presents $D$ as the closed subscheme

$$D = \operatorname{Spf} T[\![x]\!]/(f_D(x))$$

for $f_D(x) = x^n + a_{n-1} x^{n-1} + \cdots + a_0$ monic. Additionally, for any prime ideal $\mathfrak{p} \in T$ we can form the field $T_\mathfrak{p}/\mathfrak{p}$, over which the module $\mathcal{O}_D \otimes_T T_\mathfrak{p}/\mathfrak{p}$ must still be of rank $n$. It follows that

$$f_D(x) \otimes_T T_\mathfrak{p}/\mathfrak{p} \equiv x^n,$$

hence that each $a_j$ lies in the intersection of all prime ideals of $T$, hence that each $a_j$ is nilpotent.

In turn, this means that the polynomial $f_D$ is selected by a map $\operatorname{Spec} T \to \widehat{\mathbb{A}}^n$.

Conversely, given such a map, we can form the polynomial $f_D(x)$ and the divisor $D$. □

*Remark 2.2.3* This lemma effectively connects several simple dots: especially nice polynomials $f_D(x) \in \mathcal{O}_C$, their vanishing loci $D \subseteq C$, and the ideal sheaves $\mathcal{I}_D$ of functions divisible by $f$ – i.e., functions with a partially prescribed vanishing set. Basic operations on polynomials affect their vanishing loci in predictable ways, and these operations are also reflected at the level of divisor schemes. For instance, there is a unioning map

$$\mathrm{Div}_n^+ C \times \mathrm{Div}_m^+ C \to \mathrm{Div}_{n+m}^+ C,$$
$$(D_1, D_2) \mapsto D_1 \sqcup D_2.$$

At the level of ideal sheaves, we use their one-dimensionality to produce the formula

$$\mathcal{I}_{D_1 \sqcup D_2} = \mathcal{I}_{D_1} \otimes_{\mathcal{O}_C} \mathcal{I}_{D_2}.$$

Under a choice of coordinate $x$, the map at the level of polynomials is given by

$$(f_{D_1}, f_{D_2}) \mapsto f_{D_1} \cdot f_{D_2}.$$

Next, note that there is a canonical isomorphism $C \xrightarrow{\cong} \mathrm{Div}_1^+ C$. Iterating the above addition map gives a map

$$C^{\times n} \xrightarrow{\sqcup} \mathrm{Div}_n^+ C.$$

**Lemma 2.2.4** *The object $C_{\Sigma_n}^{\times n}$ exists as a formal variety, and it participates in the following triangle:*

$$\begin{array}{ccc} & & C^{\times n} \\ & \swarrow & \downarrow{\sqcup} \\ C_{\Sigma_n}^{\times n} & \xrightarrow{\cong} & \mathrm{Div}_n^+ C. \end{array}$$

*Proof* The first assertion is a consequence of Newton's theorem on symmetric polynomials: The subring of symmetric polynomials in $R[x_1, \ldots, x_n]$ is itself polynomial on generators

$$\sigma_j(x_1, \ldots, x_n) = \sum_{\substack{S \subseteq \{1, \ldots, n\} \\ |S|=j}} x_{S_1} \cdots x_{S_j},$$

and hence

$$R[\sigma_1, \ldots, \sigma_n] \subseteq R[x_1, \ldots, x_n]$$

gives an affine model of the quotient map $\widehat{\mathbb{A}}^n \to (\widehat{\mathbb{A}}^n)_{\Sigma_n}$. Picking a coordinate

## 2.2 Divisors on Formal Curves

on $C$ allows us to import this fact into formal geometry to deduce the existence of $C^{\times n}_{\Sigma_n}$. The factorization then follows by noting that the iterated $\sqcup$ map is symmetric. Finally, Remark 2.2.3 shows that the horizontal map pulls the coordinate $a_j$ back to $\sigma_j$, so the third assertion follows. □

*Remark 2.2.5* The map $C^{\times n} \to C^{\times n}_{\Sigma_n}$ is an example of a map of schemes which is surjective *as a map of sheaves*. This is somewhat subtle: For any given test ring $T$, it is not necessarily the case that $C^{\times n}(T) \to C^{\times n}_{\Sigma_n}(T)$ is surjective on $T$-points. However, for a fixed point $f \in C^{\times n}_{\Sigma_n}(T)$, we are guaranteed a flat covering $T \to \prod_j T_j$ such that there are individual lifts $\widetilde{f_j}$ of $f$ over each $T_j$.[9]

Now we use the pointing $\zeta \colon S \to C$ to interrelate divisor schemes of varying ranks. Together with the $\sqcup$ operation, $\zeta$ gives a composite

$$\mathrm{Div}_n^+ C \longrightarrow C \times \mathrm{Div}_n^+ C \longrightarrow \mathrm{Div}_1^+ C \times \mathrm{Div}_n^+ C \longrightarrow \mathrm{Div}_{n+1}^+ C,$$

$$D \longmapsto (\zeta, D) \longmapsto (\langle\zeta\rangle, D) \longmapsto \langle\zeta\rangle \sqcup D.$$

**Definition 2.2.6** We define the following variants of "stable divisor schemes":

$$\mathrm{Div}^+ C = \bigsqcup_{n \geq 0} \mathrm{Div}_n^+ C,$$

$$\mathrm{Div}_n C = \mathrm{colim}\left( \mathrm{Div}_n^+ C \xrightarrow{\langle\zeta\rangle+-} \mathrm{Div}_{n+1}^+ C \xrightarrow{\langle\zeta\rangle+-} \cdots \right),$$

$$\mathrm{Div}\, C = \mathrm{colim}\left( \mathrm{Div}^+ C \xrightarrow{\langle\zeta\rangle+-} \mathrm{Div}^+ C \xrightarrow{\langle\zeta\rangle+-} \cdots \right),$$

$$\cong \bigsqcup_{n \in \mathbb{Z}} \mathrm{Div}_n C.$$

The first of these constructions is very suggestive: It looks like the free commutative monoid formed on a set, and we might hope that the construction in formal schemes enjoys a similar universal property. In fact, all three constructions have universal properties:

**Theorem 2.2.7** (see Corollary 5.1.10) *The scheme* $\mathrm{Div}^+ C$ *models the free formal commutative monoid on the unpointed formal curve* $C$. *The scheme* $\mathrm{Div}\, C$ *models the free formal group on the unpointed formal curve* $C$. *The scheme* $\mathrm{Div}_0 C$ *simultaneously models the free formal commutative monoid and the free formal group on the* pointed *formal curve* $C$. □

---

[9] This amounts to the claim that not every polynomial can be written as a product of linear factors. For instance, the divisor on $C = \mathrm{Spf}\, \mathbb{R}[\![x]\!]$ defined by the equation $x^2 + 1$ splits as $(x+i)(x-i)$ after base-change along the flat cover $\mathrm{Spec}\, \mathbb{C} \to \mathrm{Spec}\, \mathbb{R}$.

We will postpone the proof of this theorem until later, when we've developed a theory of coalgebraic formal schemes.

*Remark 2.2.8* Given $q\colon C \to C'$ a map of formal curves over $S$ and $D \subseteq C$ a divisor on $C$, the construction of $\mathrm{Div}_n^+ C$ as a symmetric space as in Lemma 2.2.4 shows that there is a corresponding *pushforward* divisor $q_*D$ on $C'$. This can be thought of in some other ways – for instance, this map at the level of sheaves [Har77, chapter IV, exercise 2.6] is given by

$$\det q_* \mathcal{I}_D \cong (\det q_* \mathcal{O}_C) \otimes \mathcal{I}_{q_*D}.$$

We can also use Theorem 2.2.7 in the stable case: The composite map

$$C \xrightarrow{q} C' \cong \mathrm{Div}_1^+ C' \to \mathrm{Div}\, C'$$

targets a formal group scheme, and hence universality induces a map

$$q_*\colon \mathrm{Div}\, C \to \mathrm{Div}\, C'.$$

On the other hand, for a general $q$ the pullback $D \times_{C'} C$ of a divisor $D \subseteq C'$ will not be a divisor on $C$. It is possible to impose conditions on $q$ so that this is so, and in this case $q$ is called an *isogeny*. We will return to this in Appendix A.2.

Our final goal for this section is to broaden this discussion to line bundles on formal curves generally, using this nice case as a model. The main classical theorem is that line bundles, sometimes referred to as *Cartier divisors*, arise as the group-completion and sheafification of zero loci of polynomials, referred to (as above) as *Weil divisors*. In the case of a *formal* curve, sheafification has little effect, and so we seek to exactly connect line bundles on a formal curve with formal differences of Weil divisors. To begin, we need some vocabulary that connects the general case to the one studied above.

**Definition 2.2.9** (see [Vak15, section 14.2]) Suppose that $\mathcal{L}$ is a line bundle on $C$ and select a section $u$ of $\mathcal{L}$. There is a largest closed subscheme $D \subseteq C$ where the condition $u|_D = 0$ is satisfied. If $D$ is a divisor, $u$ is said to be *divisorial* and we write $\mathrm{div}\, u := D$.

**Lemma 2.2.10** (see [Vak15, exercise 14.2.E] and [Har77, proposition II.6.13]) *A divisorial section $u$ of a line bundle $\mathcal{L}$ induces an isomorphism $\mathcal{L} \cong \mathcal{I}_D$.* □

Line bundles which admit divisorial sections are therefore those that arise through our construction above, i.e., those which are controlled by the zero locus of a polynomial. In keeping with the classical inspiration, we expect generic line bundles to be controlled by the zeroes *and poles* of a rational function, and so we introduce the following class of functions:

**Definition 2.2.11** ([Str99b, definition 5.20 and proposition 5.26]) The ring of meromorphic functions on $C$, $\mathcal{M}_C$, is obtained by inverting all coordinates in $\mathcal{O}_C$.[10] Additionally, this can be augmented to a scheme $\text{Mer}(C, \mathbb{G}_m)$ of meromorphic functions on $C$ by

$$\text{Mer}(C, \mathbb{G}_m)(u \colon \text{Spec } T \to S) := (\mathcal{M}_{u^*C})^\times.$$

Thinking of a meromorphic function as the formal expansion of a rational function, we are moved to study the monoidality of Definition 2.2.9.

**Lemma 2.2.12** *If $u_1$ and $u_2$ are divisorial sections of $\mathcal{L}_1$ and $\mathcal{L}_2$ respectively, then $u_1 \otimes u_2$ is a divisorial section of $\mathcal{L}_1 \otimes \mathcal{L}_2$ and $\text{div}(u_1 \otimes u_2) = \text{div } u_1 + \text{div } u_2$.* □

**Definition 2.2.13** A *meromorphic divisorial section* of a line bundle $\mathcal{L}$ is a decompositon $\mathcal{L} \cong \mathcal{L}_1 \otimes \mathcal{L}_2^{-1}$ together with an expression of the form $u_+/u_-$, where $u_+$ and $u_-$ are divisorial sections of $\mathcal{L}_1$ and $\mathcal{L}_2$, respectively. We set $\text{div}(u_+/u_-) = \text{div } u_+ - \text{div } u_-$.

In the case of a formal curve, the fundamental theorem is that meromorphic functions, line bundles, and stable Weil divisors all essentially agree. A particular meromorphic function spans a one-dimensional $\mathcal{O}_C$-submodule sheaf of $\mathcal{M}_C$, and hence it determines a line bundle. Conversely, a line bundle is determined by local gluing data, which are exactly the data of a meromorphic function. However, it is clear that there is some overdeterminacy in this presentation: Scaling a meromorphic function by a nowhere vanishing entire function will not modify the submodule sheaf. We now make the observation that the function div gives an assignment from meromorphic functions to stable Weil divisors, which is *also* insensitive to rescaling by a nowhere vanishing function. These inputs are arranged in the following theorem:

**Theorem 2.2.14** ([Str99b, proposition 5.26], [Strb, proposition 33.4], see also [Har77, proposition II.6.11]) *In the case of a formal curve $C$, there is a short exact sequence of formal groups*

$$0 \to \underline{\text{FormalSchemes}}(C, \mathbb{G}_m) \to \text{Mer}(C, \mathbb{G}_m) \to \text{Div}(C) \to 0. \quad \square$$

## 2.3 Line Bundles Associated to Thom Spectra

In this lecture, we will exploit all of the algebraic geometry previously set up to deduce a load of topological results.

---

[10] In fact, it suffices to invert any single one [Str99b, lemma 5.21].

**Definition 2.3.1** Let $E$ be a complex-orientable theory and let $V \to X$ be a complex vector bundle over a space $X$. According to Theorem 2.0.3, the cohomology of the Thom spectrum $E^*T(V)$ forms a one-dimensional $E^*X$-module. Using Lemma 1.4.8, we construct a line bundle over $X_E$

$$\mathbb{L}(V) := \widetilde{E^*T(V)},$$

called the *Thom sheaf* of $V$.

*Remark 2.3.2* One of the main utilities of this definition is that it only uses the *property* that $E$ is complex-orientable, and it begets only the *property* that $\mathbb{L}(V)$ is a line bundle.

This construction enjoys many properties already established.

**Corollary 2.3.3** *A vector bundle $V$ over $Y$ and a map $f\colon X \to Y$ induce an isomorphism*

$$\mathbb{L}(f^*V) \cong (f_E)^*\mathbb{L}(V).$$

*There is also is a canonical isomorphism*

$$\mathbb{L}(V \oplus W) = \mathbb{L}(V) \otimes \mathbb{L}(W).$$

*Finally, this property can then be used to extend the definition of $\mathbb{L}(V)$ to virtual bundles:*

$$\mathbb{L}(V - W) = \mathbb{L}(V) \otimes \mathbb{L}(W)^{-1}.$$

*Proof* The first claim is justified by the naturality of Theorem 1.1.12, the second is justified by Lemma 1.1.8, and the last is a direct consequence of the first two. □

We use these properties to work the following example, which connects Thom sheaves with the major players from Lecture 2.1.

*Example 2.3.4* ([AHS04, section 8])    Take $\mathcal{L}$ to be the canonical line bundle over $\mathbb{C}P^\infty$. Using the same mode of argument as in Example 1.1.3, the zero-section

$$\Sigma^\infty \mathbb{C}P^\infty \xrightarrow{\simeq} T_2(\mathcal{L})$$

gives an identification

$$E^0\mathbb{C}P^\infty \supseteq \widetilde{E}^0\mathbb{C}P^\infty \xleftarrow{\simeq} \widetilde{E}^0T_2(\mathcal{L})$$

of $\widetilde{E}^0T_2(\mathcal{L})$ with the augmentation ideal in $E^0\mathbb{C}P^\infty$. At the level of Thom sheaves, this gives an isomorphism

$$\mathcal{I}(0) \xleftarrow{\simeq} \mathbb{L}(\mathcal{L})$$

## 2.3 Line Bundles Associated to Thom Spectra

of $\mathbb{L}(\mathcal{L})$ with the sheaf of functions vanishing at the origin of $\mathbb{CP}^\infty_E$. Pulling $\mathcal{L}$ back along

$$0: * \to \mathbb{CP}^\infty$$

gives a line bundle over the one-point space, which on Thom spectra gives the inclusion

$$\Sigma^\infty \mathbb{CP}^1 \to \Sigma^\infty \mathbb{CP}^\infty.$$

Stringing many results together, we can now calculate:

$$\widetilde{\pi_2 E} \cong \widetilde{E^0 \mathbb{CP}^1} \quad (S^2 \simeq \mathbb{CP}^1)$$
$$\cong \mathbb{L}(0^* \mathcal{L}) \quad \text{(Definition 2.3.1)}$$
$$\cong 0^* \mathbb{L}(\mathcal{L}) \quad \text{(Corollary 2.3.3)}$$
$$\cong 0^* \mathcal{I}(0) \quad \text{(preceding calculation)}$$
$$\cong \mathcal{I}(0)/(\mathcal{I}(0) \cdot \mathcal{I}(0)) \quad \text{(definition of } 0^* \text{ from Definition 1.4.9)}$$
$$\cong T_0^* \mathbb{CP}^\infty_E \quad \text{(Definition 2.1.2)}$$
$$\cong \omega_{\mathbb{CP}^\infty_E}, \quad \text{(proof of Lemma 2.1.24)}$$

where $\omega_{\mathbb{CP}^\infty_E}$ denotes the sheaf of invariant differentials on $\mathbb{CP}^\infty_E$. Consequently, if $k \cdot \varepsilon$ is the trivial bundle of dimension $k$ over a point, then

$$\widetilde{\pi_{2k} E} \cong \mathbb{L}(k \cdot \varepsilon) \cong \mathbb{L}(k \cdot 0^* \mathcal{L}) \cong \mathbb{L}(0^* \mathcal{L})^{\otimes k} \cong \omega_{\mathbb{CP}^\infty_E}^{\otimes k}.$$

Finally, given an $E$-algebra $f : E \to F$ (e.g., $F = E^{X_+}$), then we have

$$\widetilde{\pi_{2k} F} \cong f_E^* \omega_{\mathbb{CP}^\infty_E}^{\otimes k}.$$

Outside of this example, it is difficult to find line bundles $\mathbb{L}(V)$ which we can analyze so directly. In order to get a handle on $\mathbb{L}(V)$ in general, we now seek to strengthen this bond between line bundles and vector bundles by finding inside of algebraic topology the alternative presentations of line bundles given in Lecture 2.2. In particular, we would like a topological construction on vector bundles which produces divisors – i.e., finite schemes over $X_E$. This has the scent of a certain familiar topological construction called projectivization, and we now work to justify the relationship.

**Definition 2.3.5** Let $V$ be a complex vector bundle of rank $n$ over a base $X$. Define $\mathbb{P}(V)$, the *projectivization of $V$*, to be the $\mathbb{CP}^{n-1}$-bundle over $X$ whose fiber over $x \in X$ is the space of complex lines in the original fiber $V|_x$.

**Theorem 2.3.6** *Take $E$ to be complex-oriented. The $E$-cohomology of $\mathbb{P}(V)$ is given by the formula*

$$E^*\mathbb{P}(V) \cong E^*(X)[\![t]\!]/c(V)$$

*for a certain monic polynomial*

$$c(V) = t^n - c_1(V)t^{n-1} + c_2(V)t^{n-2} - \cdots + (-1)^n c_n(V).$$

*Proof* We fit all of the fibrations we have into a single diagram:

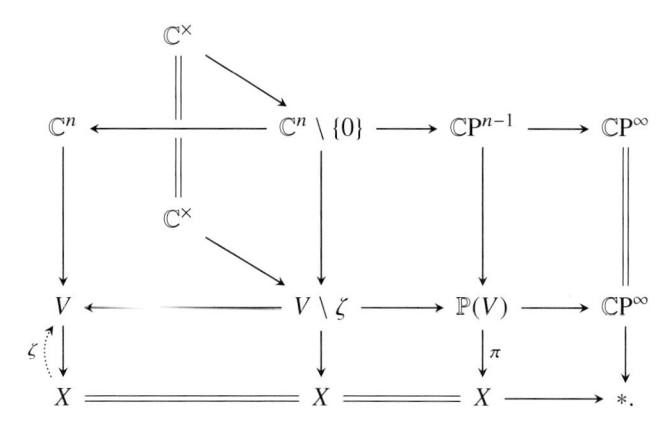

We read this diagram as follows: On the far left, there's the vector bundle we began with, as well as its zero-section $\zeta$. Deleting the zero-section gives the second bundle, a $\mathbb{C}^n \setminus \{0\}$-bundle over $X$. Its quotient by the scaling $\mathbb{C}^\times$-action gives the third bundle, a $\mathbb{C}P^{n-1}$-bundle over $X$. Additionally, the quotient map $\mathbb{C}^n \setminus \{0\} \to \mathbb{C}P^{n-1}$ is itself a $\mathbb{C}^\times$-bundle, and this induces the structure of a $\mathbb{C}^\times$-bundle on the quotient map $V \setminus \zeta \to \mathbb{P}(V)$. Thinking of these as complex line bundles, they are classified by a map to $\mathbb{C}P^\infty$, which can itself be thought of as the last vertical fibration, fibering over a point.

Note that the map on $E$-cohomology between these two last fibers is surjective. It follows that the Serre spectral sequence for the third vertical fibration is degenerate, since all the classes in the fiber must survive.[11] We thus conclude that $E^*\mathbb{P}(V)$ is a free $E^*(X)$-module on the classes $\{1, t, t^2, \ldots, t^{n-1}\}$ spanning $E^*\mathbb{C}P^{n-1}$, where $t$ encodes the chosen complex-orientation of $E$. To understand the ring structure, we need only compute $t^{n-1} \cdot t$, which must be able to be written in terms of the classes which are lower in $t$-degree:

$$t^n = c_1(V)t^{n-1} - c_2(V)t^{n-2} + \cdots + (-1)^{n-1}c_n(V)$$

for some classes $c_j(V) \in E^*X$. The main claim follows. □

---

[11] This is called the Leray–Hirsch theorem.

## 2.3 Line Bundles Associated to Thom Spectra

In coordinate-free language, we have the following corollary:

**Corollary 2.3.7** (Theorem 2.3.6 redux) *Take E to be complex-orientable. The map*

$$\mathbb{P}(V)_E \to X_E \times \mathbb{CP}^\infty_E$$

*is a closed inclusion of $X_E$-schemes, and the structure map $\mathbb{P}(V)_E \to X_E$ is free and finite of rank $n$.*[12] *It follows that $\mathbb{P}(V)_E$ is a divisor on $\mathbb{CP}^\infty_E$ considered over $X_E$, i.e.,*

$$\mathbb{P}(V)_E \in \left(\mathrm{Div}_n^+(\mathbb{CP}^\infty_E)\right)(X_E). \qquad \square$$

**Definition 2.3.8** The classes $c_j(V)$ of Theorem 2.3.6 are called the *Chern classes* of $V$ (with respect to the complex-orientation $t$ of $E$), and the polynomial $c(V) = \sum_{j=0}^n (-1)^{n-j} t^j c_{n-j}(V)$ is called the *Chern polynomial*.

The next major theorems concerning projectivization are the following:

**Corollary 2.3.9** *The sub-bundle of $\pi^*(V)$ consisting of vectors $(v, (\ell, x))$ such that $v$ lies along the line $\ell$ splits off a canonical line bundle.* $\qquad \square$

**Corollary 2.3.10** ("Splitting principle"/"Complex-oriented descent") *Associated to any n-dimensional complex vector bundle $V$ over a base $X$, there is a canonical map $f_V: Y_V \to X$ such that $(f_V)_E: (Y_V)_E \to X_E$ is finite and faithfully flat, and there is a canonical splitting into complex line bundles:*

$$f_V^*(V) \cong \bigoplus_{i=1}^n \mathcal{L}_i. \qquad \square$$

This last corollary is extremely important. Its essential algebraic content is to say that any question about characteristic classes can be checked for sums of line bundles. Specifically, because of the injectivity of $f_V^*$, any relationship among the characteristic classes deduced in $E^* Y_V$ must already be true in the ring $E^* X$. The following theorem is a consequence of this principle:

**Theorem 2.3.11** ([Swi02, theorem 16.2 and 16.10]) *Again take $E$ to be complex-oriented. The coset fibration*

$$U(n-1) \to U(n) \to S^{2n-1}$$

*deloops to a spherical fibration*

$$S^{2n-1} \to BU(n-1) \to BU(n).$$

---

[12] That this map is relative to $X_E$ is why we were free to consider only curves in the previous lecture.

*The associated Serre spectral sequence*

$$E_2^{*,*} = H^*(BU(n); E^*S^{2n-1}) \Rightarrow E^*BU(n-1)$$

*degenerates at $E_{2n}$ and induces an isomorphism*

$$E^*BU(n) \cong E^*[\![\sigma_1, \ldots, \sigma_n]\!].$$

*Now, let $V: X \to BU(n)$ classify a vector bundle $V$. Then the coefficient $c_j$ in the polynomial $c(V)$ is selected by $\sigma_j$:*

$$c_j(V) = V^*(\sigma_j).$$

*Proof sketch* The first part is a standard calculation. To prove the relation between the Chern classes and the $\sigma_j$, the splitting principle states that we can complete the map $V: X \to BU(n)$ to a square

$$\begin{array}{ccc} Y_V & \xrightarrow{\oplus_{i=1}^n \mathcal{L}_i} & BU(1)^{\times n} \\ \downarrow{f_V} & & \downarrow{\oplus} \\ X & \xrightarrow{V} & BU(n). \end{array}$$

The equation $c_j(f_V^* V) = V^*(\sigma_j)$ can be checked in $E^*Y_V$. □

We now see that not only does $\mathbb{P}(V)_E$ produce a point of $\mathrm{Div}_n^+(\mathbb{CP}_E^\infty)$, but actually the scheme $\mathrm{Div}_n^+(\mathbb{CP}_E^\infty)$ itself appears internally to topology:

**Corollary 2.3.12** *For a complex orientable cohomology theory $E$, there is an isomorphism*

$$BU(n)_E \cong \mathrm{Div}_n^+ \mathbb{CP}_E^\infty,$$

*so that maps $V: X \to BU(n)$ are transported to divisors $\mathbb{P}(V)_E \subseteq \mathbb{CP}_E^\infty \times X_E$. Selecting a particular complex orientation of $E$ begets two isomorphisms*

$$BU(n)_E \cong \widehat{\mathbb{A}}^n, \qquad \mathrm{Div}_n^+ \mathbb{CP}_E^\infty \cong \widehat{\mathbb{A}}^n,$$

*and these are compatible with the previous centered isomorphism.*[13] □

This description has two remarkable features. One is its "faithfulness": This isomorphism of formal schemes means that the entire theory of characteristic classes is captured by the behavior of the divisor scheme. The other aspect is its coherence with topological operations we find on $BU(n)$. For instance, the Whitney sum map translates as follows:

**Lemma 2.3.13** *There is a commuting square*

---

[13] See [Str99b, proposition 8.31] for a proof that recasts Theorem 2.3.11 itself in coordinate-free terms.

## 2.3 Line Bundles Associated to Thom Spectra

$$
\begin{array}{ccc}
BU(n)_E \times BU(m)_E & \xrightarrow{\oplus} & BU(n+m) \\
\| & & \| \\
\mathrm{Div}_n^+ \mathbb{CP}_E^\infty \times \mathrm{Div}_m^+ \mathbb{CP}_E^\infty & \xrightarrow{\sqcup} & \mathrm{Div}_{n+m}^+ \mathbb{CP}_E^\infty.
\end{array}
$$

*Proof* The sum map

$$BU(n) \times BU(m) \xrightarrow{\oplus} BU(n+m)$$

induces on Chern polynomials the identity [Swi02, theorem 16.2.d]

$$c(V_1 \oplus V_2) = c(V_1) \cdot c(V_2).$$

In terms of divisors, this means

$$\mathbb{P}(V_1 \oplus V_2)_E = \mathbb{P}(V_1)_E \sqcup \mathbb{P}(V_2)_E. \qquad \square$$

The following is a consequence of combining this lemma with the splitting principle:

**Corollary 2.3.14** *The map* $Y_E \xrightarrow{f_V} X_E$ *of Corollary 2.3.10 pulls* $\mathbb{P}(V)_E$ *back to give*

$$Y_E \times_{X_E} \mathbb{P}(V)_E \cong \bigsqcup_{i=1}^n \mathbb{P}(\mathcal{L}_i)_E.$$

*Interpretation* This says that the splitting principle is a topological enhancement of the claim that a divisor on a curve can be base-changed along a finite flat map where it splits as a sum of points. $\qquad \square$

The other constructions from Lecture 2.2 are also easily matched up with topological counterparts:

**Corollary 2.3.15** *There are natural isomorphisms* $BU_E \cong \mathrm{Div}_0\, \mathbb{CP}_E^\infty$ *and* $(BU \times \mathbb{Z})_E \cong \mathrm{Div}\, \mathbb{CP}_E^\infty$. *Additionally,* $(BU \times \mathbb{Z})_E$ *is the free formal group on the curve* $\mathbb{CP}_E^\infty$.[14] *These relationships are summarized in Table 2.1.* $\qquad \square$

**Corollary 2.3.16** *There is a commutative diagram*

$$
\begin{array}{ccc}
BU(n)_E \times BU(m)_E & \xrightarrow{\otimes} & BU(nm)_E \\
\| & & \| \\
\mathrm{Div}_n^+ \mathbb{CP}_E^\infty \times \mathrm{Div}_m^+ \mathbb{CP}_E^\infty & \xrightarrow{\cdot} & \mathrm{Div}_{nm}^+ \mathbb{CP}_E^\infty,
\end{array}
$$

---

[14] Similar methods to those of this section also show that the map $\mathbb{CP}_E^{n-1} \to \Omega SU(n)_E$ presents $\Omega SU(n)_E$ as the free formal group on $\mathbb{CP}_E^{n-1}$.

72                            Complex Bordism

|              | Spaces                    |                               | Schemes                   |
| ------------ | ------------------------- | ----------------------------- | ------------------------- |
| Object       | Classifies                | Object                        | Classifies                |
| $BU(n)$      | Rank $n$ vector bundles   | $\mathrm{Div}_n^+ \mathbb{CP}_E^\infty$ | Rank $n$ effective divisors |
| $\coprod_n BU(n)$ | All vector bundles   | $\mathrm{Div}^+ \mathbb{CP}_E^\infty$   | All effective divisors    |
| $BU \times \mathbb{Z}$ | Stable virtual bundles | $\mathrm{Div}\, \mathbb{CP}_E^\infty$ | Stable Weil divisors      |
| $BU \times \{0\}$ | Rank 0 stable virtual bundles | $\mathrm{Div}_0\, \mathbb{CP}_E^\infty$ | Rank 0 stable divisors |

Table 2.1  Different notions of vector bundles and their associated divisors

*where for a test algebra $f : \mathrm{Spec}\, T \to \mathrm{Spec}\, E_0$ the bottom map acts by*

$$(D_1, D_2 \subseteq f^*\mathbb{CP}_E^\infty) \mapsto (D_1 \times D_2 \subseteq f^*\mathbb{CP}_E^\infty \times f^*\mathbb{CP}_E^\infty \xrightarrow{\mu} f^*\mathbb{CP}_E^\infty),$$

*and $\mu$ is the map induced by the tensor product map $\mathbb{CP}^\infty \times \mathbb{CP}^\infty \to \mathbb{CP}^\infty$.*

*Proof* By the splitting principle, it is enough to check this on sums of line bundles. A sum of line bundles corresponds to a totally decomposed divisor, so we consider the case of a pair of such divisors $\bigsqcup_{i=1}^n \{a_i\}$ and $\bigsqcup_{j=1}^m \{b_j\}$. Referring to Definition 2.1.16, the map acts by convolution of divisors:

$$\left(\bigsqcup_{i=1}^n \{a_i\}\right)\left(\bigsqcup_{j=1}^m \{b_j\}\right) = \bigsqcup_{i,j} \{a_i +_{\mathbb{CP}_E^\infty} b_j\}. \qquad \square$$

**Corollary 2.3.17**  *The determinant map*

$$BU(n)_E \xrightarrow{B \det_E} BU(1)_E$$

*models the summation map*

$$\mathrm{Div}_n^+ \mathbb{CP}_E^\infty \xrightarrow{\sigma} \mathbb{CP}_E^\infty.$$

*Proof* This is a direct consequence of the splitting principle, the factorization

$$\begin{array}{ccc} BU(n) & \xrightarrow{B \det} & BU(1) \\ {\scriptstyle \oplus} \uparrow & \nearrow {\scriptstyle \otimes} & \\ BU(1)^{\times n}, & & \end{array}$$

and Corollary 2.3.16. $\qquad \square$

Finally, we can connect our analysis of the divisors coming from topological vector bundles with the line bundles studied at the start of the section.

**Lemma 2.3.18** ([Strb, section 19], [Str99b, definition 8.33])  *Let $\mathcal{I}(\mathbb{P}(V)_E)$ denote the ideal sheaf on $X_E \times \mathbb{CP}_E^\infty$ associated to the divisor subscheme $\mathbb{P}(V)_E$,* and let $\zeta : X_E \to X_E \times \mathbb{CP}^\infty_E$ denote the pointing of the formal curve $\mathbb{CP}^\infty_E$. There is a natural isomorphism of sheaves over $X_E$:

$$\zeta^* \mathcal{I}(\mathbb{P}(V)_E) \cong \mathbb{L}(V).$$

*Proof sketch* As employed in Example 2.3.4, a model for $\zeta^* \mathcal{I}(D)$ is

$$\zeta^* \mathcal{I}(D) = \mathcal{I}(D)/\mathcal{I}(D \sqcup 0).$$

The Thom space also has such a model: There is a cofiber sequence

$$\mathbb{P}(V) \to \mathbb{P}(V \oplus \mathrm{triv}) \to T(V).$$

Applying $E$-cohomology yields the result. □

**Theorem 2.3.19** (see Theorem 5.2.1) *A trivialization* $t \colon \mathbb{L}(\mathcal{L}) \cong \mathcal{O}_{\mathbb{CP}^\infty_E}$ *of the Thom sheaf associated to the canonical bundle induces an orientation* $MUP \to E$.

*Proof sketch* Suppose that $V$ is a rank $n$ vector bundle over $X$, and let $f \colon Y \to X$ be the space guaranteed by the splitting principle to provide an isomorphism $f^*V \cong \bigoplus_{j=1}^n \mathcal{L}_j$. The chosen trivialization $t$ then pulls back to give a trivialization of $\mathcal{I}(\mathbb{P}(f^*V)_E)$, and by finite flatness this descends to also give a trivialization of $\mathcal{I}(\mathbb{P}(V)_E)$. Pulling back along the zero section gives a trivialization of $\mathbb{L}(V)$. Then note that the system of trivializations produced this way is multiplicative, as a consequence of $\mathbb{P}(V_1 \oplus V_2)_E \cong \mathbb{P}(V_1)_E \sqcup \mathbb{P}(V_2)_E$. In the universal examples, this gives a sequence of compatible maps $MU(n) \to E$ which assemble on the colimit $n \to \infty$ to give the desired map of ring spectra. □

## 2.4 Power Operations for Complex Bordism

Our eventual goal, like in Case Study 1, is to give an algebro-geometric description of $MU_*(*)$ and of the cooperations $MU_*MU$. It is possible to approach this the same way as in Lecture 1.5, using the Adams spectral sequence [Qui69, theorem 2], [Lura, lecture 9]. However, $MU_*(*)$ is an integral algebra and so we cannot make do with working out the mod-2 Adams spectral sequence alone – we would at least have to work out the mod-$p$ Adams spectral sequence for every $p$. At odd primes $p$, there is the following unfortunate theorem:

**Theorem 2.4.1** *There is an isomorphism*

$$H\mathbb{F}_{p*}H\mathbb{F}_p \cong \mathbb{F}_p[\xi_1, \xi_2, \ldots] \otimes \Lambda[\tau_0, \tau_1, \tau_2, \ldots]$$

*with* $|\xi_j| = 2p^j - 2$ *and* $|\tau_j| = 2p^j - 1$. □

There are odd-dimensional classes in this algebra, and the *graded-commutativity* of the dual mod-$p$ Steenrod algebra means that these classes anti-commute. This prohibits us from even writing "$\mathrm{Spec}(H\mathbb{F}_{p*}H\mathbb{F}_p)$," and this is the first time we have encountered Hindrance #4 from Lecture 1.3 in the wild.

Because of this, we will not feel any guilt for taking a completely alternative approach to this calculation. This other method, due to Quillen, has as its keystone a completely different kind of cohomology operation called a *power operation*. These operations are quite technical to describe, but at their core is taking the $n$th power of a cohomology class – and hence they have a frustrating lack of properties, including failures to be additive and to be stable. Our goal in this lecture is just to define these cohomology operations, specialized to the particular setting we will need for Quillen's proof. Some remarks are in order:

- We will accomplish this as early as Remark 2.4.6, and though we proceed from there to sketch how Quillen's methods belong to a general framework, the reader should not hesitate to continue on to Lecture 2.5 when they grow bored, assured that the discussion of "Tate power operations" will not resurface.
- Additionally, unlike with Steenrod squares, the algebro-geometric interpretation of these operations will not be immediately accessible to us, and indeed their eventual reinterpretation in these terms is one of the more hard-won pursuits in this area of mathematics – the curious reader will have to wait until Appendix A.2.
- As before, because we are ultimately aiming to make a computation, we will phrase our discussion in terms of graded objects.

Power operations arise not just from taking the $n$th power of a cohomology class but from also remarking on the natural symmetry of that operation. We record this symmetry using the following technical apparatus:

**Lemma 2.4.2** ([EKMM97, theorem 1.6], [HSS00, corollary 2.2.4], [Lurc, examples 6.1.4.2 and 6.1.6.2])  *Given a spectrum $E$, its $n$-fold smash power forms a $\Sigma_n$-spectrum, in the sense that there is a natural diagram $*/\!\!/\Sigma_n \to$* SPECTRA *of $\infty$-categories which selects $E^{\wedge n}$ on objects. In the case of $E = \Sigma^\infty X$, this diagram agrees with the permutation diagram $*/\!\!/\Sigma_n \to X^{\wedge n}$ formed in the one-category* SPACES.  □

A cohomology class $f\colon \Sigma^\infty_+ X \to E$ gives rise to a morphism of $\Sigma_n$-spectra

$$f^{\wedge n}\colon (\Sigma^\infty_+ X)^{\wedge n} \to E^{\wedge n}.$$

The homotopy colimit of such a diagram is called the *homotopy orbits* of the spectrum, and this gives a natural diagram

## 2.4 Power Operations for Complex Bordism

$$\begin{array}{ccccc}
(\Sigma_+^\infty X)^{\wedge n} & \xrightarrow{f^{\wedge n}} & E^{\wedge n} & \xrightarrow{\mu} & E \\
\downarrow & & \downarrow & \nearrow_{\mu_n} & \\
(\Sigma_+^\infty X)^{\wedge n}_{h\Sigma_n} & \xrightarrow{f^{\wedge n}_{h\Sigma_n}} & E^{\wedge n}_{h\Sigma_n} & &
\end{array}$$

A ring spectrum $E$ equipped with a suite of factorizations $\mu_n$ satisfying compatibility laws embodying term collection (i.e., $x^a \cdot x^b = x^{a+b}$) and iterated exponentiation (i.e., $(x^a)^b = x^{ab}$) is called an $H_\infty$-ring spectrum [BMMS86, definition I.3.1]. The composite

$$P^{\Sigma_n}_{\text{ext}} f \colon (\Sigma_+^\infty X)^{\wedge n}_{h\Sigma_n} \xrightarrow{f^{\wedge n}_{h\Sigma_n}} E^{\wedge n}_{h\Sigma_n} \xrightarrow{\mu_n} E$$

is called the *external (total)* $\Sigma_n$-*power operation* applied to $f$, and the restriction to the diagonal subspace

$$P^{\Sigma_n} f \colon \Sigma_+^\infty X \wedge \Sigma_+^\infty B\Sigma_n \simeq (\Sigma_+^\infty X)_{h\Sigma_n} \xrightarrow{\Delta_{h\Sigma_n}} (\Sigma_+^\infty X)^{\wedge n}_{h\Sigma_n} \xrightarrow{P^{\Sigma_n}_{\text{ext}} f} E$$

is called the *(internal total)* $\Sigma_n$-*power operation* applied to $f$. We can consider it as a cohomology class lying in

$$P^{\Sigma_n} f \in E^0(X \times B\Sigma_n).$$

*Remark 2.4.3* ([BMMS86, definition IV.7.1])   The optional adjective "total" is a reference to the following variant of this construction. By choosing a class in $E_* B\Sigma_n$, thought of as a functional on $E^* B\Sigma_n$, the Kronecker pairing of the total power operation with this fixed class gives rise to a truly internal cohomology operation on $E^* X$ as in the following composite:

$$\Sigma_+^\infty X \wedge \mathbb{S} \xrightarrow{1 \wedge \sigma} \Sigma_+^\infty X \wedge \Sigma_+^\infty B\Sigma_n \wedge E \xrightarrow{P^{\Sigma_n} f \wedge 1} E \wedge E \xrightarrow{\mu} E.$$

Cohomology classes $f \in E^{-q}(X)$ lying in degrees $q \neq 0$ require more care. Passing to the representative $f \colon \Sigma_+^\infty X \to \mathbb{S}^q \wedge E$, the analogous diagram is

$$\begin{array}{ccccc}
(\Sigma_+^\infty X)^{\wedge n} & \xrightarrow{f^{\wedge n}} & (\mathbb{S}^q \wedge E)^{\wedge n} & \xrightarrow{\mu} & \mathbb{S}^{nq} \wedge E \\
\downarrow & & \downarrow & \nearrow_{\mu_{n,q}} & \\
(\Sigma_+^\infty X)^{\wedge n}_{h\Sigma_n} & \xrightarrow{f^{\wedge n}_{h\Sigma_n}} & (\mathbb{S}^q \wedge E)^{\wedge n}_{h\Sigma_n} & \longrightarrow & \mathbb{S}^{q\rho}_{h\Sigma_n} \wedge E^{\wedge n}_{h\Sigma_n},
\end{array}$$

where $\mathbb{S}^\rho$ is the representation sphere for the permutation representation of $\Sigma_n$. Such a system of factorizations is called an $H^1_\infty$-ring structure on $E$ – but in practice these are very uncommon,[15] and instead a subsystem of factorizations

---

[15] In fact, they can only appear on $H\mathbb{F}_2$-algebras [BMMS86, section VII.6.1].

for every $q \equiv 0 \pmod{d}$ is called an $H_\infty^d$-ring structure [BMMS86, definition I.4.3]. In the same way, these give rise to external and internal power operations on cohomology classes of those negative degrees which are divisible by $d$. We will mostly concern ourselves with the theory of $H_\infty$-ring spectra in this section, but we will eventually work with $H_\infty^2$-ring spectra in the intended application, and so we now describe a naturally occurring such structure on $MU$.

**Definition 2.4.4** ([Rud98, definition VII.7.4]) Suppose that $\xi\colon X \to BU(k)$ presents a complex vector bundle of rank $k$ on $X$. The $n$-fold product of this bundle gives a new bundle

$$X^{\times n} \xrightarrow{\xi^{\times n}} BU(k)^{\times n} \xrightarrow{\oplus} BU(n \cdot k)$$

of rank $nk$ on which the symmetric group $\Sigma_n$ acts. By taking the (homotopy) $\Sigma_n$-quotient of the fiber, total, and base spaces, we produce a vector bundle $\xi(\Sigma_n)$ on $X_{h\Sigma_n}^{\times n}$ participating in the diagram

$$\begin{array}{ccccc} X^{\times n} & \xrightarrow{\xi^{\times n}} & BU(k)^{\times n} & \xrightarrow{\oplus} & BU(nk) \\ \downarrow & & \downarrow & \nearrow{\widetilde{\mu}_n} & \\ X_{h\Sigma_n}^{\times n} & \xrightarrow{\xi_{h\Sigma_n}^{\times n}} & BU(k)_{h\Sigma_n}^{\times n}. & & \end{array}$$

The universal case gives the map $\widetilde{\mu}_n$.

**Lemma 2.4.5** ([Rud98, equation VII.7.3]) *There is an isomorphism*

$$T(\xi(\Sigma_n)) \simeq (T\xi)_{h\Sigma_n}^{\wedge n}.$$

*Proof* This proof is mostly a matter of having had the idea to write down the lemma to begin with. From here, we string basic properties together:

$$\begin{aligned} T(\xi(\Sigma_n)) &= T(\xi_{h\Sigma_n}^{\times n}) & \text{(definition)} \\ &= T(\xi^{\times n})_{h\Sigma_n} & \text{(colimits commute with colimits)} \\ &= T(\xi)_{h\Sigma_n}^{\wedge n}. & (T \text{ is monoidal: Lemma 1.1.8}) \end{aligned}$$

□

Applying the lemma to the universal case produces a factorization

$$MU(k)^{\wedge n} \to MU(k)_{h\Sigma_n}^{\wedge n} \to MU(nk)$$

of the unstable multiplication map, and hence a stable factorization

$$MU^{\wedge n} \to MU_{h\Sigma_n}^{\wedge n} \xrightarrow{\mu_n} MU.$$

## 2.4 Power Operations for Complex Bordism

We now enrich this to an $H_\infty^2$ structure ([JN10, section 3.2], [BMMS86, corollary VIII.5.3]): By applying Lemma 2.4.5 in reverse, the construction is

$$\mu_{n,2q} \colon \mathbb{S}_{h\Sigma_n}^{2qn} \wedge MU_{h\Sigma_n}^{\wedge n} \xrightarrow{1 \wedge \mu_n} \mathbb{S}_{h\Sigma_n}^{2qn} \wedge MU$$
$$\simeq T(\mathrm{triv}^{\oplus q}(\Sigma_n) \text{ over } B\Sigma_n) \wedge MU \quad \text{(Lemma 2.4.5)}$$
$$\simeq \Sigma_+^{2qn} B\Sigma_n \wedge MU \quad \text{(Thom isomorphism)}$$
$$\to \Sigma^{2qn} MU. \quad \text{(project to basepoint)}$$

*Remark 2.4.6* The constructions above also go through with $\Sigma_n$ replaced by a subgroup. In our application, we will work with the cyclic subgroup $C_n \leq \Sigma_n$.

We now return to the setting of a general $H_\infty$-ring spectrum $E$. Some properties of power operations are immediately visible – for instance, they are multiplicative:

$$P^{\Sigma_n}(f \cdot g) = P^{\Sigma_n}(f) \cdot P^{\Sigma_n}(g),$$

and restriction of $P^{\Sigma_n}(f)$ to the basepoint in $\Sigma_+^\infty B\Sigma_n$ yields the cup power class $f^n$. In order to state any further properties, we will need to make some extraneous observations. First, note that any map of groups $\varphi \colon G \to \Sigma_n$ gives a variation on this construction by restriction of diagrams

$$*/\!/G \xrightarrow{\varphi} *\!/\!/\Sigma_n \underset{E^{\wedge n}}{\overset{(\Sigma_+^\infty X)^{\wedge n}}{\rightrightarrows}} \text{Spectra}.$$

This construction is useful when studying composites of power operations: The group $\Sigma_k \wr \Sigma_n \subseteq \Sigma_{nk}$ acts naturally on $(E^{\wedge k})^{\wedge n} \simeq E^{\wedge nk}$, and indeed there is an equivalence

$$P^{\Sigma_n} \circ P^{\Sigma_k} = P^{\Sigma_k \wr \Sigma_n}.$$

In order to understand these modified power operations more generally, we are motivated to study such maps $\varphi$ more seriously. Some basic constructions are summarized in the following definition:

**Definition 2.4.7** ([May96, sections XI.3 and XXV.3], see also [Lurc, section 6.1.6]) Let $\varphi \colon G \to G'$ be an inclusion of finite groups and let $F$ be a $G'$-spectrum. There is a natural map of homotopy colimits $\varphi_* \colon F_{hG} \to F_{hG'}$ which induces a *restriction map* on cohomology:

$$\mathrm{Res}_G^{G'} \colon E^0 F_{hG'} \to E^0 F_{hG}.$$

The spectrum $\bigvee_{G'/G} F$ considered as a $G'$-spectrum with the diagonal $G'$-action has the property $(\bigvee_{G'/G} F)_{hG'} = F_{hG}$, and the $G'$-equivariant averaging map

$$F \xrightarrow{\bigvee_{G'/G} \mathrm{id}} \bigvee_{G'/G} F$$

passes on homotopy orbits to the *additive norm map* $N_G^{G'} : F_{hG'} \to F_{hG}$, which again induces a map on cohomology classes

$$\mathrm{Tr}_G^{G'} : E^0 F_{hG} \to E^0 F_{hG'}$$

called the *transfer map*.

The restriction and transfer maps appear prominently in the following formula, which measures the failure of the power operation construction to be additive:

**Lemma 2.4.8** ([BMMS86, corollary II.1.6], [AHS04, proposition A.5, equation 3.6])  *For cohomology classes $f, g \in E^0 X$, there is a formula*[16]

$$P^{\Sigma_n}(f+g) = \sum_{i+j=n} \mathrm{Tr}^{\Sigma_n}_{\Sigma_i \times \Sigma_j} \left( P^{\Sigma_i}(f) \cdot P^{\Sigma_j}(g) \right). \qquad \square$$

To produce binomial formulas for the modified power operations, we use the following lemma:

**Lemma 2.4.9** ([Ada78b, pp. 109–110], [HKR00, section 6.5])  *Let $G_1, G_2$ be subgroups of $G'$, and consider the homotopy pullback diagram*

$$\begin{array}{ccc} P & \longrightarrow & */\!\!/ G_1 \\ \downarrow & & \downarrow \\ */\!\!/ G_2 & \longrightarrow & */\!\!/ G'. \end{array}$$

*For any identification $P \simeq \coprod_K (*/\!\!/ K)$,*[17] *there is a push–pull interchange formula*

$$\mathrm{Res}_{G_1}^{G'} \mathrm{Tr}_{G_2}^{G'} = \sum_K \mathrm{Tr}_K^{G_1} \mathrm{Res}_K^{G_2}. \qquad \square$$

**Corollary 2.4.10**  *For a transitive subgroup $G \leq \Sigma_n$, there is a congruence*

$$P^G(f+g) \equiv P^G(f) + P^G(g) \quad (\mathrm{mod}\ \textit{transfers from proper subgroups of } G).$$

*Proof sketch*  Note that $P^G$ can be defined by means of restriction, as in $P^G = \mathrm{Res}_G^{\Sigma_n} P^{\Sigma_n}$. We can hence reuse the previous binomial formula:

$$P^G(f+g) = \mathrm{Res}_G^{\Sigma_n} P^{\Sigma_n}(f+g) = \sum_{i+j=n} \mathrm{Res}_G^{\Sigma_n} \mathrm{Tr}^{\Sigma_n}_{\Sigma_i \times \Sigma_j} \left( P^{\Sigma_i}(f) \cdot P^{\Sigma_j}(g) \right).$$

---

[16] This should be compared with the classical binomial formula $\frac{1}{n!}(x+y)^n = \sum_{i+j=n} \frac{1}{i!j!} x^i y^j$.

[17] This decomposition in terms of subgroups $K$ is *not* canonical.

## 2.4 Power Operations for Complex Bordism

In the cases $i = 0$ or $j = 0$, the transfer map is the identity operation, and we recover $P^G(g)$ and $P^G(f)$ respectively. In all the other terms, the interchange lemma lets us pull the transfer to the outside, and transitivity of $G$ guarantees this new transfer to be proper. □

Since the only operations we understand so far are stable operations, which are in particular additive, we are moved to find a target for the power operation $P^G$ which kills the ideal generated by the proper transfers (i.e., the intermediate terms in the binomial formula) yet which remains computable.

**Definition 2.4.11** ([May96, sections XXV.6 and XXI.4], [Lurc, example 6.1.6.22, definition 6.1.6.24])   Again let $F$ be a $G$-spectrum, and define its *homotopy fixed points* $F^{hG}$ to be the homotopy limit spectrum. It is possible to factor the norm map $F_{hG} \to F$ of Definition 2.4.7 through the homotopy fixed points, as in
$$F_{hG} \to F^{hG} \to F.$$
The cofiber of this first map is denoted $F^{tG}$ and is called the *Tate spectrum* of $G$. In turn, this gives rise to a notion of Tate power operation via

$$\pi_0 E_+^{\Sigma^\infty X} \xrightarrow{P^G} \pi_0 E^{(\Sigma_+^\infty X)_{hG}} = \pi_0(E_+^{\Sigma^\infty X})^{hG} \longrightarrow \pi_0(E_+^{\Sigma^\infty X})^{tG},$$
$$f \longmapsto P^G f \longmapsto P^G_{\text{Tate}} f.$$

*Example 2.4.12*   The Tate spectrum is quite destructive in the case of $G = C_p$.[18] For example, take $X$ to be a *finite* trivial $C_p$-spectrum, and let $\bigvee_{C_p} X$ be the $C_p$-indexed wedge with the free action. The norm, transfer, and restriction maps belong to the following diagram:

$$\begin{array}{ccccccc}
(X^{\vee C_p})_{hC_p} & \longrightarrow & (X^{\vee C_p})^{hC_p} & \longrightarrow & X^{\vee C_p} & \longrightarrow & (X^{\vee C_p})_{hC_p} \\
\| & & \| & & \| & & \| \\
X & \longrightarrow & X & \xrightarrow{\Delta} & X^{\vee C_p} & \xrightarrow{\mu} & X,
\end{array}$$

where on the bottom row we have identified the homotopy orbits and fixed points of the wedge. We claim that the norm map is an equivalence. This diagram is natural in $X$ and each node preserves cofiber sequences, so for finite $X$ it suffices to check the claim on the spheres $\mathbb{S}^n$. Because we are norming up from

---

[18] In fact, its job in general is to be destructive, and the failure of this example for other groups is a sign that we have chosen the "wrong" definition. For a general group $G$, the definition of "Tate construction" can be modified to kill all intermediate transfers, not just those which transfer up from the trivial subgroup – but this requires understanding *families* of subgroups [May96, section XXI.4] and the corresponding generalized norm map [May96, section XXV.6], which is more equivariant homotopy theory than we otherwise need.

the trivial group, the long horizontal composite is multiplication by $|C_p| = p$. We have thus factored $p \in \pi_0 \operatorname{End} \mathbb{S}^n = \mathbb{Z}$ as some map followed by the diagonal and fold maps, which themselves compose to $p$. This forces the norm map itself to be an equivalence, and hence the cofiber sequence

$$(X^{\vee C_p})_{hC_p} \xrightarrow{\simeq} (X^{\vee C_p})^{hC_p} \to (X^{\vee C_p})^{tC_p}$$

causes $(X^{\vee C_p})^{tC_p}$ to be contractible.

As another example of the destructiveness of the Tate operation, we have the following:

**Lemma 2.4.13** *In the case $G = C_p$, the Tate power operation is additive.*

*Proof* The image of the map $\pi_0(E^{\Sigma_+^\infty X})_{hC_p} \to \pi_0(E^{\Sigma_+^\infty X})^{hC_p}$ is the kernel of the projection to the Tate object. Since the only proper subgroup of $C_p$ is the trivial subgroup, this image contains all transfers. □

The real miracle is that these cyclic Tate power operations are not only additive, but they are even *completely* computable.

**Lemma 2.4.14** (see [Lurd, chapter IX]) *The assignment $X \mapsto \pi_0(E^{(\Sigma_+^\infty X)^{\wedge p}})^{tC_p}$ is a cohomology theory on finite spectra $X$, represented by $E^{tC_p}$.*

*Proof* The Eilenberg–Steenrod axioms are clear except for the cofiber sequence axiom, which we will boil down to checking that the assignment $X \mapsto (X^{\wedge p})^{tC_p}$ preserves cofiber sequences of finite complexes. A cofiber sequence

$$X \to Y \to Y/X$$

of pointed spaces is equivalent data to the diagram

$$X \to Y,$$

which has colimit $Y$ and which admits a filtration by distance from the initial node with filtration quotients:

$$\begin{array}{ccc} X & \longrightarrow & Y \\ \| & & \downarrow \\ X & & Y/X. \end{array}$$

By taking the $p$-fold Cartesian power of the diagram $X \to Y$, we produce a diagram shaped like a $p$-dimensional hypercube with colimit $Y^{\wedge p}$ and which again admits a filtration by distance from the initial node. The colimits of these partial diagrams give a $C_p$-equivariant filtration of $Y^{\wedge p}$ as:

### 2.4 Power Operations for Complex Bordism

$$
\begin{array}{ccccccc}
F_0 & \longrightarrow & F_1 & \longrightarrow & \cdots \longrightarrow & F_{p-1} & \longrightarrow & Y^{\wedge p} \\
\| & & \downarrow & & & \downarrow & & \downarrow \\
X^{\wedge p} & \bigvee \left(X^{\wedge(p-1)} \wedge (Y/X)\right) & & \cdots & \bigvee \left(X \wedge (Y/X)^{\wedge(p-1)}\right) & & (Y/X)^{\wedge p}.
\end{array}
$$

We now apply $(-)^{tC_p}$ to this diagram. The Tate construction carries cofiber sequences of $C_p$-spectra to cofiber sequences of spectra, so this is again a filtration diagram. In the intermediate filtration quotients, the $C_p$-action is given by freely permuting wedge factors, from which it follows that the Tate construction vanishes on these nodes. Hence, the diagram postcomposed with the Tate construction takes the form

$$
\begin{array}{ccccccc}
F_0^{tC_p} & \xrightarrow{\simeq} & F_1^{tC_p} & \xrightarrow{\simeq} & \cdots \xrightarrow{\simeq} & F_{p-1}^{tC_p} & \longrightarrow & (Y^{\wedge p})^{tC_p} \\
\| & & \downarrow & & & \downarrow & & \downarrow \\
(X^{\wedge p})^{tC_p} & * & & \cdots & * & & ((Y/X)^{\wedge p})^{tC_p}.
\end{array}
$$

Eliminating the intermediate filtration stages with empty filtration quotients, we see that this filtration is equivalent data to a cofiber sequence

$$(X^{\wedge p})^{tC_p} \to (Y^{\wedge p})^{tC_p} \to ((Y/X)^{\wedge p})^{tC_p}.$$

Repeating this proof inside of $E$-module spectra gives the desired result. □

The effect of this lemma is twofold. For one, the Tate operation is not only additive, it is even *stable*. Second, it suffices to understand the behavior of passing to the Tate construction in the case of $X = *$, i.e., the effect of the map $E^{hC_p} \to E^{tC_p}$. Since we are intending to make a particular computation, it will at this point be convenient to return to our case of interest, $E = MU$.

**Theorem 2.4.15** *There is an isomorphism*

$$\pi_* MU^{tC_n} = MU^* BC_n[x^{-1}],$$

*where $x$ is the restriction to $MU^2 BC_n$ of the canonical class $x \in MU^2(\mathbb{CP}^\infty)$.*[19]

*Proof* Consider the $C_n$-equivariant cofiber sequence

$$S(\mathbb{C}^m)_+ \to S^0 \to S^{\mathbb{C}^m},$$

where $S(\mathbb{C}^m)$ is the unit sphere inside of $\mathbb{C}^m$ and $S^{\mathbb{C}^m}$ is the one-point compactification of the $C_n$-representation $\mathbb{C}^m$. A key fact is that $S(\mathbb{C}^m)$ admits a $C_n$-equivariant cell decomposition by free cells, natural with respect to the inclusions as $m$ increases. This buys us several consequences:

---

[19] Everything we say here will actually be valid for any number $n$, not just a prime, as well as any ring spectrum $E$ under $MU$.

1. The following Tate objects vanish: $(MU \wedge S(\mathbb{C}^m)_+)^{tC_n} \simeq *$. As in Example 2.4.12, this is because the Tate construction vanishes on free $C_n$-cells.
2. We can use $\operatorname*{colim}_{m\to\infty} S(\mathbb{C}^m)_+$ as a model for $EC_n$, so that
$$MU_{hC_n} = \left(\operatorname*{colim}_{m\to\infty} MU \wedge S(\mathbb{C}^m)_+\right)_{hC_n}.$$
3. Coupling these two facts together, we get
$$MU_{hC_n} = \left(\operatorname*{colim}_{m\to\infty} MU \wedge S(\mathbb{C}^m)_+\right)_{hC_n} = \left(\operatorname*{colim}_{m\to\infty} MU \wedge S(\mathbb{C}^m)_+\right)^{hC_n}.$$
4. Pulling the fixed points functor out, this gives
$$MU^{tC_n} = \operatorname{cofib}(MU_{hC_n} \to MU^{hC_n})$$
$$= \operatorname{cofib}\left(\left(\operatorname*{colim}_{m\to\infty} MU \wedge S(\mathbb{C}^m)_+\right)^{hC_n} \to MU^{hC_n}\right)$$
$$= \left(\operatorname{cofib}\left(\operatorname*{colim}_{m\to\infty} MU \wedge S(\mathbb{C}^m)_+\right) \to MU \wedge S^0\right)^{hC_n}$$
$$= \left(\operatorname*{colim}_{m\to\infty} MU \wedge S^{\mathbb{C}^m}\right)^{hC_n}.$$

This last formula puts us in a position to calculate. The Thom isomorphism for $MU$ gives an identification $MU \wedge S^{\mathbb{C}} \simeq \Sigma^2 MU$ as $C_n$-spectra, and the map
$$(MU \wedge S^0)^{hC_n} \to (MU \wedge S^{\mathbb{C}})^{hC_n}$$
can be identified with multiplication by the Thom class:
$$MU^{\Sigma^\infty BC_n} \xrightarrow{t \cdot -} (\Sigma^2 MU)^{\Sigma^\infty BC_n}.$$

In all, this gives
$$MU^{tC_n} \simeq \operatorname*{colim}_{m\to\infty}\left((MU \wedge S^{\mathbb{C}^m})^{hC_n}\right)$$
$$\simeq \operatorname*{colim}_{m\to\infty}\left((\Sigma^{2m} MU)^{BC_n}\right) \simeq MU^{BC_n}[t^{-1}]. \qquad \square$$

**Corollary 2.4.16** *The non-additivity of the power operations associated to $MU$ is governed by t-torsion classes.* $\qquad \square$

*Remark 2.4.17* ([Qui71, equations 3.10–3.11]) The picture Quillen paints of all this is considerably different from ours. He begins by giving a different presentation of the complex cobordism groups of a manifold $M$: a complex orientation of a smooth map $Z \to M$ is a factorization
$$Z \xrightarrow{i} E \xrightarrow{\pi} M$$
through a complex vector bundle $\pi \colon E \to M$ by an embedding $i$, as well as a

complex structure on the normal bundle $v_i$. Up to suitable notions of stability (in the dimension of $E$) and homotopy equivalence (involving, in particular, isotopies of different embeddings $i$), these quotient to give cobordism classes of maps complex-oriented maps $Z \to M$. The collection of cobordism classes over $M$ of codimension $q$ is isomorphic to $MU^q(M)$ [Qui71, proposition 1.2]. Quillen's definition of the power operations is then given in terms of this geometric model: a representative $f \colon Z \to M$ of a cobordism class gives rise to another complex-oriented map $f^{\times n} \colon Z^{\times n} \to M^{\times n}$, and he defines $P_{\text{ext}}^{C_n}(f)$ to be the postcomposition with $M^{\times n} \to M_{hC_n}^{\times n}$ (after taking care that the target isn't typically a manifold). All the properties of his construction must therefore be explored through the lens of groups acting on manifolds.

*Example 2.4.18* ([Ste53a, Ste53b]) The chain model for $H\mathbb{F}_2$-homology is actually also rigid enough to define power operations, and somewhat curiously these operations automatically turn out to be additive, without passing to the Tate construction. This means that they are recognizable in terms of classical Steenrod operations: The $C_2$-construction

$$\{\Sigma^n \Sigma_+^\infty X \xrightarrow{f} H\mathbb{F}_2\} \xrightarrow{P^{C_2}} \{\Sigma^{2n} \Sigma_+^\infty X \wedge \Sigma_+^\infty BC_2 \xrightarrow{P^{C_2}(f)} H\mathbb{F}_2\}$$

gives a class in $H\mathbb{F}_2^{2n-*}(X) \otimes H\mathbb{F}_2^*(\mathbb{R}P^\infty)$, which decomposes as

$$P^{C_2}(f) = \sum_{j=0}^{2n} \mathrm{Sq}^{2n-j}(f) \otimes x^j.$$

The Adem relations [Ade52] can be extracted by studying the wreath product $\Sigma_2 \wr \Sigma_2$ and the compositional identity for power operations.

*Remark 2.4.19* ([BMMS86, theorems III.4.1–III.4.1.4, remark III.4.4]) Since the failure of the power operations to be additive was a consequence of the binomial formula, it is somewhat intuitive that modulo 2, where $(x+y)^2 \equiv x^2 + y^2$, that this operation becomes stable. In fact, more than this is true: for instance, if an $H_\infty$-ring spectrum $E$ has $\pi_0 E = \mathbb{F}_p$, it must be the case that $E$ receives a ring spectrum map from $H\mathbb{F}_p$. This fact also gives an inexplicit means to recover Lemma 1.5.8, after noting that the same methods used to endow $MU$ with an $H_\infty$-ring spectrum structure do the same for $MO$.

## 2.5 Explicitly Stabilizing Cyclic $MU$–Power Operations

Having thus demonstrated that the Tate variant of the cyclic power operation decomposes as a sum of stable operations, we are motivated to understand the available such stable operations in complex bordism. This follows quickly from our discussions in the previous few lectures. We learned in Corollary 2.3.12 that for any complex-oriented cohomology theory $E$ we have the calculation

$$E^*BU \cong E^*[\![\sigma_1, \sigma_2, \ldots, \sigma_j, \ldots]\!],$$

and we gave a rich interpretation of this in terms of divisor schemes:

$$BU_E \cong \mathrm{Div}_0\, \mathbb{CP}^\infty_E.$$

We would like to leverage the Thom isomorphism to gain a description of $E^*MU$ generally and $MU^*MU$ specifically. However, the former is *not* a ring, and although the latter is a ring, its multiplication is exceedingly complicated,[20] which means that our extremely compact algebraic description of $E^*BU$ in Corollary 2.3.12 will be of limited use. Instead, we will have to content ourselves with an $E_*$-module basis of $E^*MU$.

**Definition 2.5.1** Take $\varphi\colon MU \to E$ to be a complex-oriented ring spectrum, which presents $E^*BU$ as the subalgebra of symmetric functions inside of an infinite-dimensional polynomial algebra:

$$E^*BU \subseteq E^*BU(1)^{\times\infty} \cong E^*[\![x_1, x_2, \ldots]\!].$$

For any nonnegative multi-index $\alpha = (\alpha_1, \alpha_2, \ldots)$ with finitely many entries nonzero, there is an associated *symmetric monomial* $c_\alpha$, which is the sum of those monomials whose exponent lists contain exactly $\alpha_j$ many instances of $j$.[21] We then set $s_\alpha \in E^*MU$ to be the image of $c_\alpha$ under the Thom isomorphism of $E_*$-modules

$$E^*MU \cong E^*BU.$$

It is called the $\alpha$th *Landweber–Novikov operation* with respect to $\varphi$.

**Definition 2.5.2** In the case of the identity orientation $MU \xrightarrow{1} MU$, the resulting classes are called the *Conner–Floyd–Chern classes* and the associated cohomology operations are called the *Landweber–Novikov operations* (without further qualification).

---

[20] For a space $X$, $E^*X$ has a ring structure because $X$ has a diagonal, and $MU$ does not have a diagonal. In the special case of $E = MU$, there is a ring product coming from endomorphism composition.

[21] For example, $\alpha = (1, 0, 2, 0, \ldots)$ corresponds to the sum

$$b_\alpha = \sum_i \sum_{j \neq i} \sum_{\substack{k \neq i \\ k > j}} x_i x_j^3 x_k^3 = x_1 x_2^3 x_3^3 + x_1^3 x_2 x_3^3 + x_1^3 x_2^3 x_3 + \cdots.$$

## 2.5 Explicitly Stabilizing Cyclic MU–Power Operations

**Remark 2.5.3** For a vector bundle $V$ and a complex-oriented cohomology theory $E$, we define the *total symmetric Chern class* of $V$ by the sum

$$c_\mathbf{t}(V) = \sum_\alpha c_\alpha(V) \mathbf{t}^\alpha.$$

In the case of a line bundle $\mathcal{L}$ with first Chern class $c_1(\mathcal{L}) = x$, this degenerates to the sum

$$c_\mathbf{t}(\mathcal{L}) = \sum_{j=0}^\infty x^j t_j.$$

For a direct sum, $c_\mathbf{t}(V \oplus W)$ satisfies a Cartan formula:

$$c_\mathbf{t}(V \oplus W) = c_\mathbf{t}(V) \cdot c_\mathbf{t}(W).$$

Again specializing to line bundles $\mathcal{L}$ and $\mathcal{H}$ with first Chern classes $c_1(\mathcal{L}) = x$ and $c_1(\mathcal{H}) = y$, this gives

$$\begin{aligned}c_\mathbf{t}(\mathcal{L} \oplus \mathcal{H}) &= \left(\sum_{j=0}^\infty x^j t_j\right) \cdot \left(\sum_{k=0}^\infty y^k t_k\right) = \sum_{j=0}^\infty \sum_{k=0}^\infty x^j y^k t_j t_k \\ &= 1 + (x+y)t_1 + (xy)t_1^2 + (x^2 + y^2)t_2 + (xy^2 + x^2 y)t_1 t_2 + \cdots \\ &= 1 + c_1(U)t_1 + c_2(U)t_1^2 + \\ &\quad + (c_1^2(U) - 2c_2(U))t_2 + (c_1(U)c_2(U))t_1 t_2 + \cdots,\end{aligned}$$

where we have expanded out some of the pieces of the total symmetric Chern class in polynomials in the usual Chern classes.

**Definition 2.5.4** ([Ada95, theorem I.5.1]) Take $MU \xrightarrow{1} MU$ as the orientation, so that we are considering $MU^*MU$ and the Landweber–Novikov operations arising from the Conner–Floyd–Chern classes. These account for the *stable* operations in $MU$-cohomology, analogous to the Steenrod operations for $H\mathbb{F}_2$. They satisfy the following properties:

- $s_0$ is the identity;
- $s_\alpha$ is natural: $s_\alpha(f^*x) = f^*(s_\alpha x)$;
- $s_\alpha$ is stable: $s_\alpha(\sigma x) = \sigma(s_\alpha x)$;
- $s_\alpha$ is additive: $s_\alpha(x+y) = s_\alpha(x) + s_\alpha(y)$;
- $s_\alpha$ satisfies a Cartan formula. Define

$$s_\mathbf{t}(x) := \sum_\alpha s_\alpha(x) \mathbf{t}^\alpha := \sum_\alpha s_\alpha(x) \cdot t_1^{\alpha_1} t_2^{\alpha_2} \cdots t_n^{\alpha_n} \cdots \in MU^*(X)[\![t_1, t_2, \ldots]\!]$$

for an infinite sequence of indeterminates $t_1, t_2, \ldots$. Then

$$s_\mathbf{t}(xy) = s_\mathbf{t}(x) \cdot s_\mathbf{t}(y).$$

- Let $\xi\colon X \to BU(n)$ classify a vector bundle and let $\varphi$ denote the Thom isomorphism

$$\varphi\colon MU^*X \to MU^*T(\xi).$$

Then the Chern classes of $\xi$ are related to the Landweber–Novikov operations on the Thom spectrum by the formula

$$\varphi(c_\alpha(\xi)) = s_\alpha(\varphi(1)).$$

Having set up an encompassing theory of stable operations, we now seek to give a formula for the cyclic Tate power operation in this framework. In order to approach this, we initially set our sights on the too-lofty goal of computing the total power operation $P^{C_n}(f)$ for $f \in MU^{2q}(X)$m an $MU$-cohomology class on a finite complex $X$, without necessarily passing to the Tate construction. Because of the definition $MU = \colim_k MU(k)$ and because $P^{C_n}$ is natural under pullback, it will suffice for us to study the effect of $P^{C_n}$ on the universal classes

$$u_m\colon MU(m) \to MU,$$

after using the canonical $MU$-Thom isomorphism to reinterpret them as classes on a suspension spectrum. We begin with the canonical orientation itself:

$$u_1 = x \in h\text{Spectra}(MU(1), MU) \cong MU^2\mathbb{CP}^\infty.$$

In order to understand the effect $P^{C_n}(x)$ of the power operation on $x$, recall that $x$ is also the first Conner–Floyd–Chern class of the tautological bundle $\mathcal{L}$ on $\mathbb{CP}^\infty$, i.e.,

$$x\colon MU(1) \to MU$$

is both the Thomification of the block inclusion

$$\mathcal{L}\colon BU(1) \to BU$$

as well as its first Chern class in the canonical orientation for $MU$-theory. The $H_\infty^2$-ring spectrum construction defining $P_{\text{ext}}^{C_n}(x)$ thus fits into the following diagram:

$$\begin{array}{ccccc}
\Sigma_+^\infty BU(1) & \longrightarrow & \Sigma_+^\infty BU(1) \wedge \Sigma_+^\infty BC_n & & \\
\downarrow \Delta & & \downarrow \Delta_{hC_n} & & \\
(\Sigma_+^\infty BU(1))^{\wedge n} & \longrightarrow & \Sigma_+^\infty BU(1)^{\wedge n}_{hC_n} & \xrightarrow{\mathcal{L}(C_n)} & \Sigma_+^\infty BU(n) \\
\downarrow c_1^{\wedge n} & \mathcal{L}^{\times n} & \downarrow & P_{\text{ext}}^{C_n} \searrow & \downarrow c_n \\
(\Sigma^2 MU)^{\wedge n} & \longrightarrow & (\Sigma^2 MU)^{\wedge n}_{hC_n} & \longrightarrow & \Sigma^{2n} MU.
\end{array}$$

## 2.5 Explicitly Stabilizing Cyclic MU–Power Operations

The commutativity of the widest rectangle (i.e., the justification for the name "$c_n$" on the right-most vertical arrow) comes from the Cartan formula for Chern classes: Because $\mathcal{L}^{\oplus n}$ splits as the sum of $n$ line bundles, $c_n(\mathcal{L}^{\oplus n})$ is computed as the product of the first Chern classes of those line bundles. Second, the commutativity of the right-most square is not trivial: It is a specific consequence of how the multiplicative structure on $MU$ arises from the direct sum of vector bundles.[22] The commutativities of the other two squares comes from the natural transformation from a $C_n$-space to its homotopy orbit space.

Hence, the internal cyclic power operation $P^{C_n}(x)$ applied to the canonical coordinate $x$ is defined by the composite

$$\Sigma_+^\infty BU(1)_{hC_n} \xrightarrow{\Delta_{hC_n}} (\Sigma_+^\infty BU(1))^{\wedge n}_{hC_n} \to \Sigma^{2n} MU^{\wedge n}_{hC_n} \to \Sigma^{2n} MU,$$

which is to say

$$P^{C_n}(x) = c_n(\Delta^*_{hC_n} \mathcal{L}(C_n)).$$

We have thus reduced to computing a particular Conner–Floyd–Chern class of a particular bundle. Our next move is to transport more information from the $C_n$-equivariant bundle $\Delta^* \mathcal{L}^{\times n}$ to the bundle $\Delta^*_{hC_n} \mathcal{L}(C_n)$.

**Lemma 2.5.5** (see Definition 2.4.4)  *Given a G-equivariant bundle V over a base X on which G acts trivially, the homotopy quotient bundle $V_{hG}$ determines a vector bundle over $X_{hG} \simeq X \times BG$. This construction preserves direct sums, and it preserves tensor products when one of the factors has the trivial action.* □

We thus proceed to analyze $\Delta^*_{hC_n} \mathcal{L}(C_n)$ by first decomposing the $C_n$-equivariant bundle $\Delta^* \mathcal{L}^{\times n}$. The $C_n$-action is given by permutation of the factors, and hence we have an identification

$$\Delta^* \mathcal{L}^{\times n} \cong \mathcal{L} \otimes \pi^* \rho,$$

where $\rho$ is the permutation representation of $C_n$ (considered as a vector bundle over a point) and $\pi \colon BU(1) \to *$ is the constant map. The permutation representation for the abelian group $C_n$, also known as its regular representation, is accessible by character theory. The generating character $\chi \colon U(1)[n] \to U(1)$ gives a decomposition

$$\rho \cong \bigoplus_{j=0}^{n-1} \chi^{\otimes j}.$$

---

[22] In general, any notion of first Chern class $\Sigma_+^\infty BU(1) \to \Sigma^2 E$ gives rise to a *noncommuting* diagram of this same shape. The two composites $\Sigma_+^\infty BU(1)^{\wedge n}_{hC_n} \to \Sigma^{2n} E$ need not agree, since $\mathcal{L}(C_n)$ has no *a priori* reason to be compatible with the factorization appearing in the $H^2_\infty$-structure. They turn out to be loosely related nonetheless, and their exact relation (as well as a procedure for making them agree in certain cases) is the subject of Appendix A.2.

Applying this to our situation, we get a sequence of isomorphisms of $C_n$-equivariant vector bundles

$$\Delta^* \mathcal{L}^{\times n} \cong \mathcal{L} \otimes \pi^* \rho \cong \mathcal{L} \otimes \bigoplus_{j=0}^{n-1} \pi^* \chi^{\otimes j} \cong \bigoplus_{j=0}^{n-1} \mathcal{L} \otimes \pi^* \chi^{\otimes j}.$$

Applying Lemma 2.5.5, we recast this as a calculation of the bundle $\Delta^*_{hC_n} \mathcal{L}(C_n)$:

$$\Delta^*_{hC_n} \mathcal{L}(C_n) = \bigoplus_{j=0}^{n-1} \pi_1^* \mathcal{L} \otimes \pi_2^* \eta^{\otimes j},$$

where $\eta$ is the bundle classified by $\eta \colon BU(1)[n] \to BU(1)$ and $\pi_1, \pi_2$ are the two projections off of $BU(1) \times BC_n$.

We can use this to access $c_n(\Delta^*_{hC_n} \mathcal{L}(C_n))$. As the top Chern class of this $n$-dimensional vector bundle, we think of this as a calculation of its Euler class, which lets us lean on multiplicativity:

$$p^{C_n}(x) = c_n(\Delta^*_{hC_n} \mathcal{L}(C_n)) = e\left(\bigoplus_{j=0}^{n-1} \pi_1^* \mathcal{L} \otimes \pi_2^* \eta^{\otimes j}\right) = \prod_{j=0}^{n-1} e\left(\pi_1^* \mathcal{L} \otimes \pi_2^* \eta^{\otimes j}\right)$$

$$= \prod_{j=0}^{n-1} c_1\left(\pi_1^* \mathcal{L} \otimes \pi_2^* \eta^{\otimes j}\right) = \prod_{j=0}^{n-1} (x +_{MU} [j]_{MU}(t)).$$

Here, $x$ is still the first Conner–Floyd–Chern class of $\mathcal{L}$ and $t$ is the Euler class of $\eta$. We now try to make sense of this product expression for $c_n(\Delta^*_{hC_n} \mathcal{L}(C_n))$ by expanding it in powers of $x$ and identifying its component pieces.

**Lemma 2.5.6** *There is a series expansion in $MU^* BU(1)_{hC_n}$:*

$$p^{C_n}(x) = \prod_{j=0}^{n-1} (x +_{MU} [j]_{MU}(t)) = w + \sum_{j=1}^{\infty} a_j(t) x^j,$$

*where $a_j(t)$ is a series with coefficients in the subring $C \subseteq MU_*$ spanned by the coefficients of the natural $MU$-formal group law. The leading term*

$$w = e(\overline{\rho}) = \prod_{j=0}^{n-1} e(\eta^{\otimes j}) = \prod_{j=0}^{n-1} [j]_{MU}(e(\eta)) = (n-1)! t^{n-1} + \sum_{j \geq n} b_j t^j$$

*is the Euler class of the reduced permutation representation, and, again, the elements $b_j$ lie in the subring $C$.* □

We now turn from the first universal case to our original goal: understanding the action of $p^{C_n}$ on each of the canonical classes

$$u_m \colon MU(m) \to MU.$$

## 2.5 Explicitly Stabilizing Cyclic MU–Power Operations

We approach this via the splitting principle by first rewriting the formula in Lemma 2.5.6 in a form amenable to wrangling direct sums with the Cartan formula:

$$P^{C_n}(c_1(\mathcal{L})) = \sum_{|\alpha| \leq 1} w^{1-|\alpha|} a_\alpha(t) c_\alpha(\mathcal{L}), \qquad a_\alpha(t) = \prod_{j=0}^{\infty} a_j(t)^{\alpha_j}.$$

**Corollary 2.5.7** *There is the universal formula*

$$P^{C_n}(u_m) = \sum_{|\alpha| \leq m} w^{m-|\alpha|} a_\alpha(t) s_\alpha(u_m).$$

*Proof* Writing $\varphi$ for the canonical Thom isomorphism associated to $MU$-cohomology, the main new observation is

$$P^{C_n}(u_m) = \varphi(P^{C_n}(c_m(\xi_m))).$$

From here, we apply the splitting principle, multiplicativity of power operations, and properties of Landweber–Novikov operations:

$$= \varphi(P^{C_n}(c_1(\mathcal{L}_1) \cdots c_1(\mathcal{L}_m)))$$
$$= \varphi(P^{C_n}(c_1(\mathcal{L}_1)) \cdots P^{C_n}(c_1(\mathcal{L}_m)))$$
$$= \varphi\left(\left(\sum_{|\alpha| \leq 1} w^{1-|\alpha|} a_\alpha(t) c_\alpha(\mathcal{L}_1)\right) \cdots \left(\sum_{|\alpha| \leq 1} w^{1-|\alpha|} a_\alpha(t) c_\alpha(\mathcal{L}_m)\right)\right)$$
$$= \varphi\left(\sum_{|\alpha| \leq m} w^{m-|\alpha|} a_\alpha(t) c_\alpha(\mathcal{L}_1 \oplus \cdots \oplus \mathcal{L}_m)\right)$$
$$= \varphi\left(\sum_{|\alpha| \leq m} w^{m-|\alpha|} a_\alpha(t) c_\alpha(\xi_m)\right) = \sum_{|\alpha| \leq m} w^{m-|\alpha|} a_\alpha(t) s_\alpha(u_m). \qquad \square$$

We will use this to power the following conclusion about cohomology classes in general, starting with an observation about the fundamental class of a sphere:

**Lemma 2.5.8** (see [Rud98, corollary VII.7.14]) *For $f \in MU^{2q}(X)$ a cohomology class in a finite complex $X$, there is the suspension relation*

$$P^{C_n}(\sigma^{2m} f) = w^m \sigma^{2m} P^{C_n}(f).$$

*Proof* We calculate $P^{C_n}$ applied to the fundamental class

$$S^{2m} \xrightarrow{\iota_{2m}} T_m BU(m) \simeq \Sigma^{2m} MU(m) \xrightarrow{\Sigma^{2m} u_m} \Sigma^{2m} MU$$

by restricting the universal formula:

$$P^{C_n}(\iota_{2m}^* u_m) = \iota_{2m}^* P^{C_n}(u_m) = \iota_{2m}^*\left(\sum_{|\alpha|\le m} w^{m-|\alpha|} a_\alpha(t) s_\alpha(u_m)\right) = w^m \iota_{2m},$$

since $s_\alpha(\iota_{2m}) = 0$ for any nonzero $\alpha$, as the cohomology of $S^{2m}$ is too sparse. Because $\sigma^{2m} f = \iota_{2m} \wedge f$, we conclude the proof by multiplicativity of $P^{C_n}$. □

**Theorem 2.5.9** (see [Qui71, proposition 3.17], [Rud98, corollary VII.7.14]) *Let $X$ be a finite pointed space and let $f$ be a cohomology class*

$$f \in \widetilde{MU}^{2q}(X).$$

*For $m \gg 0$, there is a formula*

$$w^m P^{C_n}(f) = \sum_{|\alpha|\le m+q} w^{q+m-|\alpha|} a_\alpha(t) s_\alpha(f),$$

*with $t$, $w$, and $a_\alpha(t)$ as defined previously.*[23]

*Proof* We take $m$ large enough so that $f$ is represented by an unstable map

$$g\colon \Sigma^{2m} X \to T_{m+q} BU(m+q),$$

in the sense that $g$ intertwines $f$ with the universal class $u_{m+q}$ by the formula

$$g^* u_{m+q} = \sigma^{2m} f.$$

We use Lemma 2.5.8 and naturality to conclude

$$w^m \sigma^{2m} P^{C_n}(f) = P^{C_n}(\sigma^{2m} f) = P^{C_n}(g^* u_{m+q}) = g^* P^{C_n}(u_{m+q})$$

$$= g^*\left(\sum_{|\alpha|\le m+q} w^{m+q-|\alpha|} a_\alpha(t) s_\alpha(u_{m+q})\right)$$

$$= \sum_{|\alpha|\le m+q} w^{m+q-|\alpha|} a_\alpha(t) \sigma^{2m} s_\alpha(f).$$

□

Our conclusion, then, is that $P^{C_n}$ is *almost* naturally expressible in terms of the Landweber–Novikov operations, where the "*almost*" is controlled by some $w$-torsion. Our discussion of the Tate construction in the previous lecture shows that this is, in some sense, a generic phenomenon, and indeed the previous theorem can be divided by $w^m$ to recover a statement about the cyclic Tate

---

[23] The reader comparing with Quillen's paper will notice various apparent discrepancies between the statements of our theorem and of his. These are notational: He grades his cohomology functor homologically, which occasionally causes our $q$ to match his $-q$, so that his $n$ is comparable to our $m - q$.

power operation. However, we have learned the additional information that the various factors in this statement – including $w$ itself – are controlled by the formal group law "$+_{MU}$" associated to the tautological complex orientation of $MU$ and the subring $C$. This is *not* generic behavior. In the next lecture, we will discover the surprising fact that we only need to multiply by a *single w* – also highly non-generic – and the equally surprising consequences this entails for $MU_*$ itself.

## 2.6 The Complex Bordism Ring

With Theorem 2.5.9 in hand, we will now deduce Quillen's major structural theorem about $MU_*$. We will preserve the notation used in Lemma 2.5.6 and Theorem 2.5.9:

- $\overline{\rho}$ is the reduced regular representation of $C_n$, which coincides with its reduced permutation representation, and $w = e(\overline{\rho})$ is its Euler class.
- $\eta \colon BC_n \to BU(1)$ is the line bundle associated to a choice of generating character $C_n \to U(1)$, and $t = e(\eta)$ is its Euler class.
- $C$ is the subring of $MU_*$ generated by the coefficients of the formal group law associated to the identity complex-orientation.

In the course of working out the main theorem, we will want to make use of the following property of the class $t$.

**Definition 2.6.1** For $x$, a coordinate on a formal group, the *divided n-series* is determined by the formula

$$x \cdot \langle n \rangle(x) = [n](x).$$

**Lemma 2.6.2** ([Qui71, proposition 4.4]) *Let $X$ be a finite complex. If a class $\omega \in MU^*(X \times BC_n)$ satisfies $t \cdot \omega = 0$, then there exists a class $y \in MU^*(X)$ with $\omega = y \cdot \langle n \rangle_{MU}(t)$.*

*Proof sketch* Consider the following piece of the Gysin sequence for the bundle $\eta$ as viewed in the cohomology theory $MU^*(X \times -)$:

$$\widetilde{MU}^{*+1}(X \times S^1_{hC_n}) \xrightarrow{\partial} MU^*(X \times BC_n) \xrightarrow{t} MU^{*+2}(X \times BC_n).$$

Because the action of $C_n$ on the total space $S^1$ is free, we have $S^1_{hC_n} \simeq S^1$. The suspension isomorphism then rewrites the left-most term as

$$\widetilde{MU}^*(X) \xrightarrow{\langle n \rangle_{MU}(t)} MU^*(X \times BC_n) \xrightarrow{t} MU^{*+2}(X \times BC_n).$$

92                        Complex Bordism

In order to justify the image of $\partial$, one needs to unwrap the definition of the Gysin sequence [Qui71, equation 4.7, proposition 4.2], which we elect not to do. Given this, however, the lemma statement follows from exactness. □

We now turn to the main theorem.

**Theorem 2.6.3** ([Qui71, theorem 5.1])  *If $X$ has the homotopy type of a finite complex, then*

$$MU^*(X) = C \cdot \sum_{q \geq 0} MU^q(X), \qquad \widetilde{MU}^*(X) = C \cdot \sum_{q > 0} MU^q(X).$$

*Remark 2.6.4*  In what follows, the reader should carefully remember the degree conventions stemming from the formula

$$MU^*X = \pi_{-*}F(\Sigma_+^\infty X, MU).$$

The homotopy ring $MU_*$ appears in the *negative* degrees of $MU^*(*)$, but the fundamental class of $S^m$ appears in the *positive* degree $MU^m(S^m)$.

*Proof of Theorem 2.6.3*  We can immediately reduce the claim in two ways. First, it is true if and only if it is also true for reduced cohomology. Second, we are free to restrict attention just to $MU^{2*}(X)$, since we can handle the odd-degree parts of $MU^*(X)$ by suspending $X$ once. Defining

$$R^{2*} := C \cdot \sum_{q > 0} MU^{2q}(X),$$

we can thus focus on the claim

$$\widetilde{MU}^{2*}(X) \stackrel{?}{=} C \cdot \sum_{q > 0} MU^{2q}(X).$$

Noting that the claim is trivially true for all positive values of $*$, we will show this by working $p$-locally and inducting on the value of "$-*$."

Suppose that

$$R^{-2j}_{(p)} = \widetilde{MU}^{-2j}(X)_{(p)}$$

for $j < q$ and consider $x \in \widetilde{MU}^{-2q}(X)_{(p)}$. Then, for $m \gg 0$, we have

$$w^m p^{C_p}(x) = \sum_{|\alpha| \leq m-q} w^{m-q-|\alpha|} a(t)^\alpha s_\alpha x$$

$$= w^{m-q} x + \sum_{\substack{|\alpha| \leq m-q \\ \alpha \neq 0}} w^{m-q-|\alpha|} a(t)^\alpha s_\alpha x.$$

Recall that $w$ is a power series in $t$ with coefficients in $C$ and leading term

## 2.6 The Complex Bordism Ring

$(p-1)! \cdot t^{p-1}$, so that $t^{p-1} = w \cdot \theta(t)$ for some multiplicatively invertible series $\theta(t)$ with coefficients in $C$. Since $s_\alpha$ raises degree, we have $s_\alpha x \in R$ by the inductive hypothesis, and we may thus collect all those terms (as well as many factors of $\theta(t)^{-1}$) into a series $\psi_x(t) \in R_{(p)}[\![t]\!]$ to write

$$t^N(w^q P^{C_p}(x) - x) = \psi_x(t),$$

where $N = (m-q)(p-1)$.

Consider the set of possible integers $N$ for which we can write such an equation – we know that $N = (m-q)(p-1)$ works, but we now also consider values of $N$ which are not multiples of $(p-1)$. We aim to conclude that $N = 1$ is the minimum of this set, so suppose that $N$ is the minimum such value. Using Remark 2.4.3, we find that restricting this equation along the inclusion $i\colon X \to X \times BU(1)[p]$ sets $t = 0$ and yields $\psi_x(0) = 0$. It follows that $\psi_x(t) = t\varphi_x(t)$ is at least once $t$-divisible, and thus

$$t(t^{N-1}(w^q P^{C_p} x - x) - \varphi_x(t)) = 0.$$

Appealing to Lemma 2.6.2, we produce a class $y \in \widetilde{MU}^{-2q+2(N-1)}(X)_{(p)}$ with

$$t^{N-1}(w^q P^{C_p}(x) - x) = \varphi_x(x) + y\langle p\rangle(t).$$

If $N > 1$, then $y \in R_{(p)}$ for degree reasons and hence the right-hand side gives a series expansion contradicting our minimality hypothesis. So, we have $N = 1$; the class $y$ lies in the critical degree $-2q$; and the outer factor of $t^{N-1}$ is not present in the last expression.[24] Restricting along $i$ again to set $w = t = 0$ and $P^{C_p}(x) = x^p$, we obtain the equation

$$\left.\begin{array}{ll} -x & \text{if } q > 0 \\ x^p - x & \text{if } q = 0 \end{array}\right\} = \varphi_x(0) + py.$$

In the first case, where $q > 0$, it follows that $MU^{-2q}(X)_{(p)}$ is contained in $R^{-2q}_{(p)} + pMU^{-2q}(X)_{(p)}$, and since $MU^{-2q}(X)_{(p)}$ is finitely generated,[25] it follows that $MU^{-2q}(X)_{(p)} = R^{-2q}_{(p)}$. In the other case, $x$ can be rewritten as a sum of elements in $R^0_{(p)}$, elements in $p\widetilde{MU}^0(X)_{(p)}$, and elements in $(\widetilde{MU}^0(X)_{(p)})^p$. Since the ideal $\widetilde{MU}^0(X)_{(p)}$ is nilpotent, it again follows that $\widetilde{MU}^0(X)_{(p)} = R^0_{(p)}$, concluding the induction. □

---

[24] One can interpret the proof thus far as giving a bound on the amount of $w$-torsion needed to get the stability relation described in Theorem 2.5.9. Our answer is quite surprising: We have found that we need just a single $w$ (indeed, a single $t$), which isn't much stability at all!

[25] This is a consequence of $X$ having finitely many cells, $MU$ having finitely many cells in each degree, and each homotopy group of the stable sphere being finitely generated.

**Corollary 2.6.5** ([Qui71, corollary 5.2]) *The coefficients of the formal group law generate $MU_*$.*

*Proof* This is the case of Theorem 2.6.3 where we set $X = *$. □

*Remark 2.6.6* This proof actually also goes through for $MO$ as well. In that case, it's even easier, since the equation $2 = 0$ in $\pi_0 MO$ causes much of the algebra to collapse. The proof does not extend further to cases like $MSO$ or $MSp$: these bordism theories do not have associated formal group laws, and so we lose the control we had in Lecture 2.5.

Take $\mathcal{M}_{\text{fgl}}$ to be the moduli functor of formal group laws. Since a formal group law is a power series satisfying some algebraic identities, this moduli object is an affine scheme with coordinate ring $O_{\mathcal{M}_{\text{fgl}}}$. A rephrasing of Corollary 2.6.5 is that the natural map

$$O_{\mathcal{M}_{\text{fgl}}} \to MU_*$$

is *surjective*. This is reason enough to start studying $\mathcal{M}_{\text{fgl}}$ in earnest, which we take up in the next case study – but if we anachronistically assume one algebraic fact about $O_{\mathcal{M}_{\text{fgl}}}$, we can prove that the natural map is an *isomorphism*. We begin with the following topological observation about mixing complex-orientations:

**Lemma 2.6.7** ([Ada95, lemma II.6.3 and corollary II.6.5]) *Let $\varphi \colon MU \to E$ be a complex-oriented ring spectrum and consider the two orientations on $E \wedge MU$ given by*

$$\mathbb{S} \wedge MU \xrightarrow{\eta_E \wedge 1} E \wedge MU, \qquad MU \wedge \mathbb{S} \xrightarrow{\varphi \wedge \eta_{MU}} E \wedge MU.$$

*The two induced coordinates $x^E$ and $x^{MU}$ on $\mathbb{CP}^\infty_{E \wedge MU}$ are related by the formulas*

$$x^{MU} = \sum_{j=0}^{\infty} b_j^E (x^E)^{j+1} =: g(x^E),$$

$$g^{-1}(x^{MU} +_{MU} y^{MU}) = g^{-1}(x^{MU}) +_E g^{-1}(y^{MU}),$$

*where $E_* MU \cong \frac{\text{Sym}_{E_*} E_* \{\beta_0, \beta_1, \beta_2, \ldots\}}{\beta_0 = 1} \cong E_*[b_1, b_2, \ldots]$, as in Lemma 1.5.1, Corollary 1.5.2, and Corollary 2.0.4.*

*Proof* The second formula is a direct consequence of the first. For the first formula, note that because $\{\beta_j^E\}_j$ provides a homology basis for $(E \wedge MU)_* \mathbb{CP}^\infty$, we can decompose $x^{MU}$ by the Kronecker pairing:

$$x^{MU} = \sum_{j=0}^{\infty} \langle x^{MU}, \beta_j^E \rangle (x^E)^{j+1}.$$

2.6 The Complex Bordism Ring 95

Expending the definition of the pairing gives

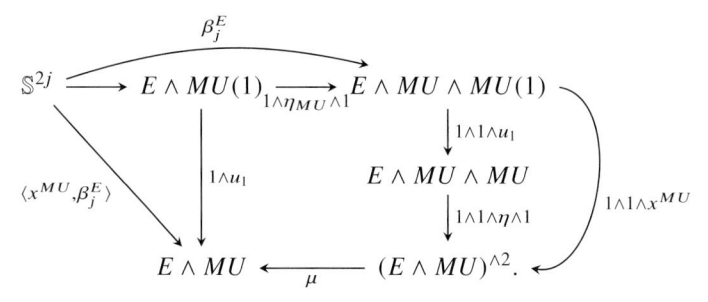

$$\begin{array}{c}
\mathbb{S}^{2j} \xrightarrow{\beta_j^E} E \wedge MU(1) \xrightarrow{1 \wedge \eta_{MU} \wedge 1} E \wedge MU \wedge MU(1) \\
\downarrow^{1 \wedge u_1} \qquad \downarrow^{1 \wedge 1 \wedge u_1} \\
\langle x^{MU}, \beta_j^E \rangle \qquad E \wedge MU \wedge MU \\
\downarrow^{1 \wedge 1 \wedge \eta \wedge 1} \qquad \downarrow^{1 \wedge 1 \wedge x^{MU}} \\
E \wedge MU \xleftarrow{\mu} (E \wedge MU)^{\wedge 2}.
\end{array}$$

By definition, the module generators $\beta_{j+1} \in E_{2(j+1)}\mathbb{C}P^\infty = E_{2j}MU(1)$ push forward along $u_1 : MU(1) \to MU$ to define the algebra generators $b_j \in E_{2j}MU$, giving the desired identification. □

**Corollary 2.6.8** ([Ada95, corollary II.6.6]) *In particular, for the orientation $MU \to H\mathbb{Z}$ we have*

$$x_1 +_{MU} x_2 = \exp^H(\log^H(x_1) + \log^H(x_2)),$$

*where* $\exp^H(x) = \sum_{j=0}^\infty b_j x^{j+1}$. □

However, one also notes that $H\mathbb{Z}_* MU = \mathbb{Z}[b_1, b_2, \ldots]$ carries the universal example of a formal group law with a logarithm – this observation follows directly from the Thom isomorphism, and so is independent of any knowledge about the ring $MU_*$. This brings us one step away from understanding $MU_*$:

**Theorem 2.6.9** (to be proven as Theorem 3.2.2) *The ring $O_{\mathcal{M}_{fgl}}$ carrying the universal formal group law is free: It is isomorphic to a polynomial ring over $\mathbb{Z}$ in countably many generators.* □

**Corollary 2.6.10** (Quillen's theorem) *The natural map $O_{\mathcal{M}_{fgl}} \to MU_*$ classifying the formal group law on $MU_*$ is an isomorphism.*

*Proof* We proved in Corollary 2.6.5 that this map is surjective. We also proved in Theorem 2.1.25 that every rational formal group law has a unique strict logarithm, i.e., the long composite on the second row

$$\begin{array}{ccccc}
O_{\mathcal{M}_{fgl}} & \longrightarrow & MU_* & \longrightarrow & (H\mathbb{Z}_* MU) \\
\downarrow & & \downarrow & \simeq & \downarrow \\
O_{\mathcal{M}_{fgl}} \otimes \mathbb{Q} & \longrightarrow & MU_* \otimes \mathbb{Q} & \longrightarrow & (H\mathbb{Z}_* MU) \otimes \mathbb{Q}
\end{array}$$

is an isomorphism. It follows from Theorem 2.6.9 that the left-most vertical map is injective, hence the top-left horizontal map is injective, hence it is an isomorphism. □

**Corollary 2.6.11** *The ring $\pi_*(MU \wedge MU)$ carries the universal example of two strictly isomorphic formal group laws. Additionally, the ring $\pi_0(MUP \wedge MUP)$ carries the universal example of two isomorphic formal group laws.*

*Proof* Combine Lemma 2.6.7 and Corollary 2.6.10. □

# Case Study 3

## Finite Spectra

Our goal in this case study is to thoroughly examine one of the techniques from Case Study 1 that has not yet resurfaced: the idea that $H\mathbb{F}_2$-homology takes values in quasicoherent sheaves over an algebro-geometric object encoding the coaction of the dual Steenrod Hopf algebra. We will find that this is quite generic: Associated to a mildly nice ring spectrum $E$, we will construct a rich algebro-geometric object $\mathcal{M}_E$, called its *context*, such that $E$-homology sends spaces $X$ to sheaves $\mathcal{M}_E(X)$ over $\mathcal{M}_E$. In still nicer situations, the sheaf $\mathcal{M}_E(X)$ tracks exactly $E_*X$ as well as the action of the $E$-analog of the dual Steenrod algebra, called the *Hopf algebroid of stable $E$-homology cooperations*. From this perspective, we reinterpret Corollary 2.6.10 as giving a presentation

$$\mathcal{M}_{MUP} \xrightarrow{\simeq} \mathcal{M}_{\mathrm{fg}},$$

where $\mathcal{M}_{\mathrm{fg}}$ is the *moduli of formal groups*. This indicates a program for studying periodic complex bordism, which we now outline.

Abstractly, one can hope to study any sheaf, including $\mathcal{M}_E(X)$, by analyzing its stalks – or, relatedly (with some luck), by analyzing its geometric fibers. The main utility of Quillen's theorem is that it gives us access to a concrete model of the context $\mathcal{M}_{MUP}$, so that we can determine where to even look. However, even this is not really enough to get off the ground: the stalks of some sheaf can exhibit nearly arbitrary behavior, and in particular there is little reason to expect the stalks of $\mathcal{M}_E(X)$ to vary nicely with $X$ or otherwise reflect its structure. In an ideal world, these stalks would themselves vary so nicely with $X$ as to form homology theories, carrying cofiber sequences to short exact sequences of fibers. In general, one might ask whether a map $f$, as in the diagram

$$\begin{array}{ccccc} \mathrm{Spec}\, R & \xrightarrow{f} & \mathcal{M}_{\mathrm{fgl}} & = & \mathcal{M}_{MUP}[0] & = & \mathrm{Spec}\, MUP_0 \\ & \searrow & \downarrow & & \downarrow & & \\ & & \mathcal{M}_{\mathrm{fg}} & = & \mathcal{M}_{MUP}, & & \end{array}$$

determines a homology theory by the formula
$$f^*\mathcal{M}_{MUP}(X) = (MUP_0 X \otimes_{MUP_0} R)^{\sim}.$$
Since $MUP_0(-)$ is already a homology theory, this is exactly the question of whether the operation $- \otimes_{MUP_0} R$ preserves exact sequences – and this is precisely what it means for $f$ to be *flat*. When flatness is satisfied, this gives the following theorem:

**Theorem 3.0.1** (Landweber)  *Given such a diagram where the diagonal arrow is flat, the functor*
$$R_0(X) := MUP_0(X) \otimes_{MUP_0} R$$
*is a 2-periodic homology theory.*

In the course of proving this theorem, Landweber additionally devised a method to recognize flat maps. Recall that a map $f\colon Y \to X$ of schemes is flat exactly when for any closed subscheme $i\colon A \to X$ with ideal sheaf $\mathcal{I}$ there is an exact sequence
$$0 \to f^*\mathcal{I} \to f^*O_X \to f^*i_*O_A \to 0.$$
Landweber classified the closed subobjects of $\mathcal{M}_{\mathbf{fg}}$, thereby giving a precise list of conditions needed to check maps for flatness.

This appears to be a moot point, however, as it is unreasonable to expect this idea to apply to computing geometric fibers: the inclusion of a geometric point is flat only in highly degenerate cases. We will see that this can be repaired: the inclusion of the formal completion of a subobject is flat in friendly situations, and so we naturally become interested in the infinitesimal deformation spaces of the geometric points $\Gamma$ on $\mathcal{M}_{\mathbf{fg}}$. If we can analyze those, then Landweber's theorem will produce homology theories called *Morava $E_\Gamma$-theories*. Moreover, if we find that these deformation spaces are *smooth*, it will follow that their deformation rings support regular sequences. In this excellent case, by taking the regular quotient we will be able to recover *Morava $K_\Gamma$-theory*, a *homology theory*, which plays the role[1] of computing the fiber of $\mathcal{M}_{MUP}(X)$ at $\Gamma$.[2]

We have thus assembled a task list:

- Describe the open and closed subobjects of $\mathcal{M}_{\mathbf{fg}}$.
- Describe the geometric points of $\mathcal{M}_{\mathbf{fg}}$.
- Analyze their infinitesimal deformation spaces.

---

[1] To be clear: $K_\Gamma(X)$ may not actually compute the literal fiber of $\mathcal{M}_{MUP}(X)$ at $\Gamma$, since the homotopical operation of quotienting out the regular sequence is potentially sensitive to torsion sections of the module $\mathcal{M}_{MUP}(X)$.

[2] Incidentally, this program has no content when applied to $\mathcal{M}_{H\mathbb{F}_2}$, as Spec $\mathbb{F}_2$ is simply too small.

These will occupy our attention for the first half of this case study. In the second half, we will exploit these homology theories $E_\Gamma$ and $K_\Gamma$, as well as their connection to $\mathcal{M}_{fg}$ and to $MUP$, to make various structural statements about the category SPECTRA. These homology theories are especially well-suited to understanding the subcategory SPECTRA$^{fin}$ of finite spectra, and we will recount several important statements in that setting. Together with these homology theories, these celebrated results (collectively called the nilpotence and periodicity theorems) form the basis of *chromatic homotopy theory*. In fact, our *real* goal in this case study is to give an introduction to the chromatic perspective that remains in line with our algebro-geometrically heavy narrative.

## 3.1 Descent and the Context of a Spectrum

In Lecture 1.4 we took for granted the $H\mathbb{F}_2$-Adams spectral sequence, which had the form
$$E_2^{*,*} = H_{gp}^*(\underline{\mathrm{Aut}}_1(\widehat{\mathbb{G}}_a); \widehat{H\mathbb{F}_2 P_0 X}) \Rightarrow \pi_* X_2^\wedge,$$
where we had already established some yoga by which we could identify the dual Steenrod coaction on $H\mathbb{F}_2 P_0 X$ with an action of $\underline{\mathrm{Aut}}\,\widehat{\mathbb{G}}_a$ on its associated quasicoherent sheaf over $\mathrm{Spec}\,\mathbb{F}_2$. Our goal in this lecture is to revise this tool to work for other ring spectra $E$ and target spectra $X$, eventually arriving at a spectral sequence with signature
$$E_2^{*,*} = H^*(\mathcal{M}_E; \mathcal{M}_E(X) \otimes \omega^{*/2}) \Rightarrow \pi_* X_E^\wedge.$$
In particular, we will encounter along the way the object "$\mathcal{M}_E$" envisioned in the introduction to this case study.

At a maximum level of vagueness, we are seeking a process by which homotopy groups $\pi_* X$ can be recovered from $E$-homology groups $E_* X$. Recognizing that $X$ can be thought of as an $\mathbb{S}$-module and $E \wedge X$ can be thought of as its base change to an $E$-module, we might be moved to double back and consider as inspiration a completely algebraic analog of the same situation. Given a ring map $f: R \to S$ and an $S$-module $N$, Grothendieck's framework of *(faithfully flat) descent* addresses the following questions:

1. When is there an $R$-module $M$ such that $N \cong S \otimes_R M = f^* M$?
2. What extra data can be placed on $N$, called *descent data*, so that the category of descent data for $f$ is equivalent to the category of $R$-modules under the map $f^*$?
3. What conditions can be placed on $f$ so that the category of descent data for any given module is always contractible, called *effectivity*?

Suppose that we begin with an $R$-module $M$ and set $N = f^*M$, so that we are certain *a priori* that the answer to the first question is positive. The $S$-module $N$ has a special property, arising from $f$ being a ring map: There is a canonical isomorphism of $(S \otimes_R S)$-modules

$$\varphi \colon S \otimes_R N =$$
$$(f \otimes 1)^* N =$$
$$((f \otimes 1) \circ f)^* M \cong ((1 \otimes f) \circ f)^* M \quad s_1 \otimes (s_2 \otimes m) \mapsto (s_1 \otimes m) \otimes s_2$$
$$= (1 \otimes f)^* N$$
$$= N \otimes_R S.$$

Equivalently, we are noticing that pulling back an $R$-module from the bottom-right corner to the top-left corner along either arm of the following pullback diagram results in an isomorphic $S \otimes_R S$-module:

$$\begin{array}{ccc} \operatorname{Spec} S \times_{\operatorname{Spec} R} \operatorname{Spec} S & \longrightarrow & \operatorname{Spec} S \\ \downarrow & & \downarrow \\ \operatorname{Spec} S & \longrightarrow & \operatorname{Spec} R. \end{array}$$

In fact, this isomorphism is compatible with further shuffles, in the sense that the following diagram commutes:[3]

$$\begin{array}{ccc} N \otimes_R S \otimes_R S & \xrightarrow[\simeq]{\varphi_{13}} & S \otimes_R S \otimes_R N \\ & \searrow\varphi_{12}\hspace{1em}\nearrow\varphi_{23} & \\ & \simeq \quad S \otimes_R N \otimes_R S, \quad \simeq & \end{array}$$

where $\varphi_{ij}$ denotes applying $\varphi$ to the $i$th and $j$th coordinates.

**Definition 3.1.1** An $S$-module $N$ equipped with such an isomorphism

$$\varphi \colon S \otimes_R N \to N \otimes_R S$$

which causes the triangle to commute is called a *descent datum for $f$*.

Descent data admit two equivalent reformulations, both of which are useful to note.

*Remark 3.1.2* ([Ami59]) The ring $C = S \otimes_R S$ admits the structure of an $S$-coring: We can use the map $f$ to produce a relative diagonal map

$$\Delta \colon S \otimes_R S \cong S \otimes_R R \otimes_R S \xrightarrow{1 \otimes f \otimes 1} S \otimes_R S \otimes_R S \cong (S \otimes_R S) \otimes_S (S \otimes_R S).$$

---

[3] The commutativity of this triangle also shows that any number of shuffles also commutes.

3.1 Descent and the Context of a Spectrum    101

The descent datum $\varphi$ on an $S$-module $N$ is equivalent to a $C$-coaction map. The $S$-linearity of the coaction map is encoded by a square

$$\begin{array}{ccc} S \otimes_R N & \xrightarrow{1 \otimes \psi} & S \otimes_R N \otimes_S (S \otimes_R S) \\ \downarrow & \searrow^{\varphi} & \downarrow \\ N & \xrightarrow{\psi} & N \otimes_S (S \otimes_R S) \rightrightarrows N \otimes_R S, \end{array}$$

and the long composite gives the descent datum $\varphi$. Conversely, given a descent datum $\varphi$ we can restrict it to get a coaction map by

$$\psi\colon N = R \otimes_R N \xrightarrow{f \otimes 1} S \otimes_R N \xrightarrow{\varphi} N \otimes_R S.$$

The coassociativity condition on the comodule is equivalent under this correspondence to the commutativity of the triangle associated to $\varphi$.

*Remark 3.1.3* ([Hov02, theorem A])   Alternatively, descent data also arise naturally as sheaves on simplicial schemes. Associated to the map $f\colon \mathrm{Spec}\, S \to \mathrm{Spec}\, R$, we can form a Čech complex

$$\mathcal{D}_f := \left\{ \mathrm{Spec}\, S \longrightarrow \begin{array}{c} \mathrm{Spec}\, S \\ \times_{\mathrm{Spec}\, R} \\ \mathrm{Spec}\, S \end{array} \rightleftarrows \begin{array}{c} \mathrm{Spec}\, S \\ \times_{\mathrm{Spec}\, R} \\ \mathrm{Spec}\, S \\ \times_{\mathrm{Spec}\, R} \\ \mathrm{Spec}\, S \end{array} \longrightarrow \cdots \right\},$$

which factors the map $f$ as

$$\mathrm{Spec}\, S \xrightarrow{\mathrm{sk}^0} \mathcal{D}_f \xrightarrow{c} \mathrm{Spec}\, R.$$

with the composite being $f$.

A quasicoherent (and Cartesian [Sta14, tag 09VK]) sheaf $\mathcal{F}$ over a simplicial scheme $X$ is a sequence of quasicoherent sheaves $\mathcal{F}[n]$ on $X[n]$ as well as, for each map $\sigma\colon [m] \to [n]$ in the simplicial indexing category inducing a map $X(\sigma)\colon X[n] \to X[m]$, a natural choice of isomorphism of sheaves

$$\mathcal{F}(\sigma)^*\colon X(\sigma)^*\mathcal{F}[m] \to \mathcal{F}[n].$$

In particular, a pullback $c^*\widetilde{M}$ gives such a quasicoherent sheaf on $\mathcal{D}_f$. By restricting attention to the first three levels we find exactly the structure of the descent datum described before. Additionally, we have a natural *Segal isomorphism*

$$\mathcal{D}_f[1]^{\times_{\mathcal{D}_f[0]}(n)} \xrightarrow{\simeq} \mathcal{D}_f[n] \quad (\text{cf. } S \otimes_R S \otimes_R S \cong (S \otimes_R S) \otimes_S (S \otimes_R S) \text{ at } n = 2),$$

102                           Finite Spectra

which shows that any descent datum (including those not arising, *a priori*, from a pullback) can be naturally extended to a full quasicoherent sheaf on $\mathcal{D}_f$.

The following theorem is the culmination of a first investigation of descent:[4]

**Theorem 3.1.4** (Faithfully flat descent [Gro71, exposé VIII])  *If $f: R \to S$ is faithfully flat, the natural assignments*

$$\text{QCoh}(\text{Spec } R) \underset{\lim}{\overset{c^*}{\rightleftarrows}} \text{QCoh}(\mathcal{D}_f)$$

*form an equivalence of categories.*

*Jumping-off point*  The basic observation in this case is that $0 \to R \to S \to S \otimes_R S$ is an exact sequence of $R$-modules.[5] This makes much of the homological algebra involved work out. □

Without the flatness hypothesis, this theorem fails dramatically and immediately. For instance, the inclusion of the closed point

$$f: \text{Spec } \mathbb{F}_p \to \text{Spec } \mathbb{Z}$$

fails to distinguish the $\mathbb{Z}$-modules $\mathbb{Z}$ and $\mathbb{Z}/p$. Remarkably, this can be to a large extent repaired by reintroducing homotopy theory and passing to derived categories – for instance, the complexes $Lf^*\widetilde{\mathbb{Z}}$ and $Lf^*\widetilde{\mathbb{Z}/p}$ become distinct as objects of $D(\text{Spec } \mathbb{F}_p)$. Our preceding discussion of descent in Remark 3.1.3 can be quickly revised for this new homotopical setting, provided we remember to derive not just the categories of sheaves but also their underlying geometric objects. Our approach is informed by the following result:

**Lemma 3.1.5** ([EKMM97, theorem IV.2.4])  *There is an equivalence of symmetric monoidal $\infty$-categories between $D(\text{Spec } R) \simeq \text{Modules}_{HR}$.* □

Hence, given a map of rings $f: R \to S$, we redefine the derived descent object to be the cosimplicial ring spectrum

$$\mathcal{D}_{Hf} := \left\{ HS \overset{\longrightarrow}{\underset{\longrightarrow}{\longleftarrow}} \wedge_{HR} \overset{\longrightarrow}{\underset{\longrightarrow}{\longleftarrow}} \begin{array}{c} \longrightarrow \\ HS \\ \wedge_{HR} \\ HS \\ \wedge_{HR} \\ HS \end{array} \overset{\longrightarrow}{\underset{\longrightarrow}{\longleftarrow}} \cdots \right\},$$

---

[4] For details and additional context, see Vistoli [Vis05, section 4.2.1]. The story in the context of Hopf algebroids is also spelled out in detail by Miller [Milb].

[5] In the language of Example 1.4.17, this says that $R$ itself appears as the cofixed points $S \square_{S \otimes_R S} R$.

### 3.1 Descent and the Context of a Spectrum

and note that an $R$-module $M$ gives rise to a cosimplicial left-$\mathcal{D}_{Hf}$-module which we denote $\mathcal{D}_{Hf}(HM)$. The totalization of this cosimplicial module gives rise to an $HR$-module receiving a natural map from $M$, and we can ask for an analog of Theorem 3.1.4.

**Lemma 3.1.6** *For $f\colon \mathbb{Z} \to \mathbb{F}_p$ and $M$ a connective complex of $\mathbb{Z}$-modules, the totalization* $\mathrm{Tot}\,\mathcal{D}_{Hf}(HM)$ *recovers the $p$-completion of $M$.*[6]

*Proof sketch* The map $H\mathbb{Z} \to H\mathbb{F}_p$ kills the ideal $(p)$, and we calculate

$$H\mathbb{F}_p \wedge_{H\mathbb{Z}} H\mathbb{F}_p \simeq H\mathbb{F}_p \vee \Sigma H\mathbb{F}_p$$

to be connective. Together, these facts show that the filtration of $\mathcal{D}_{Hf}(HM)$ gives the $p$-adic filtration on homotopy. If $\pi_*HM$ is already $p$-complete, then the reassembly map $HM \to \mathrm{Tot}\,\mathcal{D}_{Hf}(HM)$ is a weak equivalence. □

We are now close enough to our original situation that we can make the last leap: Rather than studying a map $Hf\colon HR \to HS$, we instead have the unit map $\eta\colon \mathbb{S} \to E$ associated to some ring spectrum $E$. Fixing a target spectrum $X$, we define the analog of the descent object:

**Definition 3.1.7** The *descent object* for $X$ along $\eta\colon \mathbb{S} \to E$ is the cosimplicial spectrum

$$\mathcal{D}_E(X) := \left\{ \begin{array}{c} E \\ \wedge \\ X \end{array} \xrightarrow[\eta_R]{\overset{\eta_L}{\longrightarrow}} \begin{array}{c} E \\ \wedge \\ E \\ \wedge \\ X \end{array} \xleftarrow{\mu} \xrightarrow{\Delta} \begin{array}{c} E \\ \wedge \\ E \\ \wedge \\ E \\ \wedge \\ X \end{array} \begin{array}{c} \longrightarrow \\ \longleftarrow \\ \longrightarrow \\ \longleftarrow \\ \longrightarrow \end{array} \cdots \right\}.$$

**Lemma 3.1.8** ([Lurc, theorem 4.4.2.8]) *For $E$ an $A_\infty$-ring spectrum, $\mathcal{D}_E(X)$ can be considered as a cosimplicial object in the $\infty$-category of* SPECTRA. □

**Definition 3.1.9** The *$E$-nilpotent completion* of $X$ is the totalization of this cosimplicial spectrum:

$$X_E^\wedge := \mathrm{Tot}\,\mathcal{D}_E(X).$$

---

[6] There is an important distinction between a $p$-complete module and a module over the $p$-completion. For example, $\mathbb{Q}_p$ has a (continuous!) $\mathbb{Z}_p$-module structure, but it is not $p$-complete: The identity $\mathbb{Q}_p \otimes_{\mathbb{Z}_p} \mathbb{Z}_p/p^j = 0$ inhibits its reconstruction from the associated descent data. This distinction is embedded in the formation of the derived category, but in turn this has its own wrinkles; see, for example [HS99, appendix A] and [BF15, appendix A].

It receives a natural map $X \to X_E^\wedge$, the analog of the natural map of $R$-modules $M \to c_*c^*M$ considered in Theorem 3.1.4.

*Remark 3.1.10* ([Rav84, theorem 1.12], [Bou79])  Ravenel proves the following generalization of Lemma 3.1.6. Let $E$ be a connective ring spectrum, let $J$ be the set of primes complementary to those primes $p$ for which $E_*$ is uniquely $p$-divisible, and let $X$ be a connective spectrum.[7] If each element of $E_*$ has finite order, then $X_E^\wedge = X_J^\wedge$ gives the arithmetic completion of $X$ – which we reinterpret as $\mathbb{S}_J^\wedge \to E$ being of effective descent for connective objects. Otherwise, if $E_*$ has elements of infinite order, then $X_E^\wedge = X_{(J)}$ gives the arithmetic localization – which we reinterpret as saying that $\mathbb{S}_{(J)} \to E$ is of effective descent. Finding more encompassing conditions for which descent holds is an active subject of research [Mat16], [Lurd, appendix D].

Finally, we can interrelate these algebraic and topological notions of descent by studying the coskeletal filtration spectral sequence[8] for $\pi_* X_E^\wedge$, which we define to be the *E-Adams(-Novikov) spectral sequence* for $X$. Applying the homotopy groups functor to the cosimplicial ring spectrum $\mathcal{D}_E$ gives a cosimplicial ring $\pi_* \mathcal{D}_E$, which we would like to connect with an algebraic descent object of the sort considered in Remark 3.1.3. In order to make this happen, we need two niceness conditions on $E$:

**Definition 3.1.11** (Commutativity Hypothesis)  An even-periodic ring spectrum $E$ satisfies **CH** when the ring $\pi_* E^{\wedge j}$ is commutative for all $j \geq 1$. In this case, we can form the simplicial scheme

$$\mathcal{M}_E = \operatorname{Spec} \pi_0 \mathcal{D}_E,$$

called the *context* of $E$.

**Definition 3.1.12** (Flatness Hypothesis)  An even-periodic ring spectrum $E$ satisfies **FH** when the right-unit map $E_0 \to E_0 E$ is flat. In this case, the Segal

---

[7] Even for connective ring spectra $E$, the Bousfield localization $L_E X$ does *not* have to recover an arithmetic localization of $X$ if $X$ is not connective. Take $E = H\mathbb{Z}$ and $X = KU$, which Snaith's theorem presents as $X = KU = \Sigma_+^\infty \mathbb{CP}^\infty[\beta^{-1}]$, where $\beta \colon \mathbb{CP}^1 \to \mathbb{CP}^\infty$ is the Bott element. This gives $H\mathbb{Z}_*KU = H\mathbb{Z}_*(\mathbb{CP}^\infty[\beta^{-1}]) = (H\mathbb{Z}_*\mathbb{CP}^\infty)[\beta_1^{-1}]$. We can identify the pieces in turn: Example 2.1.20 shows $\mathbb{CP}^\infty_{H\mathbb{Z}} = \widehat{\mathbb{G}}_a$, so the dual Hopf algebra $(O_{\widehat{\mathbb{G}}_a})^* = H\mathbb{Z}_*\mathbb{CP}^\infty$ is a divided polynomial ring on the class $\beta_1$. Inverting $\beta_1$ then gives $(H\mathbb{Z}_*\mathbb{CP}^\infty)[\beta_1^{-1}] = \Gamma[\beta_1][\beta_1^{-1}] = \mathbb{Q}[\beta_1^\pm]$, so that, in particular, there is a weak equivalence $H\mathbb{Z} \wedge KU \to H\mathbb{Q} \wedge KU$. The cofiber $KU \to KU \otimes \mathbb{Q} \to KU \otimes \mathbb{Q}/\mathbb{Z}$ is thus a nonzero $H\mathbb{Z}$-acyclic spectrum. You can also work this example without knowing Snaith's theorem: Since $\Sigma_+^\infty \mathbb{CP}^\infty[\beta^{-1}] \to KU$ is a map of ring spectra, $\pi_0 \Sigma_+^\infty \mathbb{CP}^\infty[\beta^{-1}]$ cannot be a rational group, as $\pi_0 KU = \mathbb{Z}$. Another entertaining consequence of this is $D(KU/p) = F(KU/p, \mathbb{S}) = 0$, since $\mathbb{S}$ is $H\mathbb{Z}$-local.

[8] The reader who would like a refresher about the construction of spectral sequences is referred to Boardman's exceptional article [Boa99].

## 3.1 Descent and the Context of a Spectrum

map
$$(E_0 E)^{\otimes_{E_0} j} \otimes_{E_0} E_0 X \to \pi_0(E^{\wedge(j+1)} \wedge X) = \pi_0 \mathcal{D}_E(X)[j]$$
is an isomorphism for all $X$. In geometric language, this says that $\mathcal{M}_E$ is valued in simplicial sets equivalent to nerves of groupoids and that
$$\mathcal{M}_E(X) := \widetilde{\pi_0 \mathcal{D}_E(X)}$$
forms a Cartesian quasicoherent sheaf over $\mathcal{M}_E$. In this sense, we have constructed a factorization

$$\begin{array}{ccc}
\text{Spectra} & \xrightarrow{E_0(-)} & \text{Modules}_{E_0} \\
{\scriptstyle \mathcal{M}_E(-)} \searrow & & \nearrow {\scriptstyle (-)[0]} \\
& \text{QCoh}(\mathcal{M}_E). &
\end{array}$$

While **CH** and **FH** are enough to guarantee that $\mathcal{M}_E$ and $\mathcal{M}_E(X)$ are well-behaved, they still do not exactly connect us with Remark 3.1.3. The main difference is that the ring of homology cooperations for $E$
$$E_0 E = \pi_0(E \wedge E) = \pi_0 \mathcal{D}_E[1]$$
is only distantly related to the tensor product $E_* \otimes_{\pi_* \mathbb{S}} E_*$ (or even $\text{Tor}_{*,*}^{\pi_* \mathbb{S}}(E_*, E_*)$). This is a trade we are eager to make, as the latter groups are typically miserably behaved, whereas $E_0 E$ is typically fairly nice. In order to take advantage of this, we enlarge our definition to match:

**Definition 3.1.13** Let $A$ and $\Gamma$ be commutative rings with associated affine schemes $X_0 = \text{Spec } A$, $X_1 = \text{Spec } \Gamma$. A *Hopf algebroid* consists of the pair $(A, \Gamma)$ together with structure maps

$$\begin{aligned}
\eta_R &: A \to \Gamma, & s &: X_1 \to X_0, \\
\eta_L &: A \to \Gamma, & t &: X_1 \to X_0, \\
\chi &: \Gamma \to \Gamma, & (-)^{-1} &: X_1 \to X_1, \\
\Delta &: \Gamma \to \Gamma \underset{A}{{}^{\eta_R}\otimes{}^{\eta_L}} \Gamma, & \circ &: X_1 \underset{X_0}{{}^t\times{}^s} X_1 \to X_1,
\end{aligned}$$

such that $(X_0, X_1)$ forms a groupoid scheme. An $(A, \Gamma)$-*comodule* is an $A$-module equipped with a $\Gamma$-comodule structure, and such a comodule is equivalent to a Cartesian quasicoherent sheaf on the nerve of $(X_0, X_1)$.[9]

*Example 3.1.14* A Hopf $k$-algebra $H$ gives a Hopf algebroid $(k, H)$. The scheme of objects $\text{Spec } k$ in the groupoid scheme is the constant scheme 0.

---

[9] An extremely enlightening discussion of the intricacies of this construction was set out by Boardman [Boa82] in the traditional language of Hopf algebroids.

**Lemma 3.1.15** *For $E$ an $A_\infty$-ring spectrum satisfying $CH$ and $FH$, the $E_2$-page of its Adams spectral sequence can be identified as*

$$E_2^{*,*} = \mathrm{Cotor}_{E_0 E}^{*,*}(E_0, E_0 X)$$
$$\cong H^*(\mathcal{M}_E; \mathcal{M}_E(X) \otimes \omega^{*/2}) \Rightarrow \pi_* X_E^\wedge,$$

*where $\omega^{n/2}$ denotes the line bundle $\omega^{n/2} = \mathcal{M}_E(\mathbb{S}^n)$.*

*Proof sketch* The homological algebra of Hopf algebras from Lecture 1.4 can be lifted almost verbatim, allowing us to define resolutions suitable for computing derived functors [Rav86, definition A1.2.3]. This includes the cobar resolution [Rav86, definition A1.2.11], which shows that the associated graded for the coskeletal filtration of $\mathcal{D}_E(X)$ is a complex computing the derived functors claimed in the lemma statement. The name $\omega^{n/2}$ given to $\mathcal{M}_E(S^n)$ is justified by Example 2.3.4. □

*Remark 3.1.16* (see Lemma 1.4.2) The sphere spectrum fails to satisfy **CH**, so the above results do not apply, but the $\mathbb{S}$-Adams spectral sequence is particularly degenerate: It consists of $\pi_* X$, concentrated on the 0-line. For any other ring spectrum $E$, the unit map $\mathbb{S} \to E$ induces a map of Adams spectral sequences whose images on the 0-line are those maps of comodules induced by applying $E$-homology to a homotopy element of $X$–i.e., the Hurewicz image of $E$.

*Remark 3.1.17* In Lemma 3.1.6, we discussed translating from the algebraic descent picture to a homotopical one, and a crucial point was how thorough we had to be: we transferred not just to the derived category $D(\mathrm{Spec}\, R)$ but we also replaced the base ring $R$ with its homotopical incarnation $HR$. In Definition 3.1.13, we have not been as thorough as possible: Both $X_0$ and $X_1$ are schemes and hence satisfy a sheaf condition individually, but the functor $(X_0, X_1)$, thought of as valued in homotopy 1-types, does not necessarily satisfy a homotopy sheaf condition. Enforcing this descent condition results in the *associated stack* [Hop, definition 8.13], denoted

$$\mathrm{Spec}\, A /\!\!/ \mathrm{Spec}\, \Gamma = X_0 /\!\!/ X_1.$$

Remarkably, this does not change the category of Cartesian quasicoherent sheaves – it is still equivalent to the category of $(A, \Gamma)$-comodules [Hop, proposition 11.6]. However, several different Hopf algebroids (such as those with maps between them inducing natural equivalences of groupoid schemes, as studied by Hovey [Hov02, theorem D], but also some with *no* such zig-zag) can give the same associated stack, resulting in surprising equivalences of comodule categories.[10] For the most part, it will not be especially relevant to us whether

---

[10] We will employ one of these surprising equivalences in Remark 3.3.19.

## 3.1 Descent and the Context of a Spectrum

we are considering the groupoid scheme or its associated stack, so we will not draw much of a distinction. For the most part, the associated stack is theoretically preferable, but the groupoid scheme is easier to make explicit.[11]

*Example 3.1.18* Most of the homology theories we will discuss satisfy **CH** and **FH**. For example, $H\mathbb{F}_2 P$ has this property: There is only one possible algebraic map $\mathbb{F}_2 \to \mathcal{A}_*$, so **FH** is necessarily satisfied. Our work in Lecture 1.3[12] thus grants us access to a description of the context for $H\mathbb{F}_2$:[13]

$$\mathcal{M}_{H\mathbb{F}_2 P} = \operatorname{Spec} \mathbb{F}_2 /\!\!/ \underline{\operatorname{Aut}} \widehat{\mathbb{G}}_a.$$

*Example 3.1.19* The context for $MUP$ is considerably more complicated, but Quillen's theorem can be equivalently stated as giving a description of it. Quillen's theorem on its face gives an equivalence $\operatorname{Spec} MUP_0 \cong \mathcal{M}_{\mathbf{fgl}}$, but in Lemma 2.6.7 we also gave a description of $\operatorname{Spec} MUP_0 MUP$: It is the moduli of pairs of formal group laws equipped with an invertible power series intertwining them. Altogether, this presents $\mathcal{M}_{MUP}$ as the moduli of formal groups:

$$\mathcal{M}_{MUP} \simeq \mathcal{M}_{\mathbf{fg}} := \mathcal{M}_{\mathbf{fgl}} /\!\!/ \mathcal{M}_{\mathbf{ps}}^{\mathrm{gpd}},$$

where $\mathcal{M}_{\mathbf{ps}} = \underline{\operatorname{End}}(\widehat{\mathbb{A}}^1)$ is the moduli of endomorphisms of the affine line (i.e., of power series) and $\mathcal{M}_{\mathbf{ps}}^{\mathrm{gpd}}$ is the multiplicative subgroup of invertible such maps. We include a picture of the $p$-localization of the $MU$-Adams $E_2$-page in Figures 3.1 and 3.2. In view of Remark 3.1.17, there is an important subtlety about the stack $\mathcal{M}_{\mathbf{fg}}$: An $R$-point is a functor on affines over $\operatorname{Spec} R$ which is locally isomorphic to a formal group, but whose local isomorphism *may not patch* to give a global isomorphism. This does not agree, a priori, with the definition of formal group given in Definition 2.1.16, where the isomorphism witnessing a functor as a formal variety was expected to be global. We will address this further in Lemma 3.2.9.[14]

*Example 3.1.20* The context for $MOP$, by contrast, is reasonably simple. Corollary 1.5.7 shows that the scheme $\operatorname{Spec} MOP_0$ classifies formal group

---

[11] Constructing the correct derived category of comodules also has subtle associated homotopical issues. Hovey gives a good reference for this in the case of a stack associated to a Hopf algebroid [Hov04].
[12] The reader should spend a moment contrasting our two approaches to the definition of $\mathcal{M}_{H\mathbb{F}_2}(X)$. Previously, we assumed that $X$ was finite type and worked with cohomology – but under exactly this assumption, the cohomology of $X$ can always be written as the homology of $Y$ for another finite spectrum satisfying $X = DY$.
[13] This is a bit glib: This gives a presentation of the *even* part of the context for $H\mathbb{F}_2 P$, and the "$\widehat{\mathbb{G}}_a$" in the formula is $\mathbb{CP}^\infty_{H\mathbb{F}_2}$ rather than $\mathbb{RP}^\infty_{H\mathbb{F}_2}$. This sleight of hand abuts the discussion of Bocksteins in Theorem 3.6.15.
[14] Naumann gives a very pleasant write-up of, among other things, the difference between $\mathcal{M}_{\mathbf{fg}}$ as a simplicial scheme and as a stack [Nau07].

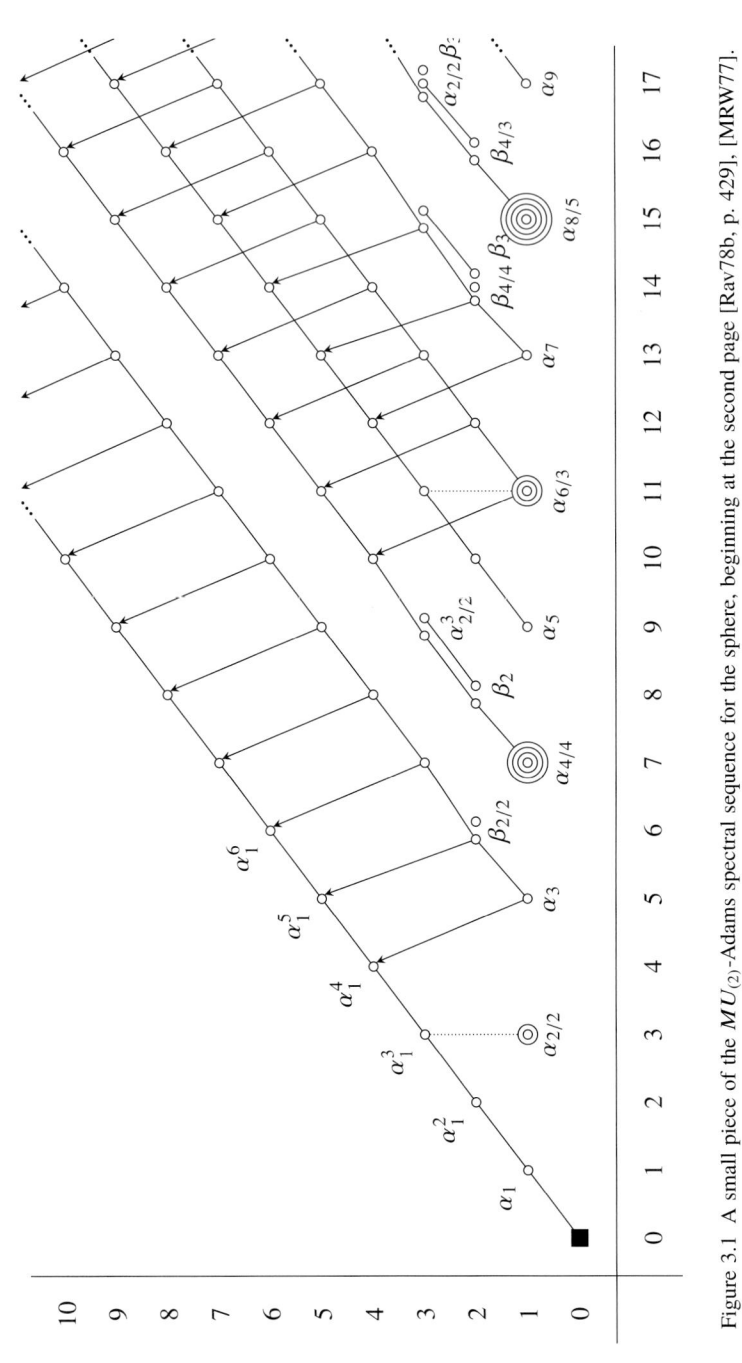

Figure 3.1 A small piece of the $MU_{(2)}$-Adams spectral sequence for the sphere, beginning at the second page [Rav78b, p. 429], [MRW77]. Northeast lines denote multiplication by $\eta = \alpha_1$, northwest lines denote $d_3$-differentials, and vertical dotted lines indicate additive extensions. Elements are labeled according to the conventions of Remark 3.6.22, and in particular $\alpha_{i/j}$ is $2^j$-torsion.

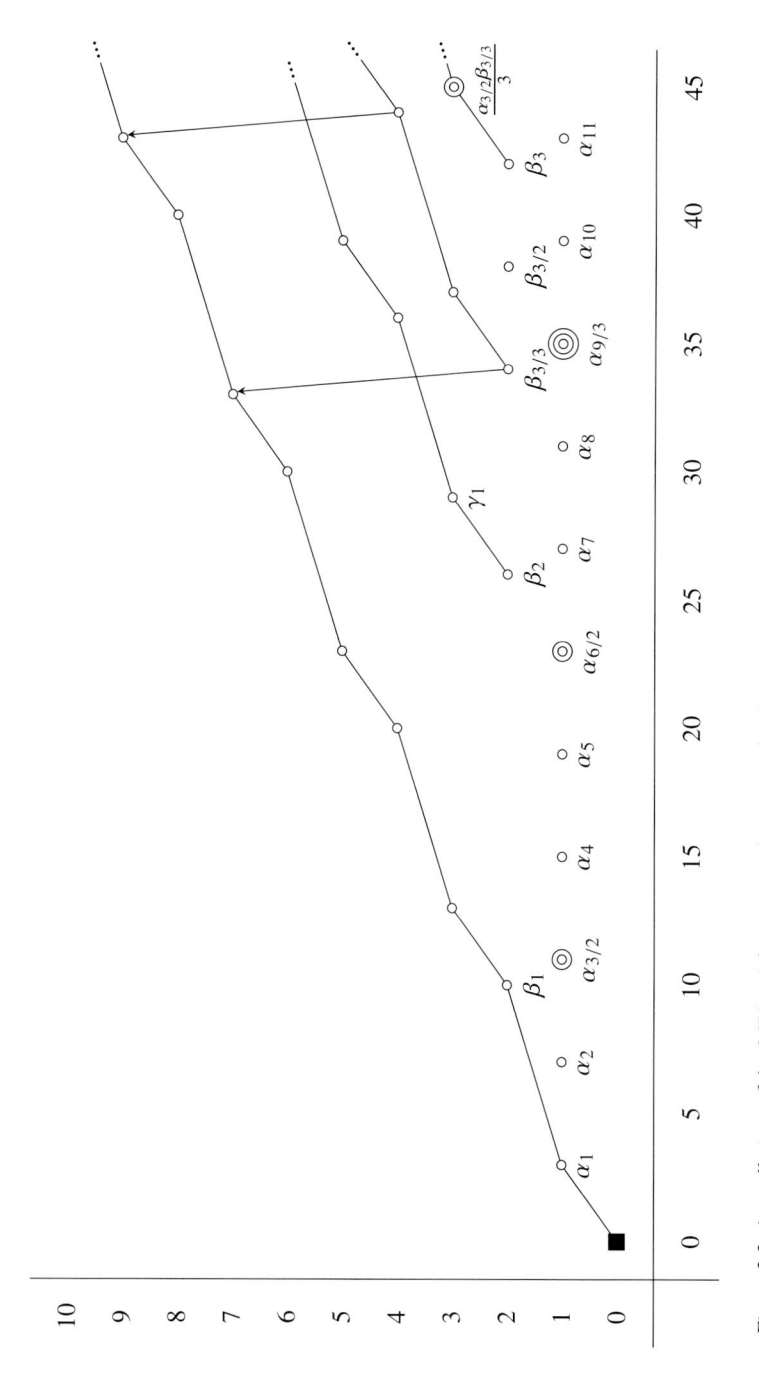

Figure 3.2 A small piece of the $MU_{(3)}$-Adams spectral sequence for the sphere, beginning at the second page [Rav86, Figure 1.2.19]. Northeast lines denote multiplication by $\alpha_1$ or by $\beta_1/\alpha_1$, and northwest lines denote $d_5$-differentials. Elements are labeled according to the conventions of Remark 3.6.22, and in particular $\alpha_{i/j}$ is $3^j$-torsion.

laws over $\mathbb{F}_2$ which admit logarithms, so that $\mathcal{M}_{MOP}$ consists of the groupoid of formal group laws with logarithms and isomorphisms between them. This admits a natural deformation-retraction to the moduli consisting just of $\widehat{\mathbb{G}}_a$ and its automorphisms, expressing the redundancy in $MOP_0(X)$ encoded in the splitting of Lemma 1.5.8.

*Remark 3.1.21* The algebraic moduli $\mathcal{M}_{MU} = (\operatorname{Spec} MU_*, \operatorname{Spec} MU_*MU)$ and the topological moduli $(MU, MU \wedge MU)$ are quite different. An orientation $MU \to E$ selects a coordinate on the formal group $\mathbb{CP}^\infty_E$, but $\mathbb{CP}^\infty_E$ itself exists independently of the orientation. Hence, while $\mathcal{M}_{MU}(E_*)$ can have many connected components corresponding to *distinct formal groups* on the coefficient ring $E_*$, the groupoid $\textsc{RingSpectra}(\mathcal{D}_{MU}, E)$ has only *one* connected component corresponding to the formal group $\mathbb{CP}^\infty_E$ intrinsic to $E$.[15,16,17]

*Remark 3.1.22* ([Hop14a, p. 5]) If $E$ is a complex-oriented ring spectrum, then the simplicial sheaf $\mathcal{M}_{MU}(E)$ has an extra degeneracy, which causes the $MU$-based Adams spectral sequence for $E$ to degenerate. In this sense, the "stackiness" of $\mathcal{M}_{MU}(E)$ is exactly a measure of the failure of $E$ to be orientable.

*Remark 3.1.23* ([Bou79, section 5], see also [Lurc, theorem 1.2.4.1], [Mil81, section 1]) It is also possible to construct an $E$-Adams spectral sequence by iteratively smashing with the fiber sequence $\overline{E} \to \mathbb{S} \to E$ to form the tower

$$\begin{array}{ccccccc}
\mathbb{S} \wedge X & \leftarrow & \overline{E} \wedge X & \leftarrow & \overline{E}^{\wedge 2} \wedge X & \leftarrow & \cdots \\
\downarrow & & \downarrow & & \downarrow & & \\
E \wedge X & & E \wedge \overline{E} \wedge X & & E \wedge \overline{E}^{\wedge 2} \wedge X & & \cdots.
\end{array}$$

This presentation makes the connection to descent much more opaque, but it does not require $E$ to be an $A_\infty$-ring spectrum.

*Remark 3.1.24* Many algebro-geometric properties lift automatically to the setting of *representable* maps: A map $f\colon \mathcal{M} \to \mathcal{N}$ is representable when for all affines $\operatorname{Spec} R \to \mathcal{N}$, the pullback $\mathcal{M} \times_\mathcal{N} \operatorname{Spec} R$ is affine. A representable such $f$ is said to have property **P** when the natural map $\mathcal{M} \times_\mathcal{N} \operatorname{Spec} R \to \operatorname{Spec} R$ has property **P** for every choice of affine over $\mathcal{N}$, giving us intrinsic notions of adjectives like open, flat, ....

---

[15] In Example 3.3.3, we will show that $\widehat{\mathbb{G}}_a$ and $\widehat{\mathbb{G}}_m$ are not isomorphic over $\operatorname{Spec} \mathbb{Z}$. It follows that the "Todd genus," which is the map $MUP_0 \to KU_0 \cong \mathbb{Z}$ induced by Example 2.1.21, *requires* $KU$ as its target and *cannot* be realized by a map $MUP \to H\mathbb{Z}$. Rationally, however, there is an isomorphism $\mathbb{Q} \otimes KU \cong \mathbb{Q} \otimes H\mathbb{Z}P$, and the resulting relation between the Todd genus and the trivial genus is known as the *Chern character*.

[16] The reader ought to compare this with the situation in explicit local class field theory, where a local number field has a preferred formal group attached to it.

[17] The precocious student might ask what functor $MU$ represents as an $E_\infty$-ring spectrum. To date, this functor has not been algebraically recognized.

## 3.2 The Structure of $\mathcal{M}_{\mathrm{fg}}$ I: The Affine Cover

In Definition 3.1.13 we gave a factorization

$$\text{Spectra} \xrightarrow{MUP_0(-)} \text{Modules}_{MUP_0}$$

with $\mathcal{M}_{MUP}(-)$ and $(-)[0]$ factoring through $\text{QCoh}(\mathcal{M}_{MUP})$,

and in Example 3.1.19 we established an equivalence

$$\varphi \colon \mathcal{M}_{MUP} \xrightarrow{\simeq} \mathcal{M}_{\mathrm{fg}}.$$

Our program, as outlined in the introduction, is to analyze this functor $\mathcal{M}_{MUP}(-)$ by postcomposing it with $(\varphi^{-1})^*$ and studying the resulting sheaf over $\mathcal{M}_{\mathrm{fg}}$. In order to perform such an analysis, we will want a firm grip on the geometry of the stack $\mathcal{M}_{\mathrm{fg}}$, and in this lecture we begin by studying the scheme $\mathcal{M}_{\mathrm{fgl}}$ as well as the natural covering map

$$\mathcal{M}_{\mathrm{fgl}} \to \mathcal{M}_{\mathrm{fg}}.$$

Additionally, Example 3.1.19 was a consequence of Corollary 2.6.10, which relied on the unproven result stated as Theorem 2.6.9, which we will now prove in this section as Theorem 3.2.2.

**Definition 3.2.1** There is an affine scheme $\mathcal{M}_{\mathrm{fgl}}$ classifying formal group laws. Begin with the scheme classifying *all* bivariate power series:

$$\text{Spec}\,\mathbb{Z}[a_{ij} \mid i,j \geq 0] \leftrightarrow \{\text{bivariate power series}\},$$

$$f \in \text{Spec}\,\mathbb{Z}[a_{ij} \mid i,j \geq 0](R) \leftrightarrow \sum_{i,j \geq 0} f(a_{ij}) x^i y^j.$$

Then, $\mathcal{M}_{\mathrm{fgl}}$ is the closed subscheme selected by the formal group law axioms in Definition 2.1.19.

This presentation of $\mathcal{M}_{\mathrm{fgl}}$ as a subscheme appears to be extremely complicated in that its ideal is generated by many hard-to-describe elements, but $\mathcal{M}_{\mathrm{fgl}}$ itself is actually not complicated at all. We will prove the following:

**Theorem 3.2.2** (Lazard's Theorem [Laz55, théorème II]) *There is a non-canonical isomorphism*

$$\mathcal{O}_{\mathcal{M}_{\mathrm{fgl}}} \cong \mathbb{Z}[t_n \mid 1 \leq n < \infty] =: L.$$

*Proof* We begin by studying simpler moduli with the intention of comparing them with the more complicated $\mathcal{M}_{\mathbf{fgl}}$. Let $U = \mathbb{Z}[b_0, b_1, b_2, \ldots]/(b_0 - 1)$ be the universal ring supporting a strict exponential

$$\exp(x) := \sum_{j=0}^{\infty} b_j x^{j+1}.$$

Because $b_0 = 1$ is invertible in this ring, this series has a formal inverse, i.e., a universal logarithm:

$$\log(x) := \sum_{j=0}^{\infty} m_j x^{j+1},$$

where the coefficients $m_j$ are rather complicated polynomials in the indeterminates $b_*$.[18] Together, these series induce a formal group law on $U$ by the conjugation formula

$$x +_u y = \exp(\log(x) + \log(y)),$$

which is in turn classified by a map $u \colon \mathcal{O}_{\mathcal{M}_{\mathbf{fgl}}} \to U$.[19] Using Theorem 2.1.25, we know that the map $u$ is a rational isomorphism, so we become interested in learning more about the behavior of the elements $m_j$, and in particular their torsion properties. Since these elements are difficult to handle precisely, we compute modulo decomposables:

$$x = \exp(\log(x))$$
$$= x + \sum_{n=1}^{\infty} m_n x^{n+1} + \sum_{n=1}^{\infty} b_n \left( x + \sum_{j=1}^{\infty} m_j x^{j+1} \right)^{n+1}$$
$$\equiv x + \sum_{n=1}^{\infty} m_n x^{n+1} + \sum_{n=1}^{\infty} b_n x^{n+1} \quad \text{(mod decomposables)},$$

hence $b_n \equiv -m_n$ (mod decomposables). Using this, we then compute

$$x +_u y = \exp(\log(x) + \log(y))$$
$$= \left( (x + y) + \sum_{n=1}^{\infty} m_n (x^{n+1} + y^{n+1}) \right)$$

---

[18] In the context of complex-oriented cohomology theories, this is called the *Miščenko logarithm*, and the coefficients $m_j$ have the simple formula $\log_\varphi(x) = \sum_{n=0}^{\infty} \frac{\varphi[\mathbb{C}P^n]}{n+1} x^{n+1}$. In trade, the coefficients $b_j$ are more mysterious.

[19] This is *not* the universal formal group law. We will soon see that some formal group laws do not admit logarithms. Rather, it is the group law associated to $H\mathbb{Z} \wedge MU$, and the map we are studying is that induced by $MU \to H\mathbb{Z} \wedge MU$ as in Lemma 2.6.7. In light of this, there is a multiplicative version of this same story: the Hattori–Stong theorem states that $MU_* \to K_*MU$ has as its image a direct summand [Ara73, Bak87].

## 3.2 The Structure of $\mathcal{M}_{\text{fg}}$ I: The Affine Cover

$$+ \sum_{n=1}^{\infty} b_n \left( (x+y) + \sum_{j=1}^{\infty} m_j (x^{j+1} + y^{j+1}) \right)^{n+1}$$

$$\equiv x + y$$

$$+ \sum_{n=1}^{\infty} -b_n(x^{n+1} + y^{n+1}) + \sum_{n=1}^{\infty} b_n(x+y)^{n+1} \quad \text{(mod decomposables)}$$

$$= x + y + \sum_{n=1}^{\infty} b_n((x+y)^{n+1} - x^{n+1} - y^{n+1}),$$

hence

$$u(a_{i(n-i)}) \equiv \binom{n}{i} b_{n-1} \quad \text{(mod decomposables)}.$$

It follows that the map $Qu$ on the indecomposable quotient has image in degree $2n$ the subgroup $T_{2n}$ generated by $d_{n+1} b_n$, where $d_{n+1} = \gcd\left(\binom{n+1}{k} \middle| 0 < k < n+1 \right)$.

Appealing to Lemma 3.2.3, select elements $t_n \in O_{\mathcal{M}_{\text{fgl}}}$ projecting to $r(d_{n+1} b_n)$ on indecomposables, and consider the induced map

$$\mathbb{Z}[t_n \mid n \geq 1] \xrightarrow{v} O_{\mathcal{M}_{\text{fgl}}} \xrightarrow{u} U.$$

The map $r$ is surjective and $Qv$ has the same image, so it is also surjective and hence $v$ is as well. Additionally, because

$$Q(uv)_{2n} \colon \mathbb{Z}\{t_n\} \to \mathbb{Z}\{d_{n+1} b_n\}$$

is a surjective map of free $\mathbb{Z}$-modules, it must also be injective. It follows that the map $uv$ of free $\mathbb{Z}$-algebras is injective, and hence $v$ itself is injective. □

We have yet to prove the following Lemma:

**Lemma 3.2.3** *There is a canonical retraction* $r \colon T_{2n} \to (QO_{\mathcal{M}_{\text{fgl}}})_{2n}$ *of* $Qu$.

In order to prove this Lemma, it will be useful to have a more moduli-theoretic interpretation of $Qu$. The graded ring $O_{\mathcal{M}_{\text{fgl}}}$ splits naturally as $\mathbb{Z}$ in degree 0 and the elements of positive degree, and using this we can extend the map $Qu$ to a ring homomorphism

$$Qu \colon O_{\mathcal{M}_{\text{fgl}}} \to \mathbb{Z} \oplus QL,$$

which factors through $T$ and projects to degree $2n$ to give a ring homomorphism

$$O_{\mathcal{M}_{\text{fgl}}} \to \mathbb{Z} \oplus T_{2n}.$$

For any abelian group $A$, a graded ring homomorphism $O_{\mathcal{M}_{\text{fgl}}} \to \mathbb{Z} \oplus \Sigma^{2n} A$ selects a formal group law with two clear properties: modulo terms of degree $n$

it equals the additive group law, and it has no terms of degree greater than $n$. The terms appearing in exactly degree $n$ take the following particular form:

**Definition 3.2.4** A *symmetric 2-cocycle of degree n* is a polynomial $f(x, y)$ satisfying

- symmetry: $f(x, y) = f(y, x)$;
- homogeneity: $f$ consists solely of terms of total degree $n$;
- 2-cocycle condition:[20]

$$f(x, y) - f(t + x, y) + f(t, x + y) - f(t, x) = 0.$$

Dropping the homogeneity condition in favor of a bounded above condition, we also have the following definition:

**Definition 3.2.5** A series satisfying the analogs of the formal group law axioms modulo terms of degree $(n + 1)$ is called a *formal n-bud*.[21]

**Lemma 3.2.6** *If $+_\varphi$ and $+'_\varphi$ are two n-buds which reduce to the same $(n - 1)$-bud, the difference $(x +_\varphi y) - (x +'_\varphi y)$ is a symmetric 2-cocycle of degree n. Conversely, given such an n-bud $+_\varphi$ and a symmetric 2-cocycle $f$, the series*

$$x +'_\varphi y := (x +_\varphi y) + f(x, y)$$

*is also an n-bud.*

*Proof* This is made explicit in the following calculation:

$$x +'_\varphi (y +'_\varphi z) = x +'_\varphi (y +_\varphi z + f(y, z))$$
$$= x +_\varphi (y +_\varphi z + f(y, z)) + f(x, y +_\varphi z + f(y, z))$$
$$\equiv x +_\varphi (y +_\varphi z) + f(y, z) + f(x, y + z) \quad (\mathrm{mod}\ (x, y)^{n+1}),$$
$$(x +'_\varphi y) +'_\varphi z = (x +_\varphi y + f(x, y)) +'_\varphi z$$
$$= (x +_\varphi y + f(x, y)) +_\varphi z + f(x +_\varphi y + f(x, y), z)$$
$$\equiv (x +_\varphi y) +_\varphi z + f(x, y) + f(x + y, z) \quad (\mathrm{mod}\ (x, y)^{n+1}),$$

resulting in the 2-cocycle condition on $f$, and symmetry of $x +'_\varphi y$ enforces the symmetry of $f$. Reading the sequence of equalities backwards shows the converse. □

*Reduction of Lemma 3.2.3 to Lemma 3.2.7* We now show that the following conditions are equivalent:

---

[20] We will justify the "2-cocycle" terminology in the course of the proof of Lemma 3.2.7.
[21] A formal $n$-bud determines a "multiplication" $(\widehat{\mathbb{A}}^1 \times \widehat{\mathbb{A}}^1)^{(n)} \to \widehat{\mathbb{A}}^{1,(n)}$. Note that this does *not* belong to a group object, since $(\widehat{\mathbb{A}}^1 \times \widehat{\mathbb{A}}^1)^{(n)} \neq \widehat{\mathbb{A}}^{1,(n)} \times \widehat{\mathbb{A}}^{1,(n)}$. This is the observation that the ideals $(x, y)^{n+1}$ and $(x^{n+1}, y^{n+1})$ are distinct.

## 3.2 The Structure of $\mathcal{M}_{fg}$ I: The Affine Cover

1. (Lemma 3.2.7) Symmetric 2-cocycles of degree $n$ are spanned by

$$c_n = \frac{1}{d_n} \cdot ((x+y)^n - x^n - y^n),$$

   where $d_n = \gcd\left(\binom{n}{k} \Big| 0 < k < n\right)$.
2. For $F$ is an $(n-1)$-bud, the set of $n$-buds extending $F$ form a torsor under addition for $R \otimes c_n$.
3. Any homomorphism $(QO_{\mathcal{M}_{fgl}})_{2n} \to A$ of additive groups factors through a homomorphism $(QO_{\mathcal{M}_{fgl}})_{2n} \to T_{2n}$.
4. (Lemma 3.2.3) There is a canonical splitting $T_{2n} \to (QO_{\mathcal{M}_{fgl}})_{2n}$.

To verify that Claims 1 and 2 are equivalent, note first that Lemma 3.2.6 shows that the set of $n$-buds in Claim 2 is a torsor for the group of such polynomials in Claim 1, and hence Claims 1 and 2 are a simultaneous assertion about the precise form of the structure group for that torsor. The equivalence of Claims 2 and 3 amounts to our moduli-theoretic interpretation of formal group laws on rings formed by square-zero extensions: A group map

$$(QO_{\mathcal{M}_{fgl}})_{2n} \to A$$

is equivalent data to a ring map

$$O_{\mathcal{M}_{fgl}} \to \mathbb{Z} \oplus A$$

with the prescribed behavior on $(QO_{\mathcal{M}_{fgl}})_{2n}$ and which sends all other indecomposables to 0. Finally, Claim 4 is the universal case of Claim 3, using the projection map $O_{\mathcal{M}_{fgl}} \to (QO_{\mathcal{M}_{fgl}})_{2n}$. Applying Claim 3 to this map yields a diagram

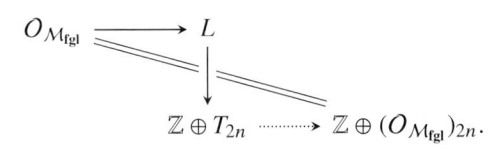

$$\begin{array}{ccc} O_{\mathcal{M}_{fgl}} & \longrightarrow & L \\ \downarrow & \searrow & \\ \mathbb{Z} \oplus T_{2n} & \cdots\cdots\to & \mathbb{Z} \oplus (O_{\mathcal{M}_{fgl}})_{2n}. \end{array}$$

$\square$

We will now verify Claim 1 computationally, thereby completing the proof of Lemma 3.2.3 (and hence Theorem 3.2.2).

**Lemma 3.2.7** (Symmetric 2-cocycle lemma [Laz55, lemme 3], see [Hop, theorem 3.1]) *Symmetric 2-cocycles of degree n are spanned by*

$$c_n = \frac{1}{d_n} \cdot ((x+y)^n - x^n - y^n),$$

*where* $d_n = \gcd\left(\binom{n}{k} \Big| 0 < k < n\right)$.

*Proof* We begin with a reduction of the sorts of rings over which we must consider the possible symmetric 2-cocycles. First, notice that only the additive group structure of the ring matters: the symmetric 2-cocycle condition does not involve any ring multiplication. Second, it suffices to show the lemma over a finitely generated abelian group, as a particular polynomial has finitely many terms and hence involves finitely many coefficients. Noticing that the lemma is true for $A \oplus B$ if and only if it's true for $A$ and for $B$, we couple these facts to the structure theorem for finitely generated abelian groups to reduce to the cases $\mathbb{Z}$ and $\mathbb{Z}/p^r$. From here, we can reduce to the prime fields: If $A \leq B$ is a subgroup and the lemma is true for $B$, it's true for $A$, so we will be able to deduce the case of $\mathbb{Z}$ from the case of $\mathbb{Q}$. Lastly, we can also reduce from $\mathbb{Z}/p^r$ to $\mathbb{Z}/p$ using an inductive Bockstein-style argument over the extensions

$$(p^r)/(p^{r+1}) \to \mathbb{Z}/p^{r+1} \to \mathbb{Z}/p^r$$

and noticing that $(p^r)/(p^{r+1}) \cong \mathbb{Z}/p$ as abelian groups. Hence, we can now freely assume that our ground object is a prime field.

We now ground ourselves by fitting symmetric 2-cocycles into a more general homological framework, hoping that we can use such a machine to power a computation. For a formal group scheme $\widehat{\mathbb{G}}$, we can form a simplicial scheme $B\widehat{\mathbb{G}}$ in the usual way:

$$B\widehat{\mathbb{G}} := \left\{ \begin{array}{c} * \\ * \leftarrow \times \rightarrow \widehat{\mathbb{G}} \leftarrow \\ * \leftarrow \times \rightarrow \widehat{\mathbb{G}} \leftarrow \times \rightarrow \cdots \\ * \leftarrow \times \rightarrow \widehat{\mathbb{G}} \leftarrow \\ * \leftarrow \times \rightarrow \\ * \leftarrow \end{array} \right\}.$$

By applying the functor $\underline{\text{FormalSchemes}}(-, \widehat{\mathbb{G}}_a)(k)$, we get a cosimplicial abelian group stemming from the group scheme structure on $\widehat{\mathbb{G}}_a$, and this gives a cochain complex of which we can take the cohomology. In the case $\widehat{\mathbb{G}} = \widehat{\mathbb{G}}_a$, the 2-cocycles in this cochain complex are *precisely* what we've been calling 2-cocycles, so we may be interested in computing $H^2$. In particular, although the elements of $H^2$ are *not* required to be symmetric, we might find symmetry to be automatically enforced if $H^2$ and $B^2$ (and hence $C^2$) are sufficiently small. Toward that end, we make the early remark that $B^2$ is indeed very small: The generators $x^k$ of $C^1$ allow us to compute generators for $B^2$ as

$$d^1(x^k) = d_k c_k.$$

## 3.2 The Structure of $\mathcal{M}_{\text{fg}}$ I: The Affine Cover

Now we turn to the computation of $H^2$. First, using the standard coordinate on $\widehat{\mathbb{G}}_a$, we can compute the above cochain complex to be

$$C^* = \left\{ (0) \xrightarrow{\partial} (x_1) \xrightarrow{\partial} (x_1, x_2) \xrightarrow{\partial} \cdots \right\},$$

where the indicated objects are ideals of $k$, $k[\![x_1]\!]$, $k[\![x_1, x_2]\!]$, and so on. The differential in this complex is given by the formula

$$(\partial f)(x_1, \ldots, x_{n+1}) = f(x_1, \ldots, x_n)$$
$$+ \sum_{j=1}^n (-1)^j f(x_1, \ldots, x_j + x_{j+1}, \ldots, x_n)$$
$$+ (-1)^{n+1} f(x_2, \ldots, x_{n+1}),$$

and one sees from this definition that it extends to the constants in $k[\![x_1, \ldots, x_n]\!]$ and that the constants form an exact complex, hence we can include them without harming the cohomology of the complex to form

$$C^* = \left\{ k \xrightarrow{\partial} k[\![x_1]\!] \xrightarrow{\partial} k[\![x_1, x_2]\!] \xrightarrow{\partial} \cdots \right\}.$$

This complex plays a recognizable role: It computes $\text{Cotor}_{k[\![x]\!]}(k, k)$, according to Corollary 1.4.5 and Definition 1.4.11. We now seek to apply theorems from homological algebra to compute this derived functor in an alternative way. First, note from Corollary 1.4.15 that there is an isomorphism

$$\text{Cotor}_{\mathcal{O}_{\widehat{\mathbb{G}}}}(k, k) \cong \text{Ext}_{\mathcal{O}_{\widehat{\mathbb{G}}}}(k, k),$$

where we have taken Ext in the category of comodules. Then, we can apply Koszul duality to thread $k$-linear duality through the formula, effectively trading comodule calculations for module calculations:

$$\text{Ext}_{\mathcal{O}_{\widehat{\mathbb{G}}}}(k, k) \cong \left( \text{Tor}_{\mathcal{O}_{\widehat{\mathbb{G}}}^*}(k, k) \right)^*.$$

These last groups, finally, are accessible by other, more efficient means.

$\mathbb{Q}$: There is a free $\mathbb{Q}[t]$-module resolution

$$\begin{array}{c} \mathbb{Q} \\ \uparrow \\ 0 \longleftarrow \mathbb{Q}[t] \xleftarrow{\cdot t} \mathbb{Q}[t] \longleftarrow 0, \end{array}$$

to which we apply $(-) \otimes_{\mathbb{Q}[t]} \mathbb{Q}$ to calculate

$$H^* \text{\underline{FormalSchemes}}(B\widehat{\mathbb{G}}_a, \widehat{\mathbb{G}}_a)(\mathbb{Q}) = \begin{cases} \mathbb{Q} & \text{when } * = 0, \\ \mathbb{Q} & \text{when } * = 1, \\ 0 & \text{otherwise.} \end{cases}$$

This means that every 2-cocycle is a coboundary, symmetric or not.

$\mathbb{F}_p$: Now we are computing Tor over a free commutative $\mathbb{F}_p$-algebra on one generator with divided powers. Such an algebra splits as a tensor of truncated polynomial algebras, and again computing a minimal free resolution results in the calculation

$H^* \text{\underline{FormalSchemes}}(B\widehat{\mathbb{G}}_a, \widehat{\mathbb{G}}_a)(\mathbb{F}_p) =$

$$= \begin{cases} \frac{\mathbb{F}_p[\alpha_k \mid k \geq 0]}{\alpha_k^2 = 0} \otimes \mathbb{F}_p[\beta_k \mid k \geq 0] & \text{when } p > 2, \\ \mathbb{F}_2[\alpha_k \mid k \geq 0] & \text{when } p = 2, \end{cases}$$

with $\alpha_k \in H^1$ and $\beta_k \in H^2$. Now that we know what to look for, we can find representatives of each of these classes:

- The class $\alpha_k$ can be represented by $x^{p^k}$, as this is a minimally divisible monomial of degree $p^k$ satisfying the 1-cocycle condition

$$x^{p^k} - (x+y)^{p^k} + y^{p^k} = 0.$$

- The 2-cohomology is concentrated in degrees of the form $p^k$ and $p^j + p^k$, corresponding to $\beta_k$ and $\alpha_j \alpha_k$. The polynomial $c_{p^k}$ is a 2-cocycle of the correct degree, and, because $d^1(x^{p^k}) = d_{p^k} c_{p^k}$ with $p \mid d_{p^k}$, it is not a 2-coboundary. We can therefore use it as a representative for $\beta_k$. (Additionally, the asymmetric class $\alpha_k \alpha_j$ is represented by $x^{p^k} y^{p^j}$.)

- Similarly, in the case $p = 2$ the exceptional class $\alpha_{k-1}^2$ is represented by $c_{2^k}(x, y)$, as this is a 2-cocycle in the correct degree which is not a 2-coboundary.

Given how few 2-coboundaries and 2-cohomology classes there are, we conclude that $c_n(x, y)$ and $x^{p^a} y^{p^b}$ give a basis for *all* of the 2-cocycles. Of these it is easy to select the symmetric ones, which agrees with our expected conclusion. □

The most important consequence of Theorem 3.2.2 is *smoothness*:

**Corollary 3.2.8** *Given a formal group law $F$ over a ring $R$ and a surjective ring map $f: S \to R$, there exists a formal group law $\widetilde{F}$ over $S$ with*

$$F = f^* \widetilde{F}.$$

## 3.2 The Structure of $\mathcal{M}_{\text{fg}}$ I: The Affine Cover

*Proof* Identify $F$ with the classifying map $\operatorname{Spec} R \to \mathcal{M}_{\text{fgl}}$. Employ an isomorphism

$$\varphi \colon \mathcal{M}_{\text{fgl}} \to \operatorname{Spec} L$$

afforded by Theorem 3.2.2, so that $\varphi \circ F$ is selected by a sequence of elements $r_n = \varphi^* F^*(t_n) \in R$. Each of these admit preimages $s_n$ through $f$, and we determine a map

$$\widetilde{\varphi \circ F} \colon \operatorname{Spec} S \to \operatorname{Spec} L$$

by the formula $\widetilde{\varphi \circ F}^*(t_j) = s_j$ and freeness of $L$. Since $\varphi$ is an isomorphism, this determines a map $\widetilde{F} = \varphi^{-1} \circ \widetilde{\varphi \circ F}$ factoring $F$. □

In order to employ Corollary 3.2.8 effectively, we will need to know when a map $\operatorname{Spec} R \to \mathcal{M}_{\text{fg}}$ classifying a local formal group (i.e., a map to the moduli stack as in Remark 3.1.17) can be lifted to a triangle

$$\begin{array}{ccc} & & \mathcal{M}_{\text{fgl}} \\ & \nearrow & \downarrow \\ \operatorname{Spec} R & \longrightarrow & \mathcal{M}_{\text{fg}}. \end{array}$$

Once this is achieved, we can apply Corollary 3.2.8 to a surjective map of rings $\operatorname{Spec} R \to \operatorname{Spec} S$ to build a second diagram

$$\begin{array}{ccc} \operatorname{Spec} S & \dashrightarrow & \mathcal{M}_{\text{fgl}} \\ \uparrow & \nearrow & \downarrow \\ \operatorname{Spec} R & \longrightarrow & \mathcal{M}_{\text{fg}}. \end{array}$$

**Lemma 3.2.9** ([Lura, proposition 11.7]) *A map $\widehat{\mathbb{G}} \colon \operatorname{Spec} R \to \mathcal{M}_{\text{fg}}$ lifts to $\mathcal{M}_{\text{fgl}}$ exactly when the Lie algebra $T_0 \widehat{\mathbb{G}}$ of $\widehat{\mathbb{G}}$ is isomorphic to $R$.*

*Proof* Certainly if $\widehat{\mathbb{G}}$ admits a global coordinate, then $T_0 \widehat{\mathbb{G}} \cong R$. Conversely, the formal group $\widehat{\mathbb{G}}$ is certainly locally isomorphic to $\widehat{\mathbb{A}}^1$ by a covering

$$i_\alpha \colon X_\alpha \to \operatorname{Spec} R$$

and isomorphisms

$$\varphi_\alpha \colon i_\alpha^* \widehat{\mathbb{G}} \cong \widehat{\mathbb{A}}^1.$$

However, *a priori*, these isomorphisms may not glue, corresponding to the potential nontriviality of the Čech 1-cocycle

$$[\varphi_\alpha] \in \check{H}^1(\operatorname{Spec} R; \mathcal{M}_{\text{ps}}^{\text{gpd}}).$$

The group scheme $\mathcal{M}_{\text{ps}}^{\text{gpd}}$ is populated by $T$-points of the form

$$\mathcal{M}_{\text{ps}}^{\text{gpd}}(T) = \left\{ t_0 x + t_1 x^2 + t_2 x^3 + \cdots \mid t_j \in T, t_0 \in T^\times \right\},$$

and it admits a filtration by the closed subschemes

$$\mathcal{M}_{\text{ps}}^{\text{gpd}, \geq N}(T) = \left\{ 1 \cdot x + t_N x^{N+1} + t_{N+1} x^{N+2} + \cdots \mid t_j \in T \right\}.$$

The associated graded of this filtration is $\mathbb{G}_m \times \mathbb{G}_a^{\times \infty}$, and hence the filtration spectral sequence shows

$$\check{H}^1(\operatorname{Spec} R; \mathcal{M}_{\text{ps}}^{\text{gpd}}) \xrightarrow{\simeq} \check{H}^1(\operatorname{Spec} R; \mathbb{G}_m),$$

as $\check{H}^1(\operatorname{Spec} R; \mathbb{G}_a) = 0$ for all affine schemes. Finally, given a choice[22] of trivialization $T_0 \widehat{\mathbb{G}} \cong R$, this induces compatible trivializations of $T_0 i_\alpha^* \widehat{\mathbb{G}}$, which we can use to rescale the isomorphisms $\varphi_\alpha$ so that their image in $\check{H}^1(\operatorname{Spec} R; \mathbb{G}_m)$ vanishes, and hence $[\varphi_\alpha]$ is induced from a class in

$$\check{H}^1(\operatorname{Spec} R; \mathcal{M}_{\text{ps}}^{\text{gpd}, \geq 1}).$$

This obstruction group vanishes. □

*Example 3.2.10* In light of Lemma 3.2.9, it is easy to produce examples of "non-coordinatizable formal groups", i.e., maps of stacks $\operatorname{Spec} T \to \mathcal{M}_{\text{fg}}$ that do not come from maps of simplicial schemes $\operatorname{Spec} T \to \mathcal{M}_{\text{fg}}$. Take $\mathcal{L}$ to be any nontrivial line bundle over $\operatorname{Spec} T$, trivialized on a cover $\bigsqcup_j U_j$, and define the functor $\widehat{\mathbb{G}}$ by setting $\widehat{\mathbb{G}}|_{U_j} = \mathcal{L}(U_j) \otimes \widehat{\mathbb{G}}_a$ and then gluing these together using the data of $\mathcal{L}$. For a concrete source of such bundles $\mathcal{L}$, one can use nontrivial fractional ideals in a Dedekind domain.

*Remark 3.2.11* The subgroup scheme $\mathcal{M}_{\text{ps}}^{\text{gpd}, \geq 1}$ is often referred to in the literature as the group of *strict isomorphisms*. There is an associated moduli of formal groups identified only up to strict isomorphism, which sits in a fiber sequence

$$\mathbb{G}_m \to \mathcal{M}_{\text{fgl}} /\!/ \mathcal{M}_{\text{ps}}^{\text{gpd}, \geq 1} \to \mathcal{M}_{\text{fg}}.$$

These appeared earlier in this lecture as well: In the proof of Theorem 3.2.2, we constructed over $L$ the universal formal group law equipped with a *strict* exponential map. The moduli of formal group laws modulo strict isomorphisms appears as the context associated to the graded version, rather than even-periodic version, of the story told so far – i.e., as a sort of non-periodic context $\mathcal{M}_{MU}$.

---

[22] Incidentally, a choice of trivialization of $T_0 \widehat{\mathbb{G}}$ exactly resolves the indeterminacy of $\log'(0)$ baked into Theorem 2.1.25.

## 3.3 The Structure of $\mathcal{M}_{\mathbf{fg}}$ II: Large Scales

We now turn to understanding the geometry of the quotient stack $\mathcal{M}_{\mathbf{fg}}$ itself, armed with two important tools: Theorem 2.1.25 and Corollary 3.2.8. We begin with a rephrasing of the former:

**Theorem 3.3.1** (see Theorem 2.1.25) *Let $k$ be any field of characteristic 0. Then $\widehat{\mathbb{G}}_a$ describes an isomorphism*

$$\operatorname{Spec} k /\!\!/ \mathbb{G}_m \xrightarrow{\cong} \mathcal{M}_{\mathbf{fg}} \times \operatorname{Spec} k. \qquad \square$$

One of our overarching tasks from the introduction to this case study is to enhance this to a classification of *all* of the geometric points of $\mathcal{M}_{\mathbf{fg}}$, including those where $k$ is a field of positive characteristic $p$:

$$\widehat{\mathbb{G}} \colon \operatorname{Spec} k \to \mathcal{M}_{\mathbf{fg}} \times \operatorname{Spec} \mathbb{Z}_{(p)}.$$

We proved this theorem in the characteristic 0 case by solving a certain differential equation, which necessitated integrating a power series, and integration is what we expect to fail in characteristic $p$. The following definition tracks *where* it fails:

**Definition 3.3.2** Let $+_\varphi$ be a formal group law over a $\mathbb{Z}_{(p)}$-algebra. Let $n$ be the largest degree such that there exists an invertible formal power series $\ell$ with

$$\ell(x +_\varphi y) = \ell(x) + \ell(y) \pmod{(x, y)^n},$$

i.e., $\ell$ is a logarithm for the $(n-1)$-bud determined by $+_\varphi$. The *$p$-height of $+_\varphi$* is defined to be $\log_p(n)$.

*Example 3.3.3* This definition is somewhat subtle, since $\ell$ is not required to integrate a truncation of the invariant differential. Nonetheless, it will be a consequence of the results below that the height of a formal group law can be determined by attempting to integrate its unit-speed invariant differential and noting the first degree where the requisite division is impossible, provided that that degree is of the form $p^d$. For example, the unit-speed invariant differential associated to the usual coordinate on the muliplicative group has the form

$$\omega(x) = (1 + x + x^2 + \cdots + x^{p-1} + \cdots) \, dx.$$

The radius of convergence of this ordinary differential equation in $\mathbb{Z}_{(p)}$ is $p$, owing to the term

$$\int x^{p-1} \, dx = \frac{x^p}{p} \notin \mathbb{Z}_{(p)}[x].$$

We thus see that $\widehat{\mathbb{G}}_m$ has height $\log_p(p) = 1$.

122                           Finite Spectra

This turns out to be a crucial invariant of a formal group law, admitting many other interesting presentations. In this lecture, investigation of this definition will lead us to a classification of the closed substacks of $\mathcal{M}_{fg}$, another of our overarching tasks. As a first step, we would like to show that this value is well-behaved in various senses – for instance, it is always an integer (or $\infty$). First, we note that this definition really depends on the formal group rather than the formal group law.

**Lemma 3.3.4** *The height of a formal group law is an isomorphism invariant, i.e., it descends to give a function*

$$\mathrm{ht} \colon \pi_0 \mathcal{M}_{fg}(T) \to \mathbb{R}_{>0} \cup \{\infty\}$$

*for any local test $\mathbb{Z}_{(p)}$-algebra $T$.*

*Proof* The series $\ell$ is a partial logarithm for the formal group law $\varphi$, i.e., an isomorphism between the formal group defined by $\varphi$ and the additive group. Since isomorphisms compose, this statement follows. □

With this in mind, we look for a more standard form for formal group laws, where height taking on integer values will hopefully be obvious. In order to give ourselves more tools to work with, and with the intention of alleviating this restriction as soon as possible, we assume that the ground ring is torsion-free – and although our true goal is to understand formal groups over positive characteristic fields, we are free to make this assumption by Corollary 3.2.8. In this setting, the most blindly optimistic standard form is as follows:

**Definition 3.3.5** (cf. [Haz12, Proposition 15.2.4])  We say that a formal group law $+_\varphi$ over a torsion-free $\mathbb{Z}_{(p)}$-algebra is *p-typical* when its invariant differential takes the form

$$\omega(x) = \sum_{j=0}^{\infty} \ell_j x^{p^j-1} \, dx.$$

**Lemma 3.3.6** ([Haz12, theorem 15.2.9])  *Every formal group law $+_\varphi$ over a torsion-free $\mathbb{Z}_{(p)}$-algebra is naturally isomorphic to a p-typical formal group law, called the p-typification of $+_\varphi$.*

*Proof* Let $\widehat{\mathbb{G}}$ be the formal group associated to $+_\varphi$, denote its inherited parameter by

$$g_0 \colon \widehat{\mathbb{A}}^1 \xrightarrow{\cong} \widehat{\mathbb{G}},$$

so that $\omega_0(x) = (g_0^* \omega)(x) = \sum_{n=0}^{\infty} a_n x^{n-1} \, dx$ is the invariant differential associated to $+_\varphi$. Our goal is to perturb this coordinate to a new coordinate $g_\infty$ so that

## 3.3 The Structure of $\mathsf{M_{fg}}$ II: Large Scales

its associated invariant differential has the form

$$\sum_{n=0}^{\infty} a_{p^n} x^{p^n-1} \, dx.$$

To do this, we introduce four operators on functions[23] $\widehat{\mathbb{A}}^1 \to \widehat{\mathbb{G}}$:

- Given $r \in R$, we can define a *homothety* by rescaling the coordinate by $r$:

$$(\theta_r^* g_0^* \omega)(x) = \omega_0(rx) = \sum_{n=1}^{\infty} (a_n r^n) x^{n-1} \, dx.$$

- For $\ell \in \mathbb{N}$, we can define a shift operator (or *Verschiebung*) by

$$(V_\ell^* g_0^* \omega)(x) = \omega_0(x^\ell) = \sum_{n=1}^{\infty} a_n \ell x^{n\ell-1} \, dx.$$

- For $\ell \in \mathbb{Z}$, we define the $\ell$–series to be

$$[\ell](g_0(x)) = \overbrace{g_0(x) +_{\widehat{\mathbb{G}}} \cdots +_{\widehat{\mathbb{G}}} g_0(x)}^{\ell \text{ times}}.$$

We can extend this definition slightly: Given an $\ell \in \mathbb{Z}_{(p)}$, we form the $\ell$-series by writing $\ell$ as a fraction $\ell = s/t$ and composing the $s$-series with the inverse of the $t$-series. The effect on the invariant differential is given by

$$([\ell]^* g_0^* \omega)(x) = \ell \omega_0(x) = \sum_{n=1}^{\infty} \ell a_n x^{n-1} \, dx.$$

- For $\ell \in \mathbb{N}$, we can define a *Frobenius operator*[24] by

$$F_\ell g_0(x) = \sum_{j=1}^{\ell} {}_{\widehat{\mathbb{G}}}\, g_0(\zeta_\ell^j x^{1/\ell}),$$

where $\zeta_\ell$ is a primitive $\ell$th root of unity. Because this formula is Galois-invariant in choice of primitive root, it actually expands to a series which lies over the ground ring (without requiring an extension by $\zeta_\ell$ or by $z^{1/\ell}$). Using the standard identity

$$\sum_{j=1}^{\ell} \zeta_\ell^{jn} = \begin{cases} \ell & \text{if } \ell \mid n, \\ 0 & \text{otherwise,} \end{cases}$$

---

[23] Unfortunately, it is standard in the literature to call these operators "curves," which does not fit well with our previous use of the term in Case Study 2.

[24] There are other definitions of the Frobenius operator which are less mysterious but less explicit. For instance, it also arises from applying the Verschiebung to the character group (or "Cartier dual") of $\widehat{\mathbb{G}}$.

we can explicitly compute the behavior of $F_\ell$ on the invariant differential:

$$(F_\ell^* g_0^* \omega)(x) = \sum_{n=1}^{\infty} a_{n\ell} x^{n-1} \, dx.$$

Stringing these together, for $p \nmid \ell$ we have

$$([1/\ell]^* V_\ell^* F_\ell^* g_0^* \omega)(x) = \sum_{n=1}^{\infty} a_{n\ell} x^{n\ell-1} \, dx.$$

Hence, we can iterate over primes $\ell \neq p$, and for two adjacent such primes $\ell' > \ell$ we consider the perturbation

$$g_{\ell'} = g_\ell -_{\widehat{G}} [1/\ell] V_\ell F_\ell g_\ell.$$

Each of these differences gives a parameter according to Theorem 2.1.10, and the first possible nonzero term appears in degree $\ell$, hence the coefficients stabilize linearly in $\ell$. Passing to the limit thus gives a new parameter $g_\infty$ on the same formal group $\widehat{G}$, but now with the $p$-typicality property. □

**Lemma 3.3.7** *If $x$ is a parameter for a formal group $\widehat{G}$ over a torsion-free $\mathbb{Z}_{(p)}$-algebra, then any other parameter $y$ admits a unique expression as*

$$y = \sum_{j=0}^{\infty} \widehat{G} \, \theta_{b_j}^* V_j^* x.$$

*Proof* Set $y = y_\infty$, and select the unit $b_1$ such that setting $b_1 x = y_1$ gives

$$y_\infty - y_1 = y_\infty - \theta_{b_1}^* V_1^* x \equiv 0 \pmod{x^{1+1}}.$$

We inductively define $y_n$ by examining

$$y_\infty - y_{n-1} \equiv b_n x^n \pmod{x^{n+1}}$$

and setting

$$y_n = y_{n-1} +_{\widehat{G}} \theta_{b_n}^* V_n^* x. \qquad \square$$

**Corollary 3.3.8** (see Example 3.3.3, and also [Lura, proposition 13.6]) *Let $x$ be some parameter for a formal group $\widehat{G}$ over a torsion-free $\mathbb{Z}_{(p)}$-algebra, and suppose that the minimal index so that the invariant differential is not integrable in that degree takes the form $p^d - 1$. Then $\widehat{G}$ has height $d$.*

*Proof* Take $x$ to be a parameter as in the statement. Although *a priori* the partial logarithm $\ell$ of Definition 3.3.2 may not be the partial integral of the

### 3.3 The Structure of $M_{fg}$ II: Large Scales

invariant differential associated to $x$, it is the partial logarithm associated to *some* parameter $y$, and we use the preceding lemma to re-express $y$ as

$$y = \sum_{j=0}^{\infty} \theta^*_{\hat{\mathbb{G}} b_j} V^*_j x.$$

Consider the invariant differential associated to $y$, expressed in terms of $x$ and $dx$, using the rules described in the proof of Lemma 3.3.6:

$$\omega_y = \sum_{j=0}^{\infty} \theta^*_{b_j} V^*_j \left( \sum_{n=0}^{\infty} a_{p^n} x^{p^n-1} dx \right) = \sum_{j=0}^{\infty} \sum_{n=0}^{\infty} j a_{p^n} b_j^{jp^n-1} x^{jp^n-1} dx.$$

The terms in the critical degree $(p^d - 1)$ take the form

$$\sum_{n=0}^{d} p^{d-n} a_{p^n} b_{p^{d-n}}^{p^d-1} x^{p^d-1} dx.$$

The term $a_{p^n}$ is assumed to have $p$-adic valuation at least $n$ for $n < d$, and hence $p^{d-n} a_{p^n}$ has $p$-adic valuation at least $d$ for $n < d$. Since in the case $n = d$ the coefficient $a_{p^d}$ has assumed to have $p$-adic valuation strictly less than $d$, this sum must also have $p$-adic valuation strictly less than $d$. In particular, this shows that the radius of convergence of the partial logarithm is at most $p^d - 1$, and the existence of $x$ itself shows that it is at least $p^d - 1$. □

**Corollary 3.3.9** *The height of a formal group law lies in* $\mathbb{N} \cup \{\infty\}$.

*Proof* A $p$-typical coordinate is guaranteed to satisfy the preconditions of Corollary 3.3.8, and Lemma 3.3.6 guarantees such coordinates. □

In the course of the prior discussion, manipulations with the $\ell$-series have played a critical role, and we have shied away from the $p$-series because of the noninvertibility of $p$ in $\mathbb{Z}_{(p)}$. As a consequence, this means that the $p$-series contains an unusual amount of information even after $p$-typification of a coordinate, and so we turn to its analysis.

**Lemma 3.3.10** ([Ara73, section 4]) *A formal group law $+_\varphi$ is $p$-typical if and only if there are elements $v_d$ with*

$$[p]_\varphi(x) = px +_\varphi v_1 x^p +_\varphi v_2 x^{p^2} +_\varphi \cdots +_\varphi v_d x^{p^d} +_\varphi \cdots.$$

*Proof sketch* Suppose first that $+_\varphi$ is $p$-typical with invariant differential

$$\omega(x) = \sum_{n=0}^{\infty} a_n x^{p^n-1} dx.$$

We can then compare the two series

$$(\theta_p^* \omega)(x) = p\,dx + \cdots,$$
$$([p]^* \omega)(x) = p\omega(x) = p\,dx + \cdots.$$

The difference is concentrated in degrees of the form $p^d - 1$, beginning in degree $p - 1$, so we can find an element $v_1$ such that

$$[p]^* \omega(x) - (\theta_p^* \omega(x) + \theta_{v_1}^* V_p \omega(x))$$

is also concentrated in degrees of the form $(p^d - 1)$ but now starts in degree $(p^2 - 1)$. Iterating this gives the equation

$$[p]^* \omega = \theta_p^* \omega + \theta_{v_1}^* V_p^* \omega + \theta_{v_2}^* V_{p^2}^* \omega + \cdots,$$

at which point we can use the invariance property of $\omega$ to deduce

$$[p]_\varphi(x) = px +_\varphi v_1 x^p +_\varphi v_2 x^{p^2} +_\varphi \cdots +_\varphi v_n x^{p^n} +_\varphi \cdots.$$

In the other direction, the invariant differential coefficients can be recursively recovered from the coefficients $v_d$ for a formal group law with $p$-typical $p$-series, using a similar manipulation. In fact, we can push this slightly further:

$$\sum_{d=0}^{\infty} p a_{p^d} x^{p^d - 1}\, dx = \sum_{d=0}^{\infty} \sum_{j=0}^{\infty} p^d a_{p^j} v_d^{p^j} x^{p^{d+j} - 1}\, dx$$
$$= \sum_{n=0}^{\infty} \left( \sum_{k=0}^{n} a_{p^k} p^{n-k} v_{n-k}^{p^k} \right) x^{p^n - 1}\, dx,$$

implicitly taking $a_1 = 1$ and $v_0 = p$. □

Lemma 3.3.10 shows that, at least in the case that the ground ring is an integral domain, the $p$-series of a $p$-typical formal group law contains exactly as much information as the logarithm itself (and hence fully determines the formal group law).[25] We make this observation precise in the following important definition and theorem:

**Definition 3.3.11** In the setting of a coordinate on a formal group over a torsion-free $\mathbb{Z}_{(p)}$-algebra, the following conditions are equivalent:

1. Its invariant differential is $p$-typical, as in Definition 3.3.5.
2. For each prime $\ell \neq p$, the Frobenius $F_\ell$ vanishes on the coordinate.

---

[25] In the complete setting, there is another formula for recovering the logarithm from the $p$-series:

$$\log_\varphi(x) = \lim_{j \to \infty} \left( p^{-j} \cdot [p^j]_\varphi(x) \right).$$

### 3.3 The Structure of $\mathcal{M}_{fg}$ II: Large Scales

3. The coordinate factors through $\varepsilon \colon \mathcal{M}_{fgl} \to \mathcal{M}_{fgl}$, where $\varepsilon$ denotes the idempotent $p$-typicalizing procedure developed in Lemma 3.3.6.

The second and third conditions do not require the ground ring to be torsion-free, and so **we now drop this standing assumption** and use the second condition to define a closed subscheme $\mathcal{M}_{fgl}^{p\text{-typ}} \subseteq \mathcal{M}_{fgl}$, the *moduli of p-typical formal group laws*.[26]

**Theorem 3.3.12** (see [Mila, proposition 5.1], [Rav86, theorem A2.2.3], and the proof of [Hop, proposition 19.10]) *The Kudo–Araki map determined by Lemma 3.3.10*

$$\mathbb{Z}_{(p)}[v_1, v_2, \ldots, v_d, \ldots] \xrightarrow{v} O_{\mathcal{M}_{fgl}^{p\text{-typ}}}$$

*is an isomorphism.*

*Proof* We begin with the image factorization definition of $\mathcal{M}_{fgl}^{p\text{-typ}}$:

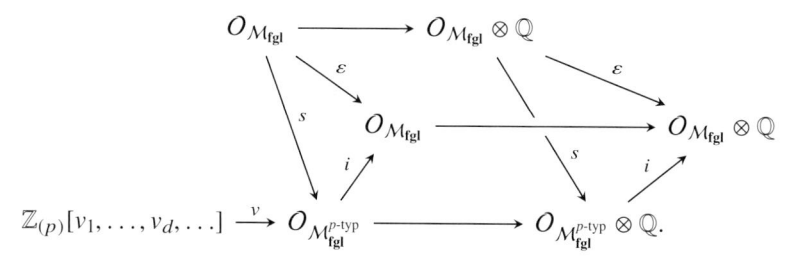

$$\begin{array}{c}
O_{\mathcal{M}_{fgl}} \longrightarrow O_{\mathcal{M}_{fgl}} \otimes \mathbb{Q} \\
\searrow^{s} \quad \nearrow^{\varepsilon} \quad \searrow^{s} \quad \nearrow^{\varepsilon} \\
O_{\mathcal{M}_{fgl}} \longrightarrow O_{\mathcal{M}_{fgl}} \otimes \mathbb{Q} \\
\searrow^{i} \quad \nearrow \quad \searrow^{i} \quad \nearrow \\
\mathbb{Z}_{(p)}[v_1, \ldots, v_d, \ldots] \xrightarrow{v} O_{\mathcal{M}_{fgl}^{p\text{-typ}}} \longrightarrow O_{\mathcal{M}_{fgl}^{p\text{-typ}}} \otimes \mathbb{Q}.
\end{array}$$

We immediately deduce that all the horizontal arrows are injections: in Theorem 3.2.2 we calculated $O_{\mathcal{M}_{fgl}}$ to be torsion-free; idempotence of $\varepsilon$ shows that $O_{\mathcal{M}_{fgl}^{p\text{-typ}}}$ is a subring of $O_{\mathcal{M}_{fgl}}$, hence it is also torsion-free; and Lemma 3.3.10 shows that $(i \circ v)(v_n)$ agrees with $pm_{p^n}$ in the module of indecomposables $Q(O_{\mathcal{M}_{fgl}} \otimes \mathbb{Q})$.

To complete the proof, we need to show that $v$ is surjective, which will follow from calculating the indecomposables in $O_{\mathcal{M}_{fgl}^{p\text{-typ}}}$ and checking that $Qv$ is surjective. Since $s$ is surjective, the map $Qs$ on indecomposables is surjective as well, and its effect can be calculated rationally. Since $(Q\varepsilon)(m_n) = 0$ for $n \neq p^d$, we have that $Q(O_{\mathcal{M}_{fgl}^{p\text{-typ}}})$ is generated by $s(t_{p^d-1})$ under an isomorphism as in Theorem 3.2.2. It follows that $Qi$ injects, hence $Qv$ must surject by the previous calculation of $Q(i \circ v)(v_n)$. □

**Corollary 3.3.13** *If $[p]_\varphi(x) = [p]_\psi(x)$ for two p-typical formal group laws $+_\varphi$ and $+_\psi$, then $+_\varphi$ and $+_\psi$ are themselves equal.* □

---

[26] The third condition selects the same functor as the second condition, but it does not as obviously show that this is a closed subscheme of $\mathcal{M}_{fgl}$ (and hence itself an affine scheme).

**Corollary 3.3.14** *For any sequence of coefficients $v_j \in R$ in a $\mathbb{Z}_{(p)}$-algebra $R$, there is a unique p-typical formal group law $+_\varphi$ with*

$$[p]_\varphi(x) = px +_\varphi v_1 x^p +_\varphi v_2 x^{p^2} +_\varphi \cdots +_\varphi v_d x^{p^d} +_\varphi \cdots.$$ □

**Corollary 3.3.15** *In the case that the ground ring $R$ is a field of positive characteristic, the height of a formal group can be taken to be the size of its p-torsion:*

$$\mathrm{ht}(\widehat{\mathbb{G}}) = \log_p \dim_R O_{\widehat{\mathbb{G}}[p]} = \log_p \dim_R (O_{\widehat{\mathbb{G}}}/[p](x)).$$

*This is the first nonzero coefficient in the expansion of its p-series.*[27] □

Finally, we exploit these results to make deductions about the geometry of $\mathcal{M}_{\mathrm{fg}} \times \mathrm{Spec}\, \mathbb{Z}_{(p)}$. There is an inclusion of groupoid-valued sheaves from $p$-typical formal group laws with $p$-typical isomorphisms to all formal group laws with all isomorphisms. Lemma 3.3.6 can be viewed as presenting this inclusion as a deformation-retraction, witnessing a natural *equivalence* of groupoids. It follows from Remark 3.1.17 that they both present the same stack. The central utility of this equivalence is that the Kudo–Araki moduli of $p$-typical formal group laws is a considerably smaller algebra than $O_{\mathcal{M}_{\mathrm{fgl}}}$, resulting in a less noisy picture of the Hopf algebroid. Our final goal in this lecture is to exploit this refined presentation in the study of invariant functions.

**Definition 3.3.16** ([Goe08, lemma 2.28])   Let $(A, \Gamma)$ be a Hopf algebroid and let $(X_0, X_1)$ be the associated groupoid scheme. A function $f \colon X_0 \to \mathbb{A}^1$ is said to be *invariant* when it is stable under isomorphism, i.e., when there is a diagram

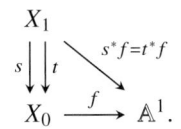

$$\begin{array}{c} X_1 \\ {}_s\downarrow\downarrow{}_t \quad \searrow{}^{s^*f = t^*f} \\ X_0 \xrightarrow{f} \mathbb{A}^1. \end{array}$$

(In terms of Hopf algebroids, the corresponding element $a \in A$ satisfies $\eta_L(a) = \eta_R(a)$.) Correspondingly, a closed subscheme $A \subseteq X_0$ determined by the simultaneous vanishing of functions $f_\alpha$ is said to be *invariant* when the vanishing condition is invariant – i.e., a point lies in the simultaneous vanishing locus if and only if its entire orbit under $X_1$ also lies in the simultaneous vanishing locus. (In terms of Hopf algebroids, the corresponding ideal $I \subseteq A$ satisfies $\eta_L(I) = \eta_R(I)$.) Finally, a *closed substack* is a substack determined by an invariant ideal of $X_0$.

---

[27] In the case of a field of characteristic 0, this definition of height is constant at 0. This convention is uniformly abided.

### 3.3 The Structure of $\mathcal{M}_{\mathbf{fg}}$ II: Large Scales

We now have the language to describe the closed substacks of $\mathcal{M}_{\mathbf{fg}} \times \operatorname{Spec} \mathbb{Z}_{(p)}$, which is equivalent to discerning all of the invariant ideals of $O_{\mathcal{M}_{\mathbf{fgl}}^{p\text{-typ}}}$.

**Corollary 3.3.17** ([Wil82, theorem 4.6 and lemmas 4.7–4.8]) *The ideal $I_d = (p, v_1, \ldots, v_{d-1})$ is invariant under the action of strict formal group law isomorphisms for all $d$. It determines the closed substack $\mathcal{M}_{\mathbf{fg}}^{\geq d}$ of formal group laws of $p$-height at least $d$.*

*Proof* Recall from Theorem 3.3.12 the Kudo–Araki isomorphism

$$\mathcal{M}_{\mathbf{fgl}}^{p\text{-typ}} \xrightarrow{\cong} \operatorname{Spec} \mathbb{Z}_{(p)}[v_1, v_2, \ldots, v_d, \ldots] =: \operatorname{Spec} V,$$

and let $+_L$ denote the associated universal $p$-typical formal group law with $p$-series

$$[p]_L(x) = px +_L v_1 x^p +_L v_2 x^{p^2} +_L \cdots +_L v_d x^{p^d} +_L \cdots.$$

Over $\operatorname{Spec} V[t_1, t_2, \ldots]$, we can form a second group law $+_R$ by conjugating $+_L$ by the universal $p$-typical coordinate transformation $g(x) = \sum_{j=0}^{\infty} t_j x^{p^j}$.[28] The corresponding $p$-series

$$[p]_R(x) = \sum_{d=0}^{\infty}{}_R \eta_R(v_d) x^{p^d}$$

determines the $\eta_R$ map of the Hopf algebroid $(V, V[t_1, t_2, \ldots])$ presenting the moduli of $p$-typical formal group laws and $p$-typical isomorphisms. We cannot hope to compute $\eta_R(v_d)$ explicitly, but modulo $p$ we can apply Freshman's Dream to the expansion of

$$[p]_L(g(x)) = g([p]_R(x))$$

to discern some information:

$$\sum_{\substack{i \geq 0 \\ j > 0}}{}_L t_i \eta_R(v_j)^{p^i} \equiv \sum_{\substack{i > 0 \\ j \geq 0}}{}_L v_i t_j^{p^i} \pmod{p}.$$

This is still inexplicit, since $+_L$ is a very complicated operation, but we can see $\eta_R(v_d) \equiv v_d \pmod{I_d}$. It follows that $I_d$ is invariant for each $d$. Additionally, the closed substack this determines are those formal groups admitting local $p$-typical coordinates for which $v_{\leq d} = 0$, guaranteeing that the height of the associated formal group is at least $d$. □

---

[28] To see that the transformation $g$ must have this form, suppose merely that it is an isomorphism of formal group laws with $p$-typical source and target, $+_L$ and $+_R$ respectively. Since $O_{\mathcal{M}_{\mathbf{fgl}}^{p\text{-typ}}}$ is a torsion-free ring, by rationalizing we may express $g$ as the composite of the rational logarithm for $+_L$, a possible scalar multiplication, and the rational exponential for $+_R$. These are each $p$-typical transformations by hypothesis, and since $p$-typicality is stable under composition, this is true of $g$.

130  Finite Spectra

What is *much* harder to prove is the following:

**Theorem 3.3.18** (Landweber's classification of invariant prime ideals [Lan75, corollary 2.4 and proposition 2.5], see [Wil82, theorem 4.9]) *The unique closed reduced substack of* $\mathcal{M}_{\text{fg}} \times \operatorname{Spec} \mathbb{Z}_{(p)}$ *of codimension d is selected by the invariant prime ideal* $I_d \subseteq O_{\mathcal{M}_{\text{fgl}}^{p\text{-typ}}}$.

*Proof sketch*   We want to show that if $I$ is an invariant prime ideal, then $I = I_d$ for some $d$. To begin, note that $v_0 = p$ is the only invariant function on $\mathcal{M}_{\text{fgl}}^{p\text{-typ}}$, hence $I$ must either be trivial or contain $p$. Then, inductively assume that $I_d \subseteq I$. If this is not an equality, we want to show that $I_{d+1} \subseteq I$ is forced. Take $y \in I \setminus I_d$; if we could show

$$\eta_R(y) = a v_d^j t^K + \text{higher-order terms}$$

for nonzero $a \in \mathbb{Z}_{(p)}$, we could proceed by primality to show that $v_d \in I$ and hence $I_{d+1} \subseteq I$. This is possible (and, indeed, this is how the full proof goes), but it requires serious bookkeeping.   □

*Remark 3.3.19*   The complementary open substack of dimension $d$ is harder to describe. From first principles, we can say only that it is the locus where the coordinate functions $p, v_1, \ldots, v_d$ do not *all simultaneously vanish*. It turns out that:

1. On a cover, at least one of these coordinates can be taken to be invertible.
2. Once one of them is invertible, a coordinate change on the formal group law can be used to make $v_d$ (and perhaps others in the list) invertible. Hence, we can use $v_d^{-1} O_{\mathcal{M}_{\text{fgl}}^{p\text{-typ}}}$ as a coordinate chart.
3. Over a further base extension and a further coordinate change, the higher coefficients $v_{d+k}$ can be taken to be zero. Hence, we can also use the quotient ring $v_d^{-1} \mathbb{Z}_{(p)}[v_1, \ldots, v_d]$ as a coordinate chart.

*Remark 3.3.20* (cf. [Str06, Section 12] and [Lura, remark 13.9])   Specialize now to the case of a field $k$ of characteristic $p$. Since the additive group law has vanishing $p$-series and is $p$-typical, a consequence of Corollary 3.3.13 is that *every* $p$-typical group law with vanishing $p$-series is exactly equal to $\widehat{\mathbb{G}}_a$, and in fact any formal group law with vanishing $p$-series $p$-typifies exactly to $\widehat{\mathbb{G}}_a$. This connects several ideas we have seen so far: the presentation of formal group laws with logarithms in Theorem 1.5.6, the presentation of the context $\mathcal{M}_{MOP}$ in Example 3.1.20, and the Hurewicz image of $MU_*$ in $H\mathbb{F}_{p*}MU$ in Corollary 2.6.8.

*Remark 3.3.21*   It's worth pointing out how strange all of this is. In Euclidean geometry, open subspaces are always top-dimensional, and closed subspaces

### 3.3 The Structure of $\mathcal{M}_{fg}$ II: Large Scales

can drop dimension. Here, proper open substacks of every dimension appear, and every nonempty closed substack is $\infty$-dimensional (albeit of positive codimension).

*Remark 3.3.22* The results of this section have several alternative forms in the literature. For instance, there is a second set of coordinates $v_d$, called the *Hazewinkel generators*, which differ substantially from the Kudo–Araki coordinates favored here, though they are just as "canonical." Different coordinate patches are useful for accomplishing different tasks, and the reader would be wise to remain flexible.[29,30]

*Remark 3.3.23* ([Laz75, Section IV.9], [Haz12, section 17.5]) The operation of $p$-typification often gives "unusual" results. For instance, we will examine the standard multiplicative formal group law of Example 2.1.26, its rational logarithm, and its rational exponential:

$$x +_{\widehat{\mathbb{G}}_m^{std}} y = x + y - xy, \quad \log_{\widehat{\mathbb{G}}_m^{std}}(x) = -\log(1-x), \quad \exp_{\widehat{\mathbb{G}}_m^{std}}(x) = 1 - \exp(-x).$$

By Lemma 3.3.6, we see that the $p$-typification of this rational logarithm takes the form

$$\log_{\widehat{\mathbb{G}}_m^{p\text{-typ}}}(x) = \sum_{j=0}^{\infty} \frac{x^{p^j}}{p^j}.$$

We can couple this to the standard exponential of the rational multiplicative group

$$\widehat{\mathbb{A}}^1 \xrightarrow{\varepsilon x} \widehat{\mathbb{G}}_m \xrightarrow{\log^{p\text{-typ}}} \widehat{\mathbb{G}}_a \xrightarrow{\exp^{std}} \widehat{\mathbb{G}}_m \xrightarrow{x} \widehat{\mathbb{A}}^1$$

to produce the $p$-typifying coordinate change from Lemma 3.3.6:

$$1 - \exp\left(-\sum_{j=0}^{\infty} \frac{x^{p^j}}{p^j}\right) = 1 - E_p(x)^{-1}.$$

This series $E_p(x)$ is known as the *Artin–Hasse exponential*, and it has the miraculous property that it is a series lying in $\mathbb{Z}_{(p)}[\![x]\!] \subseteq \mathbb{Q}[\![x]\!]$, as it is a change of coordinate series on $\widehat{\mathbb{G}}_m$ over $\operatorname{Spec} \mathbb{Z}_{(p)}$.

---

[29] In particular, it is a largely open question whether there is a (partial) coordinate patch that is compatible with the results of Appendix A.2, though there are partial results along these lines [JN10, Law17, LN12, Str99a].

[30] An unusual set of canonical generators is discussed by Lazard [Laz75, section V.10].

## 3.4 The Structure of $\mathcal{M}_{fg}$ III: Small Scales

In the previous two lectures, we analyzed the structure of $\mathcal{M}_{fg}$ as a whole: First we studied the cover

$$\mathcal{M}_{fgl} \to \mathcal{M}_{fg},$$

and then we turned to the stratification described by the height function

$$\mathrm{ht}\colon \pi_0\mathcal{M}_{fg}(T \text{ a } \mathbb{Z}_{(p)}\text{-algebra}) \to \mathbb{N} \cup \{\infty\}.$$

In this lecture, we will concern ourselves with the small-scale behaviors of $\mathcal{M}_{fg}$: its geometric points and their local neighborhoods.[31] To begin, we have all the tools in place to perform an outright classification of the geometric points.

**Theorem 3.4.1** ([Laz55, théorème IV])  *Let $\bar k$ be an algebraically closed field of positive characteristic $p$. The height map*

$$\mathrm{ht}\colon \pi_0\mathcal{M}_{fg}(\bar k) \to \mathbb{N}_{>0} \cup \{\infty\}$$

*is a bijection.*

*Proof* Surjectivity follows from Corollary 3.3.14. Namely, the *dth Honda formal group law* is the $p$-typical formal group law over $k$ determined by

$$[p]_{\varphi_d}(x) = x^{p^d},$$

and it gives a preimage for $d$. To show injectivity, we must show that every $p$-typical formal group law $\varphi$ over $\bar k$ is isomorphic to the appropriate Honda group law. Suppose that the $p$-series for $\varphi$ begins

$$[p]_\varphi(x) = ux^{p^d} +_\varphi ax^{p^{d+k}} + \cdots$$

for a unit $u$. By replacing $x$ with $u^{-1/p^d} x$, we may assume that $u = 1$. Then, we will construct a coordinate transformation $g(x) = \sum_{j=0}^\infty b_j x^{p^j}$ satisfying

$$g(x^{p^d}) \equiv [p]_\varphi(g(x)) \pmod{x^{p^{d+k}+1}}$$

$$\sum_{j=0}^\infty {}^\varphi b_j x^{p^{d+j}} \equiv \left(\sum_{j=0}^\infty {}^\varphi b_j x^{p^j}\right)^{p^d} +_\varphi a\left(\sum_{j=0}^\infty {}^\varphi b_j x^{p^j}\right)^{p^{d+k}} \pmod{x^{p^{d+k}+1}}$$

$$\sum_{j=0}^\infty {}^\varphi b_j x^{p^{d+j}} \equiv \left(\sum_{j=0}^\infty (\mathrm{Frob}^d)^*\varphi\, b_j^{p^d} x^{p^{d+j}}\right) + ax^{p^{d+k}} \pmod{x^{p^{d+k}+1}}.$$

For $g$ to be a coordinate transformation, we must have $b_0 = 1$, and because $\bar k$ is

---

[31] For an alternative perspective on much of this material, see [Str06, section 18], where the presentation connects rather tightly with our Lecture 4.4.

## 3.4 The Structure of $\mathcal{M}_{fg}$ III: Small Scales

algebraically closed we can induct on $j$ to solve for the other coefficients in the series. The coordinate for $\varphi$ can thus be perturbed so that the term $x^{p^{d+k}}$ does not appear in the $p$-series, and inducting on $d$ gives the result. □

*Remark 3.4.2* ([Str06, Remark 11.2]) We can now see see that $\pi_0 \mathcal{M}_{fg}$, sometimes called the *coarse moduli of formal groups*, is not representable by a scheme. From Theorem 3.4.1, we see that there are infinitely many points in $\pi_0 \mathcal{M}_{fg}(\mathbb{F}_p)$. From Corollary 3.2.8, we see that these lift along the surjection $\mathbb{Z} \to \mathbb{F}_p$ to give infinitely many distinct points in $\pi_0 \mathcal{M}_{fg}(\mathbb{Z})$. On the other hand, by Theorem 3.3.1 there is a single $\mathbb{Q}$-point of the coarse moduli, whereas the $\mathbb{Z}$-points of a representable functor would inject into its $\mathbb{Q}$-points. In light of Example 3.1.19, this result is no surprise to a topologist: The nonaffine nature of the moduli of formal groups is reflected by Definition 3.1.9 and Lemma 3.1.15 in the nontrivial structure of the $MU$-Adams spectral sequence.

We now turn to understanding the infinitesimal neighborhoods of these geometric points. In general, for $s\colon \operatorname{Spec} k \to X$ a closed $k$-point of a scheme, we defined in Definition 2.1.6 and Definition 2.1.7 an infinitesimal neighborhood object $X_s^\wedge$ with a lifting property

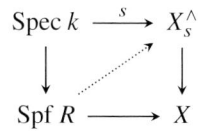

$$\begin{array}{ccc} \operatorname{Spec} k & \xrightarrow{s} & X_s^\wedge \\ \downarrow & \nearrow & \downarrow \\ \operatorname{Spf} R & \longrightarrow & X \end{array}$$

for any infinitesimal thickening $\operatorname{Spf} R$ of $\operatorname{Spec} k$. Thinking of $X$ as representing a moduli problem, a typical choice for $\operatorname{Spf} R$ is $\widehat{\mathbb{A}}_k^1$, and a map $\widehat{\mathbb{A}}_k^1 \to X$ extending $s$ is a series solution to the moduli problem with constant term $s$. In turn, $X_s^\wedge$ is the smallest object through which all such maps factor, and so we think of it as classifying Taylor expansions of solutions with constant term $s$.

For a formal group $\Gamma\colon \operatorname{Spec} k \to \mathcal{M}_{fg}$, the definition is formally similar, but actually writing it out is made complicated by Remark 3.1.17. In particular, $p\colon \operatorname{Spec} k \to X$ may not lift directly through $\operatorname{Spf} R \to X$, but instead $\operatorname{Spec} R/\mathfrak{m} \to X$ may present $p$ on a cover $i\colon \operatorname{Spec} R/\mathfrak{m} \to \operatorname{Spec} k$.

**Definition 3.4.3** ([Reza, section 2.4], see [Str97, section 6]) Define the *Lubin–Tate stack* $(\mathcal{M}_{fg})_\Gamma^\wedge$ to be the groupoid-valued functor from the category of infinitesimal thickenings of $k$,[32] which on such a thickening $R$ has objects

---

[32] We are being somewhat cavalier by using the word "stack" to refer to a groupoid-valued functor that we have *not* defined on all rings, as opposed to the usage in Definition 3.1.13 and Remark 3.1.17. The reader is advised to privilege the word "stack" with the same flexibility as the word "sheaf", which many authors freely take to have quite exotic source categories and which satisfy unusual and varied descent properties.

$$\begin{array}{ccc}
& \Gamma \longrightarrow \mathcal{M}_{\mathbf{fg}} \longleftarrow \widehat{G} & \\
& i^*\Gamma \xRightarrow{\alpha} \pi^*\widehat{G} & \\
\text{Spec } k \xleftarrow{i} & \text{Spec } R/\mathfrak{m} \xrightarrow{\pi} & \text{Spf } R,
\end{array}$$

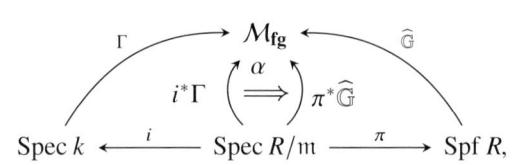

where $i$ is an inclusion of $k$ into the residue field $R/\mathfrak{m}$ and $\alpha\colon i^*\Gamma \to \pi^*\widehat{G}$ is an isomorphism of formal groups. The morphisms in the groupoid are maps $f\colon \widehat{G} \to \widehat{G}'$ of formal groups over Spf $R$ covering the identity on $i^*\Gamma$, called $\star$-*isomorphisms*.

*Remark 3.4.4* (cf. [Rezb, section 4.1]) Because the ground field $k$ is a local ring, the formal group $\Gamma\colon$ Spec $k \to \mathcal{M}_{\mathbf{fg}}$ at the special fiber always has trivializable Lie algebra, hence Lemma 3.2.9 shows that it always admits a presentation by a formal group law. In fact, any deformation $\widehat{G}\colon$ Spf $R \to \mathcal{M}_{\mathbf{fg}}$ of $\Gamma$ also has a trivializable Lie algebra, since projective modules (such as $T_0\widehat{G}$) over local rings like $R$ are automatically free (i.e., trivializable). It follows that the groupoid $(\mathcal{M}_{\mathbf{fg}})^{\wedge}_{\Gamma}(R)$ admits a presentation in terms of formal group *laws*. Starting with the pullback square of groupoids

$$\begin{array}{ccc}
(\mathcal{M}_{\mathbf{fg}})^{\wedge}_{\Gamma}(R) & \longrightarrow & \mathcal{M}_{\mathbf{fg}}(R) \\
\downarrow & & \downarrow \\
\coprod_{i\colon \text{Spec } R/\mathfrak{m} \to \text{Spec } k} \{\Gamma\} \longrightarrow \coprod_{i\colon \text{Spec } R/\mathfrak{m} \to \text{Spec } k} \mathcal{M}_{\mathbf{fg}}(k) & \longrightarrow & \mathcal{M}_{\mathbf{fg}}(R/\mathfrak{m})
\end{array}$$

and selecting formal group laws everywhere, the objects of the groupoid $(\mathcal{M}_{\mathbf{fg}})^{\wedge}_{\Gamma}(R)$ are given by diagrams

$$\begin{array}{ccccc}
(\widehat{\mathbb{A}}^1_k, +_\Gamma) & \leftarrow (\widehat{\mathbb{A}}^1_{R/\mathfrak{m}}, +_{i^*\Gamma}) & === & (\widehat{\mathbb{A}}^1_{R/\mathfrak{m}}, +_{\pi^*\widehat{G}}) \to & (\widehat{\mathbb{A}}^1_R, +_{\widehat{G}}) \\
\downarrow & & & & \downarrow \\
\text{Spec } k & \xleftarrow{i} & \text{Spec } R/\mathfrak{m} & \xrightarrow{\pi} & \text{Spf } R,
\end{array}$$

where we have required an *equality* of formal group laws over the common pullback. A morphism in this groupoid is a formal group law isomorphism $f$ over Spf $R$ which reduces to the identity over Spec $R/\mathfrak{m}$.

The main result about $(\mathcal{M}_{\mathbf{fg}})^{\wedge}_{\Gamma}$ is due to Lubin and Tate:

**Theorem 3.4.5** (Lubin–Tate theorem [LT66, theorem 3.1]) *Suppose that* ht $\Gamma < \infty$ *for $\Gamma$ a formal group over $k$ a perfect field of positive characteristic $p$. The functor $(\mathcal{M}_{\mathbf{fg}})^{\wedge}_{\Gamma}$ is valued in essentially discrete groupoids, and it is naturally equivalent to a smooth formal scheme of dimension* $(\mathrm{ht}(\Gamma) - 1)$ *over* $\mathbb{W}_p(k)$.

### 3.4 The Structure of $\mathcal{M}_{fg}$ III: Small Scales

*Remark 3.4.6* ([Zin84, theorem 4.35]) The presence of the *p-local Witt ring* $\mathbb{W}_p(k)$ is explained by its universal property: For $k$ as previously and $R$ an infinitesimal thickening of $k$, $\mathbb{W}_p(k)$ has the lifting property[33]

$$\begin{array}{ccc} \mathbb{W}_p(k) & \xrightarrow{\exists!} & R \\ \downarrow & & \downarrow \\ k & \xrightarrow{i} & R/\mathfrak{m}. \end{array}$$

For the finite perfect fields $k = \mathbb{F}_{p^d} = \mathbb{F}_p(\zeta_{p^d-1})$, the Witt ring can be computed to be $\mathbb{W}_p(\mathbb{F}_{p^d}) = \mathbb{Z}_p(\zeta_{p^d-1})$.

*Remark 3.4.7* In light of Remark 3.4.4, we can also state Theorem 3.4.5 in terms of formal group laws and their $\star$-isomorphisms. For a group law $+_\Gamma$ over a perfect field $k$ of positive characteristic, it claims that there exists a ring $X$, noncanonically isomorphic to $\mathbb{W}_p(k)[\![u_1, \ldots, u_{d-1}]\!]$, as well as a certain group law $+_{\tilde{\Gamma}}$ on this ring. The group law $+_{\tilde{\Gamma}}$ has the following property: If $+_{\widehat{G}}$ is a formal group law on an infinitesimal thickening Spf $R$ of Spec $k$ which reduces along $\pi$: Spec $R/\mathfrak{m} \to$ Spf $R$ to $+_\Gamma$, then there is a unique ring map $f: X \to R$, compatible with the identifications $X/\mathfrak{m} \cong k \cong R/\mathfrak{m}$, such that $f^*(+_{\tilde{\Gamma}})$ is $\star$-isomorphic to $+_{\widehat{G}}$, and this $\star$-isomorphism is unique.

We will spend the rest of this lecture working toward a proof of Theorem 3.4.5. We first consider a very particular sort of infinitesimal thickening: the square-zero extension $R = k[\varepsilon]/\varepsilon^2$ with pointing $\varepsilon = 0$. We are interested in two kinds of data over $R$: formal group laws $+_\Delta$ over $R$ reducing to $+_\Gamma$ at the pointing, and formal group law automorphisms $\varphi$ of $+_\Gamma$ which reduce to the identity automorphism at the pointing.

**Lemma 3.4.8** *Define*

$$\Gamma_1 = \frac{\partial(x +_\Gamma y)}{\partial x}, \qquad \Gamma_2 = \frac{\partial(x +_\Gamma y)}{\partial y}.$$

*Such automorphisms $\varphi$ are determined by series $\psi$ satisfying*

$$0 = \Gamma_1(x, y)\psi(x) - \psi(x +_\Gamma y) + \psi(y)\Gamma_2(x, y).$$

*Such formal group laws $+_\Delta$ are determined by bivariate series $\delta(x, y)$ satisfying*

$$0 = \Gamma_1(x +_\Gamma y, z)\delta(x, y) - \delta(x, y +_\Gamma z) + \delta(x +_\Gamma y, z) - \delta(y, z)\Gamma_2(x, y +_\Gamma z).$$

*Proof* Such an automorphism $\varphi$ admits a series expansion

$$\varphi(x) = x + \varepsilon \cdot \psi(x).$$

---

[33] Rings with such lifting properties are generally called *Cohen rings*. In the case that $k$ is a perfect field of positive characteristic $p$, the Witt ring $\mathbb{W}_p(k)$ happens to model a Cohen ring for $k$.

Then, we take the homomorphism property
$$\varphi(x +_\Gamma y) = \varphi(x) +_\Gamma \varphi(y)$$
$$(x +_\Gamma y) + \varepsilon \cdot \psi(x +_\Gamma y) = (x + \varepsilon \cdot \psi(x)) +_\Gamma (y + \varepsilon \cdot \psi(y))$$
and apply $\left.\frac{\partial}{\partial \varepsilon}\right|_{\varepsilon=0}$ to get
$$\psi(x +_\Gamma y) = \Gamma_1(x, y) \cdot \psi(x) + \Gamma_2(x, y) \cdot \psi(y).$$
Similarly, such a formal group law $+_\Delta$ admits a series expansion
$$x +_\Delta y = (x +_\Gamma y) + \varepsilon \cdot \delta(x, y).$$
Beginning with the associativity property
$$(x +_\Delta y) +_\Delta z = x +_\Delta (y +_\Delta z),$$
we compute $\left.\frac{\partial}{\partial \varepsilon}\right|_{\varepsilon=0}$ applied to both sides:
$$\left.\frac{\partial}{\partial \varepsilon}\right|_{\varepsilon=0} ((x +_\Delta y) +_\Delta z)$$
$$= \left.\frac{\partial}{\partial \varepsilon}\right|_{\varepsilon=0} ((((x +_\Gamma y) + \varepsilon \cdot \delta(x, y)) +_\Gamma z) + \varepsilon \cdot \delta(x +_\Gamma y, z))$$
$$= \Gamma_1(x +_\Gamma y, z) \cdot \delta(x, y) + \delta(x +_\Gamma y, z),$$
and similarly
$$\left.\frac{\partial}{\partial \varepsilon}\right|_{\varepsilon=0} (x +_\Delta (y +_\Delta z))$$
$$= \left.\frac{\partial}{\partial \varepsilon}\right|_{\varepsilon=0} ((x +_\Gamma ((y +_\Gamma z) + \varepsilon \cdot \delta(y, z))) + \varepsilon \cdot \delta(x, y +_\Gamma z))$$
$$= \Gamma_2(x, y +_\Gamma z) \cdot \delta(y, z) + \delta(x, y +_\Gamma z).$$
Equating these gives the condition in the lemma statement. □

The key observation is that these two conditions appear as cocycle conditions for the first two levels of a natural cochain complex.

**Definition 3.4.9** ([Laz97, section 3]) The deformation complex[34,35] $\widehat{C}^*(+_\Gamma; k)$ is defined by
$$k \to k[\![x_1]\!] \to k[\![x_1, x_2]\!] \to k[\![x_1, x_2, x_3]\!] \to \cdots$$

---

[34] Pieces of this complex are visible in work of Drinfeld [Dri74, section 4.A] and of Lubin–Tate [LT66], but neither actually assemble the whole complex.
[35] It would be great to have a geometric definition of this complex, i.e., one using $\Gamma$ rather than $+_\Gamma$.

### 3.4 The Structure of $\mathcal{M}_{\mathbf{fg}}$ III: Small Scales

with differential

$$(df)(x_1, \ldots, x_{n+1}) = \Gamma_1\left(\sum_{i=1}^{n}{}^{\Gamma} x_i, x_{n+1}\right) \cdot f(x_1, \ldots, x_n)$$

$$+ \sum_{i=1}^{n}(-1)^i \cdot f(x_1, \ldots, x_i +_\Gamma x_{i+1}, \ldots, x_{n+1})$$

$$+ (-1)^{n+1} \cdot \Gamma_2\left(x_1, \sum_{i=2}^{n+1}{}^{\Gamma} x_i\right) \cdot f(x_2, \ldots, x_{n+1}),$$

where we have again written

$$\Gamma_1(x, y) = \frac{\partial(x +_\Gamma y)}{\partial x}, \qquad \Gamma_2(x, y) = \frac{\partial(x +_\Gamma y)}{\partial y}.$$

The complex even knits the information together intelligently:

**Corollary 3.4.10** ([Laz97, p. 1320])  *Two extensions $+_\Delta$ and $+_{\Delta'}$ of $+_\Gamma$ to $k[\varepsilon]/\varepsilon^2$ are isomorphic if their corresponding 2-cocycles in $\widehat{Z}^2(+_\Gamma; k)$ differ by an element in $\widehat{B}^2(+_\Gamma; k)$.* □

Remarkably, we have already encountered this complex before:

**Lemma 3.4.11** ([Laz97, p. 1320])  *Write $\widehat{\mathbb{G}}$ for the formal group associated to the group law $+_\Gamma$. The cochain complex $\widehat{C}^*(+_\Gamma; k)$ is quasi-isomorphic to the cohomology cochain complex considered in the proof of Lemma 3.2.7:*

$$\widehat{C}^*(+_\Gamma; k) \to \underline{\mathrm{FormalSchemes}}(B\widehat{\mathbb{G}}, \widehat{\mathbb{G}}_a)(k)$$

$$f \mapsto \Gamma_1\left(0, \sum_{i=1}^{n}{}^{\Gamma} x_i\right)^{-1} \cdot f(x_1, \ldots, x_n).$$ □

In the course of proving Lemma 3.2.7, we computed the cohomology of this complex in the specific case of $\widehat{\mathbb{G}} = \widehat{\mathbb{G}}_a$. This is the one case where Lubin and Tate's theorem does *not* apply, since it requires $\mathrm{ht}\,\widehat{\mathbb{G}} < \infty$. Nonetheless, by filtering the multiplication on $\widehat{\mathbb{G}}$ by degree, we can use this specific calculation to get up to the one we now seek.

**Lemma 3.4.12**  *Let $\widehat{\mathbb{G}}$ be a formal group of finite height d over a field k. Then $H^1(\widehat{\mathbb{G}}; \widehat{\mathbb{G}}_a) = 0$ and $H^2(\widehat{\mathbb{G}}; \widehat{\mathbb{G}}_a)$ is a free k-vector space of dimension d.*

*Proof (after Hopkins)*  We select a $p$-typical coordinate on $\widehat{\mathbb{G}}$ of the form

$$x +_\varphi y = x + y + \mathrm{unit} \cdot c_{p^d}(x, y) + \cdots,$$

where $c_{p^d}(x, y)$ is as in one of Lazard's symmetric 2-cocycles, as in Lemma 3.2.7.

Filtering $\widehat{\mathbb{G}}$ by degree, the multiplication projects to $x +_\varphi y = x + y$ in the associated graded, and the resulting filtration spectral sequence has signature

$$[H^*(\widehat{\mathbb{G}}_a; \widehat{\mathbb{G}}_a)]_* \Rightarrow H^*(\widehat{\mathbb{G}}; \widehat{\mathbb{G}}_a),$$

where the second grading comes from the degree of the homogeneous polynomial representatives of classes in $H^*(\widehat{\mathbb{G}}_a; \widehat{\mathbb{G}}_a)$.

Because Lemma 3.2.7 gives different calculations of $H^*(\widehat{\mathbb{G}}_a; \widehat{\mathbb{G}}_a)$ for $p = 2$ and $p > 2$, we specialize to $p > 2$ for the remainder of the proof and leave the similar $p = 2$ case to the reader. For $p > 2$, Lemma 3.2.7 gives

$$[H^*(\widehat{\mathbb{G}}_a; \widehat{\mathbb{G}}_a)]_* = \left[ \frac{k[\alpha_j \mid j \geq 0]}{\alpha_j^2 = 0} \otimes k[\beta_j \mid j \geq 0] \right]_*,$$

where $\alpha_j$ is represented by $x^{p^j}$ and $\beta_j$ is represented by $c_{p^j}(x,y)$. To compute the differentials in this spectral sequence, one computes by hand the formula for the differential in the bar complex, working up to the lowest nonzero degree. For instance, to compute $d(\alpha_j)$ we examine the series

$$(x +_\varphi y)^{p^j} - (x^{p^j} + y^{p^j}) = (\text{unit}) \cdot c_{p^{d+j}}(x, y) + \cdots,$$

where we used $c_{p^d}^{p^j} = c_{p^{j+d}}$, giving $d(\alpha_j) = \beta_{j+d}$. So, we see that nothing in the 1-column of the spectral sequence is a permanent cocycle and that there are $d$ classes at the bottom of the 2-column of the spectral sequence which are not coboundaries. To conclude the lemma statement, we need only to check that they are indeed permanent cocycles. To do this, we note that they are realized as deformations, by observing

$$x +_{\text{univ}} y \cong x + y + v_j c_{p^j}(x, y) \quad (\text{mod } v_1, \ldots, v_{j-1}, (x, y)^{p^{j+1}})$$

where $+_{\text{univ}}$ is the Kudo–Araki universal $p$-typical law (see [LT66, proposition 1.1]). □

*Proof of Theorem 3.4.5 using Remark 3.4.7* Fix any group law $+_{\tilde{\Gamma}}$ over the ring $\mathbb{W}_p(k)[\![u_1, \ldots, u_{d-1}]\!]$ such that for each $j$,

$$x +_{\tilde{\Gamma}} y \cong x + y + u_j c_{p^j}(x, y) \quad (\text{mod } u_1, \ldots, u_{j-1}, (x, y)^{p^{j+1}}).$$

We will prove the claim inductively on the order of the infinitesimal neighborhood of $\operatorname{Spec} k = \operatorname{Spec} R/\mathfrak{m}$ in $\operatorname{Spf} R$:[36]

$$\operatorname{Spec} R/\mathfrak{m} \xrightarrow{j_r} \operatorname{Spec} R/\mathfrak{m}^r \xrightarrow{i_r} \operatorname{Spf} R.$$

---

[36] The reader may also be interested in a proof presented in more stacky language [Lura, reduction to theorem 21.5].

## 3.4 The Structure of $\mathcal{M}_{\text{fg}}$ III: Small Scales

Suppose that we have demonstrated the theorem for $+_{\widehat{\mathbb{G}}_{r-1}} = i^*_{r-1}(+_{\widehat{\mathbb{G}}})$, so that there is a map $\alpha_{r-1} \colon \mathbb{W}_p(k)[\![u_1, \ldots, u_{d-1}]\!] \to R/\mathfrak{m}^{r-1}$ and a strict isomorphism $g_{r-1} \colon +_{\widehat{\mathbb{G}}_{r-1}} \to \alpha^*_{r-1} +_{\widehat{\Gamma}}$ of formal group laws. The exact sequence

$$0 \to \mathfrak{m}^{r-1}/\mathfrak{m}^r \to R/\mathfrak{m}^r \to R/\mathfrak{m}^{r-1} \to 0$$

exhibits $R/\mathfrak{m}^r$ as a square–zero extension of $R/\mathfrak{m}^{r-1}$ by $M = \mathfrak{m}^{r-1}/\mathfrak{m}^r$. Then, let $\beta$ be *any* lift of $\alpha_{r-1}$ and $h$ be *any* lift of $g_{r-1}$ to $R/I^r$, and let $A$ and $B$ be the induced group laws

$$x +_A y = \beta^* \tilde{\varphi}, \qquad x +_B y = h\left(h^{-1}(x) +_{\widehat{\mathbb{G}}_r} h^{-1}(y)\right).$$

Since these both deform the group law $+_{\widehat{\mathbb{G}}_{r-1}}$, we know by Corollary 3.4.10 and Lemma 3.4.12 there exist $m_j \in M$ and $f(x) \in M[\![x]\!]$ satisfying

$$(x +_B y) - (x +_A y) = (df)(x, y) + \sum_{j=1}^{d-1} m_j c_{p^j}(x, y),$$

where $c_{p^j}(x, y)$ is the 2-cocycle associated to the cohomology 2-class $\beta_j$. The following definitions complete the induction:

$$g_r(x) = h(x) - f(x), \qquad \alpha_r(u_j) = \beta(u_j) + m_j. \qquad \square$$

*Remark 3.4.13* Our calculation $H^1(\widehat{\mathbb{G}}_\varphi; \widehat{\mathbb{G}}_a) = 0$ shows that the automorphisms $\alpha \colon \Gamma \to \Gamma$ of the special fiber induce automorphisms of the entire Lubin–Tate stack by universality. Namely, for $\Gamma \to \widetilde{\Gamma}$ the universal deformation, the precomposite

$$\Gamma \xrightarrow{\alpha} \Gamma \to \widetilde{\Gamma}$$

presents $\widetilde{\Gamma}$ as a deformation of $\Gamma$ in a different way, hence induces a map

$$\widetilde{\alpha} \colon \widetilde{\Gamma} \to \widetilde{\Gamma},$$

which by Theorem 3.4.5 is in turn induced by a map

$$\widetilde{\alpha} \colon (\mathcal{M}_{\text{fg}})^\wedge_\Gamma \to (\mathcal{M}_{\text{fg}})^\wedge_\Gamma.$$

The action is *highly* nontrivial in all but the most degenerate cases, and its study is of serious interest to homotopy theorists (see Lecture 3.6) and to arithmetic geometers (see Remark 4.4.20 as well as Appendix B.2).

*Remark 3.4.14* We also see that our analysis of infinitesimal behavior fails wildly for the case $\Gamma = \widehat{\mathbb{G}}_a$. The differential calculation in Lemma 3.4.12 is meant to give us an upper bound on the dimensions of $H^1(\Gamma; \widehat{\mathbb{G}}_a)$ and $H^2(\Gamma; \widehat{\mathbb{G}}_a)$, but this family of differentials is zero in the additive case. Accordingly, both of these vector spaces are infinite dimensional, completely prohibiting us from making any further assessment.

Having accomplished all our major goals, we close our algebraic analysis of $\mathcal{M}_{fg}$ with Figure 3.3, a diagram summarizing our results.

## 3.5 Nilpotence and Periodicity in Finite Spectra

With our analysis of $\mathcal{M}_{fg}$ complete, our first goal in this lecture is to finish the program sketched in the introduction to this case study by manufacturing those interesting homology theories connected to the functor $\mathcal{M}_{MU}(-)$. We begin by rephrasing our main tool, Theorem 3.0.1, in terms of algebraic conditions.

**Theorem 3.5.1** (Landweber Exact Functor Theorem [Lan76, corollary 2.7], [Lura, lecture 16], [Hop, theorem 21.4 and proposition 21.5]) *Let $\mathcal{F}$ be a quasicoherent sheaf over $\mathcal{M}_{fg} \times \operatorname{Spec} \mathbb{Z}_{(p)}$, thought of as a comodule $M$ for the Hopf algebroid*

$$(A, \Gamma) = (O_{\mathcal{M}_{fgl}}, O_{\mathcal{M}_{fgl}}[t_1, t_2, \ldots]).$$

*Using the Kudo–Araki map of Theorem 3.3.12, we can define the action of the infinite sequence $(p, v_1, \ldots, v_d, \ldots)$ on $M$. If this sequence is regular, then*

$$X \mapsto M \otimes_{O_{\mathcal{M}_{fgl}}} MUP_0(X)$$

*determines a homology theory.*

*Proof* Following the discussion in the introduction, note that a cofiber sequence

$$X' \to X \to X''$$

of spectra gives rise to an exact sequence

$$
\begin{array}{ccccccccc}
\cdots & \longrightarrow & \mathcal{M}_{MUP}(X') & \longrightarrow & \mathcal{M}_{MUP}(X) & \longrightarrow & \mathcal{M}_{MUP}(X'') & \longrightarrow & \cdots \\
& & \| & & \| & & \| & & \\
\cdots & \longrightarrow & \mathcal{N}' & \longrightarrow & \mathcal{N} & \longrightarrow & \mathcal{N}'' & \longrightarrow & \cdots.
\end{array}
$$

We thus see that we are essentially tasked with showing that $\mathcal{F}$ is flat, so that tensoring with $\mathcal{F}$ does not disturb the exactness of this sequence. In that case, we can then apply Brown representability to the composite functor $\mathcal{F} \otimes \mathcal{M}_{MUP}(X)$.

Flatness of $\mathcal{F}$ is equivalent to $\operatorname{Tor}_1(\mathcal{F}, \mathcal{N}) = 0$ for an arbitrary auxiliary quasicoherent sheaf $\mathcal{N}$. By our regularity hypothesis, there is an exact sequence of sheaves

$$0 \to \mathcal{F} \xrightarrow{p} \mathcal{F} \to \mathcal{F}/(p) \to 0,$$

## 3.5 Nilpotence and Periodicity in Finite Spectra

so applying $\mathrm{Tor}_*(-, \mathcal{N})$ gives an exact sequence,

$$\mathrm{Tor}_2(\mathcal{F}/(p), \mathcal{N}) \to \mathrm{Tor}_1(\mathcal{F}, \mathcal{N}) \xrightarrow{p} \mathrm{Tor}_1(\mathcal{F}, \mathcal{N}),$$

of Tor groups. The sequence gives the following sufficiency condition:

$$[\mathrm{Tor}_1(p^{-1}\mathcal{F}, \mathcal{N}) = 0, \quad \mathrm{Tor}_2(\mathcal{F}/(p), \mathcal{N}) = 0] \quad \Rightarrow \quad \mathrm{Tor}_1(\mathcal{F}, \mathcal{N}) = 0.$$

Similarly, the $v_1$-multiplication sequence gives another sufficiency condition:

$$[\mathrm{Tor}_2(v_1^{-1}\mathcal{F}/(p), \mathcal{N}) = 0, \quad \mathrm{Tor}_3(\mathcal{F}/I_2, \mathcal{N}) = 0] \quad \Rightarrow \mathrm{Tor}_2(\mathcal{F}/(p), \mathcal{N}) = 0.$$

Continuing in this fashion, for some $D \gg 0$ we would like to show

$$\mathrm{Tor}_{d+1}(v_d^{-1}\mathcal{F}/I_d, \mathcal{N}) = 0, \quad \text{(for each } d < D\text{)}$$
$$\mathrm{Tor}_{D+1}(\mathcal{F}/I_{D+1}, \mathcal{N}) = 0.$$

We begin with the first collection of conditions. They are *always* satisfied, but this requires an argument. We write $i_d \colon \mathcal{M}_{\mathrm{fg}}^{=d} \to \mathcal{M}_{\mathrm{fg}}$ for the inclusion of the substack of formal groups of height exactly $d$, which (following Remark 3.3.19) has a presentation by the Hopf algebroid

$$(v_d^{-1}A/I_d, \Gamma \otimes v_d^{-1}A/I_d).$$

We are trying to study the derived functors of

$$\mathcal{N} \mapsto (i_{d*}i_d^*\mathcal{F}) \otimes \mathcal{N} \cong i_{d*}(i_d^*\mathcal{F} \otimes i_d^*\mathcal{N}).$$

Since $i_{d*}$ is exact, we are moved to study the composite functor spectral sequence

$$\mathrm{QCoH}_{\mathcal{M}_{\mathrm{fg}}} \xrightarrow{i_d^*} \mathrm{QCoH}_{\mathcal{M}_{\mathrm{fg}}^{=d}} \xrightarrow{i_d^*\mathcal{F} \otimes -} \mathrm{QCoH}_{\mathcal{M}_{\mathrm{fg}}^{=d}}.$$

The second functor is exact, independent of $\mathcal{F}$: the geometric map

$$\Gamma_d \colon \mathrm{Spec}\, k \to \mathcal{M}_{\mathrm{fg}}^{=d}$$

is a faithfully flat cover, and $k$-modules have no nontrivial Tor. Meanwhile, the first functor has at most $d$ derived functors: $i_d^*$ is modeled by tensoring with $v_d^{-1}A/I_d$, but $A/I_d$ admits a Koszul resolution with $d$ stages and the localization $A/I_d \to v_d^{-1}A/I_d$ is flat. As $\mathrm{Tor}_{d+1}$ is beyond the length of this resolution, it is always zero.

To deal with the second condition,[37] we make a slight reduction: A generic spectrum $X$ can be written as an ind-finite spectrum, and as homology and Tor both commute with colimits, it suffices to treat the case that $X$ is itself finite. In this case, $\mathcal{N}$ is additionally *coherent*. Writing $j_{D+1} \colon \mathcal{M}_{\mathrm{fg}}^{\geq D+1} \to \mathcal{M}_{\mathrm{fg}}$

---

[37] It is common for authors to assume that $\mathcal{F}$ itself satisfies $\mathcal{F}/I_{D+1} = 0$, avoiding this part of the argument entirely. There are flat sheaves without this property – for instance, $\mathcal{O}(\mathcal{M}_{\mathrm{fgl}})$.

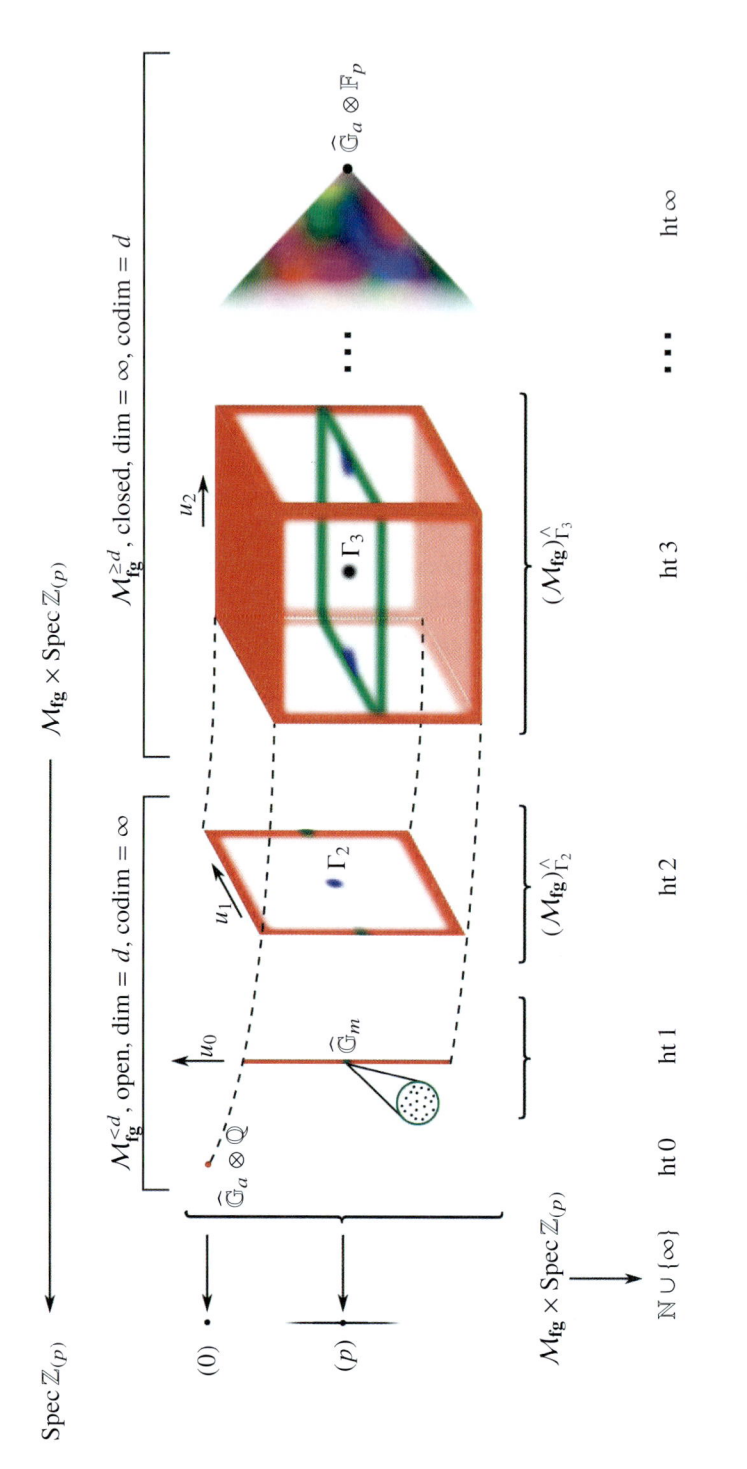

Figure 3.3 Portrait of $\mathcal{M}_{\mathbf{fg}} \times \operatorname{Spec} \mathbb{Z}_{(p)}$.

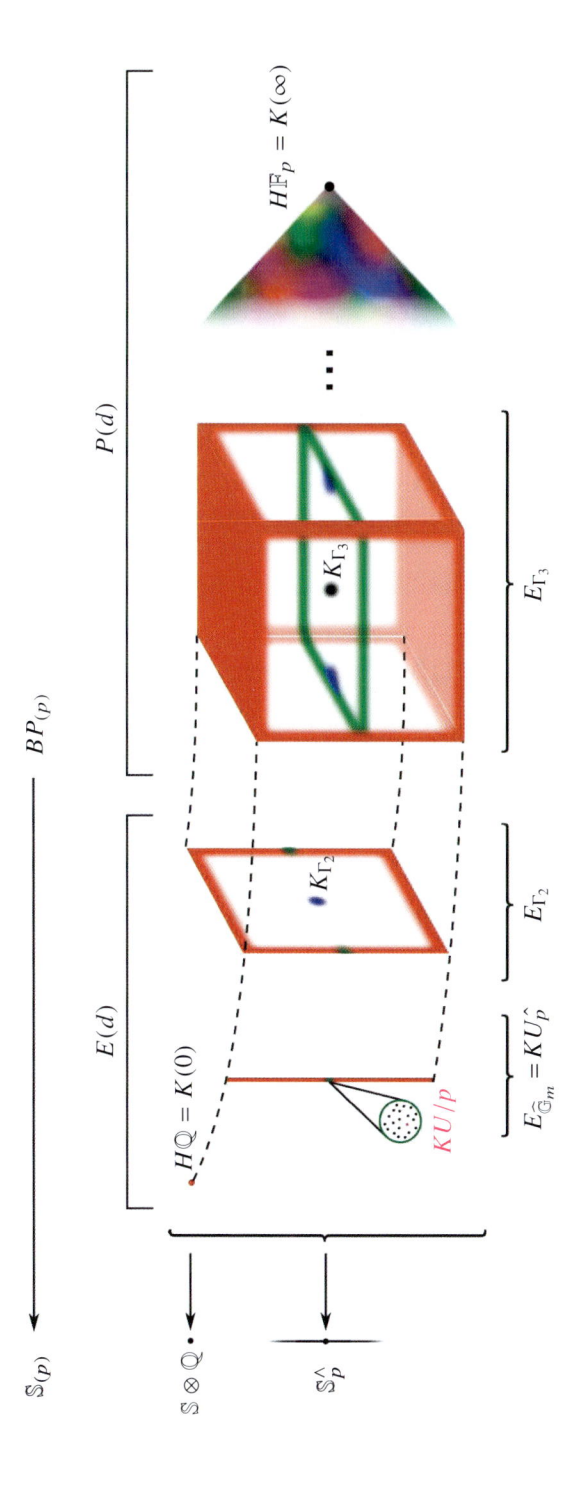

Figure 3.4 Topological realizations of various pieces of $\mathcal{M}_{\mathrm{fg}} \times \mathrm{Spec}\,\mathbb{Z}_{(p)}$.

for the inclusion of the prime closed substack, it follows from coherence and Theorem 3.3.18 that $j_{D+1}^* \mathcal{N}$ is free for large $D$, and hence

$$\mathrm{Tor}_*^{\mathcal{M}_{\mathrm{fg}}^{\geq D+1}}(j_{D+1}^* \mathcal{F}, j_{D+1}^* \mathcal{N}) = 0.$$

We proceed by downward induction: For each $d \leq D$, Mayer–Vietoris shows

$$\left[\mathrm{Tor}_*^{\mathcal{M}_{\mathrm{fg}}^{\geq d+1}}(\mathcal{F}/I_{D+1}, j_{d+1}^* \mathcal{N}) = 0, \quad \mathrm{Tor}_*^{\mathcal{M}_{\mathrm{fg}}^{=d}}(\mathcal{F}/I_{D+1}, v_d^{-1} j_d^* \mathcal{N}) = 0\right]$$

$$\Rightarrow \quad \mathrm{Tor}_*^{\mathcal{M}_{\mathrm{fg}}^{\geq d}}(\mathcal{F}/I_{D+1}, j_d^* \mathcal{N}) = 0.$$

The first statement is the inductive hypothesis, and the second statement follows again because $\Gamma_d$ is a faithfully flat cover and $k$-modules have trivial torsion. □

**Remark 3.5.2** This construction always gives an *MUP*-module spectrum in the homotopy category. If $\mathcal{F}$ comes from the pushforward of the ring of functions along a flat map $\mathrm{Spec}\, R \to \mathcal{M}_{\mathrm{fg}}$, then the resulting spectrum is a ring spectrum [HS99, theorem 2.8], and flat sheaves on it become module spectra over this ring spectrum.

**Remark 3.5.3** ([Hop, equation 20.1]) Consider a map $f: \mathrm{Spec}\, R \to \mathcal{M}_{\mathrm{fg}}$ and its associated pullback square

$$\begin{array}{ccc} P & \xrightarrow{f'} & \mathcal{M}_{\mathrm{fgl}} \\ \downarrow & & \downarrow \\ \mathrm{Spec}\, R & \xrightarrow{f} & \mathcal{M}_{\mathrm{fg}}. \end{array}$$

If $f^* \mathcal{M}_{MU}(X)$ defines a homology functor $R_0(X)$, then $f'^* \mathcal{M}_{MU}(X)$ gives a model for $(MU \wedge R)_0(X)$. In fact, $f'^* \mathcal{M}_{MU}(X)$ *always* defines a homology functor, *regardless* of the exactness of $f^*$. This observation allows us to rephrase the conditions under which $f^*$ defines a homology functor: We can always recognize $f^* \mathcal{M}_{\mathrm{fg}}(X)$ by equalizing the maps

$$s, t: (\mathcal{M}_{\mathrm{fgl}} \times \mathcal{M}_{\mathrm{ps}}^{\mathrm{gpd}}) \to \mathcal{M}_{\mathrm{fgl}}.$$

In general, forming this equalizer does not preserve exact sequences. However, in the Landweber case, where $f$ is flat and there is a lift $\widetilde{f}: \mathrm{Spec}\, R \to \mathcal{M}_{\mathrm{fgl}}$, then this equalizer becomes *forked* (i.e., we gain an extra degeneracy in the simplicial object), and the equalizer is thus guaranteed to be exact.

**Definition 3.5.4** Coupling Theorem 3.5.1 to our understanding of $\mathcal{M}_{\mathrm{fg}}$, we produce many interesting homology theories, collectively referred to as *chromatic*[38] *homology theories*:

---

[38] The elements of Figures 3.1 and 3.2 are related to each other by "$v_d$-multiplication"

## 3.5 Nilpotence and Periodicity in Finite Spectra

- Recall that the moduli of $p$-typical group laws is affine, presented in Theorem 3.3.12 by

$$O_{\mathcal{M}_{\text{fgl}}^{p\text{-typ}}} \cong \mathbb{Z}_{(p)}[v_1, v_2, \ldots, v_d, \ldots].$$

This tautologically satisfies the regularity condition in Landweber's theorem, but it can also be seen to satisfy the desired flatness condition: The moduli of $p$-typical group laws is a flat cover of its quotient by $p$-typical coordinate changes, and since the inclusion of $p$-typical group laws into all group laws induces an equivalence of stacks, the moduli of $p$-typical group laws determines a flat cover of $\mathcal{M}_{\text{fg}}$. This therefore determines a homology theory on finite spectra, called *Brown–Peterson homology*:

$$BPP_0(X) := MUP_0(X) \otimes_{MUP_0} BPP_0.$$

- A chart for the open substack $\mathcal{M}_{\text{fg}}^{\leq d}$ in terms of $\mathcal{M}_{\text{fgl}}^{p\text{-typ}}$ was given in Remark 3.3.19 by $\operatorname{Spec} \mathbb{Z}_{(p)}[v_1, v_2, \ldots, v_d^{\pm}]$. Since open maps are in particular flat, it follows that there is a homology theory $EP(d)$, called *the $d$th Johnson–Wilson homology*, defined on all spectra by

$$EP(d)_0(X) := MUP_0(X) \otimes_{MUP_0} \mathbb{Z}_{(p)}[v_1, v_2, \ldots, v_d^{\pm}].$$

- Similarly, for a formal group $\Gamma$ of height $d < \infty$ over a field $k$, we produced in Theorem 3.4.5 a chart $\operatorname{Spf} \mathbb{W}_p(k)[\![u_1, \ldots, u_{d-1}]\!]$ for its deformation neighborhood. Since inclusions of deformation neighborhoods of substacks of Noetherian stacks are flat [Mat89], there is a corresponding homology theory $E_\Gamma$, called *the (discontinuous) Morava E-theory for $\Gamma$*, determined by

$$E_{\Gamma 0}(X) := MUP_0(X) \otimes_{MUP_0} \mathbb{W}_p(k)[\![u_1, \ldots, u_{d-1}]\!][u^{\pm}].$$

In the case that $\Gamma = \Gamma_d$ is the Honda formal group of height $d$ over $\mathbb{F}_p$, the notation is often abbreviated from $E_{\Gamma_d}$ to merely $E_d$.

- Since $\mathfrak{m} = (p, u_1, \ldots, u_{d-1})$ is a regular ideal in $E_{\Gamma *}$, we can use the $MU$-module structure of $E_\Gamma$ to model the reduction to the special fiber by way of

---

(see Remark 3.6.22), and families of such elements can be selected for by inverting $v_d$, i.e., by passing to the open substack $\mathcal{M}_{\text{fg}}^{\leq d}$. The word "chromatic" here thus refers to an analogy: This localization selects certain periodic families of elements, like a bandpass filter selects certain frequencies out of a complicated radio signal.

the cofiber sequences

$$E_\Gamma \xrightarrow{p} E_\Gamma \to E_\Gamma/(p),$$

$$E_\Gamma/(p) \xrightarrow{u_1} E_\Gamma/(p) \to E_\Gamma/(p,u_1),$$

$$\vdots$$

$$E_\Gamma/I_{d-1} \xrightarrow{u_{d-1}} E_\Gamma/I_{d-1} \to E_\Gamma/I_d.$$

This determines a spectrum $K_\Gamma = E_\Gamma/I_d$, and hence determines a homology theory called *the Morava K-theory for* $\Gamma$. This is a ring spectrum with coefficients $K_{\Gamma*} = k[u^\pm]$. In the case that $\Gamma$ is the Honda $p$-typical formal group law (of height $d$, over $\mathbb{F}_p$), this spectrum is often written as $K(d)$. As an edge case, we also set $K(0) = H\mathbb{Q}$ and $K(\infty) = H\mathbb{F}_p$.[39]

- More delicately, there is a version of Morava $E$-theory that takes into account the formal topology on $(\mathcal{M}_\mathbf{fg})^\wedge_\Gamma$, called *continuous Morava E-theory*. It is defined by the pro-system $\{E_\Gamma(X)/u^I\}$, where $I$ ranges over multi-indices and the quotient is again given by cofiber sequences.

- There is also a homology theory associated to the closed substack $\mathcal{M}_\mathbf{fg}^{\geq d}$. Since $I_d = (p, v_1, \ldots, v_{d-1})$ is generated by a regular sequence on $BPP_0$, we can directly define the spectrum $PP(d)$ by a regular quotient:

$$PP(d) = BPP/(p, v_1, \ldots, v_{d-1}).$$

This spectrum does have the property $PP(d)_0 = BPP_0/I_d$ on coefficient rings, but $PP(d)_0(X) = BPP_0(X)/I_d$ *only* when $I_d$ forms a regular sequence on $BPP_0(X)$ – which is reasonably rare among the cases of interest. The relationships between these homology theories and the corresponding components of $\mathcal{M}_\mathbf{fg}$ are depicted in Figure 3.4.

*Remark 3.5.5* The "extra $P$" in some of these names is to disambiguate them from similar less-periodic objects in the literature. Other authors often prefer to take $BP$ to be a minimal wedge summand of $MU_{(p)}$ and to take the spectra $E(d)$ and $K(d)$ to be $2(p^d - 1)$-periodic (for heights $0 < d < \infty$). On the other hand, the spectra $E_\Gamma$ and $K_\Gamma$ are *usually* taken to be 2-periodic already. Since we are nearly exclusively interested in 2-periodic spectra, we insert a "$P$" into the names of the former spectra, but not to $E_\Gamma$ or $K_\Gamma$.

*Example 3.5.6* ([CF66], see Example 2.1.21)   In the case $\Gamma = \widehat{\mathbb{G}}_m$ over $k = \mathbb{F}_p$, the resulting spectra are connected to complex $K$-theory:

$$E_{\widehat{\mathbb{G}}_m} \cong KU_p^\wedge, \qquad K_{\widehat{\mathbb{G}}_m} \cong KU/p, \qquad EP(1) \cong KU_{(p)}.$$

---

[39] By Theorem 3.4.1 and Corollary 3.5.15 to follow, it often suffices to consider just these spectra $K(d)$ to make statements about all $K_\Gamma$. With more care, it even often suffices to consider $d \neq \infty$.

### 3.5 Nilpotence and Periodicity in Finite Spectra

*Remark 3.5.7* ([KLW04, section 5.2], [Rav84, corollaries 2.14 and 2.16], [Str99a, theorem 2.13]) In general, the quotient of a ring spectrum by a homotopy element does not give another ring spectrum. The most typical example of this phenomenon is that $\mathbb{S}/2$ is not a ring spectrum, since its homotopy is not 2-torsion. Most of our previous constructions do not suffer from this deficiency, with one exception: Morava $K$-theories at $p = 2$ are not commutative. Instead, there is a derivation $Q_d : K(d) \to \Sigma K(d)$ which tracks the commutativity by the relation
$$ab - ba = uQ_d(a)Q_d(b).$$
In particular, we find that $K(d)^*X$ is a commutative ring whenever $K(d)^1 X = 0$, which is often the case.

*Remark 3.5.8* ([Hop14a, section 3]) Before turning to the other half of this lecture, we feel obligated to include one out-of-place remark that the first-time reader is advised to skip. The pullback idea expressed in Remark 3.5.3 intermingles interestingly with the detection result expressed in Remark 3.1.22: There is a certain program, which is beyond the scope of this lecture, which realizes spectra associated to certain nontrivial *stacks* equipped with maps to the moduli of formal groups. For instance, the non-complex-orientable spectrum $KO$ is attached to a map
$$\operatorname{Spec} \mathbb{Z} /\!/ C_2 \to \mathcal{M}_{\mathrm{fg}}$$
which selects $\widehat{\mathbb{G}}_m$ with its complex-conjugation action. There is a sequence of non-complex-orientable ring spectra $X(n) = T(\Omega SU(n))$ which limit to $X(\infty) = MU$ and which carry "partial complex-orientations" that give a Thom isomorphism for $\mathbb{C}P^{n-1}$, i.e., they come equipped with a specified formal group law modulo terms of degree $n$. We write $\mathcal{M}_{\mathrm{fg}}^{(n)}$ for the moduli of formal groups where the isomorphisms are required to act trivially on space of $n$-jets (so that, e.g., Remark 3.2.11 concerns $\mathcal{M}_{\mathrm{fg}}^{(1)}$). The pullback of stacks

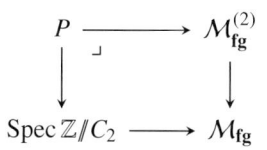

$$\begin{array}{ccc} P & \longrightarrow & \mathcal{M}_{\mathrm{fg}}^{(2)} \\ \downarrow & & \downarrow \\ \operatorname{Spec} \mathbb{Z}/\!/ C_2 & \longrightarrow & \mathcal{M}_{\mathrm{fg}} \end{array}$$

is actually affine – a witness to the fact that $X(2) \wedge KO$ is a complex-orientable ring spectrum (indeed, a variant of $KU$), even if $KO$ and $X(2)$ are not.

Having constructed these chromatic homology theories, for the rest of this lecture we pursue an example of a "fiberwise" analysis of a phenomenon in homotopy theory. First, recall the following classical theorem:

148        Finite Spectra

**Theorem 3.5.9** (Nishida's theorem [Nis73], [BMMS86, section II.2])   *Every homotopy class $\alpha \in \pi_{\geq 1} \mathbb{S}$ is nilpotent.*[40]   □

People studying $K$-theory uncovered the following related phenomenon:

**Theorem 3.5.10** ([Ada66, theorem 12.1])   *Let $M_{2n}(p)$ denote the mod-$p$ Moore spectrum with bottom cell in degree $2n$. Then there is an index $n$ and a map $v \colon M_{2n}(p) \to M_0(p)$ such that $KU_* v$ acts by multiplication by the nth power of the Bott class. The minimal such $n$ is given by the formula*

$$n = \begin{cases} p - 1 & \text{when } p \geq 3, \\ 4 & \text{when } p = 2. \end{cases}$$

□

In particular, the map $v$ cannot be nilpotent, since a null-homotopic map induces the zero map in any homology theory. Just as we took the non-nilpotent endomorphism $p \in \pi_0 \operatorname{End} \mathbb{S}$ and coned it off, we can take the endomorphism $v \in \pi_{2p-2} \operatorname{End} M_0(p)$ and cone it off to form a new spectrum called $V(1)$.[41] One can ask, then, whether the pattern continues: Does $V(1)$ have a non-nilpotent self-map, and can we cone it off to form a new such spectrum with a new such map? Can we then do that again, indefinitely? In order to study this question, we are motivated to find spectra satisfying the following condition:

**Definition 3.5.11** ([HS98, definition 4], see [DHS88, theorem 1])   A ring spectrum $E$ *detects nilpotence* if for any ring spectrum $R$ the kernel of the Hurewicz homomorphism

$$R_* \eta_E \colon \pi_* R \to E_* R$$

consists of nilpotent elements. (Setting $R = \operatorname{End}(F)$, a non-nilpotent self-map of $F$ gives an element of $\pi_* R$ which must then have nontrivial image in $E_* R$.)

This question and surrounding issues formed the basis of Ravenel's nilpotence conjectures [Rav84, section 10], which were resolved by Devinatz, Hopkins, and Smith [DHS88, HS98]. One of their two main technical achievements was to demonstrate that we already have access to a nice homology theory which detects nilpotence:

**Theorem 3.5.12** (Devinatz–Hopkins–Smith Nilpotence Theorem [DHS88, theorem 1.i])   *The spectrum $MU$ detects nilpotence.*   □

---

[40] "Nilpotent" has two equivalent interpretations here: The maps $\alpha^{\wedge n} \colon \mathbb{S}^{\wedge n} \to \mathbb{S}^{\wedge n}$ and $\alpha^{\circ n} \colon \mathbb{S} \to \mathbb{S}$ are naturally equivalent. In fact, these two kinds of nilpotence continue to agree for more general ring spectra [DHS88, introduction].

[41] The spectrum $V(1)$ is actually defined to be a finite spectrum with $BP_*V(1) \cong BP_*/(p, v_1)$. At $p = 2$ this spectrum doesn't exist and this is a misnomer. More generally, at odd primes $p$ Nave shows that $V((p+1)/2)$ doesn't exist [Nav10, theorem 1.3].

3.5 Nilpotence and Periodicity in Finite Spectra

This is a very hard theorem, and we will not attempt to prove it.[42] However, taking this as input, they are easily able to show several other interesting structural results about finite spectra. For instance, they also show that the $MU$ is the universal object which detects nilpotence, in the sense that any other ring spectrum can have this property checked stalkwise on $\mathcal{M}_{MU}$.

**Corollary 3.5.13** ([HS98, theorem 3])  *A ring spectrum $E$ detects nilpotence if and only if for all $0 \leq d \leq \infty$ and for all primes $p$, $K(d)_*E \neq 0$ (i.e., the support of $\mathcal{M}_{MU}(E)$ is not a proper substack of $\mathcal{M}_{MU}$).*

*Proof*  If $K(d)_*E = 0$ for some $d$, then the unit map $\mathbb{S} \to K(d)$ lies in the kernel of the Hurewicz homomorphism for $E$, so $E$ fails to detect nilpotence.

In the other direction, suppose that for every $d$ we have $K(d)_*E \neq 0$. Because $K(d)_*$ is a field, it follows by picking a basis of $K(d)_*E$ that $K(d) \wedge E$ is a nonempty wedge of suspensions of $K(d)$. So, for $\alpha \in \pi_*R$, if $E_*\alpha = 0$ then $(K(d) \wedge E)_*\alpha = 0$ and hence $K(d)_*\alpha = 0$. So, we need to show that if $K(d)_*\alpha = 0$ for all $n$ and all $p$ then $\alpha$ is nilpotent. Taking Theorem 3.5.12 as given, it would suffice to show merely that $MU_*\alpha$ is nilpotent. This is equivalent to showing that the ring spectrum $MU \wedge R[\alpha^{-1}]$ is contractible, which is equivalent to showing that the unit map is null:

$$\mathbb{S} \to MU \wedge R[\alpha^{-1}].$$

A nontrivial result of Johnson and Wilson shows that the following conditions are equivalent:

1. $MU_*X = 0$ holds.
2. For any $d < \infty$, $\bigvee_{j=0}^{d} K(j)_*X = 0$ and $P(d+1)_*X = 0$ hold.[43]
3. There exists a $d < \infty$ such that $\bigvee_{j=0}^{d} K(j)_*X = 0$ and $P(d+1)_*X = 0$ hold.

---

[42] In particular, this is a very hard *geometric* theorem. Its proof is phrased homotopically, but it comes down to very concrete, computational facts about such geometric objects as double-loopspaces of spheres. (One unfulfilled daydream of this book is to reencode these computations in the language of formal geometry – alas.) One way of understanding the contents of the theorem is that the spectral sequences pictured in Figures 3.1 and 3.2 have *asymptotically flat* vanishing curves, so that powers of any particular element eventually escape the populated region of the spectral sequence. (In fact, this vanishing curve is conjectured to be asymptotically equivalent to a square root function, but little about this is known.) On the other hand, we can see from Figure 3.1 that such a vanishing line is *not* initially present, due to the nonnilpotence of $\eta$ – and as a consequence, there is no analog of the nilpotence theorems for $\mathrm{QCoh}(\mathcal{M}_{\mathbf{fg}})$ itself. This is another sense in which it is a fundamentally geometric fact. It is also not even true at the $E_\infty$ page of Figure 1.2, where the "image of $J$ pattern" (the name for the "shark fins" near the main diagonal) give a family of permanent cycles along a line of slope 1.

[43] It is immediate that $MU_*X = 0$ forces $P(d+1)_*X = 0$ and $v_{d'}^{-1}P(d')_*(X) = 0$ for all $d' < d$. What's nontrivial is showing that $v_{d'}^{-1}P(d')_*(X) = 0$ if and only if $K(d')_*(X) = 0$ [Rav84, theorem 2.1.a], [JW75, section 3].

150  Finite Spectra

Taking $X = R[\alpha^{-1}]$, we have assumed that $K([0,d])_* X = 0$, but we have not made a direct assumption on $P(d+1)_* X$. However, we have

$$\operatorname*{colim}_{d} P(d+1) \simeq H\mathbb{F}_p \simeq K(\infty),$$

and we have assumed $\mathbb{S} \to K(\infty) \wedge R[\alpha^{-1}]$ to be null. By compactness of $\mathbb{S}$, that null-homotopy factors through some finite stage $P(d+1) \wedge R[\alpha^{-1}]$ for $d \gg 0$. □

Corollary 3.5.13 has the following consequence, which speaks to the primacy of both the chromatic program and these results.

**Definition 3.5.14** A ring spectrum $R$ is a *field spectrum* when every $R$-module (in the homotopy category) splits as a wedge of suspensions of $R$. (Equivalently, $R$ is a field spectrum when it has Künneth isomorphisms.)

**Corollary 3.5.15** ([HS98, proposition 1.9]) *Every field spectrum $R$ splits as a wedge of Morava's theories $K(d)$.*

*Proof* It is easy to check (as mentioned in the proof of Corollary 3.5.13) that $K(d)$ is a field spectrum.

Now, consider an arbitrary field spectrum $R$. Set $E = \bigvee_{\text{primes } p} \bigvee_{d \in [0,\infty]} K(d)$, so that $E$ detects nilpotence. The class 1 in the field spectrum $R$ is non-nilpotent, so it survives when paired with some $K$-theory $K(d)$, and hence $R \wedge K(d)$ is not contractible. Because both $R$ and $K(d)$ are field spectra, the smash product of the two simultaneously decomposes into a wedge of $K(d)$s and a wedge of $R$s. So, $R$ is a retract of a wedge of $K(d)$s, and one can show [HS98, 1.10–1.12] that any such retract is itself a wedge of $K(d)$s. □

*Remark 3.5.16* In the 2-periodic setting we've become accustomed to, the analog of Corollary 3.5.15 is that every 2-periodic field spectrum splits as a wedge of suspensions of $KP(d)$.

*Remark 3.5.17* In service of Example 3.5.6, the geometric definition of $MU$ given in Lemma 1, the edge cases of $K(0) = H\mathbb{Q}$ and $K(\infty) = H\mathbb{F}_p$, and the claimed primacy of these methods, we might wonder if there is any geometric interpretation of the field theories $K(d)$ for $0 < d < \infty$. To date, this is a completely open question and the subject of intense research.

We're now well-situated to address Ravenel's question about finite spectra and periodic self-maps. The key observation is that spectra admitting such self-maps are closed under some natural operations, leading to the following definition:

**Definition 3.5.18** A full subcategory of a triangulated category (e.g., the

## 3.5 Nilpotence and Periodicity in Finite Spectra

homotopy category of $p$-local finite spectra) is *thick* if it is closed under isomorphisms and retracts, and it has a two-out-of-three property for cofiber sequences.

Examples of thick subcategories include the following:

- The category $\mathbf{C}_d$ of $p$-local finite spectra which are $K(d-1)$-acyclic.[44] These are called "finite spectra of type at least $d$."
- The category $\mathbf{D}_d$ of $p$-local finite spectra $F$ for which there is a self-map $v : \Sigma^N F \to F$, $N \gg 0$ which induces multiplication by a unit in $K(d)$-homology and which is nilpotent in $K(\neq d)$-homology. These are called "finite spectra admitting $v_d$-self-maps."[45]

The categories $\mathbf{D}_d$ are the ones we are interested in analyzing, and we hope to identify these putative spectra $V(d)$ inside of them. Ravenel shows the following foothold interrelating the $\mathbf{C}_d$:

**Lemma 3.5.19** ([Rav84, theorem 2.11])  *For $X$ a finite complex, there is a bound*

$$\dim K(d-1)_* X \leq \dim K(d)_* X.$$

*In particular, there is an inclusion $\mathbf{C}_d \supseteq \mathbf{C}_{d+1}$.*

*Proof sketch*  One should compare this with the statement that the stalk dimension of a coherent sheaf is upper semi-continuous. In fact, this analogy gives the essentials of Ravenel's proof: One considers the ring spectrum $v_d^{-1} BP/I_{d-1}$, which admits two maps

$$(v_d^{-1} BP/I_{d-1})/v_{d-1} \longleftarrow v_d^{-1} BP/I_{d-1} \longrightarrow v_{d-1}^{-1}(v_d^{-1} BP/I_{d-1}).$$

Studying the relevant Tor spectral sequences gives the result. □

Hopkins and Smith are able to use their local nilpotence detection result, Corollary 3.5.13, to completely understand the behavior not only of the thick subcategories $\mathbf{C}_d$ but of *all* thick subcategories of $\mathrm{SPECTRA}_{(p)}^{\mathrm{fin}}$. In particular, this connects the $\mathbf{C}_d$ with the $\mathbf{D}_d$, as we will see.

**Theorem 3.5.20** ([HS98, theorem 7])  *Any thick subcategory $\mathbf{C}$ of the category of $p$-local finite spectra must be $\mathbf{C}_d$ for some $d$.*

---

[44] For instance, $K(0)$-acyclicity means that the homotopy groups of the spectrum are torsion.
[45] It is not obvious that $\mathbf{D}_d$ satisfies the cofiber extension axiom of a thick subcategory. One is required to show that for any map $f: X \to Y$ of finite complexes of type $d$, $v_d$-self-maps can be chosen on both $X$ and $Y$ which commute with $f$, which is another theorem of Hopkins and Smith [HS98, theorem 11, section 3].

152  Finite Spectra

*Proof* Since $C_d$ are nested by Lemma 3.5.19 and they form an exhaustive filtration (i.e., $C_\infty = 0$), it is thus sufficient to show that any object $X \in C$ with $X \in C_d \setminus C_{d+1}$ induces an inclusion $C_d \subseteq C$. Write $R$ for the endomorphism ring spectrum $R = F(X, X)$, and write $F$ for the fiber of its unit map:

$$F \xrightarrow{f} \mathbb{S} \xrightarrow{\eta_R} R.$$

Finally, let $Y \in C_d$ be *any* finite spectrum of type at least $d$. Our goal is to demonstrate $Y \in C$.

Now consider applying $K(n)$-homology (for *arbitrary* $n$) to the map

$$1 \wedge f \colon Y \wedge F \to Y \wedge \mathbb{S}.$$

The induced map is always zero:

- In the case that $K(n)_* X$ is nonzero, then

$$K(n)_* R = K(n)_* (X \wedge DX) = K(n)_* X \otimes (K(n)_* X)^*$$

is also nonzero. It follows that the unit map is $K(n)_* \eta_R$ is injective, and so $K(n)_* f$ is zero.

- In the case that $K(n)_* X$ is zero, then $n \le d$ and, because of the bound on type, $K(n)_* Y$ is zero as well.

By a small variant of local nilpotence detection (Corollary 3.5.13 [HS98, corollary 2.5]), it follows for $j \gg 0$ that

$$Y \wedge F^{\wedge j} \xrightarrow{1 \wedge f^{\wedge j}} Y \wedge \mathbb{S}^{\wedge j}$$

is null-homotopic. Hence, one can calculate the cofiber to be

$$\operatorname{cofib}\left(Y \wedge F^{\wedge j} \xrightarrow{1 \wedge f^{\wedge j}} Y \wedge \mathbb{S}^{\wedge j}\right) \simeq Y \wedge \operatorname{cofib} f^{\wedge j} \simeq Y \vee (Y \wedge \Sigma F^{\wedge j}),$$

so that $Y$ is a retract of this cofiber.

We now work to show that this smash product lies in the thick subcategory C of interest. First, note that it suffices to show that $\operatorname{cofib} f^{\wedge j}$ on its own lies in C: A finite spectrum (such as $Y$ or $F$) can be expressed as a finite gluing diagram of spheres, and smashing this through with $\operatorname{cofib} f^{\wedge j}$ then expresses $Y \wedge \operatorname{cofib} f^{\wedge j}$ as the iterated cofiber of maps with source and target in C. With that in mind, we will in fact show that $\operatorname{cofib} f^{\wedge k}$ lies in C for all $k \ge 1$. Consider the following smash version of the octahedral axiom: The factorization

$$F \wedge F^{\wedge(k-1)} \xrightarrow{1 \wedge f^{\wedge(k-1)}} F \wedge \mathbb{S}^{\wedge(k-1)} \xrightarrow{f \wedge 1} \mathbb{S} \wedge \mathbb{S}^{\wedge(k-1)}$$

### 3.5 Nilpotence and Periodicity in Finite Spectra

begets a cofiber sequence

$$F \wedge \operatorname{cofib} f^{\wedge(k-1)} \to \operatorname{cofib} f^{\wedge k} \to \operatorname{cofib} f \wedge \mathbb{S}^{\wedge(k-1)}.$$

Noting that the base case $\operatorname{cofib}(f) = R = X \wedge DX$ lies in C, we can inductively use the two-out-of-three property on the octahedral cofiber sequence to see that $\operatorname{cofib}(f^{\wedge k})$ lies in C for all $k$. It follows in particular that $Y \wedge \operatorname{cofib}(f^{\wedge j})$ lies in C, and using the retraction $Y$ belongs to C as well. □

Although this didn't affect the previous theorem statement, one is led by its proof to wonder whether there exist such spectra $X \in C_d \setminus C_{d+1}$ – i.e., whether these nested containments are proper. The following result lays this to rest:

**Theorem 3.5.21** (Hopkins–Smith Periodicity Theorem [HS98, proposition 5.14, theorem 9])  *A $p$-local finite spectrum is $K(d-1)$-acyclic exactly when it admits a $v_d$-self-map. Additionally, the inclusion $C_d \supsetneq C_{d+1}$ is proper.*

*Executive summary of proof*  Given the classification of thick subcategories, if a property is closed under thickness then one need only exhibit a single spectrum with the property to know that all the spectra in the thick subcategory it generates also all have that property. Inductively, they manually construct finite spectra[46] $M_0(p^{i_0}, v_1^{i_1}, \ldots, v_{d-1}^{i_{d-1}})$ for sufficiently large[47] indices $i_*$ which admit a self-map $v$ governed by a commuting square

$$\begin{array}{ccc} BP_* M_{|v_d|i_d}(p^{i_0}, v_1^{i_1}, \ldots, v_{d-1}^{i_{d-1}}) & \xrightarrow{v} & BP_* M_0(p^{i_0}, v_1^{i_1}, \ldots, v_{d-1}^{i_{d-1}}) \\ \| & & \| \\ \Sigma^{|v_d|i_d} BP_*/(p^{i_0}, v_1^{i_1}, \ldots, v_{d-1}^{i_{d-1}}) & \xrightarrow{-\cdot v_d^{i_d}} & BP_*/(p^{i_0}, v_1^{i_1}, \ldots, v_{d-1}^{i_{d-1}}). \end{array}$$

These maps are guaranteed by careful study of Adams spectral sequences. □

They therefore conclude the strongest possible positive response to Ravenel's conjectures. Not only can we continue the sequence

$$\mathbb{S}, \ \mathbb{S}/p, \ \mathbb{S}/(p, v), \ \ldots,$$

but in fact *any* finite spectrum admits an (essentially unique) interesting periodic self-map. This may be the most remarkable of the statements: Although Nishida's theorem initially led us to think of periodic self-maps as rare, they are in fact ubiquitous. Additionally, we learned that the shift[48] of this self-map is determined

---

[46] They actually first construct spectra $X_n$ with $v_n$-self-maps without attempting to constrain their $BP$-homology [HS98, theorem 4.11].

[47] We ran into the asymptotic condition $I \gg 0$ earlier, when we asserted that there is no root of the 2-local $v_1$-self-map $v \colon M_8(2) \to M_0(2)$.

[48] This is sometimes referred to as the "wavelength" in the chromatic analogy.

by the first nonvanishing $K(d)$-homology, giving an effective detection tool. Finally, all such periodicity shifts arise: For any $d$, there is a spectrum admitting a $v_d$-self-map but no $v_{<d}$-self-maps.

## 3.6 Chromatic Fracture and Convergence

In this lecture, we will couple the ideas of Lecture 3.1 to the homology theories and structure theorems described in Lecture 3.5. In particular, we have not yet exhausted Theorem 3.5.20, and for inspiration about how to utilize it, we will begin with an algebraic analog of the situation considered thus far.

For a ring $R$, the full derived category $D(\operatorname{Spec} R)$ and the derived category of perfect complexes $D^{\operatorname{perf}}(\operatorname{Spec} R)$ form examples of triangulated categories analogous to SPECTRA and SPECTRA$^{\operatorname{fin}}$. By interpreting an $R$-module as a quasicoherent sheaf over $\operatorname{Spec} R$, we can use them to probe for structure of $\operatorname{Spec} R$ – for instance, we can test whether $\widetilde{M}$ is supported over some closed subscheme $\operatorname{Spec} R/I$ by restricting the sheaf, which amounts algebraically to asking whether $M$ is annihilated by $I$. In the reverse, we can also try to discern what "closed subscheme" should mean in some arbitrary triangulated category by codifying the properties of the subcategory of $D(\operatorname{Spec} R)$ supported away from $\operatorname{Spec} R$. The key observation is this subcategory is closed under tensoring modules: If $M$ is annihilated by $I$, then $M \otimes_R N$ is also annihilated by $I$.

**Definition 3.6.1** ([Bal10, definition 1.3]) Let C be a triangulated $\otimes$-category. A thick subcategory C' ⊆ C is...

- a $\otimes$-*ideal* when $x \in$ C' forces $x \otimes y \in$ C' for any $y \in$ C;
- a *prime* $\otimes$-*ideal* when $x \otimes y \in$ C' also forces at least one of $x \in$ C' or $y \in$ C'.

Finally, define the *spectrum* of C to be its collection of prime $\otimes$-ideals. For any $x \in$ C we define a basic open $U(x) = \{$C' $\mid x \in$ C'$\}$, which altogether give a basis for a topology on the spectrum.

The basic result about this definition is that it does not miss any further conditions:

**Theorem 3.6.2** ([Bal10, proposition 8.1]) *The spectrum of $D^{\operatorname{perf}}(\operatorname{Spec} R)$ is naturally homeomorphic to the Zariski spectrum of $R$.* □

Satisfied, we apply the definition to the more difficult case of SPECTRA.

**Theorem 3.6.3** ([Bal10, corollary 9.5]) *The spectrum of* SPECTRA$^{\operatorname{fin}}_{(p)}$ *consists of the thick subcategories* $C_d$, *and* $\{C_n\}_{n=0}^d$ *are its open sets.*

### 3.6 Chromatic Fracture and Convergence

*Proof* Using Theorem 3.5.21, we can characterize $C_d$ as the kernel of $K(d-1)_*$. This shows it to be a prime $\otimes$-ideal:

$$K(d-1)_*(X \wedge Y) \cong K(d-1)_* X \otimes_{K(d-1)_*} K(d-1)_* Y$$

is zero exactly when at least one of $X$ and $Y$ is $K(d-1)$-acyclic. □

**Corollary 3.6.4** (see theorems 3.3.18, 3.5.20, and 3.5.21) *The functor*

$$\mathcal{M}_{MU}(-)\colon \text{Spectra}^{\text{fin}} \to \text{Coh}(\mathcal{M}_{MU})$$

*induces*[49] *a homeomorphism of the spectrum of* $\text{Spectra}^{\text{fin}}$ *to that of* $\mathcal{M}_{\text{fg}}$. □

The construction as we have described it falls short of completely recovering Spec $R$, as we have constructed only a topological space rather than a locally ringed space (or anything otherwise equipped locally with algebraic data, as in our functor of points perspective). The approach taken by Balmer [Bal10, section 6] is to use Tannakian reconstruction to extract a structure sheaf of local rings from the prime $\otimes$-ideal subcategories. We, however, are at least as interested in finite spectra as we are the ring spectrum $\mathbb{S}$, so we will take an approach that emphasizes module categories rather than local rings. Specifically, Bousfield's theory of homological localization allows us to lift the localization structure among open substacks of $\mathcal{M}_{\text{fg}}$ to the category Spectra as follows:

**Theorem 3.6.5** (Bousfield localization [Bou79], [Mar83, theorem 7.7]) *Let* $j\colon \text{Spec } R \to \mathcal{M}_{\text{fg}}$ *be a flat map, and let* $R_*$ *denote the homology theory associated to it by Theorem 3.0.1. There is then a diagram*

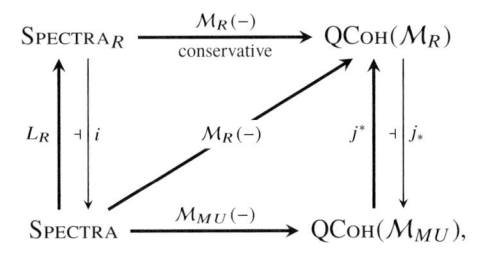

$$\begin{array}{ccc}
\text{Spectra}_R & \xrightarrow[\text{conservative}]{\mathcal{M}_R(-)} & \text{QCoh}(\mathcal{M}_R) \\
L_R \downarrow \dashv \uparrow i & \nearrow \mathcal{M}_R(-) & \uparrow j^* \dashv \downarrow j_* \\
\text{Spectra} & \xrightarrow{\mathcal{M}_{MU}(-)} & \text{QCoh}(\mathcal{M}_{MU}),
\end{array}$$

*such that* $L_R$ *is left-adjoint to* $i$, $j^*$ *is left-adjoint to* $j_*$, $i$ *and* $j_*$ *are inclusions of full subcategories,* $L_R$ *and* $j^*$ *are idempotent, the bold composites are all equal, and* $R_*$ *is conservative on* $\text{Spectra}_R$.[50] □

---

[49] This has to be interpreted delicately, as the functor $\mathcal{M}_{MU}(-)$ is not (quite) a functor of triangulated categories [Mor07a, 2.4.2].

[50] The meat of this theorem is in overcoming set-theoretic difficulties in the construction of $\text{Spectra}_R$. Bousfield accomplished this by describing a model structure on Spectra for which $R$-equivalences create the weak-equivalences.

The idea, then, is that SPECTRA$_R$ plays the topological role of the derived category of those sheaves supported on the image of the map $j$. In Definition 3.5.4, we identified several classes of interesting such maps $j$ tied to the geometry of $\mathcal{M}_{\mathrm{fg}}$. We record these special cases now:

**Definition 3.6.6** In the case $R = E_\Gamma$ of continuous Morava $E$-theory, this is a model of the inclusion of the deformation space around the point $\Gamma$, and we will denote the associated localizer by $L_\Gamma$. In the special case that $\Gamma = \Gamma_d$ is taken to be the Honda formal group, we further abbreviate the localizer by

$$\text{SPECTRA} \xrightarrow{\widehat{L}_d} \text{SPECTRA}_{\Gamma_d}.$$

In the case when $R = E(d)$ models the inclusion of the open complement of the unique closed substack of codimension $d$, we will denote the localizer by

$$\text{SPECTRA} \xrightarrow{L_d} \text{SPECTRA}_d = \text{SPECTRA}_{\mathcal{M}_{\mathrm{fg}}^{\leq d}}.$$

These localizers have a number of nice properties linking them to algebraic models.

**Lemma 3.6.7** *There are natural factorizations*

$$\mathrm{id} \to L_d \to L_{d-1}, \qquad\qquad \mathrm{id} \to L_d \to \widehat{L}_d.$$

*In particular, $L_d X = 0$ implies both $L_{d-1} X = 0$ and $\widehat{L}_d X = 0$.*

*Analogy to $j_* \vdash j^*$* The open substack of dimension $d$ properly contains both the open substack of dimension $(d - 1)$ and the infinitesimal deformation neighborhood of the geometric point of height $d$. The factorization of inclusions gives a factorization of pullback functors. □

**Lemma 3.6.8** ([Rav92, theorem 7.5.6], [Hov95, proof of lemma 2.3]) *There are equivalences*

$$L_d X \simeq (L_d \mathbb{S}) \wedge X, \qquad\qquad \widehat{L}_d X \simeq \varprojlim_I \left( M_0(v^I) \wedge L_d X \right).$$

*Analogy to $j_* \vdash j^*$* The first formula stems from $j$ an open inclusion, which has $j^* M \simeq R \otimes M$ in the algebraic setting. The second formula can be compared to the inclusion $j$ of the formal infinitesimal neighborhood of a closed subscheme, which has algebraic model $j^* M = \lim_j (R/I^j \otimes M)$.[51] □

---

[51] In keeping with our discussion of continuous Morava $E$-theory, it is also possible to consider the object $\{(M_0(v^I) \wedge L_d X)\}_I$ itself as a pro-spectrum. This is interesting to explore: Davis and Lawson have shown that setting $X = \mathbb{S}$ gives an $E_\infty$-ring pro-spectrum [DL14], even though none of the individual objects are $E_\infty$-ring spectra themselves [MNN15].

## 3.6 Chromatic Fracture and Convergence

**Lemma 3.6.9** *Let $k$ be a field of positive characteristic $p$, and let $\Gamma$ and $\Gamma'$ be two formal groups over $k$ of differing heights $0 \leq d, d', \leq \infty$. Then $K_\Gamma \wedge K_{\Gamma'} \simeq 0$.*

*Analogy to $j_* \vdash j^*$* The map classifying the formal group $\mathbb{CP}^\infty_{K_\Gamma \wedge K_{\Gamma'}}$ simultaneously factors through the maps classifying the formal groups $\mathbb{CP}^\infty_{K_\Gamma} = \Gamma$ and $\mathbb{CP}^\infty_{K_{\Gamma'}} = \Gamma'$. By Lemma 3.3.4, such a formal group must simultaneously have heights $d$ and $d'$, which forces the homotopy ring to be the zero ring.[52] □

**Lemma 3.6.10** ([Lura, lemma 23.6]) *For $d > \operatorname{ht}\Gamma$, $\widehat{L}_\Gamma L_d \simeq 0$.*

*Proof sketch* After a nontrivial reduction argument, this comes down to an identical fact: The formal group associated to $E(d) \wedge K_\Gamma$ must simultaneously be of heights at most $d$ and exactly $\operatorname{ht}\Gamma > d$, which forces the spectrum to vanish. □

**Corollary 3.6.11** $L_\Gamma E = 0$ *for any coconnective $E$, and hence*

$$L_\Gamma E = L_\Gamma(E[n, \infty))$$

*for any spectrum $E$ and any index $n$.*[53]

*Proof* Any coconnective spectrum can be expressed as the colimit of its truncations

$$\begin{array}{ccccccc} E[n,n] & \longrightarrow & E[n-1,n] & \longrightarrow & E[n-2,n] & \longrightarrow & \cdots \xrightarrow{\operatorname{colim}} E(-\infty, n] \\ \parallel & & \downarrow & & \downarrow & & \\ \Sigma^n H\pi_n E & & \Sigma^{n-1} H\pi_{n-1} E & & \Sigma^{n-2} H\pi_{n-2} E & & \cdots. \end{array}$$

Applying $L_\Gamma$ preserves this colimit diagram, but the above argument shows that $HA$ is $L_\Gamma$-acyclic for any abelian group $A$. This gives the statement about coconnective spectra, from which the general statement follows by considering the cofiber sequence

$$E[n, \infty) \to E \to E(-\infty, n).$$ □

**Corollary 3.6.12** (Chromatic fracture [Lura, proposition 23.5]) *There are homotopy pullback squares*

---

[52] Alternatively, Corollary 3.5.15 shows that $K_\Gamma \wedge K_{\Gamma'}$ simultaneously decomposes as a wedge of $K_\Gamma$s and of $K_{\Gamma'}$, which forces both wedges to be empty.

[53] A memorable slogan is that Morava $K$-theories are sensitive only to the "germ at $\infty$" [Mit05, section 3.3.3].

$$\begin{array}{ccc} L_d X & \longrightarrow & \widehat{L_d} X \\ \downarrow \lrcorner & & \downarrow \\ L_{d-1} X & \longrightarrow & L_{d-1} \widehat{L_d} X, \end{array} \qquad \begin{array}{ccc} X & \longrightarrow & \prod_p X_p^\wedge \\ \downarrow \lrcorner & & \downarrow \\ X_\mathbb{Q} & \longrightarrow & \left(\prod_p X_p^\wedge\right)_\mathbb{Q}. \end{array}$$

*Analogy to* $j_* \vdash j^*$  For the left-hand square, the inclusion of the open substack of dimension $d - 1$ into the one of dimension $d$ has as its relatively closed complement the point of height $d$. Algebraically, this gives a Mayer–Vietoris sequence with analogous terms. The right-hand square is analogous to the adèlic decomposition of abelian groups.[54] □

*Remark 3.6.13*  Corollary 3.6.12 is maybe the most useful result discussed in this lecture. It shows that a map to an $L_d$-local spectrum can be understood as a system of compatible maps to its $\widehat{L}_j$-localizations, $j \leq d$. In turn, any map into an $\widehat{L}_j$-local object factors through the $\widehat{L}_j$-localization of the source. Thus, if the source itself has chromatic properties, this often puts *very* strong restrictions on how maps can behave.

These functors and their properties listed thus far give a tight analogy between certain local categories of spectra and sheaves supported on particular submoduli of formal groups, in a way that lifts the six-functors formalism of $j_* \vdash j^*$ to the level of spectra. With this analogy in hand, however, one is led to ask considerably more complicated questions whose proofs are not at all straightforward. For instance, a useful fact about *coherent* sheaves on $\mathcal{M}_{\mathbf{fg}}$ is that they are completely determined by their restrictions to all of the open submoduli. The analogous fact about finite spectra is referred to as *chromatic convergence*:

**Theorem 3.6.14** (Chromatic convergence [Rav92, theorem 7.5.7])  *The homotopy limit of the tower*

$$\cdots \to L_d F \to L_{d-1} F \to \cdots \to L_1 F \to L_0 F$$

*recovers the p-local homotopy type of any finite spectrum* $F$.[55,56,57] □

---

[54] Whenever $L_B L_A = 0$, $L_{A \vee B}$ is the pullback of $L_A \to L_A L_B \leftarrow L_B$. Hence, this follows from Lemma 3.6.10, as well as the identifications $L_{E(d)} \simeq L_{E(d-1) \vee K(d)} \simeq L_{K(0) \vee \cdots \vee K(d)}$.

[55] Spectra satisfying this limit property are said to be *chromatically complete*, which is closely related to being *harmonic*, i.e., being local with respect to $\bigvee_{d=0}^\infty K(d)$. It is known that nice Thom spectra are harmonic [Křî94] (so, in particular, every suspension and finite spectrum), that every finite spectrum is chromatically complete, and that there exist some harmonic spectra which are not chromatically complete [Bar, section 5.1].

[56] A consequence of the fact that $p$-local finite spectra are $\bigvee_{d<\infty} K(d)$-local, there are no nontrivial maps $H\mathbb{F}_p \to F$, and in particular the Spanier–Whitehead dual of $H\mathbb{F}_p$ is null.

[57] While we're talking about $\bigvee_{d<\infty} K(d)$, the cofiber of $\bigvee_{d<\infty} K(d) \to \prod_{d<\infty} K(d)$, where we are using the $2(p^d - 1)$-periodic spectra, is concentrated in degree zero since all other

### 3.6 Chromatic Fracture and Convergence

In addition to furthering the analogy, Theorem 3.6.14 suggests a method for analyzing the homotopy groups of spheres: We could study the homotopy groups of each $L_d\mathbb{S}$ and perform the reassembly process encoded by this inverse limit. Additionally, Corollary 3.6.12 shows that this process is inductive: $L_d\mathbb{S}$ can be understood in terms of the spectrum $L_{d-1}\mathbb{S}$, the spectrum $\widehat{L}_d\mathbb{S}$, and some gluing data in the form of $L_{d-1}\widehat{L}_d\mathbb{S}$. Hence, we become interested in the homotopy of $\widehat{L}_d\mathbb{S}$, which is the target of the $E_d$-Adams spectral sequence considered in Lecture 3.1.

**Theorem 3.6.15** (Lemma 3.1.15 and Definition 3.5.4; see also Example 2.3.4, Definition 3.1.9, and Definition 3.1.13)  *The continuous $E_d$-based[58,59] Adams spectral sequence for the sphere converges strongly to $\pi_*\widehat{L}_d\mathbb{S}$. Writing $\omega$ for the line bundle on $(\mathcal{M}_{\mathbf{fg}})^\wedge_{\Gamma_d}$ of invariant differentials, we have*

$$E_2^{*,*} = H^*((\mathcal{M}_{\mathbf{fg}})^\wedge_{\Gamma_d}; \omega^{\otimes *}) \Rightarrow \pi_*\widehat{L}_d\mathbb{S}. \qquad \square$$

Our algebraic analysis from Theorem 3.4.5 and Remark 3.4.7 shows a further identification

$$\mathcal{M}_{E_d} = (\mathcal{M}_{\mathbf{fg}})^\wedge_{\Gamma_d} \simeq \widehat{\mathbb{A}}^{d-1}_{\mathbb{W}(k)} /\!\!/ \underline{\mathrm{Aut}}(\Gamma_d).$$

This computation is thus boiled down to a calculation of the cohomology of the $\mathrm{Aut}(\Gamma_d)$-representations arising via Remark 3.4.13 as the global sections of the sheaves $\omega^{\otimes *}$ (see the discussion in Example 1.4.10 and Example 1.4.17).[60] We will later deduce the following polite description of $\mathrm{Aut}\,\Gamma_d$:

---

homotopy degrees carry contributions from only finitely many factors. It follows that the cofiber is Eilenberg–Mac Lane – an unusual property.

[58] Although the $KP(d)$-Adams spectral sequence more obviously targets $\widehat{L}_d\mathbb{S}$, we have chosen to analyze the $E_d$-Adams spectral sequence above because $KP(d)$ fails to satisfy **CH**. Starting with $BPP_0BPP \cong BPP_0[t_0^\pm, t_1, t_2, \ldots]$ from Definition 3.5.4 and Corollary 3.3.17, we can calculate $EP(d)_0EP(d)$ by base-changing this Hopf algebroid: $EP(d)_0EP(d) = EP(d)_0 \otimes_{BPP_0} BPP_0BPP \otimes_{BPP_0} EP(d)_0$, which is again free over $EP(d)_0$. Since $KP(d)$ is formed from $EP(d)$ by quotienting by a regular sequence, we calculate that $KP(d)_0EP(d)$ is free over $KP(d)_0$, generated by the same summands. However, when quotienting by the regular sequence *again* to form $KP(d)_*KP(d)$, the maps in the quotient sequences act by elements in $I_d = 0$, hence introduce Bocksteins. The end result is

$$KP(d)_*KP(d) = (KP(d)_* \otimes_{BPP_*} BPP_*BPP \otimes_{BPP_*} KP(d)_*) \otimes \Lambda[\tau_0, \ldots, \tau_{d-1}],$$

where $\tau_j$ in degree 1 controls the cofiber of $E_d \xrightarrow{u_j} E_d$. In this sense, $E_d$ forms a kind of maximal deformation of $KP(d)$ as a ring spectrum, since it exhausts the collection of $KP(d)$ Bocksteins. Additionally, this behavior of odd-degree classes is quite generic: Whenever odd-primary information appears near the formal geometric picture, it seems to come from a poorly behaved homotopical quotient – some element was killed twice, say, or in some specific case a map was zero but is generically nonzero. So it is here.

[59] Baas and Madsen have calculated $H\mathbb{F}_p^*(K(n)[0, \infty)) \cong \mathcal{A}^* /\!\!/ \Lambda[Q_n]$, which also fits with this Bockstein philosophy (see also Example 1.4.20) [BM73], [HS98, proposition 4.9].

[60] A description of the *multiplicative* stable operations of $E_d$ can be found in [Str00, proposition 4], as well as most of a description of the collection of *all* stable operations as a twisted group-ring.

**Theorem 3.6.16** (see Example 4.4.13)  *For $\Gamma_d$ the Honda formal group law of height d over a perfect field k of positive characteristic p, we compute*

$$\operatorname{Aut}\Gamma_d \cong \left(\mathbb{W}_p(k)\langle S\rangle \middle/ \begin{pmatrix} Sw = w^\varphi S, \\ S^d = p \end{pmatrix}\right)^\times,$$

*where $\varphi$ denotes a lift of the Frobenius from k to $\mathbb{W}_p(k)$.*  □

*Remark 3.6.17* ([Strb, section 24], [DH95])  As a matter of emphasis, this theorem does not give any description of the $\operatorname{Aut}\Gamma_d$-*representation* structure of $\pi_* E_d$, which is very complicated (see Remark 4.4.20). Here are some basic facts about it:

- The center of $\operatorname{Aut}\Gamma_d$ is given by $\mathbb{Z}_p^\times$, which is the subgroup consisting of the multiplication-by-$n$ maps. This subgroup acts trivially on the deformation space.
- More generally, one can show that the action of $\mathbb{W}_p(k)$ extends this action. It is the stabilizer of the *canonical lift*, which is the $\mathbb{W}_p(k)$-point of Lubin–Tate space given by sending the generators $v_j$ to zero, $j \geq 1$.
- The formula for the action of $a = \sum_{j=0}^{n-1} a_j S^j$ on the tangent space is specified by the linear system

$$a^* \begin{pmatrix} u_0 \\ u_1 \\ \vdots \\ u_{d-1} \end{pmatrix} \equiv a_0 \begin{pmatrix} a_0 & 0 & \cdots & 0 \\ a_1 & a_0^p & \cdots & 0 \\ \vdots & \vdots & \ddots & \vdots \\ a_{d-1} & a_{d-2}^p & \cdots & a_0^{p^d} \end{pmatrix}^{-1} \cdot \begin{pmatrix} u_0 \\ u_1 \\ \vdots \\ u_{d-1} \end{pmatrix} \pmod{\mathfrak{m}^2}.$$

- Plenty more information can be found in the work of Devinatz and Hopkins [DH95], which is best digested after Lecture 4.4.

*Remark 3.6.18*  The arithmetically minded reader might recognize this description of $\operatorname{Aut}\Gamma_d$ as the group of units of a maximal order $\mathfrak{o}_D$ in the division algebra $D$ of Brauer–Hasse invariant $1/d$ over $k$ – another glimpse of arithmetic geometry poking through to affect stable homotopy theory.[61]

*Example 3.6.19* (Adams [HMS94, lemma 2.5])  In the case $d = 1$, the objects involved are small enough that we can compute them by hand. To begin, we have

---

[61] This finally explains our preference for using the letter "$d$" to represent the height of a formal group – the "$d$" (or, rather the "$D$") stands for "division algebra." The typical algebraic topologist writes "$n$" for the height, a trend set by Jack Morava, allegedly because "$E_n$" is a contraction of his wife's name, Ellen, and the localizer "$L_n$" a homophone. With no disrespect meant to either of them, I find "$d$" to be a considerably better mnemonic and to be less likely to conflict with other indices. "$E$" is also prone to collisions (with spectral sequences, with operads, ... ), but there is no compelling alternative.

## 3.6 Chromatic Fracture and Convergence

an isomorphism $\mathrm{Aut}(\Gamma_1) = \mathbb{Z}_p^\times$, and the action of this group on $\pi_* E_1 = \mathbb{Z}_p[u^\pm]$ is by $\gamma \cdot u^n \mapsto \gamma^n u^n$. At odd primes $p$, one computes[62]

$$H^s(\mathrm{Aut}(\Gamma_1); \pi_* E_1) = \begin{cases} \mathbb{Z}_p & \text{when } s = 0, \\ \bigoplus_{j=2(p-1)k} \mathbb{Z}_p\{u^j\}/(pku^j) & \text{when } s = 1, \\ 0 & \text{otherwise.} \end{cases}$$

This, in turn, gives the calculation[63]

$$\pi_t \widehat{L}_1 \mathbb{S}^0 = \begin{cases} \mathbb{Z}_p & \text{when } t = 0, \\ \mathbb{Z}_p/(pk) & \text{when } t = k|v_1| - 1, \\ 0 & \text{otherwise.} \end{cases}$$

With this in hand, we can compute the homotopy of the rest of the fracture square:

$$\begin{array}{ccc} \pi_* L_1 \mathbb{S} & \longrightarrow & \mathbb{Z}_p \oplus \bigoplus_{t=k|v_1|-1} \Sigma^t \mathbb{Z}_p/(pk) \\ \downarrow & & \downarrow \\ \mathbb{Q} & \longrightarrow & \mathbb{Q}_p \oplus \Sigma^{-1} \mathbb{Q}_p, \end{array}$$

from which we deduce

$$\pi_t L_1 \mathbb{S}^0 = \begin{cases} \mathbb{Z}_{(p)} & \text{when } t = 0, \\ \mathbb{Z}_p/(pk) & \text{when } t = k|v_1| - 1 \text{ and } t \neq 0, \\ \mathbb{Z}/p^\infty & \text{when } t = (0 \cdot |v_1| - 1) - 1 = -2, \\ 0 & \text{otherwise.} \end{cases}$$

*Example 3.6.20* ([Rezc, example 7.18]) We can also make an explicit chromatic analysis of the homotopy element $\eta \in \pi_1 \mathbb{S}$ studied in Lecture 1.4. As before, consider the complex $\mathbb{CP}^2 = \Sigma^2 C(\eta)$, and suppose that $\mathbb{CP}^2$ splits as $\mathbb{S}^2 \vee \mathbb{S}^4$. In this case there would be a dotted retraction in the cofiber sequence

$$\mathbb{S}^2 \xrightarrow{i} \mathbb{CP}^2 \longrightarrow \mathbb{S}^4.$$

If this were possible, we would also be able to detect the retraction after chromatic localization – so, for instance, we could consider the cohomology

---

[62] At odd primes, $p$ is coprime to the order of the torsion part of $\mathbb{Z}_p^\times$. At $p = 2$, this is not true, so the representation has infinite cohomological dimension and there is plenty of room for differentials in the ensuing $E_{\widehat{\mathbb{G}}_m}$-Adams spectral sequence.

[63] The groups $\pi_* \widehat{L}_1 \mathbb{S}$ are familiar to homotopy theorists: The Adams conjecture [Ada66] (and its solution) implies that the $J$-homomorphism $J_\mathbb{C} \colon BU \to BGL_1 \mathbb{S}$ described in Corollary 1.1.6 and Theorem 2.0.1 selects exactly these elements for nonnegative $t$.

theory $E_{\widehat{\mathbb{G}}_m} = KU_p^\wedge$ from Example 3.5.6 and test this hypothesis in $\widehat{\mathbb{G}}_m$-local homotopy. Writing $t$ for a coordinate on $\mathbb{CP}^\infty_{KU_p^\wedge}$, this cofiber sequence gives a short exact sequence on $KU_p^\wedge$-cohomology:

$$0 \longleftarrow (t)/(t)^2 \xleftarrow{i^*} (t)/(t)^3 \longleftarrow (t)^2/(t)^3 \longleftarrow 0.$$

Because $i$ is taken to be a retraction, the map $i^*$ would satisfy $i^*(t) = t$ (mod $t^2$), so that $i^*(t) = t + at^2$ for some $a$. Additionally, $i^*$ would be natural with respect to all cohomology operations on $KU_p^\wedge$. In particular, the element $(-1) \in \mathbb{Z}_p^\times \cong \operatorname{Aut}\widehat{\mathbb{G}}_m$ gives rise to an operation $\psi^{-1}$, which acts by the $(-1)$-series on the coordinate $t$. In the case that $t$ is the coordinate considered in Example 2.1.21, this gives

$$[-1](t) = -\sum_{j=1}^{\infty} t^j = -t - t^2 \pmod{t^3}.$$

We thus compute:

$$\psi^{-1} \circ i(t) = i \circ \psi^{-1}(t)$$
$$\psi^{-1}(t + at^2) = i(-t)$$
$$(-t - t^2) + a(-t - t^2)^2 = -(t + at^2)$$
$$-t + (a - 1)t^2 = -t - at^2,$$

so that we would arrive at a contradiction if the equation $2a = 1$ were insoluble. Note that this has no solution in $\mathbb{Z}_2$, so that the attaching map $\eta$ in $\mathbb{CP}^2$ is nontrivial in $\widehat{\mathbb{G}}_m$-local homotopy at the prime 2 (hence also in the global homotopy group $\pi_1 \mathbb{S}$). For $p$ odd, this equation *does* have a solution in $\mathbb{Z}_p$, and it furthermore turns out that $\eta = 0$ at odd primes. This problem also disappears if we require $i(t) = 2t + at^2$ instead, so that the above argument does not obstruct the triviality of $2\eta$ (and, indeed, Figure 1.2 shows that the relation $2\eta = 0$ holds in 2-adic homotopy).

*Example 3.6.21* ([Rezc, example 7.17 and corollary 5.12]) Take $k$ to be a perfect field of positive characteristic $p$, and take $\Gamma$ over $\operatorname{Spec} k$ to be a finite height formal group with associated Morava $E$-theory $E_\Gamma$. By smashing the unit map $\mathbb{S} \to E_\Gamma$ with the mod-$p$ Moore spectrum, we get an induced map of homotopy groups

$$h_{2n} \colon \pi_{2n} M_0(p) \to \pi_{2n} E_\Gamma/p.$$

We concluded as a consequence of Corollary 3.3.17 that there is an invariant section $v_1$ of $\omega^{\otimes(p-1)}$ on $\mathcal{M}_{\mathbf{fg}}^{\geq 1}$, and hence a preferred element of $\pi_{2(p-1)} E_\Gamma/p$

### 3.6 Chromatic Fracture and Convergence

which is natural in the choice of $\Gamma$. One might hope that these elements are the image of an element in $\pi_{2(p-1)} M_0(p)$ under the Hurewicz map $h$, and this turns out to be true: There is such an element, called $\alpha_{1/1}$. This element furthermore turns out to be $p$-torsion, meaning it extends to a map

$$\begin{array}{ccccc} \mathbb{S}^{2(p-1)} & \xrightarrow{p} & \mathbb{S}^{2(p-1)} & \xrightarrow{\text{cofib}} & M_{2(p-1)}(p) \\ & \searrow_0 & \downarrow_{\alpha_{1/1}} & \swarrow_{\nu} & \\ & & M_0(p). & & \end{array}$$

At odd primes, this turns out to be the $v_1$-self-map $v \colon M_{2(p-1)}(p) \to M_0(p)$ announced in Theorem 3.5.10 (see also [Ada66, proposition 12.7]).

This admits many generalizations. First, different powers $v_1^j$ of the section $v_1$ also give rise to homotopy elements $\alpha_{j/1} \in \pi_{2(p-1)j} M_0(p)$ (though their Hurewicz images no longer uniquely specify them). These have varying orders of divisibility, and we write $\alpha_{j/k}$ for a homotopy element satisfying $p^{k-1} \alpha_{j/k} = \alpha_{j/1}$.[64] Additionally, each Moore spectrum supports an Adams $v_1$-self-map of the form

$$v(p^j) \colon M_{|v_1| p^{j-1}}(p^j) \to M_0(p^j),$$

which collectively form commutative diagrams of the shape

$$\begin{array}{ccc} M_{p \cdot |v_1| p^{j-1}}(p^j) & \xrightarrow{v(p^j)^p} & M_0(p^j) \\ \downarrow & & \downarrow \\ M_{|v_1| p^j}(p^{j+1}) & \xrightarrow{v(p^{j+1})} & M_0(p^{j+1}). \end{array}$$

The other invariant functions described in Corollary 3.3.17 (e.g., $v_d$ modulo $I_d$) also give rise to elements in $H^*(M_{\text{fg}}^{\geq d}; \omega^{\otimes *})$, which map to the $BP$-Adams $E_2$-term and which sometimes survive the spectral sequence to give rise to homotopy elements of the generalized Moore spectra $M_0(v^I)$. Homotopy elements arising in this way are referred to as *Greek letter families* [MRW77, section 3].

*Remark 3.6.22* In the broader literature, the phrase "Greek letter elements" typically refers to the pushforward of the above elements to the homotopy groups of $\mathbb{S}$ by pinching to the top cell. This is somewhat obscuring: For instance, this significantly entangles how multiplication by $\alpha_{j/k}$ behaves.

*Remark 3.6.23* ([AA66], [Rav86, section 5.2]) The second solution of the Hopf invariant one problem, due to Adams and Atiyah, can be rephrased in this language. As discussed in Example 1.4.21, the "Hopf invariant one problem" refers to the question of whether the 1-cocycles $h_j$ in the $H\mathbb{F}_2$-Adams spectral

---

[64] The incarnation of these elements in $\widehat{\mathbb{G}}_m$-local homotopy are exactly the elements witnessed by the invariant function $u^{2(p-1)k}$ in Example 3.6.19.

164              *Finite Spectra*

sequence are permanent cycles, thereby producing stable homotopy classes. The orientation $BP \to H\mathbb{F}_2$ induces a map of Adams spectral sequences, so that if $h_j$ were to survive to detect a class, that class would also be present in the $BP$-Adams spectral sequence, and its filtration degree would be not more than that of $h_j$, which lies in degree 1. In turn, the $BP$-Adams $E_2$-term can be computed by an adèlic complex [Bei80, Hub91], which is itself the target of a Čech-type spectral sequence known internally to chromatic homotopy theory as the *chromatic spectral sequence* [MRW77, section 3], [Wil82, equation 4.60]:

$$E^2_{*,*,d} = H^*(\mathcal{M}_{\text{fg}}; v_d^{-1} BP_*/(p^\infty, v_1^\infty, \ldots, v_{d-1}^\infty)\widetilde{\phantom{)}}) \Rightarrow H^*(\mathcal{M}_{\text{fg}} \times \operatorname{Spec} \mathbb{Z}_{(p)}).$$

The 1-line of this spectral sequence is entirely governed by Example 3.6.19, and by finishing the even-primary form of that calculation, one discovers that the 2-divisibilities of the elements $\alpha_{j>4}$ are too low to produce *any* classes on the classical Adams 1-line (see Figure 3.1). It follows that the elements $h_{j>3}$ cannot be permanent cycles, lest they contribute to a contradiction.[65]

*Remark 3.6.24*  This method is nonconstructive, in the sense that we learn that the classes $h_j$ must participate in $H\mathbb{F}_2$-Adams differentials, but we do not know what those differentials are. By wholly different means [BMMS86, corollary VI.1.5], one can show for all $j$ the differential

$$d_2 h_j = h_0 h_{j-1}^2,$$

and the accident is that the right-hand side of the equation happens to be zero for $j \leq 3$. However, because we are working in characteristic 2, the existence of these differentials gives no information about the survival of the classes $h_j^2$, sometimes referred to as the "(Arf–)Kervaire invariant one problem."[66] Its recent solution by Hill, Hopkins, and Ravenel [HHR16] follows in the footsteps of previous work by Ravenel [Rav78a], and one of the crucial leaps forward is the serious employ of the theory of abelian varieties [Mor11].[67]

---

[65] As an aside, the Greek letter elements are completely well-defined on the pages of this spectral sequence. The indeterminacy in viewing them as elements of homotopy groups comes entirely from the filtration effects of this spectral sequence and the ensuing $BP$-Adams spectral sequence.

[66] Remarkably, just as the Hopf invariant one problem finds geometric application in the question of $H$-space structures on spheres, the Kervaire invariant one problem predicts the existence of certain nonsmoothable manifolds [Bro69].

[67] $\mathcal{M}_{\text{fg}}$ and the $MU$-Adams spectral sequence are not enough [MRW77, section 8.F].

# Case Study 4

## Unstable Cooperations

In Lecture 3.1 (and more broadly in Case Study 3), we codified the structure of the stable $E$-cooperations acting on the $E$-homology of a spectrum $X$, attached to it the $E$-Adams spectral sequence which approximates the stable homotopy groups $\pi_* X$, and gave algebro-geometric descriptions of the stable cooperations for some typical spectra: $H\mathbb{F}_2$, $MO$, and $MU$. We will now pursue a variation on this theme, where we consider the $E$-homology of a *space* rather than of a generic spectrum. In this case study, we will examine the theory of cooperations that arises from this set-up, called the *unstable $E$-cooperations*. This broader collection of cooperations has considerably more intricate structure than their stable counterparts, requiring the introduction of a new notion of an unstable context. With that established, we will again find that $E$-homology takes values in quasicoherent Cartesian sheaves over the unstable context, and we will again assemble an *unstable $E$-Adams spectral sequence approximating the *unstable* homotopy groups of the input space, whose $E^2$-page in favorable situations is tracked by the cohomology of the sheaf over the unstable context.

Remarkably, these unstable contexts also admit algebro-geometric interpretations. In finding the right language for this, we introduce different subclasses of cooperations, and we are also naturally led to consider *mixed cooperations* (as we did stably in Lemma 2.6.7) of the form $F_* \underline{E}_*$. The running theme is that when $E$ and $F$ are complex-orientable, there is a natural approximation map

$$\operatorname{Spec} Q^* F_* \underline{E}_* \to \underline{\operatorname{FormalGroups}}(\mathbb{CP}^\infty_F, \mathbb{CP}^\infty_E)$$

which is an isomorphism in every situation of interest. However, these isomorphisms do not appear to admit uniform proofs,[1] so we instead investigate the following cases by hand:

**Lecture 4.1** For $F = E = H\mathbb{F}_2$, we compute the full unstable dual Steenrod

---

[1] The best uniform result I can find is due to Butowiez and Turner [BT00, theorem 3.12].

algebra $H\mathbb{F}_{2*}\underline{H\mathbb{F}_{2*}}$ by means of iterated bar spectral sequences. We then pass to the additive unstable cooperations, where we show by hand that this presents the endomorphism scheme $\underline{\text{FormalGroups}}(\widehat{\mathbb{G}}_a, \widehat{\mathbb{G}}_a)$. Finally, we pass to the stable additive cooperations, and we check that our results here are compatible with the isomorphism

$$\text{Spec } H\mathbb{F}_{2*}H\mathbb{F}_2 \cong \underline{\text{Aut }\widehat{\mathbb{G}}_a}$$

presented in Lemma 1.3.5.

**Lecture 4.3** We next consider the case where $E = MU$ and where $F$ is any complex-orientable theory. We begin with the case $F = H\mathbb{F}_p$, where we can again approach the problem using iterated bar spectral sequences. The resulting computation is sufficiently nice that we can use this special case of $F = H\mathbb{F}_p$ to deduce the further case of $F = H\mathbb{Z}_{(p)}$, then $F = H\mathbb{Z}$, then $F = MU$, and then finally $F$ any complex-orientable theory.

**Lecture 4.5** Having been able to vary $F$ as widely as possible in the previous case, we then turn to trying to vary $E$. This is considerably harder, since the infinite loopspaces $\underline{E}_*$ associated to $E$ are extremely complicated and vary wildly under even "small" changes in $E$. However, in the special case of $F = H\mathbb{F}_p$, we have an incredibly powerful trick available to us: Dieudonné theory, discussed in Lecture 4.4, gives a logarithmic equivalence of categories

$$D_* : \text{GradedHopfAlgs}_{\mathbb{F}_p/}^{\geq 0, \text{fin}} \to \text{GradedDMods},$$

and applying this in the composite

$$\text{Spectra} \xrightarrow{\Omega^\infty} \text{Loopspaces}$$
$$\xrightarrow{H\mathbb{F}_{p*}} \text{GradedHopfAlgs}_{\mathbb{F}_p/}^{\geq 0, \text{fin}}$$
$$\xrightarrow{D_*} \text{GradedDMods}$$
$$\subseteq \text{GradedModules}_{\text{Cart}}$$

gives a homological functor. This means that the Dieudonné module associated to an infinite loopspace varies stably with the spectrum underlying the loopspace, which is enough leverage to settle the case where $E$ is any Landweber flat theory.

**Lecture 4.6** Finally, we settle one further case not covered by any of our generic hypotheses: $F = K_\Gamma$ and $E = H\mathbb{Z}/p^j$. Neither $K_\Gamma$ nor $H\mathbb{Z}/p^j$ is Landweber flat, but because $K_\Gamma$ is a field spectrum and because the additive group law associated to $H\mathbb{Z}/p^j$ is so simple, we can still perform the requisite iterated bar spectral sequence calculation by hand.

This last case is actually our real goal, as we are about to return to the project outlined in the Introduction. In the language of Theorem 5 from the Introduction, choosing $\Gamma$ to be the formal completion of an elliptic curve at the identity section presents the spectra $K_\Gamma$ and $E_\Gamma$ of Lecture 3.5 as the most basic examples of *elliptic spectra*. The goal of that theorem is to study $E_*BU[6, \infty)$ for $E$ an elliptic spectrum, so when proving it in Case Study 5 we will be led to consider the fiber sequences

$$BSU \to BU \to \underline{H\mathbb{Z}}_2, \qquad \underline{H\mathbb{Z}}_3 \to BU[6, \infty) \to BSU,$$

which mediate the difference between $E_*BU$ and $E_*BU[6, \infty)$ by means of $E_*\underline{H\mathbb{Z}}_2$ and $E_*\underline{H\mathbb{Z}}_3$. Thus, in our pursuit of $K_{\Gamma*}BU[6, \infty)$, we will want to have $K_{\Gamma*}\underline{H\mathbb{Z}}_*$ already in hand, as well as an algebro-geometric interpretation of it.

## 4.1 Unstable Contexts and the Steenrod Algebra

In this lecture, our goal is to codify the study of unstable cooperations, beginning with an unstructured account of how they arise. Recall that for a ring map $f\colon R \to S$, in Lecture 3.1 we studied the problem of recovering an $R$-module from an $S$-module equipped with extra data. The meaning of "extra data" that we settled on was that of *descent data*, which we phrased most enduringly as a certain cosimplicial diagram. Stripping away the commutative algebra, the only categorical formality that went into this was the adjunction

$$\text{Modules}_R \xrightleftharpoons[\text{forget}]{-\otimes_R S} \text{Modules}_S,$$

or later on, when given a ring spectrum $\eta\colon \mathbb{S} \to E$, the adjunction

$$\text{Spectra} = \text{Modules}_\mathbb{S} \xrightleftharpoons[\text{forget}]{-\wedge E} \text{Modules}_E.$$

The identification of $\text{Modules}_S$ with $T$-algebras in $\text{Modules}_R$ for the monad $T = \text{forget} \circ (-\otimes_R S)$ is the objective of *monadic descent* [Lurc, theorem 4.7.4.5]. This categorical recasting is ignorant of some of the algebraic geometry we discovered next, but it is suitable for us now as we consider the composition with a second adjunction:

$$\text{Spaces} \xrightleftharpoons[\Omega^\infty]{\Sigma^\infty} \text{Spectra} = \text{Modules}_\mathbb{S} \xrightleftharpoons[\text{forget}]{-\wedge E} \text{Modules}_E.$$

We will write $E(-)$ for the induced monad on SPACES, given by the formula
$$E(X) = \Omega^\infty(E \wedge \Sigma^\infty X) = \operatorname*{colim}_{j\to\infty} \Omega^j(\underline{E}_j \wedge X),$$
where $\underline{E}_*$ are the constituent spaces in the $\Omega$-spectrum of $E$. This space has the property $\pi_* E(X) = \widetilde{E}_{*\geq 0} X$. The monadic structure comes from the two evident natural transformations:
$$\eta\colon X \to \Omega^\infty \Sigma^\infty X \simeq \Omega^\infty(\mathbb{S} \wedge \Sigma^\infty X) \to \Omega^\infty(E \wedge \Sigma^\infty X) = E(X),$$
$$\mu\colon E(E(X)) = \Omega^\infty(E \wedge \Sigma^\infty \Omega^\infty(E \wedge \Sigma^\infty X))$$
$$\to \Omega^\infty(E \wedge E \wedge \Sigma^\infty X) \to \Omega^\infty(E \wedge \Sigma^\infty X) = E(X).$$

Just as in the stable situation, we can extract from this a cosimplicial object:

**Definition 4.1.1** The *unstable descent object* is the cosimplicial space

$$\mathcal{UD}_E(X) := \left\{ \begin{array}{c} \\ E \\ \circ \\ X \end{array} \begin{array}{c} \xrightarrow{\eta_L} \\ \xleftarrow{\mu} \\ \xrightarrow{\eta_R} \end{array} \begin{array}{c} \\ E \\ \circ \\ E \\ \circ \\ X \end{array} \begin{array}{c} \xrightarrow{} \\ \xleftarrow{} \\ \xrightarrow{\Delta} \\ \xleftarrow{} \\ \xrightarrow{} \end{array} \begin{array}{c} E \\ \circ \\ E \\ \circ \\ E \\ \circ \\ X \end{array} \begin{array}{c} \xleftarrow{} \\ \xrightarrow{} \\ \xleftarrow{} \\ \xrightarrow{} \\ \xleftarrow{} \\ \xrightarrow{} \end{array} \cdots \right\}.$$

Its totalization gives the *unstable $E$-nilpotent completion of $X$*, and the associated filtration spectral sequence on homotopy is the *unstable Adams spectral sequence*.[2] In the case that $\pi_0 \mathcal{UD}_E(S^0)$ is even, the simplicial scheme
$$\mathcal{UM}_E = \operatorname{Spec} \pi_0 \mathcal{UD}_E(S^0)$$
forms the *unstable context* for $E$, a space $X$ gives rise to a quasicoherent sheaf over $\mathcal{UM}_E$ by
$$\mathcal{UM}_E(X) = (\pi_0 \mathcal{UD}_E(X))^\sim,$$
and there is a preferred system of sheaves $\omega^{n/2} = \mathcal{UM}_E(S^n)$.

The remainder of this lecture will be spent trying to wrangle the information in Definition 4.1.1. In the stable situation, we recognized that in favorable situations the homotopy groups of the descent object formed a cosimplicial module over a certain cosimplicial ring – or, equivalently, a *Cartesian* sheaf over a certain simplicial scheme. Furthermore, we found that the simplicial scheme itself had some arithmetic meaning, and that the $E_2$-page of the descent spectral

---

[2] A reader seeking visual stimulation can find unstable analogs of figure 1.2 in [BPS92].

4.1 Unstable Contexts and the Steenrod Algebra 169

sequence computed the cohomology of this sheaf. We will find analogs of all of these results in the unstable setting, listed in order from least to most difficult.

We would first like to recognize the cosimplicial abelian group $\pi_0 \mathcal{U}\mathcal{D}_E(X)$ as a sort of comodule. In the stable case, this came from the smash product map $\mathbb{S}^0 \wedge X \to X$, as well as the lax monoidality of the functor $\mathcal{D}_E(-)$. However, to get a Segal condition by which we could identify the higher-dimensional objects in $\pi_0 \mathcal{D}_E(X)$, we had to introduce the condition **FH**.[3] The unstable situation has an analogous antecedent:

**Definition 4.1.2** ([BCM78, assumption 6.5]) An even-periodic ring spectrum $E$ is said to satisfy the **U**nstable **F**reeness **H**ypothesis, or **UFH**, if $E_0 \underline{E}_0$ is a free and even $E_0$-module.[4]

Under this condition, we again turn to studying the structure of $\mathcal{U}\mathcal{D}_E(S^0)$. If we had a Segal condition, we would expect the structure present to be determined by $\pi_0 \mathcal{U}\mathcal{D}_E(S^0)[j]$ for $j \leq 2$. The data at $j = 0$ are largely redundant:

$$\pi_0 E(S^0) = \pi_0 \underline{E}_0 = E_0$$

computes the coefficient ring of $E$. The data at $j = 1$ consist of the homology groups of the infinite loopspace associated to $E$:

$$\pi_0 E(E(S^0)) = E_0 \underline{E}_0.$$

There are three pieces of structure present here: the augmentation map

$$\varepsilon \colon E_0 \underline{E}_0 \to E_0(E) \to E_0$$

and the left- and right-unit maps $E_0 \to E_0 \underline{E}_0$. The assumption **UFH** gives us a foothold on the case $j = 2$: A choice of basis $B$ for $E_0 \underline{E}_0$ gives an isomorphism

$$E(E(S^0)) \simeq \prod_{\ell \in B}{}' \underline{E}_0 := \underset{\substack{F \subseteq B \\ F \text{ finite}}}{\operatorname{colim}} \prod_{\ell \in F} \underline{E}_0,$$

so that $\pi_0 \mathcal{U}\mathcal{D}_E(S^0)[2]$ splits as a tensor product of terms of the form $E_0 \underline{E}_0$. Here, we find a lot of structure: The addition of cohomology classes is represented by a map

$$\underline{E}_0 \times \underline{E}_0 \to \underline{E}_0,$$

---

[3] In particular, we used **FH** to invert the marked map in $E_0 X \xrightarrow{\eta_R} E_0(E \wedge X) \xleftarrow{\star} E_0 E \otimes_{E_0} E_0 X$.
[4] This helps us understand the following analogous zigzag:

$$\pi_0 E(X) \xrightarrow{\eta_R} \pi_0 E(E(X)) \xleftarrow{\mu \circ 1} \pi_0 E(E(E(X))) \xleftarrow{\text{compose}} \pi_0 E(E(S^0)) \times \pi_0 E(X).$$

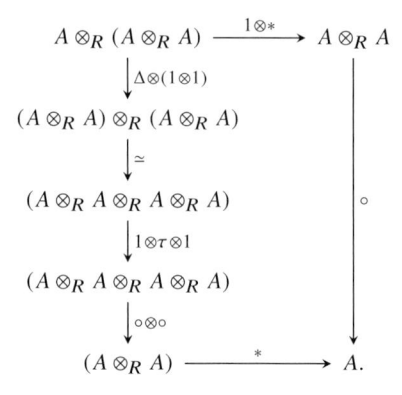

$$\begin{CD}
A \otimes_R (A \otimes_R A) @>{1 \otimes *}>> A \otimes_R A \\
@V{\Delta \otimes (1 \otimes 1)}VV @VV{\circ}V \\
(A \otimes_R A) \otimes_R (A \otimes_R A) @. \\
@V{\simeq}VV @. \\
(A \otimes_R A \otimes_R A \otimes_R A) @. \\
@V{1 \otimes \tau \otimes 1}VV @. \\
(A \otimes_R A \otimes_R A \otimes_R A) @. \\
@V{\circ \otimes \circ}VV @. \\
(A \otimes_R A) @>{*}>> A.
\end{CD}$$

Figure 4.1 The distributivity axiom for $*$ over $\circ$ in a Hopf ring.

which on $E$-homology induces a map

$$* : E_0\underline{E}_0 \otimes_{E_0} E_0\underline{E}_0 \to E_0\underline{E}_0;$$

the multiplication of cohomology classes similarly induces a map

$$\circ : E_0\underline{E}_0 \otimes E_0\underline{E}_0 \to E_0\underline{E}_0;$$

these are compatible with the images of unit classes $0, 1 \in \pi_0 E$ under the left- and right-units specified previously; there is an additive inverse map

$$\chi : E_0\underline{E}_0 \to E_0\underline{E}_0$$

compatible with the $*$-product;[5] there is a diagonal map

$$\Delta : E_0\underline{E}_0 \to E_0\underline{E}_0 \otimes_{E_0} E_0\underline{E}_0$$

induced by the diagonal map of the space $\underline{E}_0$; the maps $\chi$, $*$, and $\Delta$ imbue $E_0\underline{E}_0$ with a Hopf algebra structure; and there is a distributivity condition pictured in Figure 4.1 intertwining $*$, $\circ$, and $\Delta$.[6]

**Definition 4.1.3** ([BJW95, summary 10.46])   A *Hopf ring* is a module equipped

---

[5] When considering the graded analog of Hopf rings, the $\circ$-product obeys a skew-commutativity formula: $a \circ b = \chi^{|a| \cdot |b|}(b \circ a)$.

[6] Analogous structure also appears for aperiodic ring spectra $E$ satisfying a graded version of **UFH**, and in that case tracking through the extra grading indices is actually helpful for deciphering what these maps "feel" like.

## 4.1 Unstable Contexts and the Steenrod Algebra

with structure maps $\varepsilon$, $*$, $\circ$, $\Delta$, and $\chi$, as well as unit classes [0] and [1] for $*$ and $\circ$ respectively, subject to the axioms described above.[7,8]

These structures are enough to pin down the behavior of both $E_0 X$ and the unstable descent object:

**Definition 4.1.4** (see Lemma 1.4.4)  Let $E$ satisfy **UFH**. The associated Hopf ring $E_0\underline{E}_0$ gives rise to a comonad $G(M) = M \otimes_{E_0} E_0\underline{E}_0$, and $G$-coalgebras define comodules for the Hopf ring.

**Lemma 4.1.5** ([BCM78, theorem 6.17])  *If an even-periodic ring spectrum $E$ satisfies **UFH** and $X$ is a space with $E_* X$ a free $E_*$-module, then $E_0 X$ forms a $G$-coalgebra, the cosimplicial object $\pi_0 \mathcal{UD}_E(X)$ is its cobar resolution under $G$, and the $E_2$-page of the unstable descent spectral sequence is presented as*

$$E_2^s = R^s \text{CoalgebraS}_G(E_0, E_0 X).$$

*Proof*  For $X$ satisfying this freeness condition, there is a splitting

$$E(X) \simeq \prod_{\ell}{}' \underline{E}_{n_\ell} \simeq \prod_{\ell}{}' \underline{E}_0,$$

where the second equivalence is by even-periodicity. Since $E$ satisfies **UFH**, the Künneth spectral sequence for the comparison map $\prod_\ell E(\underline{E}_0) \to E(E(X))$ collapses to an equivalence, so that

$$\pi_0 E(E(X)) \cong \pi_0 \left( \prod_{\ell}{}' E(\underline{E}_0) \right) \cong E_0(\underline{E}_0) \otimes_{E_0} E(X).$$

Continuing inductively in this way, the levelwise homotopy of the entire unstable descent object $\mathcal{UD}_E(X)$ can be identified as the cobar complex for $G$, and the structure maps of $\pi_0 \mathcal{UD}_E(X)[\leq 2]$ endow $E_0 X$ itself with the structure of a $G$-coalgebra.  □

At this point, it is instructive to work through an extended example to understand the kinds of objects under consideration. At first appraisal, they

---

[7] Equivalently, a Hopf ring is a ring object in the category of coalgebras. This has the double-edged virtue of concision, but we will find it genuinely useful in Lecture 4.2.

[8] There are yet more pieces of topological structure available that Boardman, Johnson, and Wilson call an *enriched Hopf ring*. We omit them from the algebraic definition because they do not affect the homological algebra of modules for a Hopf ring and because they do not directly appear in the "mixed" context of Lecture 4.2. They are: an element $v \in \pi_0 E$ selects a connected component of $\underline{E}_0$, and there is an attached element $[v] \in E_0\underline{E}_0$ (i.e., the right unit structure map); a cohomology operation $r \colon E^0(-) \to E^0(-)$ induces a map $\underline{E}_0 \to \underline{E}_0$ and hence a map on $E$-homology $r_* \colon E_0\underline{E}_0 \to E_0\underline{E}_0$; and, in particular, this gives a homology suspension element $e = (e_2)_*$, where $e_2 \colon E^0(-) \to E^2(-) \cong E^0(\Omega^2-)$ witnesses the 2-periodicity of $E$.

appear so bottomlessly complicated that it must be hopeless to actually compute even the enriched Hopf ring associated to a spectrum $E$. In fact, the abundance of structure maps involved gives enough footholds that this is actually often feasible, provided we have sufficiently strong stomachs. Our example will be the aperiodic spectrum $E = H\mathbb{F}_2$ (and so we switch back into the graded setting), and the place to start is with a very old result:

**Lemma 4.1.6** *If $E$ is a spectrum with $\pi_{-1}E = 0$, then $\underline{E}_1 \simeq B\underline{E}_0$. If $E$ is a connective spectrum then $\underline{E}_q = B^q \underline{E}_0$ for $n \geq 0$.* □

We use the natural skeletal filtration on $B(-)$ to form a spectral sequence.

**Lemma 4.1.7** ([Seg70, proposition 3.2], [RW80, theorem 2.1]) *For $G$ a loopspace, there is a spectral sequence of signature*

$$E^1_{*,j} = F_*(\Sigma G)^{\wedge j} \Rightarrow F_*BG.$$

*If $G$ is a double loopspace, then the spectral sequence is one of algebras. If $F$ has Künneth isomorphisms*

$$\widetilde{F}_*((\Sigma G)^{\wedge j}) \cong \widetilde{F}_*(\Sigma G)^{\otimes_{F_*} j},$$

*then the $E^2$-page is given by*

$$E^2_{*,*} \cong \mathrm{Tor}^{F_*G}_{*,*}(F_*, F_*).$$

*If both extra conditions hold, then the spectral sequence is of Hopf algebras.* □

**Corollary 4.1.8** *If $E$ is a connective spectrum and if for all $j$ the ring spectrum $F$ has Künneth isomorphisms $\widetilde{F}_*(\underline{E}_q \wedge \underline{E}_q) \cong \widetilde{F}_*\underline{E}_q \otimes_{F_*} \widetilde{F}_*\underline{E}_q$, then there is a family of spectral sequences of Hopf algebras with signatures*

$$E^2_{*,*} \cong \mathrm{Tor}^{F_*\underline{E}_q}_{*,*}(F_*, F_*) \Rightarrow F_*\underline{E}_{q+1}.$$ □

That this spectral sequence is multiplicative for the $*$-product is useful enough, but the situation is actually much, much better than this:

**Lemma 4.1.9** ([TW80, equation 1.3], [RW80, theorem 2.2]) *Suppose further that $E$ is a ring spectrum, and denote by $E^r_{*,*}(F_*\underline{E}_q)$ the spectral sequence considered above whose $E^2$-term is constructed from Tor over $F_*\underline{E}_q$. There are maps*

$$E^r_{*,*}(F_*\underline{E}_q) \otimes_{F_*} F_*\underline{E}_{q'} \to E^r_{*,*}(F_*\underline{E}_{q+q'})$$

*which converge to the map*

$$F_*\underline{E}_{q+1} \otimes_{F_*} F_*\underline{E}_{q'} \xrightarrow{\circ} F_*\underline{E}_{q+q'+1}$$

## 4.1 Unstable Contexts and the Steenrod Algebra

on the $E^\infty$-page and which satisfy

$$d^r(x \circ y) = (d^r x) \circ y. \qquad \square$$

This lemma is incredibly useful: It means that differentials can be transported *between spectral sequences* for classes which can be decomposed as $\circ$-products. This means that the bottom spectral sequence (i.e., the case $q = 0$) exerts a large amount of control over the others – and this spectral sequence often turns out to be very computable.

We now turn to concrete computations for $E = H\mathbb{F}_2$ and $F = H\mathbb{F}_2$. To ground the induction, we will consider the first spectral sequence

$$\mathrm{Tor}_{*,*}^{H\mathbb{F}_{2*}(\mathbb{F}_2)}(\mathbb{F}_2, \mathbb{F}_2) \Rightarrow H\mathbb{F}_{2*}B\mathbb{F}_2.$$

Using $\mathbb{RP}^\infty$ as a model for $B\mathbb{F}_2$, we employ Example 1.1.15 to analyze the target of this spectral sequence: As an $\mathbb{F}_2$-module, we have an isomorphism

$$H\mathbb{F}_{2*}B\mathbb{F}_2 \cong \mathbb{F}_2\{\alpha_j \mid j \geq 0\}.$$

Using our further computation in Example 1.2.15, we can also give a presentation of the Hopf algebra structure on $H\mathbb{F}_{2*}B\mathbb{F}_2$: It is dual to the primitively generated polynomial algebra on a single class, so forms a divided power algebra on a single class which we will denote by $\alpha_{()}$. In characteristic 2, this decomposes as

$$H\mathbb{F}_{2*}B\mathbb{F}_2 \cong \Gamma[\alpha_{()}] \cong \bigotimes_{j=0}^\infty \mathbb{F}_2[\alpha_{(j)}]/\alpha_{(j)}^2,$$

where we have written $\alpha_{(j)}$ for $\alpha_{()}^{[2^j]}$ in the divided power structure.

**Corollary 4.1.10** *This* Tor *spectral sequence collapses at the $E^2$-page.*

*Proof* As an algebra, the homology $H\mathbb{F}_{2*}(\mathbb{F}_2)$ of the discrete space $\mathbb{F}_2$ is presented by a group ring, which we can identify with a truncated polynomial algebra:

$$H\mathbb{F}_{2*}(\mathbb{F}_2) \cong \mathbb{F}_2[\mathbb{F}_2] \cong \mathbb{F}_2[a]/a^{*2}, \quad \text{where } a = [1] - [0].$$

The Tor-algebra of this is then divided power on a single class:

$$\mathrm{Tor}_{*,*}^{H\mathbb{F}_{2*}(\mathbb{F}_2)}(\mathbb{F}_2, \mathbb{F}_2) = \Gamma[\alpha_{()}].$$

In order for the two computations to agree, there can therefore be no differentials in the spectral sequence. $\qquad \square$

We now summarize the rest of the induction:

**Theorem 4.1.11** $H\mathbb{F}_{2*}\underline{H\mathbb{F}_2}_q$ *is the exterior $*$-algebra on the $q$-fold $\circ$-products of the generators* $\alpha_{(j)} \in H\mathbb{F}_{2*}B\mathbb{F}_2$:

$$H\mathbb{F}_{2*}\underline{H\mathbb{F}_2}_t \cong \frac{\mathbb{F}_2[\alpha_{(j_1)} \circ \cdots \circ \alpha_{(j_q)} \mid j_1 \leq \cdots \leq j_q]}{(\alpha_{(j_1)} \circ \cdots \circ \alpha_{(j_q)})^{*2}}.$$

*Proof* Noting that the case $q = 0$ is what was proved above, make the inductive assumption that this is true for some fixed value of $q \geq 0$. The Tor groups of the associated bar spectral sequence

$$\mathrm{Tor}_{*,*}^{H\mathbb{F}_{2*}\underline{H\mathbb{F}_2}_q}(\mathbb{F}_2, \mathbb{F}_2) \Rightarrow H\mathbb{F}_{2*}\underline{H\mathbb{F}_2}_{q+1}$$

form a divided power algebra generated by the same $t$-fold $\circ$-products. An analog of a Ravenel–Wilson lemma ([RW80, lemma 9.5], [Wil82, claim 8.16]) gives a congruence

$$(\alpha_{(j_2-j_1)} \circ \cdots \circ \alpha_{(j_{q+1}-j_1)})^{[2^{j_1}]} \equiv \alpha_{(j_1)} \circ \cdots \circ \alpha_{(j_q)} \circ \alpha_{(j_{q+1})}.$$
(mod $*$-decomposables)

It follows from Lemma 4.1.9 that the differentials vanish:

$$d((\alpha_{(j_2-j_1)} \circ \cdots \circ \alpha_{(j_{q+1}-j_1)})^{[2^{j_1}]}) \equiv d(\alpha_{(j_1)} \circ \cdots \circ \alpha_{(j_q)} \circ \alpha_{(j_{q+1})})$$
(mod $*$-decomposables)

$$= \alpha_{(j_1)} \circ d(\alpha_{(j_2)} \circ \cdots \circ \alpha_{(j_{q+1})})$$
(Lemma 4.1.9)

$$= 0.$$ (inductive hyp.)

Hence, the spectral sequence collapses. To see that there are no multiplicative extensions, note that the only potentially undetermined multiplications occur as $*$-squares of exterior classes. However, the $*$-squaring map is induced by the topological map[9]

$$\underline{H\mathbb{F}_2}_q \xrightarrow{\cdot 2} \underline{H\mathbb{F}_2}_q,$$

which is already null on the level of spaces. It follows that there are no extensions and the induction holds. □

**Corollary 4.1.12** *There is an isomorphism*

$$H\mathbb{F}_{2*}\underline{H\mathbb{F}_2}_* \cong \bigotimes_q \frac{\mathbb{F}_2[\alpha_{(j_1)} \circ \cdots \circ \alpha_{(j_q)} \mid j_1 \leq \cdots \leq j_q]}{(\alpha_{(j_1)} \circ \cdots \circ \alpha_{(j_q)})^{*2}} \cong \bigotimes_q (H\mathbb{F}_{2*}\mathbb{R}P^\infty)^{\wedge q},$$

*where* $(H\mathbb{F}_{2*}\mathbb{R}P^\infty)^{\wedge q}$ *denotes the $q$th exterior power of $H\mathbb{F}_{2*}\mathbb{R}P^\infty$ as a Hopf algebra [Goe99, proposition 5.5].* □

---

[9] This is true only for the case of $E = H\mathbb{F}_p$. In Lectures 4.3 and 4.6, we will require a most robust proof that there are no extensions, which can be found in [Wil82, proof of 8.11].

### 4.1 Unstable Contexts and the Steenrod Algebra

*Remark 4.1.13* ([Wil82, theorems 8.5 and 8.11]) The odd-primary analog of this result appears in Wilson's book, where again the bar spectral sequences all collapse. The end result is

$$H\mathbb{F}_p {}_* \underline{H\mathbb{F}_p}_* \cong \frac{\bigotimes_{I,J} \mathbb{F}_p[e_1 \circ \alpha_I \circ \beta_J, \alpha_I \circ \beta_J]}{(e_1 \circ \alpha_I \circ \beta_J)^{*2} = 0, (\alpha_I \circ \beta_J)^{*p} = 0, e_1 \circ e_1 = \beta_1},$$

where we have defined the following elements:

- $e_1 \in (H\mathbb{F}_p)_1 \underline{H\mathbb{F}_p}_1$ is the homology suspension element;
- $\alpha_{(j)} \in (H\mathbb{F}_p)_{2p^j} \underline{H\mathbb{F}_p}_1$ are the analogs of the elements considered previously;
- $\beta_{(j)} \in (H\mathbb{F}_p)_{2p^j} \mathbb{CP}^\infty$ are the algebra generators of the Hopf algebra dual of the ring of functions on the formal group $\mathbb{CP}^\infty_{H\mathbb{F}_p}$ associated to $H\mathbb{F}_p$ by its natural complex orientation.

In particular, the Hopf ring is nearly *free* on these Hopf algebras, subject to the single interesting relation $e_1 \circ e_1 = \beta_{(0)}$, essentially stemming from the equivalence $S^1 \wedge S^1 \simeq \mathbb{CP}^1$.

It is now instructive to try to relate this unstable computation to the stable one from Lecture 1.3 (and, particularly, its algebro-geometric interpretation in Lemma 1.3.5). Consider the situation of cohomology operations: Each stable operation consists of a family of unstable operations intertwined by suspensions, each of which is additive and takes 0 to 0. In terms of an element $\psi \in E^*\underline{E}_j$, such an unstable operation takes 0 to 0 exactly when it lies in the augmentation ideal, and it is additive exactly when it satisfies Hopf algebra primitivity:

$$\Delta^*\psi^* = (\psi \otimes \psi)^*\Delta^*.$$

**Definition 4.1.14** For a Hopf ring $H$ with augmentation $\varepsilon$, we define the $*$-*indecomposable quotient*, a bialgebra, by $Q^*H = \ker \varepsilon/(\ker \varepsilon)^{*2}$ (see Definition 2.1.2). In the setting of unstable homology cooperations, the collection of *additive unstable cooperations* is defined as $Q^*E_*\underline{E}_j$.

We now apply this philosophy to our example:

**Corollary 4.1.15** (see Theorem 1.3.3 [Wil82, theorem 8.15])

$$\mathcal{A}_* = \mathbb{F}_2[\xi_0, \xi_1, \xi_2, \ldots]/(\xi_0 = 1).$$

*Proof* We approach via the system

$$H\mathbb{F}_2 {}_* H\mathbb{F}_2 \cong \mathrm{colim} \left( \cdots \to \widetilde{H\mathbb{F}}_{2*+j} \underline{H\mathbb{F}}_{2j} \to \widetilde{H\mathbb{F}}_{2*+(j+1)} \underline{H\mathbb{F}}_{2j+1} \to \cdots \right).$$

First, note that each map in the system factors through the $*$-indecomposables: the composite

$$\Sigma(\underline{E}_j \times \underline{E}_j) \to \Sigma \underline{E}_j \to \underline{E}_{j+1}$$

176  Unstable Cooperations

vanishes on $E$-homology for the same reason that suspension kills the cup product ([BJW95, corollary 2.18], [Swi02, proposition 13.65]). We may therefore replace this system with

$$H\mathbb{F}_{2*}H\mathbb{F}_2 \cong \text{colim}\left(\cdots \to Q^*H\mathbb{F}_{2*+j}\underline{H\mathbb{F}_2}_j \to Q^*H\mathbb{F}_{2*+(j+1)}\underline{H\mathbb{F}_2}_{j+1} \to \cdots\right).$$

Corollary 4.1.12 gives us unfettered access to the $*$-indecomposables:

$$Q^*H\mathbb{F}_{2*}\underline{H\mathbb{F}_2}_* \cong \mathbb{F}_2\{\alpha_I | I \text{ a multi-index}\}$$
$$= \mathbb{F}_2\{\alpha_{(I_0)} \circ \alpha_{(I_1)} \circ \cdots \circ \alpha_{(I_n)} | I = (I_0, \ldots, I_n) \text{ a multi-index}\}.$$

The $j$th component of this system consists of sums of monomials with $j$ factors, and the maps in the system multiply by the homology suspension element $\alpha_{(0)} = \xi_0$ to produce a monomial with $(j+1)$ factors from one with $j$ factors.[10] Writing $\xi_k = \alpha_{(k)}$, this computes the colimit to be

$$H\mathbb{F}_{2*}H\mathbb{F}_2 \cong \mathbb{F}_2[\xi_k \mid k \geq 1].$$  □

Our last goal in this lecture is to sketch a foothold that this example has furnished us with for the algebro-geometric interpretation of unstable cooperations. First, we should remark that it has been shown that there is no manifestation of the homology of a space as any kind of classical comodule [BJW95, theorem 9.4], so we are unable to directly pursue an analog of Definition 3.1.13 presenting the homology of a space as a Cartesian quasicoherent sheaf over some object. This no-go result is quite believable from the perspective of cohomology operations: We have calculated in the case of $E = H\mathbb{F}_2$ that a generic unstable cohomology operation takes the form

$$x \mapsto \sum_{S \text{ a set of multi-indices}} \left(c_S \cdot \prod_{I \in S} \text{Sq}^I(x)\right).$$

This inherently uses the multiplicative structure on $H\mathbb{F}_2^*(X)$, and the proof of the result of Boardman, Johnson, and Wilson rests entirely on the observation that decomposable elements cannot be mapped to indecomposable elements by maps of algebras, but maps of modules have no such control.[11]

However, exactly this complaint is eliminated by passing to the additive unstable cooperations: All the product terms in the above formula vanish, and the homology of a space does indeed have the structure of a comodule for this Hopf algebra. Still in the setting of our running example $E = H\mathbb{F}_2$, this makes

---

[10] As a reminder, the graded unstable element $e \in H_1S^1$ contributes to stable degree $1 - 1 = 0$.
[11] It is probably still possible to treat this carefully enough to cast the whole of unstable operations (and, in particular, the comonad $G$) into algebro-geometric language.

### 4.1 Unstable Contexts and the Steenrod Algebra

$H\mathbb{F}_{2*}(X)$ into a Cartesian quasicoherent sheaf for the simplicial scheme[12]

$$\mathcal{U}\mathcal{M}_{H\mathbb{F}_2 P} \simeq \operatorname{Spec} \mathbb{F}_2 \mathop{/\!\!/} \operatorname{Spec} Q^* H\mathbb{F}_2 P_0 \underline{H\mathbb{F}_2 P_0}.$$

In this specific example, we can even identify what this simplicial scheme is. Using Lemma 1.3.5, we have already made the identification

$$\operatorname{Spec} H\mathbb{F}_2 P_0 H\mathbb{F}_2 P \cong \underline{\operatorname{Aut}} \widehat{\mathbb{G}}_a$$

$$\left(f \colon \mathbb{F}_2[\xi_0^{\pm}, \xi_1, \ldots] \to R\right) \mapsto \left(x \mapsto \sum_{j=0}^{\infty} f(\xi_j) x^{2^j}\right),$$

The even-periodic analog of Corollary 4.1.12 presents $\operatorname{Spec} \mathcal{A} P_0$ as the open subscheme of $\operatorname{Spec} \mathbb{F}_2[\xi_0, \xi_1, \ldots]$ with $\xi_0$ inverted.[13] Hence, we see that a compatible name for the inertia group in the unstable context for $H\mathbb{F}_2 P$ is

$$\operatorname{Spec} Q^* H\mathbb{F}_2 P_0 \underline{H\mathbb{F}_2 P_0} \cong \underline{\operatorname{End}} \widehat{\mathbb{G}}_a$$

$$(f \colon \mathbb{F}_2[\xi_0, \xi_1, \ldots] \to R) \mapsto \left(x \mapsto \sum_{j=0}^{\infty} f(\xi_j) x^{2^j}\right).$$

Some of the complexity here was eliminated by the smallness of $\operatorname{Spec} H\mathbb{F}_2 P_0$. For a general ring spectrum $E$, we also have to account for $\operatorname{Spec} E_0$, but the end result is similar to that of Definition 3.1.13:

**Lemma 4.1.16** *For a ring spectrum $E$ satisfying* **UFH**, *the additive unstable cooperations form rings of functions on the objects and morphisms of a category scheme $\mathcal{U}\mathcal{M}_E$, and the $E$-homology of a space $X$ forms a* Cartesian *quasicoherent sheaf $\mathcal{U}\mathcal{M}_E(X)$ over its nerve.* □

Although it seems like we have lost a lot of information in passing to $*$-indecomposables, in many cases this is actually enough to recover everything.

**Definition 4.1.17** ([BCM78, assumptions 7.1 and 7.7]) We say that a ring spectrum $E$ satisfying **UFH** furthermore satisfies the **Unstable Generation Hypothesis**, or **UGH**, when the following conditions all hold:

1. The module of primitives $PE_0\underline{E_0}$ is $E_0$-free.
2. The composite $PE_0\underline{E_0} \to E_0\underline{E_0} \to E_0 E$ is injective.

---

[12] This identification is somewhat subtle: We are using the fact that passing to homotopy groups of this simplicial scheme factors through passing through the $*$-indecomposable quotient; see Lemma 4.1.18.

[13] Note that $Q^* H\mathbb{F}_{2*}\underline{H\mathbb{F}_{2*}}$ does *not* form a Hopf algebra, essentially because it is missing a version of $\chi$ that inverts $\circ$-multiplication. It is remarkable (and related to Corollary 5.1.10) that inverting the homology suspension element automatically produces an antipode.

3. Writing $S$ for the cofree (nonunital) coalgebra functor, $E_0\underline{E}_0 \to SPE_0\underline{E}_0$ is an isomorphism.

**Lemma 4.1.18** ([BCM78, lemma 7.5])   *Let $E$ be a ring spectrum satisfying **UGH**, and let $G$ be the comonad from Lemma 4.1.5. The composite functor $U = PG$ extends from a functor on free $E_0$-modules to all $E_0$-modules by using two-stage free resolutions and enforcing exactness, and the result is a comonad. Coalgebras for this comonad are exactly comodules for the bialgebra of additive unstable cooperations.*  □

**Corollary 4.1.19** ([BCM78, remark 7.8])   *If $E$ satisfies **UGH** and $X$ is a space with $E_0 X = SN$ for some connective free $E_0$-module $N$, then the unstable $E$-Adams $E_2$-term is computed by*
$$E_2^s = \mathrm{Ext}^s_{\mathrm{COALGEBRAS}_U}(E_0, PE_0 X).$$

*Proof*   Under **UGH**, we have a factorization
$$\mathrm{COALGEBRAS}_G(E_0, -) = \mathrm{COALGEBRAS}_U(E_0, P(-))$$
and the injective objects intertwine to give a composite functor spectral sequence
$$E_2^{r,s} = \mathrm{Ext}^r_{\mathrm{COALGEBRAS}_U}(E_0, R^s_{\mathrm{COALGEBRAS}_G} P(M)) \Rightarrow \mathrm{Ext}^{r+s}_{\mathrm{COALGEBRAS}_G}(E_0, M).$$
If $M = E_0 X = SN$ for some connective free $E_0$-module $N$, then
$$R^{q>0}_{\mathrm{COALGEBRAS}_G} P(E_0 X) = 0,$$
hence the composite functor spectral sequence collapses, and
$$\mathrm{Ext}^s_{\mathrm{COALGEBRAS}_U}(E_0, PE_0 X)$$
computes the unstable Adams $E_2$-term as claimed.   □

*Remark 4.1.20* (see Remark 1.3.9)   As in the stable case, the Hopf ring associated to a ring spectrum satisfying **UFH** splits off a factor of $B\mathbb{N}$ that tracks the grading, and $\omega^{*/2}$ can be thought of as a family of character sheaves for $\mathbb{N}$. Passing to the stable context covers the localization $B\mathbb{N} \to B\mathbb{G}_m$.

*Remark 4.1.21* ([Bau14])   Tilman Bauer has studied some of the algebraic geometry associated to unstable cohomology operations, which he gave a model for in terms of *formal plethories*.

## 4.2  Algebraic Mixed Unstable Cooperations

For simplicity, we return to the stable setting of Lecture 3.1 for a moment. For an arbitrary spectrum $X$ and ring spectrum $E$, the completion $X_E^\wedge$ is typically a quite

## 4.2 Algebraic Mixed Unstable Cooperations

poor approximation to $X$ itself. Though this can be partially mediated by placing hypotheses on $X$, the approximation can always be improved by "enlarging" the ring spectrum involved – for instance, selecting a second ring spectrum $F$ and forming the completion $X^\wedge_{E \vee F}$ at the wedge. This has the following factorization property

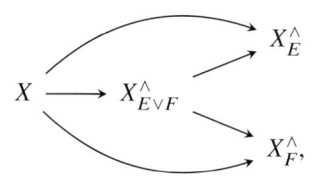

$$\begin{array}{c} X \longrightarrow X^\wedge_{E \vee F} \end{array} \begin{array}{c} \nearrow X^\wedge_E \\ \searrow X^\wedge_F, \end{array}$$

so that homotopy classes visible in either of $X^\wedge_E$ or $X^\wedge_F$ are therefore also visible in the homotopy of $X^\wedge_{E \vee F}$. Now consider the descent object $\mathcal{D}_{E \vee F}(X)$ and its layers $\mathcal{D}_{E \vee F}(X)[n]$:

$$\begin{aligned} \mathcal{U}\mathcal{D}_{E \vee F}(X)[n] &= (E \vee F)^{\wedge(n+1)} \wedge (X) \\ &\simeq (E^{\wedge(n+1)} \wedge X) \vee (F^{\wedge(n+1)} \wedge X) \\ &\quad \vee \bigvee_{\substack{i+j=n+1 \\ i \neq 0 \neq j}} (E^{\wedge i} \wedge F^{\wedge j} \wedge X)^{\vee \binom{n}{i,j}}. \end{aligned}$$

In the edge cases of $i = 0$ or $j = 0$, we can identify the descent objects $\mathcal{D}_E(X)$ and $\mathcal{D}_F(X)$ as sub-cosimplicial objects of $\mathcal{D}_{E \vee F}(X)$. The role of the cross-terms at the end of the expression is to prevent the completion at $E \vee F$ from double-counting the parts of $X$ already simultaneously visible to the completions at $E$ and at $F$ – i.e., the cross-terms handle communication between $E$ and $F$.[14]

There is a similar (but algebraically murkier) story for the unstable descent object formed at a wedge of two ring spectra. Let $X$ now be a space, and consider the first two layers of $\mathcal{U}\mathcal{D}_{E \vee F}(X)$:

$$\begin{aligned} \mathcal{U}\mathcal{D}_{E \vee F}(X)[0] &= (E \vee F)(X) \\ &= E(X) \times F(X), \\ \mathcal{U}\mathcal{D}_{E \vee F}(X)[1] &= (E \vee F)(E(X) \times F(X)) \\ &= E(E(X) \times F(X)) \times F(E(X) \times F(X)). \end{aligned}$$

Restricting attention to just the first factor, $E(E(X) \times F(X))$, its homotopy receives a map

$$\pi_0 E(E(X)) \otimes_{E_0} \pi_0 E(F(X)) \to \pi_0 (E(E(X) \times F(X)),$$

---

[14] From the perspective of spectral schemes, you might think of the descent object for $E \vee F$ as that coming from the joint cover $\{\mathbb{S} \to E, \mathbb{S} \to F\}$, and these cross-terms correspond to the scheme-theoretic intersection of $E$ and $F$ over $\mathbb{S}$.

and if $E$ has Künneth isomorphisms then the induced map off of the tensor product is an equivalence. Again, we can identify the $E(E(X))$ part of this expression as belonging to $\mathcal{UD}_E(X)[1]$, and there is a cross-term $E(F(X))$ accounting for the shared information with $F$. The other term also contains information present in $\mathcal{UD}_F(X)[1]$ and a cross-term $F(E(X))$ accounting for shared information with $E$. In order to understand how these cross-terms affect the reconstruction process, we are thus drawn to the following objects:

$$O(\mathcal{UM}_{E \vee F}(S^n)[1]) \leftarrow \pi_0 F(E(S^n)) = F_0 \underline{E}_n.$$

As $n$ ranges, these again form a Hopf ring:

**Definition 4.2.1** We will refer to $F_0(\underline{E}_0)$ as the *Hopf ring of mixed unstable cooperations* (from $F$ to $E$) or the *topological Hopf ring* (from $F$ to $E$).

We thus set about trying to understand the Hopf rings $F_0(\underline{E}_0)$ in general. In our computational example in Lecture 4.1, we found that the topological Hopf ring $H\mathbb{F}_{2*}(\underline{H\mathbb{F}_2}_*)$ modeled endomorphisms of the additive formal group after passing to a suitable quotient, and we will take this as inspiration to construct an algebraic model which approximates the topological Hopf ring.

We approach this problem in stages. To start, note that homotopy elements both of $F$ and of $E$ can be used to contribute elements to the topological Hopf ring: An element $f \in F_0$ begets a natural element $f \in F_0 \underline{E}_0$, and an element $e \in E^0 = \pi_0 \underline{E}_0$ begets an element $[e] \in F_0 \underline{E}_0$ by Hurewicz. The interaction of these rings $F_0$ and $E^0$ is captured in the following definition:

**Definition 4.2.2** ([RW80, p. 706])  Let $R$ and $S$ be graded rings. The *Hopf ring-ring* $R[S]$ forms a Hopf ring over $R$: As an $R$-module, it is free and generated by symbols $[s]$ for $s \in S$, and the ring structure on $S$ is promoted up a level to become the Hopf ring operations. Explicitly, the Hopf ring structure maps $*$, $\circ$, $\chi$, and $\Delta$ are determined by the formulas

$$R[S] \otimes_R R[S] \xrightarrow{*} R[S] \qquad [s] * [s'] = [s + s'],$$

$$R[S] \otimes_R R[S] \xrightarrow{\circ} R[S] \qquad [s] \circ [s'] = [s \cdot s'],$$

$$R[S] \xrightarrow{\chi} R[S] \qquad \chi[s] = [-s],$$

$$R[S] \xrightarrow{\Delta} R[S] \otimes_R R[S] \qquad \Delta[s] = [s] \otimes [s],$$

$$R[S] \xrightarrow{\varepsilon} R \qquad \varepsilon[s] = 1.$$

**Lemma 4.2.3** ([RW80, p. 706])  *There are natural maps of Hopf rings*

$$F_0[E^0] \to F_0(\underline{E}_0) \to F_0[E^0]$$

*augmenting the topological Hopf ring over the Hopf ring-ring.* □

## 4.2 Algebraic Mixed Unstable Cooperations

Supposing that $E$ and $F$ are complex-orientable, we now seek to involve their formal groups. The construction we are about to undertake is a variation on the proof of Lemma 2.1.4, which is itself a variation of a more general result in the theory of formal schemes:

**Lemma 4.2.4** ([Str99b, proposition 2.94]) *Let $X$ and $Y$ be schemes over $S = \operatorname{Spec} R$, such that $O_X$ forms a finite and free $R$-module. There is then a mapping scheme $M$, such that points $f \in M(A)$ naturally biject with maps $f \colon X \times_S \operatorname{Spec} A \to Y \times_S \operatorname{Spec} A$ of $A$-schemes.* □

The mode of proof of this result is to form the symmetric $R$-algebra on the $R$-module $O_Y \otimes_R O_X^*$, then quotient by the relations encoding multiplicativity of functions and unitality of multiplication. These are the same steps we will take to form a Hopf ring embodying homomorphisms of formal groups $\mathbb{CP}^\infty_F \to \mathbb{CP}^\infty_E$.

**Definition 4.2.5** (see [RW77, lemma 1.12 and equation 1.17]) For a coaugmented $R$-coalgebra $A$ and an $S$-algebra $B$, the *free relative Hopf $R[S]$-ring* $A_{R[S]}[B]$ is generated under the Hopf ring operations by symbols $a[b]$ for $a \in A$ and $b \in B$, thought of as "$a \circ [b]$." These are subject to the following rules:

1. For $\Delta(a) = \sum_j a'_j \otimes a''_j$, we have
$$\Delta(a[b]) = \sum_j (a'_j[b] \otimes a''_j[b]).$$

2. Accordingly, we have
$$a[b' + b''] = \sum_j a'_j[b'] * a''_j[b''], \quad a[b'b''] = \sum_j a'_j[b'] \circ a''_j[b''].$$

3. Thinking of the antipode as $\chi(h) = [-1] \circ h$ gives
$$\chi(a[b]) = a[-b].$$

4. Lastly, multiplication by zero gives
$$a[0] = \eta\varepsilon(a)[0].$$

Noting the similarity of the second and fourth relations to those imposed in Lemma 4.2.4, there are two additional families of relations we might consider in the presence of Hopf algebra structures on $A$ and $B$:

5. The dual to the fourth relation is
$$(\eta(1))[b] = [\eta\varepsilon(b)].$$

182    Unstable Cooperations

6. The dual to the second relation is more complicated. For $a', a'' \in A$ and for $b \in B$ with diagonal given by $\Delta(b) = \sum_{j=1}^{n} b'_j \otimes b''_j$, the dual relation is then

$$(a'a'')[b] = \underset{j}{*} \sum_{k} a'_{jk}[b'_j] \circ a''_{jk}[b''_j],$$

where $\Delta^{n-1}a' = \sum_k a'_{1k} \otimes \cdots \otimes a'_{jk} \otimes \cdots \otimes a'_{nk}$, and similarly for $a''_{jk}$.

Imposing these additional relations, we call the resulting quotient Hopf ring the *Kronecker Hopf ring-ring*, and we denote it by $A^{\circlearrowright}_{R[S]}[B]$.

**Lemma 4.2.6**  *There is a natural map*

$$(F_0\mathbb{CP}^\infty)^{\circlearrowright}_{F_0[E^0]}[E^0\mathbb{CP}^\infty] \to F_0(\underline{E}_0).$$

*Proof*  For any space $X$, we construct a Kronecker-type pairing

$$\langle -, - \rangle \colon F_0 X \times E^0 X \to F_0(\underline{E}_0)$$

as follows: Given a class $f \in \pi_0 F(X)$ and a class $e \colon X \to \underline{E}_0$, we can compose the two to produce an element $e_*(f) \in \pi_0 F(\underline{E}_0)$.[15] This pairing is "bilinear" in the following senses:

$$\langle a' + a'', b \rangle = \langle a', b \rangle + \langle a'', b \rangle, \qquad \langle f \cdot a, b \rangle = f \cdot \langle a, b \rangle,$$
$$\langle a, b' + b'' \rangle = \sum_j \langle a'_j, b' \rangle * \langle a''_j, b'' \rangle, \qquad \langle a, e \cdot b \rangle = [e] \circ \langle a, b \rangle.$$

Universality of the free relative Hopf ring thus gives a Hopf ring map $(F_0 X)_{F_0[E^0]}[E^0 X] \to F_0(\underline{E}_0)$. Specializing to $X = \mathbb{CP}^\infty$,[16] the factorization of this map through the indicated Hopf ring quotient follows the duality property of this enhanced Kronecker pairing. Namely, the four maps and their associated diagrams pictured in Figure 4.2 respectively witness the relations

$$\langle \Delta_* a, b' \otimes b'' \rangle = \langle a, \Delta^*(b' \otimes b'') \rangle, \qquad \langle \mu_*(a' \otimes a''), b \rangle = \langle a' \otimes a'', \mu^* b \rangle,$$
$$\langle \varepsilon_* 1, b \rangle = \langle 1, \varepsilon^* b \rangle, \qquad \langle \eta_* 1, b \rangle = \langle 1, \eta^* b \rangle.$$

The Kronecker pairings relate to the Künneth isomorphisms for $F_0(\mathbb{CP}^\infty \times \mathbb{CP}^\infty)$ and $E^0(\mathbb{CP}^\infty \times \mathbb{CP}^\infty)$ by the product formula

$$\langle a' \otimes a'', b' \otimes b'' \rangle = \langle a', b' \rangle \circ \langle a'', b'' \rangle.$$

Hence, writing $\Delta_* a = \sum_j a'_j \otimes a''_j$ and $\mu^* b = \sum_j b'_j \otimes b''_j$, these relations become

---

[15] This map was considered by Goerss [Goe99, proposition 10.2], who cites Strickland as inspiration.
[16] In fact, any $H$-space $X$ produces an algebraic approximation, but it is the case $X = \mathbb{CP}^\infty$ where the map tends to be an isomorphism.

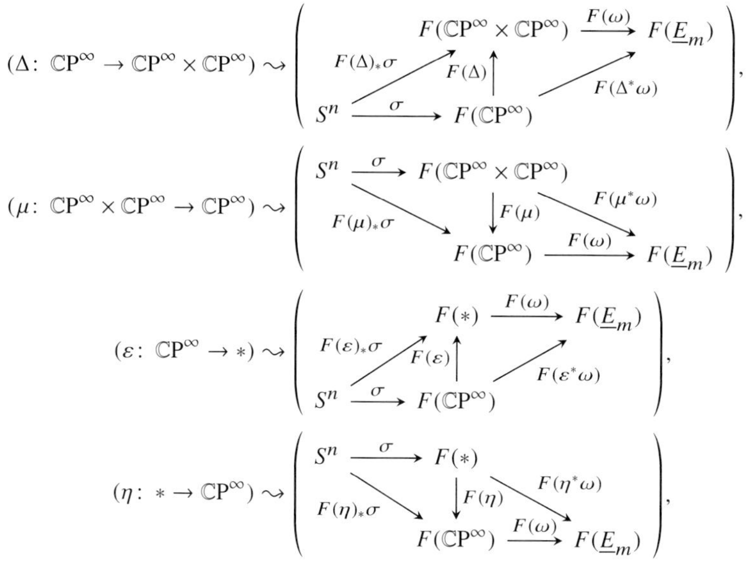

## 4.2 Algebraic Mixed Unstable Cooperations

$$(\Delta\colon \mathbb{CP}^\infty \to \mathbb{CP}^\infty \times \mathbb{CP}^\infty) \rightsquigarrow \left( \begin{array}{c} F(\mathbb{CP}^\infty \times \mathbb{CP}^\infty) \xrightarrow{F(\omega)} F(\underline{E}_m) \\ F(\Delta)_*\sigma \nearrow \quad \uparrow F(\Delta) \quad \nearrow F(\Delta^*\omega) \\ S^n \xrightarrow{\sigma} F(\mathbb{CP}^\infty) \end{array} \right),$$

$$(\mu\colon \mathbb{CP}^\infty \times \mathbb{CP}^\infty \to \mathbb{CP}^\infty) \rightsquigarrow \left( \begin{array}{c} S^n \xrightarrow{\sigma} F(\mathbb{CP}^\infty \times \mathbb{CP}^\infty) \\ F(\mu)_*\sigma \searrow \quad \downarrow F(\mu) \quad \searrow F(\mu^*\omega) \\ F(\mathbb{CP}^\infty) \xrightarrow{F(\omega)} F(\underline{E}_m) \end{array} \right),$$

$$(\varepsilon\colon \mathbb{CP}^\infty \to *) \rightsquigarrow \left( \begin{array}{c} F(*) \xrightarrow{F(\omega)} F(\underline{E}_m) \\ F(\varepsilon)_*\sigma \nearrow \quad \uparrow F(\varepsilon) \quad \nearrow F(\varepsilon^*\omega) \\ S^n \xrightarrow{\sigma} F(\mathbb{CP}^\infty) \end{array} \right),$$

$$(\eta\colon * \to \mathbb{CP}^\infty) \rightsquigarrow \left( \begin{array}{c} S^n \xrightarrow{\sigma} F(*) \\ F(\eta)_*\sigma \searrow \quad \downarrow F(\eta) \quad \searrow F(\eta^*\omega) \\ F(\mathbb{CP}^\infty) \xrightarrow{F(\omega)} F(\underline{E}_m) \end{array} \right),$$

Figure 4.2 Four topological Kronecker pairing relations.

exactly the four equations

$$\sum_j (a'_j[b'] \circ a''_j[b'']) = a[b'b''], \quad (a'a'')[b] = \circledast \sum_j \sum_k a'_{jk}[b'_j] \circ a''_{jk}[b''_j],$$

$$(\eta(1))[b] = [\eta\varepsilon(b)], \quad a[\eta(1)] = \eta\varepsilon(a)[\eta(1)]. \qquad \square$$

The main theme of this case study is that this induced map off of the quotient is very often an isomorphism (and, in turn, that the theory of formal groups also captures everything about the theory of unstable cooperations). Because we will be carrying this algebraic model around with us, we pause to give it a name.

**Definition 4.2.7** For $F$ and $E$ ring spectra, we define their *algebraic Hopf ring* $A\!A(F, E)$ (or *algebraic approximation*) by

$$A\!A(F, E) = (F_0\mathbb{CP}^\infty) \overset{\circlearrowleft}{\underset{F_0[E^0]}{}} [E^0\mathbb{CP}^\infty].$$

**Lemma 4.2.8** ([RW77, theorem 3.8], [Wil82, theorem 9.7]) *Choosing complex orientations of $E$ and $F$, define the following quantities: $\beta_j$ is dual to the jth* power of the coordinate in $F^*\mathbb{CP}^\infty$, $\beta(s)$ denotes the formal sum $\beta(s) = \sum_j \beta_j x^j$,

$$\beta(s +_F t) = \sum_n \beta_n[1] \left( \sum_{i,j} a_{ij}^F s^i t^j \right)^n,$$

$$\beta(s) +_{[E]} \beta(t) = \mathop{\text{\Large$*$}}_{i,j} \left( \beta_0[a_{ij}^E] \circ \left( \sum_k \beta_k[1] s^k \right)^{\circ i} \circ \left( \sum_\ell \beta_\ell[1] t^\ell \right)^{\circ j} \right).$$

*There is a natural isomorphism of Hopf rings*

$$\mathcal{A}\!\mathcal{A}(F, E) \cong \frac{(F_0 \mathbb{CP}^\infty)_{F_0[E^0]}[E^0]}{\beta(s +_F t) = \beta(s) +_{[E]} \beta(t)},$$

*where the equality is imposed term-by-term on the Hopf ring.*[17]

*Proof sketch* The orientations of $E$ and $F$ give rise to classes $x^j \in E^0 \mathbb{CP}^\infty$ and $\beta_k \in F_0 \mathbb{CP}^\infty$, and hence classes $\beta_k[x^j] \in \mathcal{A}\!\mathcal{A}(F, E)$. Two of the core relations imposed on this Hopf ring give us two useful identities:

1. The relation

$$\beta_k[x^0] = \varepsilon(\beta_k)[1] = \begin{cases} \beta_0[1] & \text{if } k = 0, \\ 0 & \text{if } k \neq 0 \end{cases}$$

   eliminates all elements of this form except $\beta_0[x^0] = 1$.[18]

2. The relation

$$\beta_k[x^{j+1}] = \sum_{k'+k''=k} \beta_{k'}[x^j] \circ \beta_{k''}[x]$$

   lets us rewrite these terms as $\circ$-products of terms of lower $j$-degree and no larger $k$-degree.

By consequence, the remaining terms are all sums of $\circ$-products of terms of the form $\beta_k[x]$, so that imposing these relations produces a surjection

$$(F_0 \mathbb{CP}^\infty)_{F_0[E^0]}[E^0] \to \mathcal{A}\!\mathcal{A}(F, E).$$

The remaining assertion is now a matter of imposing the fourth core relation, i.e., a matter of calculating the behavior of

$$\mathbb{CP}^\infty \times \mathbb{CP}^\infty \xrightarrow{\mu} \mathbb{CP}^\infty \xrightarrow{x} \underline{E}_0$$

---

[17] This relation is often referred to as the Ravenel–Wilson relation.

[18] The dual relation

$$\beta_0[x^j] = \beta_0[\varepsilon(x^j)] = \begin{cases} \beta_0[1] & \text{if } j = 0, \\ \beta_0[0] & \text{if } j \neq 0 \end{cases}$$

also cuts down the space of elements, but is not relevant here.

## 4.2 Algebraic Mixed Unstable Cooperations

in two different ways: using the effect of $\mu$ in $F$-homology and pushing forward in $x$, or using the effect of $\mu$ in $E$-cohomology and pushing forward along the Hurewicz map $\mathbb{S} \to F$. □

We now turn to practicing functorial algebraic geometry with Hopf rings, defining the *Hopf ring spectrum* of a Hopf ring $H$ using the formula

$$(\mathrm{Sph}\, H)(T) = \mathrm{HopfRings}(H, T).$$

Our goal for the remainder of this lecture will be to give a (partial) description of the functor $\mathrm{Sph}\, \mathcal{A}(F, E)$. In order to gain traction on this project, we seek a link with the notions of algebraic geometry that we have introduced so far, and we can find one in the form of a comparison of definitions. The description of Hopf rings given in Definition 4.1.3 was very manual, but it admits a repackaging in terms of the category theory used to define rings and ring spectra. Just as abelian groups form the abelian group objects in SETS, commutative Hopf $R$-algebras form the abelian group objects in COALGEBRAS$_R$. Both of these categories admit interesting monoidal structures capturing bilinearity: The category of abelian groups acquires a tensor product, and the category of Hopf algebras acquires the ⊠-product [Goe99, section 5]. Commutative algebras for the tensor monoidal structure on abelian groups define commutative rings, and commutative algebras for the ⊠-product define Hopf rings.

We now seek a functor HOPFALGEBRAS$_R \to$ MODULES$_R$ which is compatible with the two monoidal structures. Indeed, we have already brushed up against one in the previous lecture:

**Lemma 4.2.9** ([Goe99, proposition 6.1])   *The functor*

$$Q^* \colon \mathrm{HopfAlgebras}_R \to \mathrm{Modules}_R$$

*is strongly monoidal:*

$$Q^*(H \boxtimes_R H') = Q^*H \otimes_R Q^*H'.$$ □

In particular, we learn that $Q^*$ induces a compatible functor

$$Q^* \colon \mathrm{HopfRings}_R \to \mathrm{Algebras}_{R/},$$

so that we might build our program around it. The key observation is:

**Corollary 4.2.10**   *Both functors $Q^*$ admit right-adjoints as in the diagram*

$$\begin{array}{ccc}
\mathrm{HopfRings}_R & \xrightleftharpoons[i]{Q^*} & \mathrm{Algebras}_R \\
\downarrow & & \downarrow \\
\mathrm{HopfAlgebras}_R & \xrightleftharpoons[i]{Q^*} & \mathrm{Modules}_R.
\end{array}$$

*Proof* We begin at the simpler level of Hopf algebras and abelian groups. The situation here is clarified by breaking it into two components, unexpected from the perspective of the decomposition of definitions above:

$$\text{HopfAlgebras}_R \underset{F}{\overset{U}{\rightleftarrows}} \text{Algebras}_{R/} \underset{j}{\overset{Q^*}{\rightleftarrows}} \text{Modules}_R.$$

The presence of the forgetful functor $U$ records that the antipode and diagonal play no role in $Q^*$. The functor $j$ sends an $R$-module to its square-zero extension, and together these form an adjoint pair. The right-adjoint to the forgetful functor is considerably more complex but nonetheless exists [Fox93]. Category theory then lifts us to the setting of Hopf rings: The right-adjoint to a strongly monoidal functor is lax monoidal, which is enough to preserve algebra objects. □

As a consequence, we deduce that there is a class of test Hopf rings where the Hopf ring spectrum behaves like an object in ordinary algebraic geometry:

$$(\operatorname{SpH} H)(iA) = \text{HopfRings}(H, iA) = \text{Rings}(Q^*H, A) = (\operatorname{Spec} Q^*H)(A).$$

Before turning to $\operatorname{SpH} \mathcal{A\!A}(F, E)$ itself, we note that $i$ is an embedding, using the following sequence of generally useful results:

**Lemma 4.2.11** *For $H$ a Hopf ring and $x \in H$, write $\langle x \rangle = x - \varepsilon(x) \cdot [0]$ for the corresponding element of $\ker \varepsilon$. In the $*$-indecomposable quotient, there are the formulas*

$$\langle x \rangle \varepsilon(y) + \varepsilon(x) \langle y \rangle = \langle x * y \rangle, \qquad \langle x \rangle \circ \langle y \rangle = \langle x \circ y \rangle.$$

*Proof* Modulo $*$-decomposables, we can write

$$0 \equiv \langle x \rangle * \langle y \rangle = x * y - x\varepsilon(y) - \varepsilon(x)y + \varepsilon(x)\varepsilon(y)[0]$$
$$= \langle x * y \rangle - \langle x \rangle \varepsilon(y) - \varepsilon(x) \langle y \rangle.$$

We can also directly calculate

$$\langle x \rangle \circ \langle y \rangle = x \circ y - \varepsilon(x)\varepsilon(y)[0] - \varepsilon(x)\varepsilon(y)[0] + \varepsilon(x)\varepsilon(y)[0] = \langle x \circ y \rangle. \quad \square$$

**Corollary 4.2.12** *There is an isomorphism $Q^*R[S] \cong S$.*

*Construction* The inverse to the map $x \mapsto \langle [x] \rangle$ is given by

$$c \colon Q^*R[S] \to S,$$
$$\sum_j r_j([s_j] - [0]) \mapsto \sum_j r_j \otimes s_j. \quad \square$$

**Corollary 4.2.13** *The functor $i$ is an embedding.*

*Proof* From the perspective of $i$, the Hopf ring-ring $R[O_{\mathbb{A}^1}]$ plays the role of an affine line:

### 4.2 Algebraic Mixed Unstable Cooperations 187

$$(\text{SpH } R[O_{\mathbb{A}^1}])(iA) = (\text{Spec } O_{\mathbb{A}^1})(A) = \mathbb{A}^1(A) = A.$$

We see that we can functorially recover $A$ from $iA$. □

Finally, we turn to $Q^*\mathcal{A}\mathcal{A}(F, E)$ directly, with the goal of explaining the phenomenon uncovered in Lecture 4.1, where we passed to the $*$-indecomposables to find the classical ring of functions on the endomorphism scheme of $\widehat{\mathbb{G}}_a$.

**Corollary 4.2.14** *For complex-orientable $F$ and $E$, we have*[19]

$$\text{Spec } Q^*\mathcal{A}\mathcal{A}(F, E) \cong \underline{\text{FormalGroups}}(\mathbb{C}P_F^\infty, \mathbb{C}P_E^\infty).$$

*Proof* This is a matter of calculating $Q^*\mathcal{A}\mathcal{A}(F, E)$, which is possible to do coordinate-freely (using Lemma 4.2.6 and [Str99b, proposition 6.15]), but it is at least as clear to just give in and pick coordinates. Making such a choice and using Lemma 4.2.11, we abbreviate $\langle \beta_0[a_{ij}^E] \rangle$ to $\langle a_{ij}^E \rangle$ and $\beta_j[1]$ to $\beta_j$ to form

$$\mathop{*}_{i,j}\left(\langle a_{ij}^E \rangle \circ \left(\sum_k \beta_k s^k\right)^{\circ i} \circ \left(\sum_\ell \beta_\ell t^\ell\right)^{\circ j}\right) \equiv \sum_{i,j} a_{ij}^E \left(\sum_k \beta_k s^k\right)^i \left(\sum_\ell \beta_\ell t^\ell\right)^j,$$

(in $Q^*$)

from which it follows that

$$Q^*\mathcal{A}\mathcal{A}(F, E) = (F_* \otimes E_*)[\beta_0, \beta_1, \beta_2, \ldots]/(\beta(s +_F t) = \beta(s) +_E \beta(t)) \ . \quad □$$

*Remark 4.2.15* Using the equivalence $\mathbb{C}P^1 \simeq \mathbb{S}^2$, the homology suspension element $e_2$ is modeled by $\beta_1$. It follows immediately that the stable algebraic approximation $\text{Spec}(Q^*\mathcal{A}\mathcal{A}(F, E))[\beta_1^{-1}]$ models the scheme of formal group isomorphisms $\underline{\text{FormalGroups}}(\mathbb{C}P_F^\infty, \mathbb{C}P_E^\infty)^{\text{gpd}}$.

*Remark 4.2.16* In the unmixed case of $E = F$, as we saw in the computational example in Lecture 4.1, the algebraic Hopf ring $\mathcal{A}\mathcal{A}(E, E)$ picks up an extra diagonal corresponding to the composition of formal group endomorphisms of $\mathbb{C}P_E^\infty$, and the resulting pair $(\text{Spec } E_0, \underline{\text{End}}(\mathbb{C}P_E^\infty))$ forms a category scheme. This observation has been fully expanded in plain language by Boardman, Johnson, and Wilson [BJW95, section 10]. In the mixed case, these schemes act by pre- and post-composition on the classical part of the mixed algebraic Hopf ring $\text{Spec } Q^*\mathcal{A}\mathcal{A}(F, E)$, and in fact these actions appear as part of the structure maps in the unstable context $\mathcal{U}\mathcal{M}_{E \lor F}$ described at the beginning of this lecture. This description is also compatible with pulling back to the stable context $\mathcal{M}_{E \lor F}$: It is exactly the inclusion of the simplicial subobject consisting of the formal group isomorphisms and automorphisms. However, these induced structures *at the level of mixed Hopf rings themselves* seem under-studied in the current literature (although, see [HH95, remark 2.6]).

---

[19] Hill and Hopkins have described the behavior of SpH $\mathcal{A}\mathcal{A}(H\mathbb{Z}, MU)$ on (graded) Hopf rings not in the image of $i$ [HHR16, definition 1.22].

## 4.3 Unstable Cooperations for Complex Bordism

**Convention:** We will write $H$ for $H\mathbb{F}_p$ for the duration of this lecture.[20]

Our theme for the rest of this case study is that the comparison map

$$\mathcal{A}\mathcal{A}(F, E) \to F_*\underline{E}_*$$

of Lemma 4.2.6 is often an isomorphism. In this lecture, we begin by investigating the very modest and concrete setting of $F = H = H\mathbb{F}_p$ and $E = BP$, simply because it is the least complicated choice after the unstable Steenrod algebra: The spectrum $H$ has Künneth isomorphisms, and the formal group law associated to $BP$ has a very understandable role. Our goal is to prove the following theorem:

**Theorem 4.3.1** ([RW77, theorem 4.2]) *The natural homomorphism*

$$\mathcal{A}\mathcal{A}(H, BP) \to H_*\underline{BP}_{2*}$$

*is an isomorphism. (In particular, $H_*\underline{BP}_{2*}$ is even-concentrated.)*

Again, because we are working through a computation, we rest on the graded form of homology. With this tool available, this result is proved by a fairly elaborate counting argument: The rough idea is to show that the topological Hopf ring is polynomial, the comparison map is surjective, and the degrees arrange themselves so that the map then has no choice but to be an isomorphism. Our first move will thus be to produce an upper bound for the size of the source Hopf ring, so that surjectivity can be used to compare it with the size of the algebraic approximation.

Crucially, polynomiality will often let us replace the full Hopf ring by its ring of $*$-indecomposables. To begin, recall the following consequence of Corollary 4.2.12:

**Corollary 4.3.2** *As an algebra under the $\circ$-product,*

$$Q^*H_*[BP^*] \cong \mathbb{F}_p[\langle v_n \rangle \mid n \geq 1]. \qquad \square$$

From Lemma 4.2.8, we now know that $Q^*\mathcal{A}\mathcal{A}(H, BP)$ is generated by $\langle v_n \rangle$ for $n \geq 1$ and $\beta_j \in H_{2j}\underline{BP}_2$, $j \geq 0$. In fact, $p$-typicality shows [RW77, lemma 4.14] that it suffices to consider $\beta_{p^d} = \beta_{(d)}$ for $i \geq 0$. Altogether, this gives a secondary comparison map

$$A := \mathbb{F}_p[\langle v_n \rangle, \beta_{(d)} \mid n > 0, d \geq 0] \twoheadrightarrow Q^*\mathcal{A}\mathcal{A}(H, BP).$$

Although this map is surjective it is not an isomorphism, as these elements are subject to the following relation:

---

[20] In fact, $\mathbb{F}_p$ can be replaced by any field $k$ of positive characteristic $p$.

### 4.3 Unstable Cooperations for Complex Bordism

**Lemma 4.3.3** ([RW77, lemma 3.14], [Wil82, theorem 9.13])  *Take n to be an integer, and set $I = (\langle p \rangle, \langle v_1 \rangle, \langle v_2 \rangle, \ldots)$. In $Q^*\!A\!A(H, BP)/I^{\circ 2} \circ Q^*\!A\!A(H, BP)$, we have the relation*

$$\sum_{i=1}^{n} \langle v_i \rangle \circ \langle \beta_{(n-i)}^{\circ p^i} \rangle \equiv 0.$$

*Proof*  Since the group law on $\mathbb{C}P_H^\infty$ is additive, the Ravenel–Wilson relation for the Hopf ring $p$-series specializes to $[p]_{[BP]}(\beta(s)) = \beta(ps)$.[21] From Lemma 3.3.10, we deduce the relation

$$[p]_{BP}(s) \equiv \sum_{j \geq 0} v_j s^{p^j} \quad (\text{mod } (p, v_1, v_2, \ldots)^2).$$

These combine to give

$$\beta_0 = \beta(0) = \beta(ps) = [p]_{[BP]}(\beta(s)) \equiv \underset{j \geq 0}{\text{\Large$*$}}[v_j] \circ \beta(s)^{\circ p^j} \quad (\text{mod } I^{\circ 2}).$$

Passing to $Q^*$, we join Lemma 4.2.11 to the identity $[p] \circ \beta(s) \equiv \beta_0$ to deduce

$$0 \equiv \sum_{j > 0} \langle v_j \rangle \circ \langle \beta(s)^{\circ p^j} \rangle.$$

The coefficient of $s^{p^n}$ gives the identity claimed.  □

Let $r_n$, the $n$th relation, denote the same sum taken in $A$ instead:

$$r_n := \sum_{i=1}^{n} \langle v_i \rangle \circ \beta_{(n-i)}^{\circ p^i} \in A.$$

The lemma then shows that the image of $r_n$ in $Q^*\!A\!A(H, BP)$ is $\circ$-decomposable. Our stated goal is to show that these relations cut $A$ down to exactly the right size, and this task would be easiest if the quotient were by a regular ideal.

**Lemma 4.3.4** ([RW77, lemma 4.15])  *The sequence $(r_1, r_2, \ldots) \in A$ is regular.*

*Proof*  Our approach is intricate but standard. We seek to show that $J = (r_1, r_2, \ldots, r_n)$ is regular for every $n$, and we accomplish this by interpolation. Fixing a particular $n$, define the intermediate ideals

$$J_j = (r_n, r_{n-1}, \ldots, r_{n-j+1}),$$

as well as the intermediate rings

$$A_j = A/(\beta_{(0)}, \ldots, \beta_{(n-j-1)}), \qquad B_j = \beta_{(n-j)}^{-1} A_j.$$

---

[21] We are very sorry for the collision of $[p]_{BP}$ the $p$-series, $[p]_{[BP]}$ the fancy Hopf ring $p$-series, and $[p]$ the symbol in the Hopf ring induced from the element $p \in BP_0$. The fancy $p$-series won't linger for long, and we will always differentiate them with the relevant subscript.

Noting that $A_n = A$ and $J_n = J$, we will inductively show that $J_j$ is a regular ideal of $A_j$. The case $j = 1$ is simple: $J_1$ is a nonzero principal ideal in a ring without zero divisors, so it must be regular.

Assume the inductive result holds below some index $j$. In the quotient sequence
$$0 \to \Sigma^{|\beta_{(n-j)}|} A_j \xrightarrow{\beta_{(n-j)}} A_j \to A_{j-1} \to 0,$$
the degree shift in the multiplication map (and induction on degree) shows that if $J_{j-1}$ is regular on $A_{j-1}$, then $J_{j-1}$ is automatically regular on $A_j$. If we additionally prove that $J_{j-1}$ is prime in $A_j$ and that $r_{n-j+1} \neq 0$ in the quotient, then $A_j / J_{j-1}$ would be an integral domain, multiplication by $r_{n-j+1}$ would be injective, and we would be done. In the degree $|r_{n-j+1}|$ of interest, there is an isomorphism $(A)_{|r_{n-j+1}|} \cong (A_j / J_{j-1})_{|r_{n-j+1}|}$, and hence $r_{n-j+1} \neq 0$ as desired.

We thus turn to primality. Note first that $J_{j-1}$ is automatically prime in $B_j$, since $B_j$ is a polynomial $\mathbb{F}_p[\beta^{\pm}_{(n-j)}]$-algebra and each of the generators of $J_{j-1}$ is one of these polynomial generators of $B_j$. Suppose for contradiction that $J_{j-1}$ is not prime in $A_j$, as witnessed by some elements $x, y \notin J_{j-1}$ satisfying $xy \in J_{j-1}$. Since $J_{j-1}$ *is* prime in $B_j$, after perhaps trading $x$ and $y$ there is some minimum $k > 0$ such that
$$\beta^{\circ k}_{(n-j)} \circ x \in J_{j-1}.$$
We may as well assume $k = 1$, which we can arrange by tucking the stray factors of $\beta_{(n-j)}$ into $x$. Invoking the generators of $J_{j-1}$, we thus have an equation
$$\beta_{(n-j)} \circ x = \sum_{i=1}^{j-1} a_i \circ r_{n-i+1}$$
with $a_i \in A_j$ not all divisible by $\beta_{(n-j)}$. In fact, by moving elements onto the left-hand side, we can assume that if $a_i \neq 0$ then $a_i \notin J_{i-1}$. In $A_{j-1}$, this equation becomes
$$0 = \sum_{i=1}^{j-1} a_i \circ r_{n-i+1}$$
with $a_i$ not all in $J_{i-1}$. This is the desired contradiction, since $J_{j-1}$ is regular in $A_{j-1}$ by inductive hypothesis. □

**Corollary 4.3.5** *Set*
$$c_{i,j} = \dim_{\mathbb{F}_p} Q^* \mathcal{A}(H, BP)_{(2i,2j)}, \qquad d_{i,j} = \dim_{\mathbb{F}_p} \mathbb{F}_p[\langle v_n \rangle, \beta_{(0)}]_{2i,2j}.$$
*Then $c_{i,j} \leq d_{i,j}$ and $d_{i,j} = d_{i+2, j+2}$.*

4.3 Unstable Cooperations for Complex Bordism     191

*Proof* We have seen that $c_{i,j}$ is bounded by the $\mathbb{F}_p$-dimension of
$$\left[\mathbb{F}_p[\langle v_n\rangle, \beta_{(d)} \mid d \geq 0, n \geq 0]/(r_1, r_2, \ldots)\right]_{i,j}.$$
But, since this ideal is regular and $|r_j| = |\beta_{(j)}|$, this is the same value as $d_{i,j}$. The other relation among the $d_{i,j}$ follows from multiplication by $\beta_{(0)}$, with $|\beta_{(0)}| = (2, 2)$. □

We now turn to showing that this estimate is *sharp* and that the secondary comparison map is *surjective*, and hence an isomorphism, using the bar spectral sequence. Recalling that the bar spectral sequence converges to the homology of the *connective* delooping, let $\underline{BP}'_{2*}$ denote the connected component of $\underline{BP}_{2*}$ containing $[0_{2*}]$. We will then demonstrate the following theorem inductively:

**Theorem 4.3.6** ([RW77, induction 4.18])   *The following hold for all values of the induction index $k$:*

1. *$Q^*H_{\leq 2(k-1)}\underline{BP}'_{2*}$ is generated by $\circ$-products of the $\langle v_n \rangle$ and $\beta_{(j)}$.*
2. *$H_{\leq 2(k-1)}\underline{BP}'_{2*}$ is isomorphic to a polynomial algebra in this range.*
3. *For $0 < i \leq 2(k-1)$, we have $d_{i,j} = \dim_{\mathbb{F}_p} Q^*H_i\underline{BP}_{2j}$.*

Before addressing the theorem, we show that this finishes our calculation:

*Proof of Theorem 4.3.1, assuming Theorem 4.3.6 for all $k$*   Recall that we are considering the natural map
$$\mathcal{A}\!\mathcal{A}(H, BP) \to H_*\underline{BP}_{2*}.$$
The first part of Theorem 4.3.6 shows that this map is a surjection. The third part of Theorem 4.3.6 together with our counting estimate shows that the induced map
$$Q^*\mathcal{A}\!\mathcal{A}(H, BP) \to Q^*H_*\underline{BP}_{2*}$$
is an isomorphism. Finally, the second part of Theorem 4.3.6 says that the original surjective map, before passing to $*$-indecomposables, targets a polynomial algebra and is an isomorphism on indecomposables, hence must be an isomorphism as a whole. □

*Proof of Theorem 4.3.6*   The infinite loopspaces in $\underline{BP}_{2*}$ are related by
$$\Omega^2 \underline{BP}'_{2(*+1)} = \underline{BP}_{2*},$$
so we will use two bar spectral sequences to extract information about $\underline{BP}'_{2(*+1)}$ from $\underline{BP}_{2*}$. Since we have assumed that $H_{\leq 2(k-1)}\underline{BP}_{2*}$ is polynomial in the indicated triangular range near zero, we know that in the first spectral sequence
$$E^2_{*,*} = \mathrm{Tor}^{H_*\underline{BP}_{2*}}_{*,*}(\mathbb{F}_p, \mathbb{F}_p) \Rightarrow H_*\underline{BP}_{2*+1}$$

the $E^2$-page is, in the same range, exterior on generators in Tor-degree 1 and topological degree one higher than the generators in the polynomial algebra. Since differentials lower Tor-degree, the spectral sequence is multiplicative, and there are no classes on the 0-line, it collapses in the range $[0, 2k - 1]$. Additionally, since all the generating classes are in odd topological degree, there are no algebra extension problems, and we conclude that $H_*\underline{BP}_{2*+1}$ is indeed exterior up through degree $(2k - 1)$.

We now consider the second bar spectral sequence

$$E^2_{*,*} = \mathrm{Tor}_{*,*}^{H_*\underline{BP}_{2*+1}}(\mathbb{F}_p, \mathbb{F}_p) \Rightarrow H_*\underline{BP}_{2(*+1)'}.$$

The Tor algebra of an exterior algebra is divided power on a class of topological dimension one higher. Since these classes are now all in even degrees, the spectral sequence collapses in the range $[0, 2k]$. Additionally, these primitive classes are related to the original generating classes by double suspension, i.e., by forming the $\circ$-product with $\beta_{(0)}$. This shows the first inductive claim on the *primitive classes* through degree $2k$, and we must argue further to deduce our generation result for $x^{[p^j]}$ of degree $2k$ with $j > 0$. By inductive assumption, we can write

$$x = \sum_I \langle y_I \rangle \circ \beta_{(0)}^{\circ I_0} \circ \beta_{(1)}^{\circ I_1} \circ \cdots,$$

and it suffices to treat the case where the sum has just one term. Aiming to understand $x^{[p^j]}$, one might be divinely inspired to consider the element

$$z := \langle y_I \rangle \circ \beta_{(j)}^{\circ I_0} \circ \beta_{(j+1)}^{\circ I_1} \circ \cdots.$$

This element $z$ isn't equal to $x^{[p^j]}$ on the nose, but the coproduct of $z - x^{[p^j]}$ can be manually calculated to in lower filtration degree, so that $z \equiv x^{[p^j]}$ holds modulo filtration degree in the bar spectral sequence. Since $z$ itself is defined as a $\circ$-product, the full inductive claim follows.

The remaining thing to do is to use the size bounds: The only way that the map

$$\mathcal{A}(H, BP) \to H_*\underline{BP}_{2*}$$

could be surjective is if there were multiplicative extensions in the spectral sequence joining $x^{[p]}$ to $x^p$. Granting this, we see that the module ranks of the algebra itself and of its indecomposables are exactly the right size to be a free (i.e., polynomial) algebra, and hence this must be the case by Milnor–Moore. □

We have actually accomplished quite a lot in proving Theorem 4.3.1, as this forms the input to an Atiyah–Hirzebruch spectral sequence.

**Corollary 4.3.7** (Ravenel–Wilson Theorem [RW77, corollary 4.7]) *For any* complex-orientable cohomology theory $E$, the natural approximation maps give isomorphisms of Hopf rings[22]

$$\mathcal{AA}(E, MU) \xrightarrow{\cong} E_*\underline{MU}_{2*}, \qquad \mathcal{AA}(E, BP) \xrightarrow{\cong} E_*\underline{BP}_{2*}.$$

*Proof* The case of $E = H\mathbb{F}_p$ is Theorem 4.3.1. Since $H\mathbb{F}_{p*}\underline{BP}_{2*}$ is even, it follows that $H\mathbb{Z}_{(p)*}\underline{BP}_{2*}$ is torsion-free on a lift of a basis; that $H\mathbb{Z}_{(p)*}\underline{MU}_{2*}$ is also torsion-free on a lift of a basis; and that $H\mathbb{Z}_*\underline{MU}_{2*}$ is as well. Next, these properties trivialize the Atiyah–Hirzebruch spectral sequence governing $MU_*\underline{MU}_{2*}$, so the theorem holds in this case too. Finally, using naturality of the Atiyah–Hirzebruch spectral sequence, given a complex-orientation $MU \to E$ we deduce that the spectral sequence

$$E_* \otimes H_*(\underline{MU}_{2*}; \mathbb{Z}) \cong E_* \otimes_{MU_*} MU_*\underline{MU}_{2*} \Rightarrow E_*\underline{MU}_{2*}$$

collapses, and similarly for the case of $BP$. □

This is an impressively broad claim: the loopspaces $\underline{MU}_{2*}$ are quite complicated, and that any general statement can be made about them is remarkable. That this fact follows from a calculation in $H\mathbb{F}_p$-homology and some niceness observations is meant to showcase the density of $\mathbb{CP}_H^\infty \cong \widehat{\mathbb{G}}_a$ inside of $\mathcal{M}_\mathbf{fg}$.[23],[24]

*Remark 4.3.8* The analysis of the first bar spectral sequence in the proof of Theorem 4.3.6 also gave us a description of $H_*\underline{BP}_{2*+1}$, which is not directly visible to $\mathcal{AA}(H, BP)$. Namely, the Hopf ring $H_*\underline{BP}_*$ can be presented as

$$H_*\underline{BP}_* \xleftarrow{\cong} \mathcal{AA}(H, BP)[e]/(e^{\circ 2} = \beta_{(0)}),$$

with $e$ of degree $(1, 1)$. Additionally, analyzing the cohomological bar spectral sequence (and noting that the dual of a divided power algebra is a polynomial algebra) shows that each $H_*\underline{BP}_{2*}$ forms a *bipolynomial Hopf algebra* – i.e., both it and its dual are polynomial algebras. These bipolynomial algebras also play a critical role in the next two sections.

*Remark 4.3.9* ([Cha82], [Wil82, section 10]) There is an alternative proof that $H_*\underline{BP}_{2j}$ forms a bipolynomial Hopf algebra for each choice of $j$ that makes no

---

[22] In the case $E = MU$, we actually have brushed against this before: The formulas leading to Lemma 2.5.6 look suspiciously like formal group homomorphisms with prescribed kernels. We explore this observation more seriously in Appendix A.2.
[23] Equivalently, the convergence of Postnikov towers.
[24] It is worth pointing out that the success of this calculation is misleading as to how difficult unstable calculations can be. For instance, the Johnson–Yosimura conjecture is that for a space $X$ and $x \in BP_n(X)$ a $BP$-homology class, $x$ is $v_n$-torsion-free [Wil82, p. 37]. Even related stable conjectures are unsettled: For $\mathbb{S}^m \to X \to Y$ a cofibration of finite complexes with $BP_*(X)$ of homological dimension less than $n$, it is conjectured that the homological dimension of $BP_*(Y)$ is less than $n + 1$ [JW73, conjecture 6.8]; and if $0 \neq y \in BP\langle n \rangle_s(X)$ is a $v_n$-torsion-element of a space $X$, it is conjectured that $s < \dim(X)$ [JW73, question 6.11].

reference to Hopf rings. It proceeds along very similar lines, as it also studies the iterated bar spectral sequence, but it proceeds entirely by counting: The elements in the spectral sequence are never given explicit names, and hence it gives no real hope of understanding the functor $\mathrm{SpH}\, H_*\underline{BP}_{2*}$. By contrast, the Ravenel–Wilson method can be used to give an explicit enumeration of these classes [RW77, section 5]. Our presentation here is something of a compromise.

*Remark 4.3.10* The identification of the $p$-local and mod-$p$ homology and cohomology of $\underline{BP}_{2k}$ as a bipolynomial Hopf algebra was first accomplished by Wilson in his Ph.D. thesis [Wil73, theorem 3.3]. He deduces quite a lot of interesting results from this observation. For instance, each bipolynomial Hopf algebra can be shown to split as a tensor product of indecomposable such [Wil73, proposition 3.5], and this splitting is reflected by a splitting of $\underline{BP}_{2k}$ into a product of indecomposable $H$-spaces.

Remarkably, these indecomposable spaces can themselves be identified. For each $n$ there is a ring spectrum $BP\langle n\rangle$ over $BP$ with homotopy presented by the subalgebra $\pi_*BP\langle n\rangle = \mathbb{Z}_{(p)}[v_1,\ldots,v_n]$. This spectrum is *not* uniquely specified, a reflection of the algebraic failure of the ideal $(v_{n+1}, v_{n+2}, \ldots)$ to be invariant, and so this resists formal-geometric interpretation (see, however, [AL17], [LN12], [Str99a]). Nonetheless, using Steenrod module techniques, Wilson [Wil75, section 6] shows that every simply connected $p$-local $H$-space with torsion-free homology and ($p$-local ordinary) homology splits into a product of spaces $Y_k$, and that $Y_k = \underline{BP\langle n\rangle}_k$ for $|v_n| < k(p-1) \leq |v_{n+1}|$.

In particular, the spaces $\underline{BP\langle n\rangle}_k$ in these bands *are* independent of choice of parent spectrum $BP\langle n\rangle$, and all $p$-local $H$-spaces satisfying these freeness properties are automatically infinite loopspaces – extremely surprising results.

## 4.4 Dieudonné Modules

Our goal in this lecture is to give a compact presentation of formal groups based on the following observation: The category of commutative cocommutative Hopf algebras of finite type over a ground field $k$ forms an abelian category. It follows abstractly that this category admits a presentation as a full subcategory of the module category for some (possibly noncommutative) ring, but in fact this ring and the assignment from a group scheme to linear algebraic data can both be described explicitly. This is the subject of *Dieudonné theory*, and we will give an overview of some of its main results, including two different presentations of the equivalence.[25]

---

[25] Emphasis on "*some of its results*." Dieudonné theory is an enormous subject with many interesting results both internal and connected to arithmetic geometry and the theory of abelian varieties. We will explore almost none of this.

4.4 Dieudonné Modules 195

In the first presentation, we follow notes by Weinstein [Wei11, lecture 1]. Begin with a one-dimensional formal group $\widehat{\mathbb{G}}$ over a ring $A$, and recall that we have previously been interested in the invariant differentials $\omega_{\widehat{\mathbb{G}}} \subseteq \Omega^1_{\widehat{\mathbb{G}}/A}$ on $\widehat{\mathbb{G}}$. As explored in Theorem 2.1.25, when $A$ is a $\mathbb{Q}$-algebra such differentials give rise to logarithms through integration. On the other hand, if $A$ has positive characteristic $p$ then there is a potential obstruction to integrating terms with exponents congruent to $-1 \pmod{p}$, and in Lecture 3.3 we used this to lead us to the notion of $p$-height. We now explore a third twist on this set-up, recalling that $\Omega^1_{\widehat{\mathbb{G}}/A}$ forms the first level of the *algebraic de Rham complex* $\Omega^*_{\widehat{\mathbb{G}}/A}$. The group operation and two projections

$$\mu, \pi_1, \pi_2 \colon \widehat{\mathbb{G}} \times \widehat{\mathbb{G}} \to \widehat{\mathbb{G}}$$

induce maps

$$\mu^*, \pi_1^*, \pi_2^* \colon C^1_{dR}(\widehat{\mathbb{G}}/A) \to C^1_{dR}(\widehat{\mathbb{G}} \times \widehat{\mathbb{G}}/A).$$

The translation invariant differentials are exactly those in the kernel of $\mu^* - \pi_1^* - \pi_2^*$, as considered at the chain level. We can weaken this to request only that the difference be *exact*, or zero at the level of cohomology of the de Rham complex.

**Definition 4.4.1** The *cohomologically translation invariant differentials* constitute the $A$-submodule $PH^1_{dR}(\widehat{\mathbb{G}}/A) \subseteq H^1_{dR}(\widehat{\mathbb{G}}/A)$ defined as the kernel of $\mu^* - \pi_1^* - \pi_2^*$.[26]

*Example 4.4.2* ([Kat81, lemma 5.1.2]) Consider the case that $A$ is torsion-free, and set $K = A \otimes \mathbb{Q}$ so that $A \to K$ is an injection. In this case the differentiation map $xA[\![x]\!] \to A[\![x]\!]$ is an injection and integration of power series is possible in $K$, so we can re-express first the definition of $H^1_{dR}$ and second the conditions on our algebraic differentials in the following diagram of exact rows:

$$
\begin{array}{ccccccccc}
0 & \to & \left\{\begin{array}{c}\text{integrals}\\\text{with } A\\\text{coeffs}\end{array}\right\} & \longrightarrow & \left\{\begin{array}{c}\text{all formal integrals}\\\text{of differentials}\\\text{defined over } A\end{array}\right\} & \longrightarrow & \left\{\begin{array}{c}\text{missing}\\\text{integrals}\end{array}\right\} & \to & 0 \\
 & & \| & & \| & & \| & & \\
0 & \longrightarrow & xA[\![x]\!] & \longrightarrow & \{f \in xK[\![x]\!] \mid df \in A[\![x]\!]\,dx\} & \xrightarrow{d} & H^1_{dR}(\widehat{\mathbb{G}}/A) & \to & 0 \\
 & & \| & & \uparrow & & \uparrow & & \\
0 & \longrightarrow & xA[\![x]\!] & \longrightarrow & \left\{f \in xK[\![x]\!] \;\middle|\; \begin{array}{c} df \in A[\![x]\!]\,dx,\\ \delta f \in A[\![x,y]\!]\end{array}\right\} & \xrightarrow{d} & PH^1_{dR}(\widehat{\mathbb{G}}/A) & \to & 0,
\end{array}
$$

where $x$ is a coordinate on $\widehat{\mathbb{G}}$, and $\delta$ is defined by $\delta f = (\mu^* - \pi_1^* - \pi_2^*)f$.

---
[26] The symbol "$P$" here denotes the primitives. Using the identity $\mu^* = \Delta$, the equation $\Delta(x) = x \otimes 1 + 1 \otimes x$ holds if and only if $x$ lies in the kernel of this operator.

The flatness condition is not satisfied when working over a perfect field of positive characteristic $p$ – our favorite setting in Lecture 3.3 and Case Study 3 more generally – and without it we cannot make the identifications in the example. However, de Rham cohomology has the following remarkable lifting property (which we have written here after specializing to $H^1_{dR}$):

**Theorem 4.4.3** (Poincaré Lemma [Kat81, key lemma 5.1.3]) *Let $A$ be a $p$-local torsion-free ring, and let $f_1(x), f_2(x) \in xA[\![x]\!]$ be power series without constant term. If $f_1 \equiv f_2 \pmod{p}$, then for any differential $\omega \in A[\![x]\!]dx$ the difference $f_1^*(\omega) - f_2^*(\omega)$ is exact.*

*Proof* Write $\omega = dg$ for $g \in K[\![x]\!]$, and write $f_2 = f_1 + p\Delta$. Then

$$\int \left( f_2^*\omega - f_1^*\omega \right) = g(f_2) - g(f_1) = g(f_1 + p\Delta) - g(f_1)$$

$$= \sum_{n=1}^{\infty} \frac{(p\Delta)^n}{n!} g^{(n)}(f_1).$$

Since $g' = \omega$ has coefficients in $A$, so does the iterated derivative $g^{(n)}$ for all $n$; the fraction $p^n/n!$ lies in $\mathbb{Z}_{(p)}$; and hence the entire sum lies in the $\mathbb{Z}_{(p)}$-algebra $A$. □

**Corollary 4.4.4** ($H^1_{dR}$ is "crystalline") *If $f_1, f_2 \colon V \to V'$ are maps of pointed formal varieties which agree mod $p$, then they induce the same map on $H^1_{dR}$.* □

**Corollary 4.4.5** ([Kat81, theorem 5.1.4]) *Any map $f \colon \widehat{\mathbb{G}}' \to \widehat{\mathbb{G}}$ of pointed varieties which is a group homomorphism mod $p$ induces a map*

$$f^* \colon PH^1_{dR}(\widehat{\mathbb{G}}/A) \to PH^1_{dR}(\widehat{\mathbb{G}}'/A).$$

*Additionally, if $f_1$, $f_2$, and $f_3$ are three such maps satisfying*

$$f_3 \equiv f_1 + f_2 \in \text{FormalGroups}(\widehat{\mathbb{G}}'/p, \widehat{\mathbb{G}}/p),$$

*then $f_3^* = f_1^* + f_2^*$ as maps $PH^1_{dR}(\widehat{\mathbb{G}}/A) \to PH^1_{dR}(\widehat{\mathbb{G}}'/A)$.* □

In the case that $k$ is a *perfect* field, recall from Remark 3.4.6 that the ring $\mathbb{W}_p(k)$ of $p$-typical Witt vectors on $k$ is simultaneously torsion-free and universal among nilpotent thickenings of the residue field $k$. This emboldens us to make the following definition:[27]

---

[27] There is a better definition one might hope for, which instead assigns to each potential thickening and lift a "Dieudonne module," and then works to show that they all arise as base-changes of this universal one. This is possible and technically superior to the approach we are taking here ([Kat81, theorem 5.1.6], [Mes72, chapter 4], [Gro74]).

## 4.4 Dieudonné Modules

**Definition 4.4.6** [Kat81, section 5.5] Let $k$ be a perfect field of characteristic $p > 0$, and let $\widehat{\mathbb{G}}_0$ be a (one-dimensional) formal group over $k$. Then, choose a lift $\widehat{\mathbb{G}}$ of $\widehat{\mathbb{G}}_0$ to $\mathbb{W}_p(k)$, and define the *(contravariant) Dieudonné module* of $\widehat{\mathbb{G}}_0$ by $D^*(\widehat{\mathbb{G}}_0) := PH^1_{dR}(\widehat{\mathbb{G}}/\mathbb{W}_p(k))$.

*Remark 4.4.7* This is independent of choice of lift up to coherent isomorphism. Given any other lift $\widehat{\mathbb{G}}'$ of $\widehat{\mathbb{G}}_0$ to $\mathbb{W}_p(k)$, we can find *some* power series – not necessarily a group homomorphism – covering the identity on $\widehat{\mathbb{G}}_0$. Corollary 4.4.4 then shows that this map induces a *canonical* isomorphism between the two potential definitions of $D^*(\widehat{\mathbb{G}}_0)$.

Next, note that the module $D^*(\widehat{\mathbb{G}}_0)$ carries some natural operations:

- Arithmetic: $D^*(\widehat{\mathbb{G}}_0)$ is a $\mathbb{Z}_p^\wedge$-module, with the action by $\ell \in \mathbb{Z} \subseteq \mathbb{Z}_p^\wedge$ corresponding to multiplication-by-$\ell$ internal to $\widehat{\mathbb{G}}_0$.
- Frobenius: Modulo $p$, the map $x \mapsto x^p$ determines a homomorphism $\widehat{\mathbb{G}}_0 \to \mathrm{Frob}^*\widehat{\mathbb{G}}_0$, in turn inducing a map $F \colon D^*(\widehat{\mathbb{G}}_0) \to D^*(\varphi^*\widehat{\mathbb{G}}_0)$. This map is sometimes referred to as $\varphi$-*semilinear*, owing to the identification $D^*(\varphi^*\widehat{\mathbb{G}}_0) \cong \mathbb{W}_p(k) \otimes^\varphi_{\mathbb{W}_p(k)} D^*(\widehat{\mathbb{G}}_0)$ and hence the formula
$$F(\alpha v) = \alpha^\varphi F(v),$$
where $\varphi$ is a lift of the Frobenius on $k$ to $\mathbb{W}_p(k)$.
- Verschiebung: Inspired by Lemma 3.3.6, we seek a Verschiebung operator $V$ satisfying $VF = p$. Our explicit formula for $F$ lets us guess such a map:
$$V \colon \sum_{n=1}^\infty a_n x^{n-1} dx \mapsto \sum_{n=1}^\infty a_{pn}^{\varphi^{-1}} x^{n-1} dx.$$
It satisfies $VF = p$ and anti-semilinearity: $aV(v) = V(a^\varphi v)$.

With this, we come to the main theorem of this lecture:

**Theorem 4.4.8** ([Gro74, théorème 4.2], [Dem86, sections III.8–III.9])  *For $k$ a perfect field of characteristic $p$, the (de Rham–Dieudonné) functor $D^*$ determines a contravariant equivalence of categories between smooth one-dimensional formal groups $\widehat{\mathbb{G}}_0$ over $k$ of finite $p$-height and Dieudonné modules, which are modules $M$ over the ring*
$$\mathrm{Cart}_p = \mathbb{W}_p(k)\langle F, V\rangle \left/ \left( \begin{array}{c} FV = VF = p, \\ Fw = w^\varphi F, \\ wV = Vw^\varphi \end{array} \right)\right.$$
*that are additionally free and finite rank over $\mathbb{W}_p(k)$.* □

*Remark 4.4.9* It is possible to read off several invariants of the formal group via its Dieudonné module. For example, the $\mathbb{W}_p(k)$-rank of $D^*(\widehat{\mathbb{G}}_0)$ computes the height of $\widehat{\mathbb{G}}_0$. Additionally, the quotient $D^*(\widehat{\mathbb{G}}_0)/FD^*(\widehat{\mathbb{G}}_0)$ is canonically isomorphic to the cotangent space $T_0^*\widehat{\mathbb{G}}_0 \cong \omega_{\widehat{\mathbb{G}}_0}$, so that

$$\dim \widehat{\mathbb{G}}_0 = \dim_k D^*(\widehat{\mathbb{G}}_0)/FD^*(\widehat{\mathbb{G}}_0).$$

*Example 4.4.10* Using Remark 4.4.9, we see that the Dieudonné module associated to a formal group $\Gamma$ of height 1 must take the form

$$D^*(\Gamma) = \mathbb{W}_p(k)\{x\} \bigg/ \begin{pmatrix} Fx = p(u^{-1})^\varphi x^\varphi \\ Vx = ux^{-\varphi} \end{pmatrix}$$

for a constant $u$, referred to as the *modulus* (see [Mor89, theorem 4.2.2]). In the case of $\Gamma = \widehat{\mathbb{G}}_m$, this constant is simply $u = 1$, owing essentially to the formula $[p](x) = x^p$ for the standard coordinate on the multiplicative formal group.

*Example 4.4.11* We also give a kind of non-example: $\widehat{\mathbb{G}}_a$ is *not* a finite-height formal group, and its Dieudonné module is correspondingly strangely behaved:

$$D^*(\widehat{\mathbb{G}}_a) = k[\![F]\!]\{x\}/(V = 0).$$

*Example 4.4.12* (see Example 1.2.9) Dieudonné theory admits an extension to finite group schemes over $k$ as well, and the appropriate quotient of the Dieudonné module of a formal group agrees with the Dieudonné module associated to the appropriate subscheme:

$$D^*(\widehat{\mathbb{G}}_0[p^j]) = D^*(\widehat{\mathbb{G}}_0)/p^j.$$

For example, this gives

$$D^*(\widehat{\mathbb{G}}_m[p]) = k\{x\} \bigg/ \begin{pmatrix} F(cx) = 0, \\ V(cx) = c^{-\varphi}x \end{pmatrix}.$$

Specializing to $k = \mathbb{F}_2$, we extract the subgroup scheme $\alpha_2$ as the finite Dieudonné quotient module $D^*(\alpha_2) = \mathbb{F}_2\{x\} \leftarrow D^*(\widehat{\mathbb{G}}_a)$ of the Dieudonné module associated to $\widehat{\mathbb{G}}_a$. We can now verify the four claims from Example 1.2.9:

- *The group scheme $\alpha_2$ has the same underlying structure ring as $\mu_2 = \mathbb{G}_m[2]$ but is not isomorphic to it.* There are now several ways to see this, the simplest of which is that the Verschiebung operator acts nontrivially on $D^*(\mu_2)$ but wholly trivially on $D^*(\alpha_2)$.[28]

---

[28] See also [Strb, example 8.5].

## 4.4 Dieudonné Modules

- *There is no commutative group scheme G of rank four such that $\alpha_2 = G[2]$.* Suppose that $G$ were such a group scheme, so that $D^*(G)/2$ would give $D^*(\alpha_2)$. It can't be the case that $D^*(G)$ has only 2-torsion, since then this quotient would be a null operation, so it must be the case that $D^*(G) = \mathbb{Z}/4\{x\}$. The action of both $F$ and $V$ on $x$ must vanish after quotienting by 2, so it must be the case that $Fx = 2cx$ and $Vx = 2dx$ for some constants $c$ and $d$ – but this violates $VFx = 2x$.

- *If $E/\mathbb{F}_2$ is the supersingular elliptic curve, then there is a short exact sequence*

$$0 \to \alpha_2 \to E[2] \to \alpha_2 \to 0.$$

*However, this short exact sequence doesn't split (even after making a base change).* One definition of *supersingular* is that $\widehat{E}$ is a formal group of height 2. The claim then follows from calculating the action of $F$ and $V$ to get a short exact sequence of Dieudonné modules:

$$0 \to \mathbb{F}_2\{Fx\}/(F, V) \to \mathbb{F}_2\{x, Fx\}\bigg/\begin{pmatrix} F^2 x = 0, \\ V = 0 \end{pmatrix} \to \mathbb{F}_2\{x\}/(F, V) \to 0.$$

The exact sequence is split as $\mathbb{F}_2$-modules, but not as Dieudonné modules.

- *The subgroups of $\alpha_2 \times \alpha_2$ of rank two are parameterized by $\mathbb{P}^1$.* The Dieudonné module of the product is quickly computed:

$$D^*(\alpha_2 \times \alpha_2) = D^*(\alpha_2) \oplus D^*(\alpha_2) = \mathbb{F}_2\{x_1, x_2\}\bigg/\begin{pmatrix} F = 0, \\ V = 0 \end{pmatrix}.$$

An inclusion of a rank 2 subgroup scheme corresponds to a projection of this Dieudonné module onto a one-dimensional quotient module, and the ways to choose the kernel of this projection encompass a $\mathbb{P}^1$.

*Example 4.4.13* (see Theorem 3.6.16) We can also use Dieudonné theory to compute the automorphism group of a fixed Honda formal group, which is information we wanted back in Lecture 3.6. Take $\Gamma_d$ to be Honda formal group law of height $d$ over a perfect field $k$ containing $\mathbb{F}_{p^d}$. Its Dieudonné module is

$$D^*(\Gamma_d) = \mathrm{Cart}_p/(\mathrm{Cart}_p \cdot (p - F^d)),$$

where the relation stems from the defining $p$-series: $[p]_{\Gamma_d}(x) = x^{p^d}$. In general, the endomorphism ring of a quotient $R$-module $M = R/I$ is canonically isomorphic to the $I$-torsion in $R/I$; in our case, this gives

$$\mathrm{End}\,\Gamma_d \cong \mathbb{W}_p(\mathbb{F}_{p^d})\langle F \rangle\bigg/\begin{pmatrix} Fw = w^\varphi F, \\ F^d = p \end{pmatrix}$$

200    Unstable Cooperations

and hence

$$\operatorname{Aut}\Gamma_d \cong \left(\mathbb{W}_p(\mathbb{F}_{p^d})\langle F\rangle \middle/ \left(\begin{array}{c} Fw = w^\varphi F, \\ F^d = p \end{array}\right)\right)^\times.$$

*Remark 4.4.14* ([Kat81, theorem 5.2.1])  There is also a relationship between this representation of the Dieudonné functor and the deformation theory of formal groups from Lecture 3.4: A class $[f(x)\,\mathrm{d}x] \in D^*(\widehat{\mathbb{G}}_0)$ begets a class in $e(f) \in \operatorname{Ext}^1(\widehat{\mathbb{G}}, \widehat{\mathbb{G}}_a)$ given by the cobar 1-cocycle $f(x +_{\widehat{\mathbb{G}}} y) - f(x) - f(y)$. In fact, this assignment is surjective, and the additional information lost in the kernel is a trivialization of the Lie algebra extension

$$0 \longrightarrow \operatorname{Lie}(\widehat{\mathbb{G}}_a) \longrightarrow \operatorname{Lie}(E) \longrightarrow \operatorname{Lie}(\widehat{\mathbb{G}}) \longrightarrow 0$$

associated to the group scheme extension $E$ classified by $e(f)$. Studying the universal extension of $\widehat{\mathbb{G}}$ by an additive formal group is an exceedingly interesting thing to do ([Gro74, section V.4], [MM74], [HG94a, section 11]).

Having gotten some feel for the behavior and the usefulness of the Dieudonné functor, we now turn our attention to an alternative presentation of it. We will not have to worry about lifts to $\mathbb{W}_p(k)$ in this part, so we take $\widehat{\mathbb{G}}$ itself to be a formal group over a perfect field $k$ of positive characteristic $p$. Cartier's *functor of curves* is defined by the formula

$$C\widehat{\mathbb{G}} = \textsc{FormalSchemes}(\widehat{\mathbb{A}}^1, \widehat{\mathbb{G}}).$$

This is, again, a kind of relaxing of familiar data from Lie theory, taken from a different direction: Rather than studying just the exponential curves, $C\widehat{\mathbb{G}}$ tracks all possible curves. In Lecture 3.3, we considered three kinds of operations on a given curve $\gamma\colon \widehat{\mathbb{A}}^1 \to \widehat{\mathbb{G}}$:

- Homothety: For a scalar $a \in k$ we define $\theta_a \gamma(t) = \gamma(a \cdot t)$, which is designed to restrict to $k \cong \operatorname{End} T_0 \widehat{\mathbb{A}}^1$ on the linear part of the parameterization domain.
- Arithmetic: Given two curves $\gamma_1$ and $\gamma_2$, we can use the group law on $\widehat{\mathbb{G}}$ to define $\gamma_1 +_{\widehat{\mathbb{G}}} \gamma_2$. Moreover, given $\ell \in \mathbb{Z}$, the $\ell$-fold sum in $\widehat{\mathbb{G}}$ gives an operator

$$\ell \cdot \gamma = \overbrace{\gamma +_{\widehat{\mathbb{G}}} \cdots +_{\widehat{\mathbb{G}}} \gamma}^{\ell \text{ times}}.$$

Coupling this $\mathbb{Z}$-action to the action by homotheties and employing the completeness of power series rings, these together complete to give an action by a $p$-complete ring that reduces modulo $p$ to the factor of $k$ which acts by homotheties. Using Remark 3.4.6, this entails that the module of curves carries an action by $\mathbb{W}_p(k)$.

4.4 Dieudonné Modules

- Verschiebung: given an integer $n \geq 1$, we define $V_n \gamma(t) = \gamma(t^n)$.
- Frobenius: given an integer $n \geq 1$, we define

$$F_n \gamma(t) = \sum_{i=1}^{n} {}^{\widehat{\mathbb{G}}} \gamma(\zeta_n^i t^{1/n}),$$

where $\zeta_n$ is an $n$th root of unity. (This formula is invariant under permuting the root of unity chosen, so determines a curve defined over the original ground ring by Galois descent.)

**Definition 4.4.15** (see Lemma 3.3.6)  A curve $\gamma$ on a formal group is *p*-typical when $F_n \gamma = 0$ for $n \neq p^j$. Write $D_* \widehat{\mathbb{G}} \subseteq C \widehat{\mathbb{G}}$ for the subset of *p*-typical curves.

**Lemma 4.4.16** ([Zin84, equation 4.13])  *In the case that the base ring is p-local, $C\widehat{\mathbb{G}}$ splits as a sum of copies of $D_* \widehat{\mathbb{G}}$. The operation of p-typification gives a natural section $C\widehat{\mathbb{G}} \to D_* \widehat{\mathbb{G}}$, given by the same formula as in Lemma 3.3.6.* □

This construction also plays the role of a Dieudonné functor:

**Theorem 4.4.17** ([Zin84, theorems 3.5 and 3.28])  *For $k$ a perfect field of characteristic p, the (Cartier–Dieudonné) functor $D_*$ determines a covariant equivalence of categories between smooth one-dimensional formal groups over $k$ of finite p-height and Dieudonné modules, which again are modules over the Cartier ring that are additionally free and finite rank over $\mathbb{W}_p(k)$.* □

**Theorem 4.4.18** ([MM74, section II.15])  *There is a natural isomorphism*

$$(D_* \widehat{\mathbb{G}}_0)^* \cong D^* \widehat{\mathbb{G}}_0,$$

*where $F$ and $V$ interchange rules through the linear dual.*[29] □

*Remark 4.4.19* ([Gro74, chapitre VI])  The contravariant Dieudonné functor described above has a natural extension by choosing lifts over other pro-Artinian $k$-algebras, like the Lubin–Tate moduli stack of Definition 3.4.3. The resulting network of objects most naturally organizes into a sheaf over the *crystalline site*, but it is possible in this setting to re-express such a sheaf as a quasicoherent sheaf over the Lubin–Tate stack which is equipped with a flat connection, and it is additionally acted upon by the familiar operators $F$ and $V$.

*Remark 4.4.20* (see [Mor85, section 2.3])  Dieudonné theory gives rise to an important function called the *period map*, which we will not have direct use for in this book but which we would be negligent to omit. Although the crystalline nature of the cohomology group $H^1_{dR}$ makes our definition of $D^*$ invariant

---

[29] This is a remarkably difficult theorem to prove, see Appendix B.2.

of choice of lift, the underlying chain complex is *not* invariant of choice of lift. In particular, the subsheaf of honestly invariant differentials $\omega_{\widehat{\mathbb{G}}}$ selects an interesting one-dimensional vector subspace of $PH^1_{dR}(\widehat{\mathbb{G}})$ [Mor85, section 2.3]. Thinking of $\widehat{\mathbb{G}}$ as a point in $(\mathcal{M}_{\mathbf{fg}})^\wedge_{\widehat{\mathbb{G}}_0}(\mathbb{W}_p(k))$, this observation gives rise to an interesting function

$$\pi_{GH}\colon (\mathcal{M}_{\mathbf{fg}})^\wedge_{\widehat{\mathbb{G}}_0}(\mathbb{W}_p(k)) \to \mathbb{P}(D^*(\widehat{\mathbb{G}}_0)),$$

$$\widehat{\mathbb{G}} \mapsto [\omega_{\widehat{\mathbb{G}}} \subseteq D^*(\widehat{\mathbb{G}}_0)].$$

This map has incredibly good properties. It is equivariant for the action of $\operatorname{Aut}\widehat{\mathbb{G}}_0$ [HG94b, theorem 1], and with enough work one can use this to extract explicit (recursive) formulas expressing the action ([DH95], [Strb, section 24], [HG94a, section 22]), bringing some relief to the problem of Remark 3.6.17. Also, in a suitable context it becomes an étale morphism with identifiable fibers ([HG94b, theorem 1], [HG94a, sections 23–24]), which allows one to give an explicit formula for the dualizing sheaf [HG94b, corollary 3] with direct applications to topology [HG94b, theorem 6]:

$$\mathcal{M}_{E_\Gamma}(\Sigma^{-d^2-d}\mathbb{I}_{\mathbb{Q}/\mathbb{Z}}) = \Omega^{d-1}_{(\mathcal{M}_{\mathbf{fg}})^\wedge_\Gamma} = \omega^{\otimes d}[\det],$$

where the *Brown–Comenetz dualizing object* $\mathbb{I}_{\mathbb{Q}/\mathbb{Z}}$ is defined by the property

$$\mathbb{I}^0_{\mathbb{Q}/\mathbb{Z}}(E) = \textsc{AbelianGroups}(\pi_0 E, \mathbb{Q}/\mathbb{Z}).$$

In Figure 4.3 we sketch the period map for the values $p = 2$, $d = 2$, $O_A = \mathbb{W}_p(\mathbb{F}_{p^d})$, and $\Gamma = \Gamma_2$ the Honda formal group. This map has the following list of properties ([HG94a, appendix 25], [Yu95]):

- The center of the $\mathbb{W}(\mathbb{F}_{p^2})$-points of Lubin–Tate space corresponds to the canonical lift, which is the formal group that further acquires an $O_A$-module structure [LT65]. It has 2-series $[2](x) = 2x + x^{p^2}$.
- There are three nontrivial points in $\widehat{\mathbb{G}}[2]$: $\alpha$, $\beta$, and $\alpha + \beta$. Quotienting by them gives three points at order $1/(p+1)$, the first bunch of *quasicanonical lifts*, which have partial formal $O_A$-module structures [Gro86].
- At each quasicanonical point, you can form three quotients: Two of them make the situation "worse" (i.e., raise the order of the quasicanonical point), and one of them makes the situation "better" (i.e., lowers the order), which is a kind of witness to the identification $\widehat{\mathbb{G}}/\widehat{\mathbb{G}}[2] \cong \widehat{\mathbb{G}}$.
- The map $\pi_{GH}$ sends the canonical lift to $0 = [1 : 0]$. The first-order quasicanonical points are targets of isogenies of degree $p$ from the canonical point which reduce to the Frobenius isogeny on the special fiber. The Frobenius

*4.5 Ordinary Cooperations for Landweber Flat Theories* 203

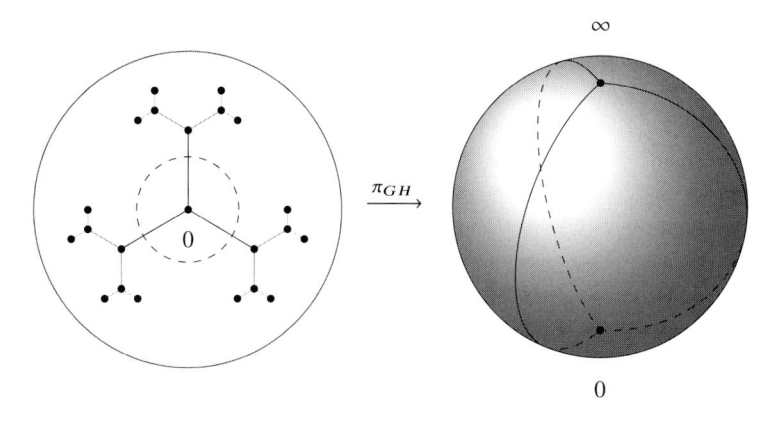

Figure 4.3 The period map at $n = 2$, $p = 2$.

isogeny can be shown to act by $F_{\text{can}} = \begin{bmatrix} 0 & p \\ 1 & 0 \end{bmatrix}$ on $\mathbb{P}^1$, and hence the first-order quasicanonical points are sent to $\infty = [0 : 1]$ under $\pi_{GH}$. This pattern continues: the second-order quasicanonical points are sent to 0, and so on.
- Out to order $1/p$, $\pi_{GH}$ is injective.
- The group $\mathbb{F}_{p^2}^\times \leq \operatorname{Aut} \widehat{\mathbb{G}}$ acts by rotation on $\mathbb{P}^1$.
- These quasicanonical points are the ones with nontrivial stabilizers under the action by $\operatorname{Aut} \Gamma_d$ – all the other points belong to free orbits. The canonical lift has the largest stabilizer of all.

*Remark 4.4.21* It is also possible to build versions of Dieudonné theory over still more exotic rings. The most successful such version is Zink's theory of Dieudonné displays [Zin02], which has found some application in algebraic topology [Law10].

## 4.5 Ordinary Cooperations for Landweber Flat Theories

**Convention: We will write $H$ for $H\mathbb{F}_p$ for the duration of the lecture.**

Our goal in this lecture is to put Dieudonné modules to work for us in algebraic topology. The executive summary of Dieudonné theory is that it gives a *linear* (or *logarithmic*) presentation of the theory of Hopf algebras. From the perspective of algebraic topology, functors sending cofiber sequences to exact sequences in a linear category are precisely homology functors. Tying these two ideas together, if we can find a functor that sends exact sequences of spaces (or

204  *Unstable Cooperations*

spectra) to exact sequences of Hopf algebras, we can post-compose it with a suitable version of the Dieudonné functor to get a homology functor and hence a *spectrum*.

To meet algebraic topology in its natural setting, it will be useful to also have a version of Dieudonné theory that is well-adapted to working with *graded* Hopf algebras.[30] This falls out of an extended study of the covariant Dieudonné module: It can be shown that the functor $D_*$ is representable by the *p-typical Witt formal group* $\widehat{\mathbb{W}}_p$,[31,32] i.e.,

$$
\begin{array}{ccc}
C_*\widehat{\mathbb{G}} = \text{FormalSchemes}(\mathbb{A}^1, \widehat{\mathbb{G}}) = \text{FormalGroups}(\widehat{\mathbb{W}}, \widehat{\mathbb{G}}) \\
\uparrow & \uparrow & \uparrow \\
D_*\widehat{\mathbb{G}} = \text{FormalSchemes}(\mathbb{A}^1, \widehat{\mathbb{G}})^{p\text{-typ}} = \text{FormalGroups}(\widehat{\mathbb{W}}_p, \widehat{\mathbb{G}}).
\end{array}
$$

The Witt formal scheme has the additional miraculous property that it is dualizable: There is a coWitt formal scheme $\widehat{C\mathbb{W}}_p$ with

$$\text{FormalGroups}(\widehat{\mathbb{W}}_p, \widehat{\mathbb{G}}) \cong \text{FormalGroups}(\widehat{\mathbb{G}}, \widehat{C\mathbb{W}}_p)$$
$$\cong \text{HopfAlgebras}(O_{\widehat{C\mathbb{W}}_p}, O_{\widehat{\mathbb{G}}}).$$

We are thus moved to form graded versions of the coWitt Hopf algebra. More precisely, the following theorem says that there are graded versions of the coWitt Hopf algebra that give a sequence of projective generators for the category of connected graded Hopf algebras over $\mathbb{F}_p$:

**Theorem 4.5.1** ([Sch70, section 3.2], [GLM93, proposition 1.6])  *Let $S(n)$ denote the free graded-commutative Hopf algebra over $\mathbb{F}_p$ on a single generator in degree $n > 0$. There is a projective cover $H(n) \twoheadrightarrow S(n)$, described as follows:*

- *If either of the following conditions hold...*
  - *$p = 2$ and $n = 2^m k$ for $2 \nmid k$ and $m > 0$, or*

---

[30] Another useful observation is that for a Hopf algebra arising as the ordinary homology of an infinite loopspace, its degree-zero part is always group-like.

[31] An extremely pleasant perspective on the construction of the Witt scheme is presented by Lazard [Laz75, chapter III].

[32] A coordinate $x$ on $\mathbb{CP}_E^\infty$ induces a sequence of isomorphisms

$$\text{FormalGroups}(BU_E, \widehat{\mathbb{G}}) \cong \text{FormalSchemes}(\mathbb{CP}_E^\infty, \widehat{\mathbb{G}})$$
$$\stackrel{x}{\cong} \text{FormalSchemes}(\widehat{\mathbb{A}}^1, \widehat{\mathbb{G}}) = C\widehat{\mathbb{G}},$$

which presents $E^*BU$ as the Witt Hopf algebra. However, this isomorphism is not especially interesting [Strb, remark 18.1]: For one, it is *highly* dependent upon the choice of coordinate, but the far right-hand object has no dependence on $\mathbb{CP}_E^\infty$, and so the operations we have been studying – formal group auto- and endomorphisms of $\mathbb{CP}_E^\infty$, mainly – do not act, and this isomorphism cannot be equivariant in any useful sense.

### 4.5 Ordinary Cooperations for Landweber Flat Theories 205

- $p \neq 2$ and $n = 2p^m k$ for $p \nmid k$ and $m > 0$, then $H(n) = \mathbb{F}_p[x_0, x_1, \ldots, x_k]$ with the Witt vector diagonal, i.e., the diagonal is arranged so that the elements $w_j = x_0^{p^j} + px_1^{p^{j-1}} + \cdots + x_j$ are primitive.
- Otherwise, $H(n) = S(n)$ is the identity. □

**Corollary 4.5.2** ([Sch70, section 5]) *The category* $\text{GradedHopfAlgs}_{\mathbb{F}_p/}^{>0,\text{fin}}$ *of finite-type graded connected Hopf algebras under* $\mathbb{F}_p$ *is a full subcategory of modules for the ring*

$$\bigoplus_{n,m} \text{GradedHopfAlgs}(H(n), H(m)).$$

*Construction* This is a general nonsense consequence of having found a set of projective generators. The functor presenting the inclusion is

$$M \mapsto \bigoplus_{n=0}^{\infty} \text{GradedHopfAlgs}(H(n), M),$$

and since this functor is corepresented its endomorphisms are encoded by the indicated ring. □

We would also like to give a set of conditions, analogous to the technical conditions appearing in the previous two presentations, which select this full subcategory out from all possible modules over this endomorphism ring.

**Definition 4.5.3** ([GLM93, p. 116]) Let $\text{GradedDMods}$ denote the category of graded abelian groups $M$ equipped with maps[33]

$$V \colon M_{pn} \to M_n, \qquad F \colon M_n \to M_{pn}$$

satisfying

1. $M_{<1} = 0$.
2. If $n$ is odd, then $pM_n = 0$.
3. The composites are controlled by $FV = p$ and $VF = p$.[34]

*Remark 4.5.4* Suppose that $n$ is even, written at odd primes in the form $n = 2p^m k$ with $p \nmid k$ or at $p = 2$ in the form $n = 2^m k$ with $2 \nmid k$ at $p = 2$. Then, combining the above relations, we get the torsion condition $p^{m+1} M_n = F^{m+1} V^{m+1} M_n = 0$.

---
[33] Here $n$ is required to be even if $p \neq 2$.
[34] These come from $H(n) \subseteq H(pn)$ and the map $H(pn) \to H(n)$ sending $x_n$ to $x_{n-1}^p$.

**Theorem 4.5.5** ([Sch70, section 5], [GLM93, theorem 1.11])  *The functor*

$$D_* \colon \mathrm{GradedHopfAlgs}_{\mathbb{F}_p/}^{>0,\mathrm{fin}} \to \mathrm{GradedDMods},$$

$$D_*(H) = \bigoplus_n D_n(H) = \bigoplus_n \mathrm{GradedHopfAlgs}_{\mathbb{F}_p/}^{>0,\mathrm{fin}}(H(n), H)$$

*is an exact equivalence of categories. Moreover, $D_*H(n)$ is characterized by the equation*

$$\mathrm{GradedDMods}(D_*H(n), M) = M_n. \qquad \square$$

Having produced our desired graded Dieudonné theory, we now need some topological input. We are by now well aware that the homology of an $H$-space forms a Hopf algebra, and the Serre spectral sequence for a fibration of $H$-spaces $F \to E \to B$ gives a spectral sequence of Hopf algebras:

$$E_2^{*,*} = H^*B \otimes H^*F \Rightarrow H^*E.$$

The following result of Goerss, Lannes, and Morel says that in the case that the fibration is one of *infinite loopspaces*, we have the exactness property we need:

**Theorem 4.5.6** ([GLM93, lemma 2.8])  *Let $X \to Y \to Z$ be a cofiber sequence of spectra. Then, provided $n > 1$ satisfies $n \not\equiv \pm 1 \pmod{2p}$, there is an exact sequence*

$$D_n H_* \Omega^\infty X \to D_n H_* \Omega^\infty Y \to D_n H_* \Omega^\infty Z. \qquad \square$$

This theorem is not especially easy to prove: one works very directly with unstable modules over the Steenrod algebra, the bar spectral sequence, and Postnikov decomposition of infinite loopspaces. We refer the reader to the paper directly, as we have been unable to find a useful improvement upon or even summary of the results presented there. Nonetheless, granting this theorem, we use Brown representability to draw the following consequence:

**Corollary 4.5.7** ([GLM93, theorem 2.1, remark 2.9])  *For $n > 1$ an integer satisfying $n \not\equiv \pm 1 \pmod{2p}$,[35] there exists a spectrum $B(n)$, called the nth Brown–Gitler spectrum, which satisfies*

$$B(n)_n X \cong D_n H_* \Omega^\infty X. \qquad \square$$

We will now use the $B(n)$ spectra to analyze the Hopf rings arising from unstable cooperations. Our intention is to prove the following:

---

[35] As convention, when $n \equiv \pm 1 \pmod{2p}$ we set $B(n) := B(n-1)$, and $B(0) := \mathbb{S}^0$.

## 4.5 Ordinary Cooperations for Landweber Flat Theories

**Theorem 4.5.8** (see Lemma A.2.12)  *For $F = H$ and $E$ a Landweber flat homology theory, the comparison map*

$$\mathcal{A}\!\mathcal{A}(H, E) \to H_*\underline{E}_{2*}$$

*is an isomorphism of Hopf rings.*

We have previously computed that the comparison map

$$\mathcal{A}\!\mathcal{A}(H, BP) \to H_*\underline{BP}_{2*}$$

is an isomorphism. In order to make use of this statement now, we must reimagine it in terms of Dieudonné theory. In order to do that, we again have to reimagine some of Dieudonné theory itself, as our description of it is concerned with *Hopf algebras* rather than *Hopf rings*. Recall that a Hopf ring is an algebra object

$$\circ \colon A_* \boxtimes A_* \to A_*,$$

where "$\boxtimes$" is a kind of graded tensor product of externally graded Hopf algebras ([HT98, proposition 2.6], [BL07, definition 2.2], [Goe99, section 5]). Since $D_*$ gives an equivalence of categories between internally graded Hopf algebras and internally graded Dieudonné modules, we should be able to find an analogous formula for the tensor product of Dieudonné modules.

**Definition 4.5.9**  [Goe99, p. 154] The naive tensor product $M \otimes N$ of Dieudonné modules $M$ and $N$ receives the structure of a $\mathbb{Z}_p\langle V \rangle$-module, where $V(x \otimes y) = V(x) \otimes V(y)$. We define the *tensor product of Dieudonné modules*[36] by

$$M \boxtimes N = \frac{\mathbb{Z}_p\langle F, V \rangle}{(VF = p)} \otimes_{\mathbb{Z}_p[V]} (M \otimes N) \Big/ \left( \begin{array}{c} 1 \otimes Fx \otimes y = F \otimes x \otimes Vy, \\ 1 \otimes x \otimes Fy = F \otimes Vx \otimes y \end{array} \right).$$

**Lemma 4.5.10** ([Goe99, corollary 8.14])  *The natural map*

$$D_*(M) \boxtimes D_*(N) \to D_*(M \boxtimes N)$$

*is an isomorphism.* □

**Definition 4.5.11**  For a ring $R$, a *Dieudonné $R$-algebra* $A_*$ is an externally graded Dieudonné module equipped with an $R$-action and a unital multiplication

$$\circ \colon A_* \boxtimes A_* \to A_*.$$

---

[36] This definition is specialized to $\mathbb{F}_p$ and $\mathbb{W}_p(\mathbb{F}_p) = \mathbb{Z}_p$, where we don't have to worry about Frobenius semilinearity.

*Example 4.5.12* ([Goe99, proposition 10.2]) Inspired by Lemma 4.2.8 and our interest in $H_*\underline{E}_*$, for a complex-oriented homology theory $E$ we define its *algebraic Dieudonné $E_*$-algebra* by

$$R_E = E_*[b_1, b_2, \ldots] \Big/ \Big( b(s+t) = b(s) +_E b(t) \Big),$$

where $V$ is multiplicative, $V$ fixes $E_*$, and $V$ satisfies $Vb_{pj} = b_j$.[37] This is the Dieudonné-theoretic analog of the algebraic approximation $\mathcal{A}\mathcal{A}(H, E)$. We also write $D_E = \{D_{2m}H_*\underline{E}_{2n}\}$ for the even part of the topological Dieudonné algebra, and these come with natural comparison maps

$$R_E \to D_E \to D_*H_*\underline{E}_{2*}.$$

**Theorem 4.5.13** (Dieudonné theoretic form of Theorem 4.5.8 [Goe99, theorem 11.7]) *Restricting attention to the even parts, the maps*

$$R_E \to D_E \to D_*H_*\underline{E}_{2*}$$

*are isomorphisms for $E$ Landweber flat.*

*Proof* In Corollary 4.3.7, we showed that these maps are isomorphisms for $E = BP$. However, the right-hand object can be identified via Brown–Gitler juggling:

$$D_n H_* \underline{E}_{2j} = B(n)_n \Sigma^{2j} E = E_{2j+n} B(n).$$

It follows that if $E$ is Landweber flat, then the middle and right terms are determined by change-of-base from the respective $BP$ terms. Finally, the left term commutes with change-of-base by its algebraic definition, and the theorem follows. □

*Remark 4.5.14* Goerss's original proof of Theorem 4.5.13 [Goe99] involved a lot more work, essentially because he didn't want to assume Theorem 4.3.1 or Corollary 4.3.7. Instead, he used the fact that $\Sigma_+^\infty \Omega^2 S^3$ is a regrading of the ring spectrum $\bigvee_n B(n)$, together with knowledge of $BP_*\Omega^2 S^3$.

*Remark 4.5.15* ([Goe99, proposition 11.6, remark 11.4]) The Dieudonné algebra framework also makes it easy to add in the odd part after the fact. Namely, suppose that $E$ is a torsion-free ring spectrum and suppose that $E_*B(2n)$ is even for all $n$. In this setting, we can verify the purely topological version of this statement: the map

$$D_E[e]/(e^2 - b_1) \to D_*H_*\underline{E}_*$$

---

[37] If $E_*$ is torsion-free, then this determines the behavior of $F$ by $FV = p$.

### 4.5 Ordinary Cooperations for Landweber Flat Theories

is an isomorphism. To see this, note that because

$$E_{2n-2k-1}B(2n) \to D_{2n}H_*\underline{E}_{2k+1}$$

is surjective and $E_{2n-2k-1}B(2n)$ is assumed zero, the group $D_{2n}H_*\underline{E}_{2k+1}$ vanishes as well. A bar spectral sequence argument shows that $D_{2n+1}H_*\underline{E}_{2k+2}$ is also empty [Goe99, lemma 11.5.1]. Hence, the map on even parts

$$(D_E[e]/(e^2 - b_1))_{*,2n} \to (D_*H_*\underline{E}_*)_{*,2n}$$

is an isomorphism, and we need only show that

$$D_*H_*\underline{E}_{2n} \xrightarrow{e \circ -} D_*H_*\underline{E}_{2n+1}$$

is an isomorphism as well. Since we have $e(Fx) = F(Ve \circ x) = 0$ generally and $D_*A/FD_*A \cong Q^*A$ for a Hopf algebra $A$, we see that $e$ kills decomposables and suspends indecomposables:

$$e \circ D_*H_*\underline{E}_{2n} = \Sigma Q H_*\underline{E}_{2n}.$$

This is also what happens in the bar spectral sequence, and the claim follows. In light of Theorem 4.5.13, this means that for Landweber flat $E$, the comparison isomorphism can be augmented to a further isomorphism

$$R_E[e]/(e^2 - b_1) \to D_*H_*\underline{E}_*.$$

*Remark 4.5.16* ([HH95]) The results of this lecture are accessed from a different perspective by Hopkins and Hunton, essentially by forming a tensor product of Hopf rings and showing that Landweber flatness induces a kind of flatness with respect to the Hopf ring tensor product as well.

Before moving on, we prove a sequence of small results which make these spectra $B(n)$ somewhat more tangible, though we advise the reader that these do not bear any further on our investigation of unstable cooperations.

**Lemma 4.5.17** ([GLM93, lemma 3.2]) *The spectrum $B(n)$ is connective and $p$-complete.*

*Proof* First, rearrange:

$$\pi_k B(n) = B(n)_n \mathbb{S}^{n-k} = D_n H_* \Omega^\infty \Sigma^\infty \mathbb{S}^{n-k}.$$

If $k < 0$, $n$ is below the connectivity of $\Omega^\infty \mathbb{S}^{n-k}$ and hence this vanishes. The second assertion follows from the observation that $H\mathbb{Z}_*B(n)$ is an $\mathbb{F}_p$-module, followed by an Adams spectral sequence argument. To see the assertion about

210                  *Unstable Cooperations*

being an $\mathbb{F}_p$-module, restrict to the case $n \not\equiv \pm 1 \pmod{2p}$ and calculate

$$H\mathbb{Z}_k B(n) = B(n)_n \Sigma^{n-k} H\mathbb{Z}$$
$$= D_n H_* K(\mathbb{Z}, n-k)$$
$$= [H(n), H_* K(\mathbb{Z}, n-k)]_n$$
$$= [Q^* H_* K(\mathbb{Z}, n-k)]_n. \qquad \square$$

We can use a similar trick to calculate the cohomology groups $H^*B(n)$:

**Definition 4.5.18** ([GLM93, example 3.6])  Let $G(n)$ denote the free unstable $\mathcal{A}^*$-module on a degree $n$ generator.[38] For $M$ an unstable $\mathcal{A}^*$-module,

$$\text{MODULES}_{\mathcal{A}^*}(G(n), M) = M_n.$$

**Theorem 4.5.19** ([GLM93, proof of theorem 3.1])  *There is an isomorphism*

$$H^* B(n) \cong \Sigma^n (DG(n))^*.$$

*Proof*  We restrict attention to $n \not\equiv \pm 1 \pmod p$, where we can use Corollary 4.5.7 directly. Start, as before, by addressing the dual problem of computing the mod-$p$ homology:

$$H_k B(n) = B(n)_n \Sigma^{n-k} H = D_n H_* K(\mathbb{F}_p, n-k).$$

The unstable module $G(n)$ also enjoys a universal property in the category of stable $\mathcal{A}^*$-modules, by passing to the maximal unstable submodule $\Omega^\infty M$ of a stable module $M$:

$$\text{MODULES}_{\mathcal{A}^*}(G(n), M) \cong [\Omega^\infty M]_n.$$

Hence, we can continue our computation:

$$H_k B(n) = D_n H_* K(\mathbb{F}_p, n-k)$$
$$= \text{MODULES}_{\mathcal{A}^*}(G(n), \Sigma^{n-k} \mathcal{A}_*)$$
$$= \text{MODULES}_{\mathbb{F}_p}(G(n)_{n-k}, \mathbb{F}_p).$$

We learn immediately that the $H_* B(n)$ spectrum, which we already knew to

---

[38] This module admits a presentation as

$$G(n) = \begin{cases} \Sigma^n \mathcal{A} / \{\beta^\varepsilon P^i \mid 2pi + 2\varepsilon > n\}\mathcal{A} & \text{if } p > 2, \\ \Sigma^n \mathcal{A} / \{\text{Sq}^i \mid 2i > n\}\mathcal{A} & \text{if } p = 2. \end{cases}$$

The Spanier–Whitehead dual of this right-module, $DG(n)$, is given by

$$\Sigma^n (DG(n))^* = \begin{cases} \mathcal{A}/\mathcal{A}\{\chi(\beta^\varepsilon P^i) \mid 2pi + 2\varepsilon > n\} & \text{if } p > 2, \\ \mathcal{A}/\mathcal{A}\{\chi \text{Sq}^i \mid 2i > n\} & \text{if } p = 2. \end{cases}$$

be bounded-below and $p$-complete, is a finite spectrum. It follows that a unique Spanier–Whitehead predual exists and agrees with the Spanier–Whitehead dual; the statement then follows from the previous equality. □

Lastly, for a *space* $X$, we definitionally have that $H_*X$ forms an unstable module over the Steenrod algebra, i.e., $\Omega^\infty H_*X = H_*X$. This has the following direct sequence (with minor fuss at the bad indices $n \equiv \pm 1 \pmod{p}$):

**Lemma 4.5.20** ([GLM93, lemma 3.3])  *For $X$ a space, there is a natural surjection $B(n)_n X \to H_n X$.*

*Remark 4.5.21* ([Coh81])  These properties uniquely characterize the spectra $B(n)$ as *Brown–Gitler spectra*, which arise in many other settings in stable homotopy theory. For instance, the stable James splitting for $\Omega^2 S^3$ is given by

$$\Sigma^\infty_+ \Omega^2 \Sigma^2 S^1 \simeq \bigvee_{j=0}^{\infty} \Sigma^j B(\lfloor j/2 \rfloor).$$

## 4.6 Cooperations among Geometric Points on $\mathcal{M}_{\mathrm{fg}}$

Our discussion of unstable cooperations has touched on each of the families of chromatic homology theories described in Definition 3.5.4 except one: the Morava $K$- and $E$-theories. Our final goal before moving on to other subjects is to describe some of the mixed unstable cooperations for $(K_\Gamma)_* \underline{K_{\Gamma'}}_*$. In complete generality, this seems like a difficult problem: Our algebraic model is rooted in formal group homomorphisms, and we have not proven any theorems about the moduli of such for arbitrary finite-height formal groups. However, the landscape brightens considerably in the case where we pick $\Gamma' = \widehat{\mathbb{G}}_a$, as this is the sort of calculation we considered in Lemmas 3.4.11 and 3.4.12. In light of this, we specialize $\Gamma'$ to $\widehat{\mathbb{G}}_a$ (and hence $K_{\Gamma'}$ to an Eilenberg–Mac Lane spectrum $H$), and we abbreviate $K_\Gamma$ to just $K$.

As with all the other major results of this case study, our approach will rest on the bar spectral sequence

$$\mathrm{Tor}^{K_*\underline{H}_q}_{*,*}(K_*, K_*) \Rightarrow K_* \underline{H}_{q+1}.$$

The analysis of this spectral sequence was first accomplished by Ravenel and Wilson [RW80], but has since been re-envisioned by Hopkins and Lurie [HL, section 2]. In order to give an effective analysis of this spectral sequence in line with the theme of this book, we will endeavor to give algebro-geometric interpretations of its input and its output, beginning with the case $q = 0$ and

$H = H(S^1[p^j])$ for some $1 \le j \le \infty$. This task itself begins with giving just *algebraic* descriptions of the input and output. For $j < \infty$, we have essentially already computed the output by other means:

**Theorem 4.6.1** ([RW80, theorem 5.7], [HL, proposition 2.4.4]) *There is an isomorphism*
$$BS^1[p^j]_K \cong BS^1_K[p^j].$$

*Proof* The circle bundle $S^1 \to BS^1[p^j] \to BS^1$ has associated Gysin sequence

$$K^*BS^1 \xrightarrow{-\smile [p^j](x)} K^*BS^1$$
$$\partial \nwarrow \swarrow$$
$$K^*BS^1[p^j],$$

where $x$ is any choice of coordinate on $BS^1_K \cong \Gamma$. Any $p$-series for $\Gamma$ takes the form $[p](x) = cx^{p^d} + \cdots$ for $c$ a unit. Such an element is not a zero divisor, so $\partial$ vanishes, and this presents $K^*BS^1[p^j]$ as the quotient ring
$$K^*BS^1[p^j] \cong K^*BS^1/[p^j](x). \qquad \square$$

*Remark 4.6.2* In the proof of the homological statement dual to Theorem 4.6.1, there is a corresponding exact sequence of Hopf algebras

$$K_*BS^1 \xleftarrow{-\frown [p^j](x)} K_*BS^1$$
$$\partial \searrow \nearrow$$
$$K_*BS^1[p^j]$$

where again $\partial = 0$ and hence $K_*(BS^1[p^j])$ is presented as the kernel of the map "cap with $[p^j](x)$." We will revisit this duality in the next case study.

With this in hand, the analysis of the bar spectral sequence proceeds very much analogously to the example of the unstable dual Steenrod algebra of Lecture 4.1. We will analyze what *must* happen in the bar spectral sequence
$$\mathrm{Tor}^{K_*S^1[p^j]}_{*,*}(K_*, K_*) \Rightarrow K_*BS^1[p^j]$$
in order to reach the conclusion of Theorem 4.6.1. In the input to this spectral sequence, the ground algebra is given by a noncanonical isomorphism
$$K_*\underline{H(S^1[p^j])}_0 \cong K_*\underline{H\mathbb{Z}/p^j}_0 = K_*[[1]]/([1]^{p^j} - 1) = K_*[[1]-[0]]/\langle[1]-[0]\rangle^{p^j}.$$
The Tor-algebra for this truncated polynomial algebra $K_*[\alpha_0]/\alpha_0^p$ is then given by the formula
$$\mathrm{Tor}^{K_*[\alpha_0]/\alpha_0^{p^j}}_{*,*}(K_*, K_*) = \Lambda[\alpha'_0] \otimes \Gamma[\alpha_0],$$

## 4.6 Cooperations among Geometric Points on $\mathcal{M}_{\mathrm{fg}}$

the combination of an exterior algebra and a divided power algebra.[39] We know which classes are supposed to survive this spectral sequence, and hence we know where the differentials must be [RW80, section 8]:

$$d\left(\alpha_{()}^{[p^{jd}]}\right) = \alpha'_{()},$$

$$\Rightarrow d\left(\alpha_{()}^{[i+p^{jd}]}\right) = \alpha'_{()} \cdot \alpha_{()}^{[i]}.$$

The spectral sequence collapses after this differential. In the case $1 < j < \infty$, there are some hidden multiplicative extensions in the spectral sequence, but these too are all determined by already knowing the multiplicative structure on $K_* \underline{H(S^1[p^j])}_1$.

However, the case of $j = \infty$ is a bit different, beginning with the following:

**Lemma 4.6.3** *For $q \geq 1$, $K(\mathbb{Q}/\mathbb{Z}_{(p)}, q)$ and $K(\mathbb{Z}, q + 1)$ are $p$-adically equivalent.*

*Proof* This is a consequence of the fiber sequence

$$K(\mathbb{Q}, q) \to K(\mathbb{Q}/\mathbb{Z}_{(p)}, q) \to K(\mathbb{Z}_{(p)}, q + 1).$$

The first term has vanishing mod-$p$ homology, forcing the $H\mathbb{F}_p$-Serre spectral sequence of the fibration to collapse and for the edge homomorphism to be an isomorphism. Similarly, the map $K(\mathbb{Z}, q + 1) \to K(\mathbb{Z}_{(p)}, q + 1)$ is an equivalence on mod-$p$ homology. □

**Remark 4.6.4** Thinking of $K(\mathbb{Z}, q + 1)$ as $B^q S^1$, one can also think of this theorem as giving a $p$-adic equivalence between $B^q(S^1[p^\infty])$ and $B^q S^1$ – i.e., the prime-to-$p$ parts of $S^1$ do not matter for $p$-adic homotopy theory.

We use this to continue the analysis of the case $q = 1$ and $j = \infty$, where the lemma gives $B(S^1[p^\infty]) = \mathbb{C}P^\infty$. The bar spectral sequence of interest then takes the form

$$\operatorname{Tor}_{*,*}^{K_* S^1}(K_*, K_*) \Rightarrow K_* \mathbb{C}P^\infty.$$

The input algebra $K_* S^1$ is exterior on a single generator in odd degree, and so its Tor-algebra is linearly dual to a power series algebra on a single generator in even degree. Since all of its input is even, this spectral sequence collapses immediately.[40]

There is a lot of structure visible in this collection of spectral sequences, as considered simultaneously. Without further inspection, the spectral sequence at

---

[39] In the work of Ravenel and Wilson [RW80, lemma 6.6], the elements $\alpha_{()}$ and $\alpha'_{()}$ are identified as a *transpotence* and a *homology suspension*, respectively.

[40] Thinking of this as the limiting "$j \to \infty$" instance of the family of examples above is a great opportunity to meditate on the role of odd-dimensional classes in homotopy theory.

$j = \infty$ records that $\mathbb{C}P_K^\infty$ is a formal variety, and the spectral sequences at the finite values $1 \le j < \infty$ encode in their differentials the behavior of the map $p^j: \Gamma \to \Gamma$ on functions – indeed, this appears to be the entire job of $\alpha'_0$. Lastly, we notice that the $E_\infty$ page of each finite-range spectral sequence includes into the spectral sequence at $j = \infty$, and moreover this filtration is exhaustive: every term in the $j = \infty$ spectral sequence appears at some $j < \infty$ stage. Since this last property is about the $E_\infty$ pages, it is really a property of the formal group $\Gamma$, which we record in a definition:

**Definition 4.6.5** ([Gro74, definition 4.1]) A *p-divisible group*[41] of height $d$ over a field is a system $\mathbb{G}_j$ of finite group schemes satisfying $\dim O(\mathbb{G}_j) = p^{jd}$, as well as maps $i_k^j: \mathbb{G}_k \to \mathbb{G}_j$ for $k < j$ which belong to exact sequences

$$0 \to \mathbb{G}_k \xrightarrow{i_k^j} \mathbb{G}_j \xrightarrow{p^k} \mathbb{G}_j.$$

**Definition 4.6.6** A *p*-divisible group is said to be *connected* when its constituent subgroups $\mathbb{G}_j$ are infinitesimal thickenings of $\operatorname{Spec} k$. An example of this is the sequence of torsion subgroups $\widehat{\mathbb{G}}[p^j]$ of a formal group $\widehat{\mathbb{G}}$. An example of a *p*-divisible group which is *not* connected is the sequence of constant group schemes $\widehat{\mathbb{G}}_j = S^1[p^j]$.

**Lemma 4.6.7** ([Gro74, section 6.7]) *Over a perfect field of positive characteristic p, a connected p-divisible group is equivalent to a smooth formal group of finite height.*

*Correspondence* The maps in both directions are easy: A *p*-divisible group is sent to its colimit, and a formal group of finite height is sent to its system of $p^j$-torsion subgroups. In both directions there is something mild to check: that the colimit gives a formal variety, and that the system of $p^j$-torsion subgroups has the indicated exactness properties. □

*Remark 4.6.8* Using the extension of Dieudonné theory to finite group schemes described in Example 4.4.12, the Dieudonné modules of a connected *p*-divisible group $\mathbb{G}$ and its associated formal group $\widehat{\mathbb{G}}$ belong to a short exact sequence:

$$0 \to D_*(\widehat{\mathbb{G}}) \to \mathbb{Q} \otimes D_*(\widehat{\mathbb{G}}) \to \operatorname*{colim}_j D_*(\mathbb{G}_j) \to 0.$$

We will soon see that these interrelations among the bar spectral sequences for the different Eilenberg–Mac Lane spaces, as well as the special behavior of the spectral sequence at $j = \infty$, are generic phenomena in $q$. We record the steps in our upcoming induction in the following theorem:

---

[41] Some like to call these *Barsotti–Tate groups*, which is probably the better name, since "*p*-divisible group" does not communicate these subtle extra properties.

### 4.6 Cooperations among Geometric Points on $\mathcal{M}_{\mathrm{fg}}$

**Theorem 4.6.9** ([HL, theorems 2.4.11–2.4.13]) *The following claims give a complete description of the Morava K-theory schemes associated to the Eilenberg–Mac Lane spaces $\underline{HS^1[p^j]}_q$.*

1. *The formal scheme $(\underline{H(S^1[p^\infty])}_q)_K$ is a formal variety of dimension $\binom{d-1}{q-1}$.*
2. *Suppose that $(\underline{H(S^1[p^\infty])}_{q-1})_K$ is a p-divisible formal group of height $\binom{d}{q-1}$ and dimension $\binom{d-1}{q-1}$, that $(\underline{H(S^1[p])}_{q-1})_K$ models its p-torsion, and that the cup product induces an isomorphism*

$$\theta^{q-1}\colon \mathbb{Q}/\mathbb{Z}_{(p)} \otimes D(\underline{H(S^1[p^\infty])}_1)^{\wedge(q-1)} \to D(\underline{H(S^1[p^\infty])}_{q-1}),$$

   *where $D(G)$ denotes the Dieudonné module associated to $K_0(G)$, where*

$$K_*(G) = K_0(G) \otimes_k K_*$$

   *is a p-divisible Hopf algebra. The same claims are then true with $q-1$ replaced everywhere by $q$.*
3. *Consider the model $\mathbb{Q}/\mathbb{Z}_{(p)} \cong S^1[p^\infty]$ for the p-primary part of the circle group. Suppose that for each $j$, the short exact sequence of groups*

$$0 \to \frac{1/p^j \cdot \mathbb{Z}_{(p)}}{\mathbb{Z}_{(p)}} \to \frac{\mathbb{Q}}{\mathbb{Z}_{(p)}} \to \frac{\mathbb{Q}}{1/p^j \cdot \mathbb{Z}_{(p)}} \to 0$$

   *induces a short exact sequence of group schemes upon applying $(\underline{H(-)}_{q-1})_K$. The sequence of group schemes under $(\underline{H(-)}_q)_K$ is then also short exact.*

*Proof of Part 1* This claim turns out to be entirely algebraic and a matter of being able to compute $H^*(\mathbb{G}; \widehat{\mathbb{G}}_a)$ for $\mathbb{G}$ a connected p-divisible group. This is expressed in the main algebraic result of Hopkins and Lurie:

**Theorem 4.6.10** ([HL, theorem 2.2.10]) *Let $\mathbb{G}$ be a p-divisible group over a perfect field k of positive characteristic p. There is then an isomorphism*

$$H^*(\mathbb{G}; \widehat{\mathbb{G}}_a) \cong \mathrm{Sym}^*(\Sigma H^1(\mathbb{G}[p], \widehat{\mathbb{G}}_a)),$$

*where "$\Sigma$" indicates that the classes are taken to lie in degree 2.* □

**Lemma 4.6.11** ([HL, remark 2.2.5]) *If $\mathbb{G}$ is a connected p-divisible group of height d and of dimension n as a formal variety, then*

$$\mathrm{rank}\left(H^1(\mathbb{G}[p]; \widehat{\mathbb{G}}_a)\right) = d - n.$$ □

*Remark 4.6.12* In the case where $\mathbb{G} = \Gamma$ is the original height $d$ formal group of dimension 1, this computes $H^*(\Gamma; \widehat{\mathbb{G}}_a)$ to be a power series algebra on $(d-1)$ generators. This is precisely what we found by hand in Lemma 3.4.12.

Returning to the task at hand, we assume inductively that $(\underline{H(S^1[p^\infty])}_{q-1})_K$ is a connected $p$-divisible group of height $\binom{d}{q-1}$ and dimension $\binom{d-1}{q-1}$. Since the input to the bar spectral sequence is computed by formal group cohomology ([Laz97], [HL, example 2.3.5], Proof of Lemma 3.2.7), it follows that the instance computing $K^*\underline{H(S^1[p^\infty])}_q$ has as its $E_2$-page an even-concentrated power series algebra of dimension

$$\binom{d}{q-1} - \binom{d-1}{q-1} = \binom{d-1}{q}.$$

The spectral sequence therefore collapses at this page, so that $(\underline{H(S^1[p^\infty])}_q)_K$ is a formal variety of the dimension claimed.

*Proof of Part 2, with a gap* The other claims in Part 2 are formal after we check that $\theta^q$ is an isomorphism, since the $p$-power-torsion structure of

$$(\underline{H(S^1[p^\infty])}_q)_K$$

can be read off from its Dieudonné module, as can its height. We introduce notation to analyze this statement: Set $M$ to be the Dieudonné module associated to $K_*\mathbb{CP}^\infty$, i.e.,

$$M = D(\underline{H(S^1[p^\infty])}_1).$$

In the following diagram

$$\begin{array}{ccccccccc} 0 & \longrightarrow & M^{\wedge q} & \longrightarrow & \mathbb{Q} \otimes M^{\wedge q} & \longrightarrow & \mathbb{Q}/\mathbb{Z}_{(p)} \otimes M^{\wedge q} & \longrightarrow & 0 \\ & & \downarrow V & & \downarrow V & & \downarrow V & & \\ 0 & \longrightarrow & M^{\wedge q} & \longrightarrow & \mathbb{Q} \otimes M^{\wedge q} & \longrightarrow & \mathbb{Q}/\mathbb{Z}_{(p)} \otimes M^{\wedge q} & \longrightarrow & 0 \end{array}$$

the middle map is an isomorphism. This forces $V$ to be a surjective endomorphism of $M^{\wedge q} \otimes \mathbb{Q}/\mathbb{Z}_{(p)}$, and the snake lemma shows that there is an isomorphism

$$\ker(V \colon \mathbb{Q}/\mathbb{Z}_{(p)} \otimes M^{\wedge q} \to \mathbb{Q}/\mathbb{Z}_{(p)} \otimes M^{\wedge q}) \cong \mathrm{coker}(V \colon M^{\wedge q} \to M^{\wedge q}).$$

Picking any coordinate $x$ and considering it as an element in the curves model of the Dieudonné functor, we see that the right-hand side is spanned by elements $x \wedge V^{\wedge I}x$, and hence the left-hand side has $k$-vector-space dimension $\binom{d-1}{q}$.

By very carefully studying the bar spectral sequence, one can learn that $\theta^q$ induces a surjection[42]

$$\ker V|_{\mathbb{Q}/\mathbb{Z}_{(p)} \otimes M^{\wedge q}} \to \ker V|_{D(\underline{H(S^1[p^\infty])}_q)}.$$

---

[42] The proof of this is quite complicated, and it rests on a pairing between the spectral sequence for $q = 1$ and the spectral sequence at $q - 1$, mapping to the spectral sequence for $q$. Remarkably, this same pairing is the main tool that powers the original approach of Ravenel and Wilson [RW80] and of our approach to the unstable dual Steenrod algebra in Lecture 4.1 (see Lemma 4.1.9). There, their program is to fix $j = 1$ and inductively analyze $q$ using this same pairing, then use

### 4.6 Cooperations among Geometric Points on $M_{\mathrm{fg}}$

In fact, since these two have the same rank, $\theta^q|_{\ker V}$ is an isomorphism on these subspaces. This is enough to conclude that $\theta^q$ is an injection: Since the action of $V$ is locally nilpotent, if $\theta^q$ ever failed to be an injection then we could apply $V$ enough times to get an example of a nontrivial element in $\ker V|_{\mathbb{Q}/\mathbb{Z}_{(p)} \otimes M^{\wedge q}}$ mapping to zero. Finally, to show that $\theta^q$ is surjective, we again use the local nilpotence of $V$ to filter $\mathbb{Q}/\mathbb{Z}_{(p)} \otimes M^{\wedge q}$ by the subspaces $\ker V^\ell$, $\ell \geq 1$, and it is then possible (though we omit the proof) to use our understanding of $\ker V$ to form preimages.

*Proof of Part 3, mostly omitted* This proof is quite complicated, but it is, in spirit, a generalization of the observation at $q = 1$ that the role of the odd-degree classes in the bar spectral sequence is to pair up with those classes in the image of the $[p^j]$-map. In fact, their main assertion is:

> We give notation for the following rings:
> 
> $$A = K_0 \underline{H(S^1[p^\infty])}_{q-1}, \quad A' = K_0 \underline{H(S^1[p^j])}_{q-1}, \quad R = K_0 \underline{H(S^1[p^\infty])}_q.$$
> 
> Let $x' \in E_2^{1,0}$ be an element, and let $y' \in \mathfrak{m}_R$ satisfy
> 
> $$y' = \psi(x') \otimes v \in \mathrm{Ext}_A^2 \otimes_k \pi_2 K \cong \mathfrak{m}_R / \mathfrak{m}_R^2.$$
> 
> Suppose that the Hopf algebra homomorphism $[p^j] \colon R \to R$ carries $y'$ to an element $y \in \mathfrak{m}_R^r$, and let $x \in E_2^{2r,2r-2}$ denote the image of $y$ under the composite
> 
> $$\mathfrak{m}_R^r/\mathfrak{m}_R^{r+1} \cong \mathrm{Ext}_A^{2r} \otimes_k \pi_{2r} K \to \mathrm{Ext}_{A'}^{2r} \otimes_k \pi_{2r} K = E_2^{2r,2r} \xrightarrow{-v^{-1}} E_2^{2r,2r-2}.$$
> 
> Then $x$ and $x'$ survive to the $(2r-1)$th page of the bar spectral sequence, and there we have
> 
> $$d_{2r-1} x' = x.$$

From here, it is a matter of *carefully* pairing elements (see [HL, p. 60]).  □

*Remark 4.6.13* Theorem 4.6.9 admits a restatement purely in terms of Hopf algebras, although Dieudonné theory was essential in its proof. The cup product gives a natural map[43]

$$K_* \underline{H(S^1[p^j])}_1^{\wedge q} \to K_* \underline{H(S^1[p^j])}_q.$$

The main result of this section is that this map is an isomorphism for all $q$, and indeed that the map from the free exterior *Hopf ring* maps isomorphically to the topological Hopf ring.

---

these base cases to ground a strong induction on $j$ and $q$, and then finally to fix $q$ and take the limit as $j \to \infty$.

[43] Note that this alternation condition becomes dramatically more complicated in the case that the formal group law and its formal group inverse series become more complicated than that of $\widehat{\mathbb{G}}_a$.

*Remark 4.6.14* ([HL, section 3], [Hed14], [Hed15])   Because $K^*\underline{H\mathbb{Z}/p^j}_q$ is even, you can hope to augment this to a calculation of $E^*\underline{H\mathbb{Z}/p^j}_q$ for $E = E_\Gamma$ the associated continuous Morava $E$-theory. This is indeed possible, and the analogous formula is true at the level of Hopf algebras:

$$E_*\underline{H(S^1[p^j])}_q \cong E_*\underline{H(S^1[p^j])}_1^{\wedge q}.$$

However, the attendant algebraic geometry is quite complicated: You either need a form of Dieudonné theory that functions over $\mathcal{M}_{E_\Gamma}$ (and then attempt to drag the previous proof through that setting), or you need to directly confront what "alternating power of a $p$-divisible group" means at the level of $p$-divisible groups (and forego all of the time-saving help afforded to you by Dieudonné theory).

*Remark 4.6.15* (see Lemma 3.6.9)   Notice that if we let the $q$-index tend to $\infty$ in $K_*\underline{H}_{q+1}$, we get the $K$-homology of a point. This is another way to see that the stable cooperations $K_*H$ vanish, meaning that the *only* information present comes from unstable cooperations.

*Remark 4.6.16*   Although the method of starting with the $j = \infty$ case and deducing from it the $1 \leq j < \infty$ cases is due to Hopkins and Lurie, the observation that the spectral sequence at $j = \infty$ is remarkably simple had already been made by Ravenel and Wilson [RW80, theorem 12.3], [RWY98, theorem 8.1.3].

*Remark 4.6.17*   Theorem 4.6.9 has a statement in the language of Lecture 4.2. Abbreviating $H = H(S^1[p^j])$, rather than forming the algebraic approximation

$$\mathcal{A}\mathcal{A}(K, H) = K_*(\mathbb{C}P^\infty)^{\bigcirc}_{K_*[H^*]}[H^*\mathbb{C}P^\infty]$$

as usual, we form the modified version

$$\mathcal{A}\mathcal{A}_{p^j}(K, H) = K_*(\underline{H}_1)^{\bigcirc}_{K_*[H^*]}[H^*\underline{H}_1].$$

Using the same unstable Kronecker pairing, this supports a natural map to $K_*\underline{H}_*$, and a summary of this lecture's results is that it is an isomorphism. This auxiliary approximation has two interesting features: It makes use of *odd*-dimensional Eilenberg–Mac Lane spaces, and it is an elaborate name for the alternating Hopf ring on the Hopf algebra $K_*\underline{H}_1$.

# Case Study 5

## The $\sigma$-Orientation

By this point, we have seen a great many ways that algebraic geometry exerts control over the behavior of homotopy theory, stable and unstable. The goal in this case study is to explore a setting where algebraic geometry is itself tightly controlled: Whereas the behavior of formal groups is quite open-ended, the behavior of *abelian varieties* is comparatively strict. We import this strictness into algebraic topology by studying complex-orientable cohomology theories $E$ which have been tagged with an auxiliary abelian variety $A$ and an isomorphism $\varphi \colon \mathbb{CP}^\infty_E \cong A_0^\wedge$. In the case that $A$ is an elliptic curve, this is our definition of an *elliptic cohomology theory*. The idea, then, is not that this puts serious constraints on the formal group $\mathbb{CP}^\infty_E$ (although it does place some), but rather that the theory of abelian varieties endows $A$, and hence $A_0^\wedge$, with various bits of preferred data. This is the tack we take to construct a *canonical $MU[6, \infty)$-orientation* of $E$: For any complex-orientable $E$, we identify the collection of such ring maps with "$\Theta$-structures on $\mathbb{CP}^\infty_E$"; a basic theorem about abelian varieties endows the elliptic curve $A$ with a canonical such structure; and altogether this yields the desired orientation for an elliptic spectrum.

Making the identification of $MU[6, \infty)$-orientations with $\Theta$-structures requires real work, but many of the stepping stones are now in place. We begin with a technical section about especially nice formal schemes, called *coalgebraic*, and we use this to finally give the proof, announced back in Theorem 2.2.7, that the scheme of stable Weil divisors on a formal curve presents the free formal group on that curve. With that out of the way and with $MU[6, \infty)$ in mind as the eventual goal, we then summarize the behavior of the part of the Postnikov tower for complex bordism that we *do* understand – the cases of $MUP$ and $MU$ – and use this to make an analysis of $MSU$. In particular, we rely heavily on the results from Case Studies 2 and 3 to understand the co/homological behaviors of $BU \times \mathbb{Z}$, $BU$, $MUP$, and $MU$, which we employ as traction in understanding $BSU$ and $MSU$.

The coherence of all of these statements gives us very explicit target theorems to aim for in our study of $MU[6, \infty)$, but we are forced to approach them from a different vantage point: Whereas we can prove a splitting principle for $SU$-bundles, the analogous statement for $U[5, \infty)$-bundles does not appear to admit a direct proof. Consequently, the proofs of the other structure theorems for $BU[6, \infty)$ and $MU[6, \infty)$ are made considerably more complicated because we have to work with our splitting principle hands tied. Instead, our main tools are the results developed in Case Study 4, which give us direct access to the co/homology of the layers of the Postnikov tower. When the dust of all this settles, we will have arrived at a very satisfying and complete theory of $MU[6, \infty)$-orientations, applicable to an arbitrary complex-orientable cohomology theory.

The reader gifted with an exceptional attention span will recall from the Introduction that we were *really* interested in $MString$-orientations, and that our interest in $MU[6, \infty)$-orientations was itself only a stepping stone. We close this case study with an analysis of this last setting, where we finally yield and place more hypotheses on $E$ – a necessity for gaining calculational access to co/homological behavior of objects like $BString$, which lie outside of the broader complex-orientable story.

We also give a short résumé on the theory of elliptic curves in Lecture 5.5, extracting the smallest possible subset of their theory that we will need here.

## 5.1 Coalgebraic Formal Schemes

We will now finally address a point that we have long let slide: In the first third of this book we were primarily interested in the formal scheme associated to the *cohomology* of a space, but in the second third we were primarily interested in a construction converting the *homology* of a spectrum to a sheaf over a context. Our goal for this lecture is to, when possible, put these two variances on an even footing. Our motivation for putting this lingering discrepancy to rest is more technical than aesthetic: We have previously wanted access to certain colimits of formal schemes (e.g., in Theorem 2.2.7). While such colimits are generally forbidding, similarly to colimits of manifolds, we will in effect produce certain conditions under which they are accessible.

For $E$ a ring spectrum and $X$ a space, the diagonal map $\Delta \colon X \to X \times X$ induces a multiplication map on $E$-cohomology via the composite

$$E^*X \otimes_{E^*} E^*X \xrightarrow{\text{Künneth}} E^*(X \times X) \xrightarrow{E^*\Delta} E^*X.$$

## 5.1 Coalgebraic Formal Schemes

Dually, applying $E$-homology, we have a pair of maps

$$E_*X \xrightarrow{E_*\Delta} E_*(X \times X) \xleftarrow{\text{Künneth}} E_*X \otimes_{E_*} E_*X,$$

where, remarkably, the Künneth map goes the wrong way to form a composite. In the case where that map is an isomorphism, the long composite induces the structure of an $E_*$-coalgebra on $E_*X$. In the most generous case that $E$ is a field spectrum (in the sense of Corollary 3.5.15), the Künneth map is always invertible and, moreover, $E^*X$ is functorially the linear dual of $E_*X$. By taking our definition of the formal scheme (associated to a space) and dualizing it, we are motivated to consider the following purely algebraic construction:

**Definition 5.1.1** Let $C$ be a coalgebra over a field $k$. We define a functor

$\operatorname{Sch} C \colon \text{Algebras}_{k/} \to \text{Sets}$,

$$T \mapsto \left\{ f \in C \otimes T \,\middle|\, \begin{array}{l} \Delta f = f \otimes f \in (C \otimes T) \otimes_T (C \otimes T), \\ \varepsilon f = 1 \end{array} \right\}.$$

**Lemma 5.1.2** *For a field $k$ and a $k$-algebra $A$ which is finite dimensional as a $k$-module, there is a natural isomorphism* $\operatorname{Spec} A \cong \operatorname{Sch} A^*$.

*Proof sketch* A point $f \in (\operatorname{Sch} A^*)(T) \subseteq A^* \otimes T$ gives rise to a $k$-module map $f_* \colon A \to T$, which the extra conditions in the formation of $(\operatorname{Sch} C)(T)$ force to be a ring homomorphism. The finiteness assumption is present exactly so that $A$ is its own double-dual, giving an inverse assignment. □

If we drop the finiteness assumption, then this comparison proof fails entirely. Indeed, the multiplication on $A$ gives rise only to maps

$$A^* \to (A \otimes_k A)^* \leftarrow A^* \otimes_k A^*,$$

which is not enough to make $A^*$ into a $k$-coalgebra. However, if we start instead with a $k$-coalgebra $C$ of infinite dimension, the following result is very telling:

**Lemma 5.1.3** ([Dem86, p. 12], [Mic03, appendix 5.3], [HL, remark 1.1.8]) *For $C$ a coalgebra over a field $k$, any finite-dimensional $k$-linear subspace of $C$ can be finitely enlarged to a subcoalgebra of $C$. Accordingly, taking the colimit gives a canonical equivalence*

$$\operatorname{Ind}(\text{Coalgebras}_k^{\text{fin}}) \xrightarrow{\simeq} \text{Coalgebras}_k.$$ □

This result allows us to leverage our duality lemma pointwise: For an arbitrary $k$-coalgebra, we break it up into a lattice of finite $k$-coalgebras, and take their linear duals to get a reversed lattice of finite $k$-algebras. Altogether, this indicates that $k$-coalgebras generally want to model *formal schemes*.

**Corollary 5.1.4** *For $C$ a coalgebra over a field $k$ expressed as a colimit $C = \operatorname{colim}_k C_k$ of finite subcoalgebras, there is an equivalence*

$$\operatorname{Sch} C \cong \{\operatorname{Spec} C_k^*\}_k = \operatorname{Spf} C^*.$$

*This induces a* covariant *equivalence of categories*

$$\textsc{Coalgebras}_k \cong \textsc{FormalSchemes}_{/k}.$$

*This equivalence translates between the product of formal schemes, the tensor product of pro-algebras, and the tensor product of coalgebras.* □

This covariant algebraic model for formal schemes is very useful. For instance, this equivalence makes the following calculation trivial:

**Lemma 5.1.5** (see Lemma 1.5.1, Theorem 2.2.7, and Corollary 2.3.15) *Select a coalgebra $C$ over a field $k$ together with a pointing $k \to C$. Write $M$ for the coideal $M = C/k$. The free formal commutative monoid on the pointed formal scheme $\operatorname{Sch} k \to \operatorname{Sch} C$ is given by*

$$F(\operatorname{Sch} k \to \operatorname{Sch} C) = \operatorname{Sch} \operatorname{Sym}^* M.$$

*Writing $\Delta c = \sum_j \ell_j \otimes r_j$ for the diagonal on $C$, the diagonal on $\operatorname{Sym}^* C$ is given by*

$$\Delta(c_1 \cdots c_n) = \sum_{j_1,\ldots,j_n} (\ell_{1,j_1} \cdots \ell_{n,j_n}) \otimes (r_{1,j_1} \cdots r_{n,j_n}). \qquad \Box$$

It is unfortunate, then, that when working over a ring rather than a field, Lemma 5.1.3 fails [Mic03, appendix 5.3]. Nonetheless, it is possible to bake into the definitions the machinery needed to get a good-enough analog of Corollary 5.1.4.

**Definition 5.1.6** ([Str99b, definition 4.58])  Let $C$ be an $R$-coalgebra which is free as an $R$-module. A basis $\{x_j\}$ of $C$ is said to be a *good basis* when any finite subcollection of $\{x_j\}$ can be finitely enlarged to a subcollection that spans a subcoalgebra. The coalgebra $C$ is itself said to be *good* when it admits a good basis. A formal scheme $X$ is said to be *coalgebraic* when it is isomorphic to $\operatorname{Sch} C$ for a good coalgebra $C$.

*Example 5.1.7*  The formal scheme $\widehat{\mathbb{A}}^n$ is coalgebraic. Beginning with the presentation

$$\widehat{\mathbb{A}}^n = \operatorname{Spf} R[\![x_1,\ldots,x_n]\!] = \operatorname*{colim}_J \operatorname{Spec} R[x_1,\ldots,x_n]/(x_1^{j_1},\ldots,x_n^{j_n}),$$

write $A_J$ for the algebra on the right-hand side. Each $A_J$ is a free $R$-module,

## 5.1 Coalgebraic Formal Schemes

and we write
$$C_J = A_J^* = R\{\beta_K \mid K < J\}$$
for the dual coalgebra, with
$$\beta_K(x^L) = \begin{cases} 1 & \text{if } K = L, \\ 0 & \text{otherwise.} \end{cases}$$
The elements $\beta_K$ form a good basis for the full coalgebra $C = \operatorname{colim}_J C_J$: Any finite collection of them $\{\beta_K\}_{K \in \mathcal{K}}$ is contained inside any $C_J$ satisfying $K < J$ for all $K \in \mathcal{K}$. As an additional consequence, all formal varieties are coalgebraic.

The main utility of this condition is that it gives us access to colimits of formal schemes:

**Theorem 5.1.8** ([Str99b, proposition 4.64])  *Suppose that*
$$F \colon \mathrm{I} \to \mathrm{Coalgebras}_R$$
*is a colimit diagram of coalgebras such that each object in the diagram, including the colimit point, is a good coalgebra. Then*
$$\mathrm{Sch} \circ F \colon \mathrm{I} \to \mathrm{FormalSchemes}$$
*is a colimit diagram of formal schemes.* □

For an example of the sort of constructions that become available via this theorem, we prove the following corollary by analyzing the symmetric power of coalgebras:

**Corollary 5.1.9** ([Str99b, example 4.65 and proposition 6.4])  *When a formal scheme $X$ is coalgebraic, the symmetric power $X^{\times n}_{\Sigma_n}$ exists. In fact, $\coprod_{n \geq 0} X^{\times n}_{\Sigma_n}$ models the free formal commutative monoid on $X$. Given an additional pointing $\operatorname{Spec} R \to X$, the colimit of the induced system*
$$\operatorname{colim}\left(\cdots \to X^{\times n}_{\Sigma_n} = \operatorname{Spec} R \times X^{\times n}_{\Sigma_n} \to X \times X^{\times n}_{\Sigma_n} \to X^{\times (n+1)}_{\Sigma_{n+1}} \to \cdots \right)$$
*models the free formal commutative monoid on the pointed formal scheme.*

*Proof sketch*  The main points entirely mirror the case over a field: The symmetric power coalgebra construction gives models for $X^{\times n}_{\Sigma_n}$, the total symmetric power Hopf algebra gives a model for the free formal commutative monoid, and the stabilization against the pointing is modeled by inverting an element in the symmetric algebra. In each case, choosing a good basis for the coalgebra underlying $X$ yields choices of good bases for the coalgebras arising from these constructions, essentially because their elements are crafted out of finite combinations of the elements of the original. □

In the specific case that $\operatorname{Spec} R \to X$ is a pointed formal *curve*, we can prove something more:

**Corollary 5.1.10** ([Str99b, proposition 6.12])  *For* $\operatorname{Spec} R \to X$ *a pointed formal curve, the free formal commutative monoid is automatically an abelian group.*

*Proof sketch*  The main idea is that the coalgebra associated to a formal curve admits an increasing filtration $F_k$ so that the reduced diagonal

$$\overline{\Delta} = \Delta - (1 \otimes \eta) - (\eta \otimes 1)$$

reduces filtration degree:

$$\overline{\Delta}|_{F_k} : F_k \to \sum_{\substack{i,j>0 \\ i+j=k}} F_i \otimes F_j.$$

In turn, the symmetric algebra on the coalgebra associated to a formal curve inherits enough of this filtration that one can iteratively solve for a Hopf algebra antipode. □

We now reconnect this algebraic discussion with the algebraic topology that spurred it.

**Lemma 5.1.11**  *If $E$ and $X$ are such that $E_* X$ is an $E_*$-coalgebra and*

$$E^* X \cong \operatorname{Modules}_{E_*}(E_* X, E_*),$$

*then there is an equivalence*

$$\operatorname{Sch} E_* X \cong X_E.$$

*Proof*  We have defined $X_E$ to have the formal topology induced by the compactly generated topology of $X$, and this same topology can also be used to write $\operatorname{Sch} E_* X$ as the colimit of finite $E_*$-coalgebras. □

*Example 5.1.12* (see Theorem 4.6.1 and Remark 4.6.2)  For a Morava $K$-theory $K_\Gamma$ associated to a formal group $\Gamma$ of finite height, we have seen that there is an exact sequence of Hopf algebras

$$K_\Gamma^0(BS^1) \xrightarrow{[p^j]} K_\Gamma^0(BS^1) \to K_\Gamma^0(BS^1[p^j]),$$

presenting $(BS^1[p^j])_K$ as the $p^j$-torsion formal subscheme $BS_K^1[p^j]$. The Hopf algebra calculation also holds in $K$-homology, where there is instead the exact sequence

$$(K_\Gamma)_0 B(S^1[p^j]) \to (K_\Gamma)_0 BS^1 \xrightarrow{(-)^{*p^j}} (K_\Gamma)_0 BS^1$$

## 5.1 Coalgebraic Formal Schemes

presenting $(K_\Gamma)_0 B(S^1[p^j])$ as the $p^j$-order $*$-nilpotence in the middle Hopf algebra. Applying Sch to this last line covariantly converts this second statement about Hopf algebras to the corresponding statement above about the associated formal schemes – i.e., the behavior of the homology Hopf algebra is a covariant avatar of the behavior of the formal schemes.

The example above, where the space in question is an $H$-space, also spurs us to consider a certain "wrong-way" operation. We have seen that the algebra structure of the $K$-cohomology of a space and the coalgebra structure of the $K$-homology of the same space contain equivalent data: They both give rise to the same formal scheme. However, in the case of a commutative $H$-space, the $K$-homology and $K$-cohomology give *commutative and cocommutative Hopf algebras*. Hence, in addition to considering the coalgebraic formal scheme $\mathrm{Sch}(K_\Gamma)_0 B(S^1[p^j])$, we can also consider the affine scheme $\mathrm{Spec}(K_\Gamma)_0 B(S^1[p^j])$. This, too, should contain identical information, and this is the subject of Cartier duality.

**Definition 5.1.13** ([Str99b, sections 6.3–6.4])  The *Cartier dual* of a commutative finite group scheme $G$ is defined by the formula

$$DG = \textsc{GroupSchemes}(G, \mathbb{G}_m),$$

itself a finite group scheme. More generally, the Cartier dual of a commutative *coalgebraic* formal group $\widehat{\mathbb{G}}$ can also be defined by

$$D\widehat{\mathbb{G}} = \textsc{GroupSchemes}(\widehat{\mathbb{G}}, \mathbb{G}_m).$$

**Lemma 5.1.14** ([Str99b, proposition 6.19])  *Let $\widehat{\mathbb{G}}$ be a coalgebraic formal group over a formal scheme $X$, and write $\mathbb{H} = \mathrm{Spec}\, O^*_{\widehat{\mathbb{G}}}$ for the group scheme associated to its dual Hopf algebra. Cartier duality then has the effects $D\widehat{\mathbb{G}} = \mathbb{H}$ and $D\mathbb{H} = \widehat{\mathbb{G}}$.*

*Proof*  We show that the first two objects, $D\widehat{\mathbb{G}}$ and $\mathbb{H}$, represent the same object. A point $f \in D\widehat{\mathbb{G}}(t\colon \mathrm{Spec}\, T \to S)$ is specified by a function

$$f\colon t^*\widehat{\mathbb{G}} \to t^*(\mathbb{G}_m \times X).$$

The map $f$ is equivalent to a map of Hopf algebras $f^*\colon T[u^\pm] \to O_{\widehat{\mathbb{G}}} \otimes_{O_X} T$, which is determined by its value $f^*(u) \in O_{\widehat{\mathbb{G}}} \otimes_{O_X} T$, which must satisfy the two relations $\Delta(f^*u) = f^*u \otimes f^*u$ and $\varepsilon(f^*u) = 1$. Invoking linear duality, $f^*u$ can also be considered as an element of $\textsc{Modules}_{O_X}(O^*_{\widehat{\mathbb{G}}}, T)$, and the two relations on $f^*u$ show that it lands in the subset

$$f^*u \in \textsc{Algebras}_{O_X/}(O^*_{\widehat{\mathbb{G}}}, T) \subseteq \textsc{Modules}_{O_X}(O^*_{\widehat{\mathbb{G}}}, T).$$

This assignment is invertible, and the proof is entirely similar for $D\mathbb{H} \cong \widehat{\mathbb{G}}$. □

*Remark 5.1.15* ([Dem86, p. 72]) The Dieudonné module of the Cartier dual of a formal group is *also* described by linear duality: It is the linear dual of the Dieudonné module of the original formal group. Hence, the covariant and contravariant Dieudonné modules described in Lecture 4.4 can be taken to be related by Cartier duality.

*Remark 5.1.16* Cartier duality intertwines the homological and cohomological schemes assigned to a commutative $H$-space. When such a commutative $H$-space $X$ has free and even $E$-homology, there is an isomorphism

$$D(\mathrm{Spf}\, E^0 X) = DX_E = \underline{\mathrm{GroupSchemes}}(X_E, \mathbb{G}_m) \cong \mathrm{Spec}\, E_0 X.$$

## 5.2 Special Divisors and the Special Splitting Principle

Starting in this lecture, after our extended interludes on chromatic homotopy theory and cooperations, we return to thinking about bordism orientations directly. To begin, we will summarize the various perspectives already adopted in Case Study 2 when we were studying complex orientations of ring spectra.

1. Definition 2.0.2: A complex-orientation of $E$ is, definitionally, a map $MUP \to E$ of ring spectra in the homotopy category.
2. Theorem 2.3.19: A complex-orientation of $E$ is also equivalent to a multiplicative system of Thom isomorphisms for complex vector bundles. Such a system is determined by its value on the universal line bundle $\mathcal{L}$ over $\mathbb{CP}^\infty$. We can also phrase this algebro-geometrically: Such a Thom isomorphism is the data of a trivialization of the Thom sheaf $\mathbb{L}(\mathcal{L})$ over $\mathbb{CP}^\infty_E$.
3. Lemma 2.6.7: Ring spectrum maps $MUP \to E$ induce on $E$-homology maps $E_0 MUP \to E_0$ of $E_0$-algebras. This, too, can be phrased algebro-geometrically: These are elements of $(\mathrm{Spec}\, E_0 MUP)(E_0)$.

We can summarize our main result about these perspectives as follows:

**Theorem 5.2.1** ([AHS01, example 2.53]) *Take $E$ to be* even-periodic *(hence complex-orientable, see Remark 2.1.15). The functor*

$$\mathrm{AffineSchemes}_{/\,\mathrm{Spec}\, E_0} \to \mathrm{Sets},$$

$$(\mathrm{Spec}\, T \xrightarrow{u} \mathrm{Spec}\, E_0) \mapsto \left\{ \textit{trivializations of } u^* \mathbb{L}(\mathcal{L}) \textit{ over } u^* \mathbb{CP}^\infty_E \right\}$$

*is isomorphic to the affine scheme* $\mathrm{Spec}\, E_0 MUP$. *Moreover, the $E_0$-points of this scheme biject with ring spectrum maps* $MUP \to E$.

## 5.2 Special Divisors and the Special Splitting Principle 227

*Proof summary* The equivalence between (1) and (3) – i.e., between complex-orientations and $E_0$-points of $\operatorname{Spec} E_0 MUP$ – follows from calculating that $E_0 MUP$ is a free $E_0$-module, so that there is a collapse in the universal coefficient theorem. Then, the equivalence between (1) and (2) follows from the splitting principle for complex line bundles, which says that the first Chern class of $\mathcal{L}$ – i.e., a trivialization of $\mathbb{L}(\mathcal{L})$ – determines the rest of the map $MUP \to E$. □

An analogous result holds for ring spectrum maps $MU \to E$ and the line bundle $1 - \mathcal{L}$, and it is proven in an analogous way. In particular, we will want a version of the splitting principle for virtual vector bundles of virtual rank 0. Given a finite complex $X$ and such a rank 0 virtual vector bundle, write $\tilde{V}\colon X \to BU$ for the classifying map. Because $X$ is a finite complex, there exists an integer $n$ so that $\tilde{V} = -(n \cdot 1 - V)$ for an honest rank $n$ vector bundle $V$ over $X$. Using Corollary 2.3.10, the splitting $f^*V \cong \bigoplus_{i=1}^{n} \mathcal{L}_i$ over $Y$ gives a presentation of $\tilde{V}$ as

$$\tilde{V} = -(n \cdot 1 - V) = -\bigoplus_{i=1}^{n}(1 - \mathcal{L}_i).$$

Crucially, we have organized this sum *entirely in terms of bundles classified by BU*, as each bundle $1 - \mathcal{L}_i$ itself has the natural structure of a rank 0 virtual vector bundle. This version of the splitting principle, together with our extended discussion of formal geometry, begets the following analog of the previous result:

**Theorem 5.2.2** ([AHS01, example 2.54], see also [AS01, lemma 6.2]) *Take $E$ to be even-periodic. The functor*

$$\textsc{AffineSchemes}_{/\operatorname{Spec} E_0} \to \textsc{Sets},$$

$$(\operatorname{Spec} T \xrightarrow{u} \operatorname{Spec} E_0) \mapsto \left\{ \textit{trivializations of } u^*\mathbb{L}(1 - \mathcal{L}) \textit{ over } u^*\mathbb{CP}_E^\infty \right\}$$

*is isomorphic to the affine scheme* $\operatorname{Spec} E_0 MU$. *Moreover, the $E_0$-points of this scheme biject with ring spectrum maps* $MU \to E$. □

In Lecture 2.3, we preferred to think of the cohomology of a Thom spectrum as a sheaf over the formal scheme associated to its base space. This extra structure has not evaporated in the homological context – it just takes a different form.

**Lemma 5.2.3** *For $\xi\colon G \to BGL_1\mathbb{S}$ a group map, the Thom spectrum $T\xi$ is a $(\Sigma_+^\infty G)$-cotorsor.*

*Construction* The Thom isomorphism $T\xi \wedge T\xi \simeq T\xi \wedge \Sigma_+^\infty G$ composes with the unit map $\mathbb{S} \to T\xi$ to give the *Thom diagonal*

$$T\xi \to T\xi \wedge \Sigma_+^\infty G. \qquad \square$$

Applying Spec $E_0(-)$, the Thom diagonal is translated into the structure of a free and transitive action map

$$\operatorname{Spec} E_0 T(\xi) \times \operatorname{Spec} E_0 G \to \operatorname{Spec} E_0 T(\xi).$$

This construction is natural in the formation of $G$ or $\xi$, and so we are also moved to specialize to the cases of $G = \mathbb{Z} \times BU$ and $G = BU$ and to understand Spec $E_0 G$ in those contexts. Again, this is a matter of chaining together results we have already proven:

$$\begin{aligned}
\operatorname{Spec} E_0(\mathbb{Z} \times BU) &= D((\mathbb{Z} \times BU)_E) & \text{(Remark 5.1.16)} \\
&= D(\operatorname{Div} \mathbb{CP}_E^\infty) & \text{(Corollary 2.3.15)} \\
&= \underline{\operatorname{FormalGroups}}(\operatorname{Div} \mathbb{CP}_E^\infty, \mathbb{G}_m) & \text{(Definition 5.1.13)} \\
&= \underline{\operatorname{FormalSchemes}}(\mathbb{CP}_E^\infty, \mathbb{G}_m), & \text{(Corollary 5.1.9)}
\end{aligned}$$

and similarly

$$\operatorname{Spec} E_0(BU) = \underline{\operatorname{FormalSchemes}}_{*/}(\mathbb{CP}_E^\infty, \mathbb{G}_m)$$

is the subscheme of those maps sending the identity point of $\mathbb{CP}_E^\infty$ to the identity point of $\mathbb{G}_m$. Such functions can be identified with trivializations of the trivial sheaf over $\mathbb{CP}_E^\infty$, and the action map induced by the Thom diagonal belongs to a commuting square

$$\begin{array}{ccc}
\operatorname{Spec} E_0 MU \\
\times & \longrightarrow & \operatorname{Spec} E_0 MU \\
\operatorname{Spec} E_0 BU \\
\| & & \| \\
\{\operatorname{triv}^{\operatorname{ns}} \text{ of } \mathbb{L}(1-\mathcal{L}) \downarrow \mathbb{CP}_E^\infty\} \\
\times & \longrightarrow & \{\operatorname{triv}^{\operatorname{ns}} \text{ of } \mathbb{L}(1-\mathcal{L}) \otimes 1 \downarrow \mathbb{CP}_E^\infty\}. \\
\{\operatorname{triv}^{\operatorname{ns}} \text{ of } 1 \downarrow \mathbb{CP}_E^\infty\}
\end{array}$$

*Remark 5.2.4* ([AHS01, corollary 2.30 and theorem 2.50]) The topological maps

$$BU \to \mathbb{Z} \times BU, \qquad MU \to MUP$$

## 5.2 Special Divisors and the Special Splitting Principle 229

induce recognizable algebro-geometric maps upon application of Spec $E_0(-)$. The comparison map

$$(\mathrm{Spec}\, E_0(\mathbb{Z} \times BU))(T) \to \underline{\mathrm{FORMALSCHEMES}}(\mathbb{CP}_E^\infty, \mathbb{G}_m)$$

reads off the image of a map

$$f\colon E_0(\mathbb{Z} \times BU) \cong E_0[b_0^\pm, b_1, \ldots] \to T$$

as the components of a function $\sum_j f(b_0^j b_j) x^j \in T \otimes O_{\mathbb{CP}_E^\infty}$, whereas the comparison map for $BU$ reads off the image of a map

$$g\colon E_0(BU) \cong E_0[b_0^\pm, b_1, b_2, \ldots]/(b_0 = 1) \to T$$

as the components of a function $\sum_j g(b_j) x^j$, effecting a normalizing division by $b_0$, itself the image of $\mathbb{CP}_E^0 \subseteq \mathbb{CP}_E^\infty$ in $\mathbb{G}_m$. Geometrically, this gives the commuting square

$$\begin{array}{ccc}
\mathrm{Spec}\, E_0(\mathbb{Z} \times BU) & \longrightarrow & \mathrm{Spec}\, E_0(BU) \\
\| & & \| \\
\underline{\mathrm{FORMALSCHEMES}}(\mathbb{CP}_E^\infty, \mathbb{G}_m) & \longrightarrow & \underline{\mathrm{FORMALSCHEMES}}_{*/}(\mathbb{CP}_E^\infty, \mathbb{G}_m)
\end{array}$$

$$f(t) \mapsto f(t)/f(0).$$

At the level of Thom spectra, these identifications are controlled by the Chern classes associated to these bundles, and we now briefly summarize their relationship. The spaces $\mathbb{Z} \times BU$ and $BU$ are the zeroth and second spaces in the $\Omega$-spectrum for connective complex $K$-theory, and since connective complex $K$-theory is complex-orientable, we have $kU^*(\mathbb{CP}^\infty) = \mathbb{Z}[\beta][\![c_1]\!]$. Inside this ring there is the relation

$$\beta c_1 = (1 - \mathcal{L}).$$

The pieces of this topological formula have recognizable provenances: $\beta$ is the restriction of the tautological bundle on $\mathbb{CP}^\infty$ to $S^2 \simeq \mathbb{CP}^1$, $(1 - \mathcal{L})$ governs the theory of Chern classes for an $MUP$-orientation, and $c_1$ governs the theory of Chern classes for an $MU$-orientation. Applying our conversions to algebro-geometric terms (i.e., the previous theorems as well as Example 2.3.4), this same equation says that the trivialization $f$ of $\mathbb{L}(u^*\mathcal{L})$, corresponding to a point in $(\mathrm{Spec}\, E_0 MUP)(T)$, is sent to the trivialization $f'(0)/f$ of $\mathbb{L}(u^*(1-\mathcal{L}))$, corresponding to the induced point in $(\mathrm{Spec}\, E_0 MU)(T)$.

This last remark indicates a direction of possible generalization to the other spaces in the $\Omega$-spectrum for connective complex $K$-theory, which have the following polite description:

**Lemma 5.2.5** *There is an equivalence*
$$\underline{kU}_{2k} = BU[2k, \infty).$$

*Proof* Consider the element $\beta^k \in kU_* = \mathbb{Z}[\beta]$. The source of the induced map $\beta^k : \Sigma^{2k} kU \to kU$ is $2k$-connective, and hence there is a factorization
$$\Sigma^{2k} kU \to kU[2k, \infty) \to kU.$$
Then, the structure of the homotopy ring $kU_*$ shows that this is an equivalence: Every class of degree at least $2k$ can be uniquely written as a $\beta^k$-multiple.[1] Applying $\Omega^\infty$ gives the desired statement:
$$\underline{kU}_{2k} = \Omega^\infty \Sigma^{2k} kU \simeq \Omega^\infty kU[2k, \infty) = BU[2k, \infty). \qquad \square$$

The next space and Thom spectrum in the sequence are thus $BSU$ and $MSU$. These are approachable using methods we have developed so far, and so we set their description as our goal for the remainder of this lecture. Our jumping off point for that story will be, again, a partial extension of the splitting principle.

**Lemma 5.2.6** *Let $X$ be a finite complex, and let $\tilde{V} \colon X \to BU$ classify a virtual vector bundle of rank $0$ over $X$. Suppose that there is a factorization $\tilde{V} \colon X \to BSU$ of $\tilde{V}$ through $BSU$. Then, there is a space $f \colon Y \to X$, where $f_E \colon Y_E \to X_E$ is finite and flat, as well as a collection of line bundles $\mathcal{H}_j$, $\mathcal{H}'_j$ expressing a $BSU$-internal decomposition*
$$\tilde{V} = -\bigoplus_{j=1}^{n} (1 - \mathcal{H}_j)(1 - \mathcal{H}'_j).$$

*Proof* Begin by using Corollary 2.3.10 on $V$ to get an equality of $BU$-bundles
$$f^* \tilde{V} = V' + \mathcal{L}_1 + \mathcal{L}_2 - n \cdot 1.$$
Adding $(1 - \mathcal{L}_1)(1 - \mathcal{L}_2)$ to both sides, this gives
$$f^* \tilde{V} + (1 - \mathcal{L}_1)(1 - \mathcal{L}_2) = V' + \mathcal{L}_1 + \mathcal{L}_2 + (1 - \mathcal{L}_1)(1 - \mathcal{L}_2) - n \cdot 1$$
$$= V' + \mathcal{L}_1 \mathcal{L}_2 - (n - 1) \cdot 1.$$
By thinking of $(1 - \mathcal{L}_j)$ as an element of $kU^2(Y) = [Y, BU]$, we see that the product element
$$(1 - \mathcal{L}_1)(1 - \mathcal{L}_2) \in kU^4(Y) = [Y, BSU]$$

---
[1] Similarly, there is an equivalence $\underline{kO}_{8k} = BO[8k, \infty)$, and this *does not hold* for indices which are not precise multiples of 8.

## 5.2 Special Divisors and the Special Splitting Principle 231

has the natural structure of a $BSU$-bundle and hence so does the sum on the left-hand side.[2] The right-hand side is the rank 0 virtualization of a rank $(n-1)$ vector bundle, hence succumbs to induction. Finally, because $SU(1)$ is the trivial group, there are no nontrivial complex line bundles with structure group $SU(1)$, grounding the induction. □

From this, we would like to directly conclude an equivalence between trivializations of the Thom sheaf $\mathbb{L}((\mathcal{L}_1 - 1)(\mathcal{L}_2 - 1)) \downarrow (\mathbb{CP}^\infty)_E^{\times 2}$ and multiplicative maps $MSU \to E$, but we are not quite yet ready to do so. Certainly an $MSU$-orientation of $E$ gives such a trivialization, but it is not clear that all possible trivializations of that universal Thom sheaf give consistent trivializations of other Thom sheaves – that is, the decomposition in Lemma 5.2.6 may admit unexpected symmetries which, in turn, place requirements on our universal trivialization so that these symmetric decompositions all result in the same restricted trivialization.[3]

*Example 5.2.7*   There is an equivalence of $SU$-bundles

$$(\mathcal{L}_1 - 1)(\mathcal{L}_2 - 1) \cong (\mathcal{L}_2 - 1)(\mathcal{L}_1 - 1).$$

Correspondingly, the trivializations of $\mathbb{L}((\mathcal{L}_1 - 1)(\mathcal{L}_2 - 1))$ which respect this twist are the *symmetric* sections.

*Example 5.2.8*   There is an equivalence of $SU$-bundles

$$(1 - 1)(\mathcal{L}_2 - 1) \cong 0.$$

Correspondingly, the trivializations of $\mathbb{L}((1 - \mathcal{L}_1)(1 - \mathcal{L}_2))$ which respect this degeneracy are the *rigid* sections, meaning they trivialize the Thom sheaf of the trivial bundle using the trivial section 1.

*Example 5.2.9*   There is another less obvious symmetry, inspired by our use of the product map

$$kU^2(Y) \otimes kU^2(Y) \to kU^4(Y)$$

in the course of the proof. There is also a product map

$$kU^2(Y) \otimes kU^0(Y) \otimes kU^2(Y) \to kU^4(Y).$$

Taking one of our splitting summands $(1 - \mathcal{L}_1)(1 - \mathcal{L}_2)$ and acting by some

---

[2] In the language of the previous case study, we are making use of a certain Hopf ring ∘-product on $\underline{kU}_{2*}$.
[3] By contrast, our splitting principle for ordinary complex vector bundles was completely deterministic, since a given isomorphism class of line bundles tautologically admits no other expression as an isomorphism class of line bundles.

bundle $\mathcal{H} \in kU^0(Y)$ gives

$$(1 - \mathcal{L}_1)\mathcal{H}(1 - \mathcal{L}_2) = (1 - \mathcal{L}_1)\mathcal{H}(1 - \mathcal{L}_2)$$
$$(\mathcal{H} - \mathcal{L}_1\mathcal{H})(1 - \mathcal{L}_2) = (1 - \mathcal{L}_1)(\mathcal{H} - \mathcal{H}\mathcal{L}_2)$$
$$(1 - \mathcal{L}_1\mathcal{H})(1 - \mathcal{L}_2)$$
$$-(1 - \mathcal{H})(1 - \mathcal{L}_2) = (1 - \mathcal{L}_1)(1 - \mathcal{H}\mathcal{L}_2)$$
$$- (1 - \mathcal{L}_1)(1 - \mathcal{H}).$$

This "$kU^0$-linearity" is sometimes called a "2-cocycle condition," in reference to the similarity with the formula in Definition 3.2.4.

We would like to show that these observations suffice, as in the following version of Theorems 5.2.1 and 5.2.2:

**Theorem 5.2.10** ([AHS01, theorem 2.50]) *The functor*

$$\{\operatorname{Spec} T \xrightarrow{u} \operatorname{Spec} E_0\} \to \left\{ \begin{array}{c} \text{trivializations of } u^*\mathbb{L}((1 - \mathcal{L}_1)(1 - \mathcal{L}_2)) \\ \text{over } u^*(\mathbb{CP}^\infty)_E^{\times 2} \text{ which are} \\ \text{symmetric, rigid, and } kU^0\text{-linear} \end{array} \right\}$$

*is isomorphic to the affine scheme* $\operatorname{Spec} E_0 MSU$. *Moreover, the $E_0$-points of this scheme biject with ring spectrum maps* $MSU \to E$.

In pursuit of this, we will show rather manually that $BSU_E$ represents an object subject to exactly such symmetries, hence $\operatorname{Spec} E_0 BSU$ represents the scheme of such symmetric functions, and finally conclude that $\operatorname{Spec} E_0 MSU$ represents the scheme of such symmetric trivializations. The place to begin is with a Serre spectral sequence:

**Lemma 5.2.11** ([AS01, lemma 6.1], see also [AS01, proposition 6.5]) *The Postnikov fibration*

$$BSU \to BU \xrightarrow{B\det} BU(1)$$

*induces a short exact sequence of Hopf algebras*

$$E^0 BSU \leftarrow E^0 BU \xleftarrow{c_1 \leftarrow c_1} E^0 BU(1).$$ □

An equivalent statement is that there is a short exact sequence of formal group schemes

$$\begin{array}{ccccc} BSU_E & \longrightarrow & BU_E & \xrightarrow{B\det} & BU(1)_E \\ \| & & \| & & \| \\ \operatorname{SDiv}_0 \mathbb{CP}^\infty_E & \longrightarrow & \operatorname{Div}_0 \mathbb{CP}^\infty_E & \xrightarrow{\text{sum}} & \mathbb{CP}^\infty_E, \end{array}$$

## 5.2 Special Divisors and the Special Splitting Principle

where the scheme "$\mathrm{SDiv}_0 \, \mathbb{CP}^\infty_E$" of *special divisors* is defined to parameterize those divisors which vanish under the summation map. However, whereas the map $BU(1)_E \to BU_E$ has an identifiable universal property – it presents $BU_E$ as the universal formal group on the pointed curve $BU(1)_E$ – the description of $BSU_E$ as a scheme of special divisors does not bear much immediate resemblance to a free object on the special divisor $(1-[a])(1-[b])$ classified by

$$(\mathbb{CP}^\infty)^{\times 2}_E \xrightarrow{(1-\mathcal{L}_1)(1-\mathcal{L}_2)_E} BSU_E \to BU_E = \mathrm{Div}_0 \, \mathbb{CP}^\infty_E.$$

Our task is thus exactly to justify this statement.

**Definition 5.2.12**  If it exists, let $C_2\widehat{\mathbb{G}}$ denote the symmetric square of $\mathrm{Div}_0 \, \widehat{\mathbb{G}}$, thought of as a module over the ring scheme $\mathrm{Div} \, \widehat{\mathbb{G}}$. This scheme has the property that a formal group homomorphism $\varphi \colon C_2\widehat{\mathbb{G}} \to H$ is equivalent data to a symmetric function $\psi \colon \widehat{\mathbb{G}} \times \widehat{\mathbb{G}} \to H$ satisfying a rigidity condition ($\psi(0,0) = 0$) and a 2-cocycle condition as in Example 5.2.9.[4]

**Theorem 5.2.13** (Ando, Hopkins, and Strickland, unpublished)  $\mathrm{SDiv}_0 \, \widehat{\mathbb{G}}$ *is a model for* $C_2\widehat{\mathbb{G}}$.

*Proof sketch*  Consider the map

$$\widehat{\mathbb{G}} \times \widehat{\mathbb{G}} \to \mathrm{Div}_0 \, \widehat{\mathbb{G}},$$
$$(a,b) \mapsto (1-[a])(1-[b])$$

for which there is a factorization of formal schemes

$$\widehat{\mathbb{G}} \times \widehat{\mathbb{G}}$$
$$\downarrow \qquad \searrow$$
$$\mathrm{SDiv}_0 \, \widehat{\mathbb{G}} \xrightarrow{\ker} \mathrm{Div}_0 \, \widehat{\mathbb{G}} \xrightarrow{\sigma} \widehat{\mathbb{G}}$$

because

$$\sigma((1-[a])(1-[b])) = (a+b) - a - b + 0 = 0.$$

A homomorphism $\varphi \colon \mathrm{SDiv}_0 \, \widehat{\mathbb{G}} \to H$ pulls back to a function $\psi \colon \widehat{\mathbb{G}} \times \widehat{\mathbb{G}} \to H$ satisfying the properties of Definition 5.2.12 as follows:

- To check rigidity, we have

$$\psi(a,0) = \varphi((1-[a])(1-[0])) = \varphi((1-[a])(1-1)) = \varphi(0) = 0.$$

---

[4] The short exact sequence $0 \to I \to \mathbb{Z}[G] \to \mathbb{Z} \to 0$ gives a purely algebraic analog of the short exact sequence $0 \to \mathrm{Div}_0 \, \widehat{\mathbb{G}} \to \mathrm{Div} \, \widehat{\mathbb{G}} \to \mathbb{Z} \to 0$. The algebro-geometric results of this lecture (in particular, Theorem 5.2.13) have direct analogs in the setting of group-rings, which the reader may find more instructive and concrete.

- To check symmetry, we have

$$\psi(a,b) = \varphi((1-[a])(1-[b])) = \varphi((1-[b])(1-[a])) = \psi(b,a).$$

- To check $kU^0$-linearity, we have

$$\begin{aligned}\psi(ac,b) - \psi(c,b) &= \varphi((1-[a][c])(1-[b])) - \varphi((1-[c])(1-[b])) \\ &= \varphi((1-[a][c])(1-[b]) - (1-[c])(1-[b])) \\ &= \varphi((1-[a])(1-[c][b]) - (1-[a])(1-[c])) \\ &= \varphi((1-[a])(1-[c][b])) - \varphi((1-[a])(1-[c])) \\ &= \psi(a,cb) - \psi(a,c).\end{aligned}$$

The other direction is more obnoxious, so we give only a sketch. Begin by selecting a function $\psi \colon \widehat{\mathbb{G}} \times \widehat{\mathbb{G}} \to H$, then mimic the construction in Lemma 5.2.6. Expanding the definition of $\mathrm{Div}_0\,\widehat{\mathbb{G}}$, we are moved to consider the object $\widehat{\mathbb{G}}^{\times k}$, where we define a map

$$\widehat{\mathbb{G}}^{\times k} \to H,$$

$$(a_1,\ldots,a_k) \mapsto -\sum_{j=2}^{k} \psi\left(\sum_{i=1}^{j-1} a_i, a_j\right).$$

This gives a compatible system of symmetric maps, and hence altogether this gives a map $\tilde{\varphi}\colon \mathrm{Div}_0\,\widehat{\mathbb{G}} \to H$ off of the colimit. In general, this map is not a homomorphism, but it is a homomorphism when restricted to

$$\varphi\colon \mathrm{SDiv}_0\,\widehat{\mathbb{G}} \to \mathrm{Div}_0\,\widehat{\mathbb{G}} \xrightarrow{\tilde{\varphi}} H.$$

Finally, one checks that any homomorphism $\mathrm{SDiv}_0\,\widehat{\mathbb{G}} \to H$ of formal groups restricting to the zero map $\widehat{\mathbb{G}} \times \widehat{\mathbb{G}} \to H$ was already the zero homomorphism by rewriting a point in $\mathrm{SDiv}_0\,\widehat{\mathbb{G}}$ as a sum as in Lemma 5.2.6. This gives the desired identification of $\mathrm{SDiv}_0\,\widehat{\mathbb{G}}$ with the universal property of $C_2\widehat{\mathbb{G}}$. □

**Corollary 5.2.14** *There is an isomorphism*

$$\mathrm{Spec}\,E_0BSU = \left\{\begin{array}{c} \text{functions } f\colon u^*(\mathbb{CP}_E^\infty)^{\times 2} \to \mathbb{G}_m \\ \text{which are symmetric, rigid, and } kU^0\text{-linear} \end{array}\right\}.$$

*Proof* Follow the sequence of isomorphisms

$$\begin{aligned}\mathrm{Spec}\,E_0BSU &= D(BSU_E) & \text{(Remark 5.1.16)}\\ &= D(\mathrm{SDiv}_0\,\mathbb{CP}_E^\infty) & \text{(Lemma 5.2.11)}\\ &= D(C_2\mathbb{CP}_E^\infty) & \text{(Theorem 5.2.13)}\\ &= \underline{\mathrm{FormalGroups}}(C_2\mathbb{CP}_E^\infty, \mathbb{G}_m), & \text{(Definition 5.1.13)}\end{aligned}$$

## 5.2 Special Divisors and the Special Splitting Principle 235

and then use the universal property in Definition 5.2.12. □

In order to lift this analysis to Spec $E_0 MSU$, we again appeal to the torsor structure. At this point, it will finally be useful to introduce some notation:

**Definition 5.2.15** ([AHS01, definition 2.39]) For a sheaf $\mathcal{L}$ over a formal group $\widehat{\mathbb{G}}$, we introduce the schemes

$$C^0(\widehat{\mathbb{G}}_E; \mathcal{L})(T) = \{\text{triv}^{ns} \text{ of } u^*\mathcal{L} \downarrow u^*\widehat{\mathbb{G}}\},$$

$$C^1(\widehat{\mathbb{G}}_E; \mathcal{L})(T) = \left\{\text{triv}^{ns} \text{ of } u^*\left(\frac{e^*\mathcal{L}}{\mathcal{L}}\right) \downarrow u^*\widehat{\mathbb{G}} \text{ which are rigid}\right\}$$

$$C^2(\widehat{\mathbb{G}}_E; \mathcal{L})(T) = \left\{\begin{array}{c} \text{triv}^{ns} \text{ of } u^*\left(\frac{e^*\mathcal{L} \otimes \mu^*\mathcal{L}}{\pi_1^*\mathcal{L} \otimes \pi_2^*\mathcal{L}}\right) \downarrow u^*\widehat{\mathbb{G}}^{\times 2} \\ \text{which are rigid, symmetric, and } kU^0\text{-linear} \end{array}\right\}.$$

Thus far, we have established the following families of isomorphisms:

Cohomological formal schemes: $(\mathbb{Z} \times BU)_E \cong C_0 \mathbb{CP}_E^\infty,$

$$BU_E \cong C_1 \mathbb{CP}_E^\infty,$$

$$BSU_E \cong C_2 \mathbb{CP}_E^\infty,$$

Homological schemes: $\text{Spec } E_0(\mathbb{Z} \times BU) \cong C^0(\mathbb{CP}_E^\infty; \mathbb{G}_m),$

$$\text{Spec } E_0(BU) \cong C^1(\mathbb{CP}_E^\infty; \mathbb{G}_m),$$

$$\text{Spec } E_0(BSU) \cong C^2(\mathbb{CP}_E^\infty; \mathbb{G}_m),$$

Orientation schemes: $\text{Spec } E_0(MUP) \cong C^0(\mathbb{CP}_E^\infty; \mathcal{I}(0)),$

$$\text{Spec } E_0(MU) \cong C^1(\mathbb{CP}_E^\infty; \mathcal{I}(0)),$$

where we have abusively abbreviated the sheaf of functions on $\mathbb{CP}_E^\infty$ to $\mathbb{G}_m$. In order to fill in the missing piece, we exploit the torsor structure on Thom spectra discussed earlier.

**Lemma 5.2.16** ([AHS01, theorem 2.50]) *There is a commuting square*

$$\begin{array}{ccc}
\text{Spec } E_0 MSU & & \\
\times & \longrightarrow & \text{Spec } E_0 MSU \\
\text{Spec } E_0 BSU & & \\
\downarrow & & \downarrow \\
C^2(\mathbb{CP}_E^\infty; \mathcal{I}(0)) & & \\
\times & \longrightarrow & C^2(\mathbb{CP}_E^\infty; \mathcal{I}(0)), \\
C^2(\mathbb{CP}_E^\infty; \mathbb{G}_m) & &
\end{array}$$

*where the horizontal maps are the action maps defining torsors, and the vertical maps are those described previously.*

*Proof sketch*  Recall the isomorphism $T(\mathcal{L} \downarrow \mathbb{CP}^\infty) \simeq \Sigma^\infty \mathbb{CP}^\infty$. The main point of this claim is that the Thom diagonal for $MU[2k, \infty)$ restricts to the Thom diagonal for $\mathbb{CP}^\infty$, which agrees with its diagonal as a space:

$$\begin{array}{ccc} (\Sigma^\infty \mathbb{CP}^\infty)^{\wedge k} & \xrightarrow{\Delta} & (\Sigma^\infty \mathbb{CP}^\infty)^{\wedge k} \wedge \Sigma^\infty_+ (\mathbb{CP}^\infty)^{\times k} \\ \downarrow & & \downarrow \\ MU[2k, \infty) & \xrightarrow{\Delta} & MU[2k, \infty) \times BU[2k, \infty). \end{array}$$

The diagonal at the level of $(\mathbb{CP}^\infty)^{\times k}$ is responsible for the cup product, so that the classes in the cohomology of projective space which induce the maps

$$\operatorname{Spec} E_0 MU[2k, \infty) \to C^k(\mathbb{CP}^\infty_E; \mathcal{I}(0)),$$
$$\operatorname{Spec} E_0 BU[2k, \infty) \to C^k(\mathbb{CP}^\infty_E; \mathbb{G}_m)$$

literally multiply together to give the description of the action. This multiplication of sections is exactly the action claimed in the model. □

*Proof of Theorem 5.2.10*  The claim of this theorem is that the map

$$\operatorname{Spec} E_0 MSU \to C^2(\mathbb{CP}^\infty_E; \mathcal{I}(0))$$

studied above is an isomorphism. By Lemma 5.2.16 this is a map of torsors over a fixed base, hence is automatically an isomorphism. □

*Remark 5.2.17* ([AS01, lemma 6.4])  We can also analyze the map

$$\operatorname{Spec} E_0 BSU \to \operatorname{Spec} E_0 BU$$

in terms of these models of functions to $\mathbb{G}_m$. Again, the analysis passes through a computation in connective $K$-theory, using the identification

$$kU^*(\mathbb{CP}^\infty)^{\times 2} = \mathbb{Z}[\beta][\![x_1, x_2]\!],$$

where $x_1 = \pi_1^* x$ and $x_2 = \pi_2^* x$ are the Chern classes associated to the tautological bundle pulled back along projections to the first and second factors

$$\pi_1 \colon (\mathbb{CP}^\infty)^{\times 2} \to \mathbb{CP}^\infty \times *, \qquad \pi_2 \colon (\mathbb{CP}^\infty)^{\times 2} \to * \times \mathbb{CP}^\infty.$$

Inside of this ring, we have the equations

$$\begin{aligned} \beta^2 x_1 x_2 &= (1 - \mathcal{L}_1)(1 - \mathcal{L}_2) \\ &= (1 - \mathcal{L}_1) - (1 - \mathcal{L}_1 \mathcal{L}_2) + (1 - \mathcal{L}_2) \\ &= \beta \left( \pi_1^*(x) - \mu^*(x) + \pi_2^*(x) \right), \end{aligned}$$

where $\mu \colon \mathbb{CP}^\infty \times \mathbb{CP}^\infty \to \mathbb{CP}^\infty$ is the tensor product map. Since the orientation schemes are governed as torsors over these base schemes, we automatically get a description

## 5.2 Special Divisors and the Special Splitting Principle

$$\operatorname{Spec} E_0 MU \longrightarrow \operatorname{Spec} E_0 MSU,$$

$$f(x) \longmapsto \frac{f(x_1) \cdot f(x_2)}{f(x_1 +_{\mathbb{C}P_E^\infty} x_2)}$$

as a section of

$$\pi_1^* \left( \frac{e^* \mathcal{I}(0)}{\mathcal{I}(0)} \right) \otimes \pi_2^* \left( \frac{e^* \mathcal{I}(0)}{\mathcal{I}(0)} \right) \otimes \left( \mu^* \left( \frac{e^* \mathcal{I}(0)}{\mathcal{I}(0)} \right) \right)^{-1} = \frac{e^* \mathcal{I}(0) \otimes \mu^* \mathcal{I}(0)}{\pi_1^* \mathcal{I}(0) \otimes \pi_2^* \mathcal{I}(0)}.$$

*Remark 5.2.18* ([AHS01, remark 2.32]) The published proofs of Ando, Hopkins, and Strickland differ substantially from the account given here. The primary difference is that "$C_2\widehat{\mathbb{G}}$" is not even mentioned, essentially because it is a fair amount of technical work to show that such a scheme even exists (especially in the case to come of $BU[6, \infty)$). On the other hand, it is very easy to demonstrate the existence of its Cartier dual: This is a scheme parameterizing certain bivariate power series subject to certain algebraic conditions, hence exists for the same reason that $\mathcal{M}_{\mathsf{fgl}}$ existed (see Definition 3.2.1). The compromise for this is that they then have to analyze the scheme $\operatorname{Spec} E_0 BSU$ directly, through considerably more computational avenues. This is not too high of a price: The analysis of the $BU[6, \infty)$ case turns out to be primarily computational anyhow, so this manner of approach is inevitable.

*Remark 5.2.19* Our definition of the scheme $C_2\widehat{\mathbb{G}}$ was by the formula

$$C_2\widehat{\mathbb{G}} = \operatorname{Sym}^2_{\operatorname{Div}\widehat{\mathbb{G}}} \operatorname{Div}_0 \widehat{\mathbb{G}},$$

where we are thinking of $\operatorname{Div}_0 \widehat{\mathbb{G}} \subseteq \operatorname{Div} \widehat{\mathbb{G}}$ as the augmentation ideal inside of an augmented ring. The formal schemes $\operatorname{Div} \widehat{\mathbb{G}}$ and $\operatorname{Div}_0 \widehat{\mathbb{G}}$ are the formal schemes associated by $E$-theory to the infinite loopspaces underlying $kU$ and $\Sigma^2 kU$ respectively. Remarking that $BSU$ is the infinite loopspace underlying $\Sigma^4 kU$, we arrive at the analogous topological formula

$$\Sigma^4 kU = (\Sigma^2 kU) \wedge_{kU} (\Sigma^2 kU).$$

*Remark 5.2.20* One consequence of our analysis is that the theory of $MSU$-orientations for a complex-orientable theory is not especially different from the theory of $MU$-orientations. Explicitly, we have found a short exact sequence

$$0 \to BSU_E \to BU_E \to BU(1)_E \to 0,$$

the first two terms of which are bipolynomial group schemes, and which altogether has a canonical splitting at the level of products of formal schemes. This splitting is *not* a map of group schemes, and so it does *not* survive to give a splitting of the Cartier dual sequence

$$0 \leftarrow \operatorname{Spec} E_0 BSU \leftarrow \operatorname{Spec} E_0 BU \leftarrow \operatorname{Spec} E_0 BU(1) \leftarrow 0,$$

but this sequence is nonetheless *noncanonically* split. It follows from our analysis above that any $MSU$-orientation of a complex-orientable theory arises as the restriction of an $MU$-orientation (where the space of available $MU$-orientation preimages is a torsor for Spec $E_0 BU(1)$).

## 5.3 Chromatic Analysis of $BU[6, \infty)$

We now embark on an analysis of $MU[6, \infty)$-orientations in earnest. As in the case of $MSU$, it is fruitful to first study the behavior of vector bundles with structure map lifted through $\underline{kU}_6 = BU[6, \infty)$ and to analyze the schemes $BU[6, \infty)_E$ and Spec $E_0 BU[6, \infty)$. In the previous case, we studied a particular bundle

$$\Pi_2 \colon \mathbb{CP}^\infty \times \mathbb{CP}^\infty \xrightarrow{(1-\mathcal{L}_1)(1-\mathcal{L}_2)} BSU,$$

which controlled much of the geometry through our splitting principle for $BSU$-bundles, recorded as Lemma 5.2.6. Analogously, we can construct a naturally occurring such bundle as the product

$$\Pi_3 \colon \mathbb{CP}^\infty \times \mathbb{CP}^\infty \times \mathbb{CP}^\infty \xrightarrow{(1-\mathcal{L}_1)(1-\mathcal{L}_2)(1-\mathcal{L}_3)} BU[6, \infty),$$

but the proof of Lemma 5.2.6 falls apart almost immediately – there does not appear to be a splitting principle for bundles lifted through $BU[6, \infty)$. This is quite worrying, and it dampens our optimism across the board: about the behavior of $\Pi_3$ exerting enough control over $BU[6, \infty)$, about the existence of "$C_3 \widehat{\mathbb{G}}$," and about $C_3 \mathbb{CP}^\infty_E$ serving as a good model for $BU[6, \infty)_E$.

*Nevertheless*, we will show that this algebraic model is still accurate by complete topological calculation. Our approach is divided between two fronts.

1. If we specialize to a particularly nice cohomology theory – such as $E = E_\Gamma$ a Morava $E$-theory – then we can use our extensive body of knowledge about finite height formal groups and their relationship to algebraic topology in order to force nice behavior into the story. This should be thought of as an exploratory step: If there is a general statement to be found, it will be visible in this particularly algebro-geometric setting, where we can maybe algebraically compute fully enough to determine what it is.
2. If we specialize to a particularly simple formal group – such as $\widehat{\mathbb{G}}_a$ and its associated cohomology theory $H\mathbb{F}_p$ – then we can use our talent for performing computations in algebraic *topology* to completely exhaust the problem. This should be thought of as the "actual" proof: As in Lecture 4.3, we will show that successfully transferring the guess result from Morava

## 5.3 Chromatic Analysis of $BU[6, \infty)$

$E$-theory to the setting of ordinary cohomology entails the result for *any* complex-orientable cohomology theory.

In this lecture, we will pursue the first avenue. We begin by setting $\Gamma$ to be a formal group of finite $p$-height of a field $k$ of positive characteristic $p$, and we let $E = E_\Gamma$ denote the associated continuous Morava $E$-theory. Our main technical tool will be the Postnikov fibration

$$\underline{H\mathbb{Z}_3} \to BU[6, \infty) \to BSU,$$

and our main goals are to construct a model sequence of formal schemes, then show that $E$-theory is well-behaved enough that the formal schemes it constructs exactly match the model.

In the previous setting of $MSU$, we gained indirect access to the algebraic model $C_2\widehat{\mathbb{G}}$ by separately proving that it was modeled by $\mathrm{SDiv}_0\,\widehat{\mathbb{G}}$ and that this had a good comparison map to $BSU_E$. This time, since we do not have access to $C_3\widehat{\mathbb{G}}$ or anything like it, we proceed by much more indirect means, along the lines of Remark 5.2.18: We know that $C^3(\widehat{\mathbb{G}}; \mathbb{G}_m)$ exists as an affine scheme, since we can explicitly construct it as a closed subscheme of the scheme of trivariate power series, and so we seek a map

$$\mathrm{Spec}\, E_0 BU[6, \infty) \to C^3(\mathbb{CP}_E^\infty; \mathbb{G}_m)$$

that does not pass through any intermediate cohomological construction. Our main tool for accomplishing this is as follows:

**Definition 5.3.1** A map $f: X \to Y$ of spaces induces a map $f_E: X_E \to Y_E$ of formal schemes. In the case that $Y$ is a commutative $H$-space and $Y_E$ is connected, we can construct a map according to the composite

$$\begin{array}{c}
X_E \times \underline{\mathrm{GroupSchemes}}(Y_E, \mathbb{G}_m) \dashrightarrow \widehat{\mathbb{A}}^1 \\
\| \qquad\qquad\qquad\qquad\qquad\qquad\qquad \simeq \uparrow \\
X_E \times \underline{\mathrm{FormalGroups}}(Y_E, \widehat{\mathbb{G}}_m) \xrightarrow{f_E \times 1} Y_E \times \underline{\mathrm{FormalGroups}}(Y_E, \widehat{\mathbb{G}}_m) \xrightarrow{\mathrm{ev}} \widehat{\mathbb{G}}_m.
\end{array}$$

This is called *the adjoint map*, and we write $\widehat{f}$ for any of the above versions of this map, whether valued in $\widehat{\mathbb{G}}_m$, $\mathbb{G}_m$, or $\widehat{\mathbb{A}}^1$. It encodes equivalent information to the $E_0$-linear map

$$E_0 \to E_0 Y \widehat{\otimes}_{E_0} E^0 X$$

dual to the map $E_0 X \to E_0 Y$ induced on $E$-homology.

*Remark 5.3.2* This construction converts many properties of $f$ into corresponding properties of this adjoint element. For instance:

- It is natural in the source: For $f\colon X \to Y$ and $g\colon W \to X$, we have
$$\widehat{fg} = \widehat{f} \circ (g_E \times 1)\colon W_E \times D(Y_E) \to \mathbb{G}_m.$$
- It is natural in the target: For $f\colon X \to Y$ and $h\colon Y \to Z$ a map of $H$-spaces, we have
$$\widehat{hf} = \widehat{f} \circ (1 \times D(h_E))\colon X_E \times D(Z_E) \to \mathbb{G}_m.$$
- It converts sums of classes to products of maps to $\mathbb{G}_m$.

*Example 5.3.3* Recall the vector bundle $\Pi_2$ lifted through $BSU$, defined at the beginning of this lecture and of great interest to us in Lecture 5.2. The adjoint to the classifying map of $\Pi_2$ is a map of formal schemes
$$\widehat{\Pi}_2\colon (\mathbb{CP}_E^\infty)^{\times 2} \times \operatorname{Spec} E_0 BSU \to \mathbb{G}_m,$$
which passes through the exponential adjunction to become a map
$$\operatorname{Spec} E_0 BSU \to \underline{\operatorname{FormalSchemes}}((\mathbb{CP}_E^\infty)^{\times 2}, \mathbb{G}_m).$$
Because the adjoint construction preserves properties of the class $\Pi_2$, we learn that this map factors through the closed subscheme
$$\operatorname{Spec} E_0 BSU \xdashrightarrow{\widehat{\Pi}_2} C^2(\mathbb{CP}_E^\infty; \mathbb{G}_m) \longrightarrow \underline{\operatorname{FormalSchemes}}((\mathbb{CP}_E^\infty)^{\times 2}, \mathbb{G}_m)$$
of symmetric, rigid functions satisfying $kU^0$-linearity. By careful manipulation of divisors in Theorem 5.2.13, we showed an isomorphism
$$BSU_E \cong \operatorname{SDiv}_0 \mathbb{CP}_E^\infty,$$
which on applying Cartier duality showed the factorized map
$$\operatorname{Spec} E_0 BSU \to C^2(\mathbb{CP}_E^\infty; \mathbb{G}_m)$$
to be an isomorphism.

*Example 5.3.4* Similarly, we have defined a cohomology class
$$\Pi_3 = (\mathcal{L}_1 - 1)(\mathcal{L}_2 - 1)(\mathcal{L}_3 - 1) \in kU^6(\mathbb{CP}^\infty)^{\times 3} = [(\mathbb{CP}^\infty)^{\times 3}, BU[6, \infty)].$$
As above, its adjoint induces a map (which we abusively also denote by $\widehat{\Pi}_3$)
$$\widehat{\Pi}_3\colon \operatorname{Spec} E_0 BU[6, \infty) \to C^3(\mathbb{CP}_E^\infty; \mathbb{G}_m),$$
where $C^3(\mathbb{CP}_E^\infty; \mathbb{G}_m)$ is the scheme of $\mathbb{G}_m$-valued trivariate functions on $\mathbb{CP}_E^\infty$ satisfying symmetry, rigidity, and $kU^0$-linearity.[5]

---

[5] If $C_3 \mathbb{CP}_E^\infty := \operatorname{Sym}^3_{\operatorname{Div} \mathbb{CP}_E^\infty} \operatorname{Div}_0 \mathbb{CP}_E^\infty$ were to exist, this scheme $C^3(\mathbb{CP}_E^\infty; \mathbb{G}_m)$ would be its Cartier dual.

## 5.3 Chromatic Analysis of $BU[6, \infty)$

We also have the following analog of the compatibility results recorded as Remarks 5.2.4 and 5.2.17 of the previous section:

**Lemma 5.3.5** ([AS01, lemma 7.1], [AHS01, proposition 2.27, corollary 2.30]) *There is a commutative square*

$$\begin{array}{ccc} \operatorname{Spec} E_0 BSU & \longrightarrow & \operatorname{Spec} E_0 BU[6, \infty) \\ \downarrow \widehat{\Pi}_2 & & \downarrow \widehat{\Pi}_3 \\ C^2(\mathbb{CP}_E^\infty; \widehat{\mathbb{G}}_m) & \xrightarrow{\delta} & C^3(\mathbb{CP}_E^\infty; \widehat{\mathbb{G}}_m), \end{array}$$

*where the map*[6] $\delta$ *is specified by*

$$\delta(f)(x_1, x_2, x_3) := \frac{f(x_1, x_3) f(x_2, x_3)}{f(x_1 +_E x_2, x_3)}.$$

*Proof* As in the proofs of Remarks 5.2.4 and 5.2.17, this is checked by performing a calculation in $kU$-cohomology of projective space, where we have the relation

$$\begin{aligned} \Pi_3 &= (1 - \mathcal{L}_1)(1 - \mathcal{L}_2)(1 - \mathcal{L}_3) \\ &= ((1 - \mathcal{L}_1) - (1 - \mathcal{L}_1 \mathcal{L}_2) + (1 - \mathcal{L}_2))(1 - \mathcal{L}_3) \\ &= ((\pi_1 \times 1)^* - (\mu \times 1)^* + (\pi_2 \times 1)^*)((1 - \mathcal{L}_1)(1 - \mathcal{L}_3)) \\ &= ((\pi_1 \times 1)^* - (\mu \times 1)^* + (\pi_2 \times 1)^*) \Pi_2. \end{aligned}$$ □

Thus far, we have constructed the solid maps in the following commutative diagram:

$$\begin{array}{ccccc} \operatorname{Spec} E_0 BSU & \longrightarrow & \operatorname{Spec} E_0 BU[6, \infty) & \longrightarrow & \operatorname{Spec} E_0 \underline{H\mathbb{Z}}_3 \\ \cong \downarrow \widehat{\Pi}_2 & & \downarrow \widehat{\Pi}_3 & & \cong \downarrow \\ C^2(\mathbb{CP}_E^\infty; \widehat{\mathbb{G}}_m) & \xrightarrow{\delta} & C^3(\mathbb{CP}_E^\infty; \widehat{\mathbb{G}}_m) & \xdashrightarrow{e} & \underline{\mathrm{FormalGroups}}((\mathbb{CP}_E^\infty)^{\wedge 2}, \widehat{\mathbb{G}}_m), \end{array}$$

where $\widehat{\mathbb{G}}^{\wedge n}$ denotes the exterior $n$th power of $\widehat{\mathbb{G}}$, the left-most vertical map is an isomorphism by Corollary 5.2.14, and right-most vertical map is an isomorphism by combining Remark 4.6.14 with Remarks 4.6.8 and 5.1.16. We would like to promote this diagram to an isomorphism of short exact sequences, and to do so we need to finish constructing the sequences themselves – we need a horizontal map $e$ making the diagram commute.

The idea behind the construction of $e$ is to pretend that $\widehat{\Pi}_3$ is an isomorphism,

---

[6] Despite its name and its formula, this map $\delta$ does not really belong to a cochain complex from our perspective. *All* of the functions we are considering, no matter how many inputs they take, are always subject to a 2-cocycle condition.

so that we could completely detect $e$ by comparing the image of the identity on Spec $E_0BU[6, \infty)$ through $\widehat{\Pi}_3$ to the image of the identity through

$$\text{Spec } E_0BU[6,\infty) \to \text{Spec } E_0\underline{H\mathbb{Z}_3} \to \text{FormalGroups}((\mathbb{CP}_E^\infty)^{\wedge 2}, \mathbb{G}_m).$$

Using our calculation that $(\mathbb{CP}_E^\infty)^{\wedge 2}$ is a $p$-divisible group, we see that we can further restrict attention to the torsion subgroups

$$(\mathbb{CP}_E^\infty)^{\wedge 2}[p^j] = (BS^1[p^j]_E)^{\wedge 2}$$

which filter it, corresponding to analyzing the bundle classified by the restriction

$$BS^1[p^j]^{\wedge 2} \xrightarrow{\mu} \underline{HS^1[p^j]_2} \xrightarrow{\beta_j} \underline{H\mathbb{Z}_3} \xrightarrow{\gamma} \underline{kU_6}.$$

Using the abbreviation $B_j = BS^1[p^j]$, our summary goal is to find an express description of a map $d$ making the following square commute:

$$\begin{array}{ccc} B_j \wedge B_j & \longrightarrow & \mathbb{CP}^\infty \wedge \mathbb{CP}^\infty \\ {\scriptstyle \beta_j\mu(\alpha\wedge\alpha)} \downarrow & & \downarrow {\scriptstyle d} \\ \Sigma^3 H\mathbb{Z} & \xrightarrow{\gamma} & \Sigma^6 kU, \end{array}$$

where we have quietly replaced spaces by their suspension spectra, and where $\beta_j\mu(\alpha \wedge \alpha)$ denotes the composite

$$B_j^{\wedge 2} \xrightarrow{\alpha\wedge\alpha} (\Sigma H\mathbb{Z}/p^j)^{\wedge 2} \xrightarrow{\mu} \Sigma^2 H\mathbb{Z}/p^j \xrightarrow{\beta_j} \Sigma^3 H\mathbb{Z}.$$

Our strategy is to extend this putative square to a map of putative cofiber sequences

$$\begin{array}{ccccccc} (\mathbb{CP}^\infty)^{\wedge 2}/B_j^{\wedge 2} & \xrightarrow{\Delta} & \Sigma B_j \wedge B_j & \longrightarrow & \Sigma\mathbb{CP}^\infty \wedge \mathbb{CP}^\infty & \longrightarrow & \Sigma(\mathbb{CP}^\infty)^{\wedge 2}/(B_j)^{\wedge 2} \\ {\scriptstyle f}\downarrow & & {\scriptstyle \beta_j\mu(\alpha\wedge\alpha)}\downarrow & & \downarrow{\scriptstyle d} & & \downarrow{\scriptstyle f} \\ \Sigma^4 kU & \xrightarrow{\sigma} & \Sigma^4 H\mathbb{Z} & \xrightarrow{\gamma} & \Sigma^7 kU & \longrightarrow & \Sigma^5 kU, \end{array}$$

and thereby trade the task of constructing $d$ for the task of constructing $f$. This is a gain because $\sigma \colon kU \to H\mathbb{Z}$, the standard $kU$-orientation of $H\mathbb{Z}$, is a considerably simpler map to understand than $\gamma$.

**Lemma 5.3.6** ([AS01, section 5]) *Make the definitions*

- $x \colon \mathbb{CP}^\infty \to \Sigma^2 kU$ *is the $kU$-Euler class for* $(1 - \mathcal{L})$.
- $u \colon T(\mathcal{L}^{\otimes p^j}) \to kU^2$ *is the $kU$-Thom class for* $T(\mathcal{L}^{\otimes p^j}) = \mathbb{CP}^\infty/B_j$.
- $A_1$ *is the projection*

$$\frac{\mathbb{CP}^\infty \wedge \mathbb{CP}^\infty}{B_j \wedge B_j} \to \frac{\mathbb{CP}^\infty \wedge \mathbb{CP}^\infty}{B_j \wedge \mathbb{CP}^\infty} = (\mathbb{CP}^\infty/B_j) \wedge \mathbb{CP}^\infty = T(\mathcal{L}^{\otimes p^j}) \wedge \mathbb{CP}^\infty.$$

### 5.3 Chromatic Analysis of $BU[6, \infty)$

- Similarly, $A_2$ is the swapped projection $(\mathbb{C}P^\infty)^{\wedge 2}/B_j^{\wedge 2} \to \mathbb{C}P^\infty \wedge T(\mathcal{L}^{\otimes p^j})$.

Setting $f = \mu(u \wedge x)A_1 - \mu(x \wedge u)A_2$ gives the desired commuting square:

$$\sigma \circ f = \beta_j \mu(\alpha \wedge \alpha) \circ \Delta.$$

*Proof* The idea is to gain control of the cohomology group $H\mathbb{Z}^4((\mathbb{C}P^\infty)^{\wedge 2}, B_j^{\wedge 2})$ by Mayer–Vietoris, which is rendered complicated by our simultaneous use of the cofiber sequence $B_j \to \mathbb{C}P^\infty \to T(\mathcal{L}^{\otimes p^j})$ in *two* factors of a smash product. Toward this end, consider the maps

$$B_1\colon B_j \wedge T(\mathcal{L}^{\otimes p^j}) \to (\mathbb{C}P^\infty)^{\wedge 2}/B_j^{\wedge 2}, \quad B_2\colon T(\mathcal{L}^{\otimes p^j}) \wedge B_j \to (\mathbb{C}P^\infty)^{\wedge 2}/B_j^{\wedge 2},$$

which have cofibers $A_1$ and $A_2$ respectively. Direct calculation [AS01, lemma 5.6] shows that $(\ker B_1^*) \cap (\ker B_2^*)$ is torsion-free, so if we can identify $B_1^*(\beta_j \mu \circ \Delta)$ and $B_2^*(\beta_j \mu \circ \Delta)$, we will be most of the way there. We pick $B_1$ to consider, as $B_2$ is similar, and we start computing, beginning with

$$B_1^*(\beta_j \mu(\alpha \wedge \alpha) \circ \Delta) = \beta_j \mu(\alpha \wedge \alpha) \circ \Delta B_1.$$

Writing $\delta\colon T(\mathcal{L}^{\otimes p^j}) \to \Sigma B_j$ for the going-around map, we have

$$\begin{array}{ccc} \Sigma B_j^{\wedge 2} & \xleftarrow{1 \wedge \delta} & B_j \wedge T(\mathcal{L}^{\otimes p^j}) \\ & \Delta \nwarrow & \downarrow B_1 \\ & & (\mathbb{C}P^\infty)^{\wedge 2}/B_j^{\wedge 2}, \end{array}$$

and hence

$$\beta_j \mu(\alpha \wedge \alpha) \circ \Delta B_1 = \beta_j \mu(\alpha \wedge \alpha) \circ (1_B \wedge \delta)$$
$$= \beta_j \mu(\alpha \wedge \alpha \delta).$$

The maps $\alpha$ and $\delta$ appear in the following map of cofiber sequences:

$$\begin{array}{ccccccc} B & \xrightarrow{j} & P & \xrightarrow{q} & T & \xrightarrow{\delta} & \Sigma B \\ \downarrow \alpha & & \downarrow y & & \downarrow w & & \downarrow \alpha \\ \Sigma H\mathbb{Z}/p^j & \xrightarrow{\beta_j} & \Sigma^2 H\mathbb{Z} & \xrightarrow{p^j} & \Sigma^2 H\mathbb{Z} & \xrightarrow{\rho} & \Sigma^2 H\mathbb{Z}/p^j, \end{array}$$

where $y$ is the standard Euler class in $H^2 \mathbb{C}P^\infty$ and the first block commutes because the bottom row is the stabilization of the top row; $w$ is the Thom class associated to $T(\mathcal{L}^{\otimes p^j})$ and the middle block commutes because it witnesses the $H\mathbb{Z}$-analog of Theorem 4.6.1; and the last block commutes because $[B, \Sigma^2 H\mathbb{Z}] = 0$ and because the other two do. In particular, an application of the right-most block gives

$$\beta_j \mu(\alpha \wedge \alpha \delta) = \beta_j \mu(\alpha \wedge \rho w).$$

Using the fact that $\beta_j$ is the cofiber of the ring map $\rho$, there is a juggle

$$\beta_j \mu(\alpha \wedge \rho w) = \mu(\beta_j \alpha \wedge w),$$

and then we use the first block in the above map of cofiber sequences to conclude

$$\mu(\beta_j \alpha \wedge w) = \mu(yj \wedge w).$$

Finally, we can use this to guess a formula for our desired map $f$: We set

$$f = \mu(u \wedge x)A_1 + \mu(x \wedge u)A_2,$$

because, for instance,

$$\begin{aligned}B_1^*(\sigma f) &= \sigma(\mu(u \wedge x)A_1 + \mu(x \wedge u)A_2)B_1 \\ &= \sigma\mu(x \wedge u)(j \wedge 1),\end{aligned}$$

where we used $A_1 B_1 = 0$ and $A_2 B_1 = (j \wedge 1)$, a calculation similar to the calculation involving $\delta$ earlier in the proof. Then, because $\sigma \colon kU \to H\mathbb{Z}$ sends Euler classes to Euler classes, we have

$$\begin{aligned}\sigma\mu(x \wedge u)(j \wedge 1) &= \mu(y \wedge w)(j \wedge 1) \\ &= \mu(yj \wedge w).\end{aligned}$$

Hence, we have crafted a class $f$ with $\sigma f - \beta_j \mu(\alpha \wedge \alpha) \in (\ker B_1^*) \cap (\ker B_2^*)$.

What remains is to show that this class is torsion, hence identically zero. Half of this is obvious: $p^j \beta_j \mu(\alpha \wedge \alpha) = 0$, since $p^j \beta_j = 0$ on its own. For $p^j \sigma f$, we make an explicit calculation:

$$\begin{aligned}p^j \sigma f &= p^j(\mu(w \wedge y)A_1 - \mu(y \wedge w)A_2) \\ &= \mu(w \wedge p^j y)A_1 - \mu(p^j y \wedge w)A_2 \\ &= \mu(w \wedge q^* w)A_1 - \mu(q^* w \wedge w)A_2 \\ &= \mu(w \wedge w) \circ ((1 \wedge q)A_1 - (q \wedge 1)A_2) = 0. \quad \square\end{aligned}$$

The upshot of all of this is that we have our desired calculation of the map $e$:

**Corollary 5.3.7** ([AS01, lemma 5.4 and subsection "Modelling $d_n(L_1, L_2)$"]) *There is a commuting triangle*

$$\begin{array}{ccc}(\Sigma^\infty BS^1[p^j])^{\wedge 2} & & \\ {\scriptstyle \beta_j}\downarrow & \searrow^{d_j} & \\ H\mathbb{Z}_3 & \xrightarrow{\gamma} & kU_6,\end{array}$$

## 5.3 Chromatic Analysis of $BU[6, \infty)$

where $d_j$ classifies the bundle

$$d_j = \sum_{k=1}^{p^j-1} \left((1-\mathcal{L}_1)(1-\mathcal{L}_1^{\otimes k})(1-\mathcal{L}_2) - (1-\mathcal{L}_1)(1-\mathcal{L}_2^{\otimes k})(1-\mathcal{L}_2)\right).$$

*Proof* We return to our putative map of cofiber sequences, and in particular to the right-most block

$$\begin{array}{ccc} \mathbb{CP}^\infty \wedge \mathbb{CP}^\infty & \xrightarrow{r} & (\mathbb{CP}^\infty)^{\wedge 2}/B_j^{\wedge 2} \\ {\scriptstyle d}\big\downarrow & & \big\downarrow{\scriptstyle f} \\ \Sigma^6 kU & \xrightarrow{\beta} & \Sigma^4 kU. \end{array}$$

This expresses $d$ in terms of $f$ in the cohomology ring $kU^*(\mathbb{CP}^\infty)^{\times 2}$, a by-now familiar situation. Namely, we have

$$\begin{aligned}\beta d &= (\mu(u \wedge x)A_1 - \mu(x \wedge u)A_2)r \\ &= \mu(u \wedge x)(q \wedge 1) - \mu(x \wedge u)(1 \wedge q) \\ &= \mu(q^*u \wedge x) - \mu(x \wedge q^*u).\end{aligned}$$

At this point, we need to make an actual identification: $u$ is a Thom class associated to the line bundle $\mathcal{L}^{\oplus p^j}$, hence $q^*u$ is its associated Euler class, which we compute in terms of $x$ to be $q^*u = [p^j]_{\mathbb{CP}^\infty_{kU}}(x)$, where the $n$-series on $\mathbb{CP}^\infty_{kU}$ expressed in terms of the coordinate $x$ is given by

$$[n]_{\mathbb{CP}^\infty_{kU}}(x) = \beta^{-1}(1-(1-\beta x)^n).$$

We use this formula to continue the calculation:

$$\begin{aligned}\mu(q^*u \wedge x) - \mu(x \wedge q^*u) &= [p^j]_{\mathbb{CP}^\infty_{kU}}(x_1) \cdot x_2 - x_1 \cdot [p^j]_{\mathbb{CP}^\infty_{kU}}(x_2) \\ &= x_1 x_2 \left(\frac{1-(1-\beta x_1)^{p^j}}{\beta x_1} - \frac{1-(1-\beta x_2)^{p^j}}{\beta x_2}\right) \\ &= \sum_{k=1}^{p^j-1}(x_1[k](x_1)x_2 - x_1[k](x_2)x_2).\end{aligned}$$  $\square$

We take this as inspiration for an algebraic definition:

**Definition 5.3.8** Let $\widehat{\mathbb{G}}$ be a connected $p$-divisible group of dimension 1. Given a point $f \in C^3(\widehat{\mathbb{G}}; \mathbb{G}_m)(T)$, we construct the function

$$e_{p^j}(f) \colon \widehat{\mathbb{G}}[p^j]^{\wedge 2} \to \mathbb{G}_m,$$

$$e_{p^j}(f) \colon (x_1, x_2) \mapsto \prod_{k=1}^{p^j} \frac{f(x_1, kx_1, x_2)}{f(x_1, kx_2, x_2)}.$$

As $j$ ranges, this assembles into a map

$$e\colon C^3(\widehat{\mathbb{G}};\mathbb{G}_m) \to \underline{\text{FormalGroups}}(\widehat{\mathbb{G}}^{\wedge 2},\mathbb{G}_m),$$

called the *Weil pairing* associated to $f$.

By design, the map $e$ participates in a commuting square with

$$\text{Spec } E_0 BU[6,\infty) \to \text{Spec } E_0 \underline{H\mathbb{Z}}_3,$$

so that this fills out the map of sequences we were considering before we got involved in this analysis of vector bundles. What remains, then, is to assemble enough exactness results to apply the 5-lemma.

**Lemma 5.3.9** ([AS01, lemma 7.2]) *For $\widehat{\mathbb{G}}$ a connected $p$-divisible group of dimension 1, the map $\delta\colon C^2(\widehat{\mathbb{G}};\mathbb{G}_m) \to C^3(\widehat{\mathbb{G}};\mathbb{G}_m)$ is injective.*

*Proof* Being finite height means that the multiplication-by-$p$ map of $\widehat{\mathbb{G}}$ is fppf-surjective. The kernel of $\delta$ consists of symmetric, biexponential maps $\widehat{\mathbb{G}}^{\times 2} \to \mathbb{G}_m$.[7] By restricting such a map $f$ to

$$f\colon \widehat{\mathbb{G}}[p^j] \times \widehat{\mathbb{G}} \to \mathbb{G}_m,$$

we can calculate

$$f(x, p^j y) = f(p^j x, y) = f(0, y) = 1.$$

But since $p^j$ is surjective on $\widehat{\mathbb{G}}$, every point on the right-hand side can be so written (after perhaps passing to a flat cover of the base), so at every left-hand stage the map is trivial. Finally, $\widehat{\mathbb{G}} = \text{colim}_j \widehat{\mathbb{G}}[p^j]$, so this filtration is exhaustive and we conclude that the kernel is trivial. □

**Lemma 5.3.10** ([AS01, lemma 7.3]) *More generally, the following sequence is exact*

$$0 \to C^2(\widehat{\mathbb{G}};\mathbb{G}_m) \xrightarrow{\delta} C^3(\widehat{\mathbb{G}};\mathbb{G}_m) \xrightarrow{e} \underline{\text{FormalGroups}}(\widehat{\mathbb{G}}^{\wedge 2},\mathbb{G}_m).$$

*Remarks on proof* The previous lemma demonstrates exactness at the first node. Showing that $e \circ \delta = 0$ is simple enough, but constructing preimages of $\ker e$ through $\delta$ is hard work. The main tool, again, is $p$-divisibility: Given a

---

[7] The condition $f \in \ker \delta$ gives $f(x, y + z) = f(x, y)f(x, z)$, so that the $kU^0$-linearity condition becomes redundant:

$$\frac{f(x,y)f(t,x+y)}{f(t+x,y)f(t,x)} = \frac{f(x,y)[f(t,x)f(t,y)]}{[f(t,y)f(x,y)]f(t,x)} = 1.$$

point $(g_1, g_2) \in \widehat{\mathbb{G}}[p^j]^{\wedge 2}$, over some flat base extension we can find $g_2'$ satisfying $p^j g_2' = g_2$. With significant effort, the assignment

$$(g_1, g_2) \mapsto \{e_{p^j}(f)(g_1, g_2')^{-1}\}$$

as $j$ ranges can be shown to be independent of the choices $g_2'$. Furthermore, if $e(f) = 1$, it even determines an element of $C^2(\widehat{\mathbb{G}}; \mathbb{G}_m)$. □

Luckily, the remaining bit of topology is very easy:

**Lemma 5.3.11** ([AS01, lemma 7.5]) *The top row of the main diagram is a short exact sequence of group schemes.*

*Proof* Consider the sequence of homology Hopf algebras, before applying Spec. Since the integral homology of $BSU$ and the $E$-homology of $\underline{H\mathbb{Z}}_3$ are both free and even, the Atiyah–Hirzebruch spectral sequence for $E_* BU[6, \infty)$ collapses to their tensor product over $E_*$. □

**Corollary 5.3.12** ([AS01, theorem 1.4]) *The map*

$$\widehat{\Pi}_3 \colon \operatorname{Spec} E_0 BU[6, \infty) \to C^3(\mathbb{CP}^\infty_E; \mathbb{G}_m)$$

*is an isomorphism.*

*Proof* This is now a direct consequence of the 5-lemma. □

*Remark 5.3.13* (see Theorem 5.4.1) We will soon show that $H_* BU[6, \infty)$ is also free and even. The proof of Lemma 5.3.11 thus also shows that the $E$-theory of $\underline{kU}_8$ fits into a similar short exact sequence.

*Remark 5.3.14* ([AS01, corollary 7.6]) The topological input to the 5-lemma also gave us a purely algebraic result for free: the map $e$ is a *surjective* map of group schemes.

## 5.4 Analysis of $BU[6, \infty)$ at Infinite Height

**Convention: We will write $H$ for $H\mathbb{F}_p$ for the duration of this lecture.**

Motivated by our success at analyzing the schemes $\operatorname{Spec}(E_\Gamma)_0 BU[6, \infty)$ associated to $BU[6, \infty)$ through Morava $E$-theory, we move on to considering the scheme constructed via ordinary homology. As usual, we expect this to be harder: The formal group associated to ordinary homology is not $p$-divisible, and this causes many sequences which are short exact from the perspective of Morava $E$-theory to go awry. Instead, we will have to examine the problem more directly – luckily, the extremely polite formal group law associated to $\widehat{\mathbb{G}}_a$

and the strong effects of the grading together make computations accessible. We also expect the reward to be greater: As in Corollary 4.3.7, we will be able to use a successful analysis of the ordinary homology scheme to give a description of the complex-orientable homology schemes, no matter what complex-orientable homology theory we use.

As in the $p$-divisible case, our framework is a map of sequences

$$\begin{array}{ccccc} \operatorname{Spec} H_*BSU & \longrightarrow & \operatorname{Spec} H_*BU[6,\infty) & \longrightarrow & \text{"}\operatorname{Spec} H_*\underline{H\mathbb{Z}}_3\text{"} \\ \downarrow & & \downarrow & & \downarrow \\ C^2(\widehat{\mathbb{G}}_a;\mathbb{G}_m) & \longrightarrow & C^3(\widehat{\mathbb{G}}_a;\mathbb{G}_m) & \longrightarrow & \underline{\operatorname{FormalGroups}}(\widehat{\mathbb{G}}_a^{\wedge 2},\mathbb{G}_m). \end{array}$$

Our task, as then, is to discern as much about these nodes as possible, as well as any exactness properties of the two sequences.[8]

We begin with the topological sequence. The Serre spectral sequence

$$E_2^{*,*} = H^*BSU \otimes H^*\underline{H\mathbb{Z}}_3 \Rightarrow H^*BU[6,\infty)$$

gives us easy access to the middle node, and we will recount the case of $p = 2$ in detail. In this case, the spectral sequence has $E_2$-page

$$E_2^{*,*} = H\mathbb{F}_2^*BSU \otimes H\mathbb{F}_2^*\underline{H\mathbb{Z}}_3$$
$$\cong \mathbb{F}_2[c_2, c_3, \ldots] \otimes \mathbb{F}_2\left[\operatorname{Sq}^{2^n} \operatorname{Sq}^{2^{n-1}} \cdots \operatorname{Sq}^2 \iota_3 \mid n \geq 1\right].$$

Because the target is 6-connective, we must have the differential $d_4 \iota_3 = c_2$, which via the Kudo transgression theorem spurs the family of differentials

$$d_{4+I_+} \operatorname{Sq}^I \iota_3 = \operatorname{Sq}^I c_2.$$

This necessitates understanding the action of the Steenrod operations on the cohomology of $BSU$, which is due to Wu [May99, section 23.6]:[9]

$$\operatorname{Sq}^{2^j} \cdots \operatorname{Sq}^4 \operatorname{Sq}^2 c_2 \equiv c_{1+2^j} \quad \text{(mod decomposables)}.$$

Accounting for the squares of classes left behind, this culminates in the following calculation:

**Theorem 5.4.1** *There is an isomorphism*

$$H\mathbb{F}_2^*BU[6,\infty) \cong \frac{H\mathbb{F}_2^*BU}{(c_j \mid j \neq 2^k + 1, j \geq 3)} \otimes \mathbb{F}_2[\iota_3^2, (\operatorname{Sq}^2 \iota_3)^2, \ldots].$$

---

[8] The quotes indicate that the right-hand topological node does not even make sense: $H_*\underline{H\mathbb{Z}}_3$ is not even-concentrated, and we do not understand the algebraic geometry of spaces whose homology is not even-concentrated. This is quite troubling – but we will press on for now.

[9] The reader might enjoy reviewing some of the formulas from Lecture 1.5 and proving their $BU$- and $BSU$-analogs, which entail this one.

## 5.4 Analysis of $BU[6, \infty)$ at Infinite Height

*Further remarks on proof*   More generally, there is an isomorphism

$$H\mathbb{F}_2^* \underline{kU}_{2k} \cong \frac{H\mathbb{F}_2^* BU}{(c_j \mid \sigma_2(j-1) < k-1)} \otimes \mathrm{Op}[\mathrm{Sq}^3 \iota_{2k-1}],$$

where $\sigma_2$ is the dyadic digital sum and "Op" denotes the Steenrod–Hopf-subalgebra of $H\mathbb{F}_2^* \underline{H\mathbb{Z}}_{2k-1}$ generated by the indicated class ([Sin68], [Sto63]). Stong specialized to $p = 2$ and carefully applied the Serre spectral sequence to the fibrations

$$\underline{kU}_{2(k+1)} \to \underline{kU}_{2k} \to \underline{H\mathbb{Z}}_{2k}.$$

Singer worked at an arbitrary prime and used the Eilenberg–Moore spectral sequence for the fibrations

$$\underline{H\mathbb{Z}}_{2k-1} \to \underline{kU}_{2(k+1)} \to \underline{kU}_{2k}.$$

Both used considerable knowledge of the interaction of these spectral sequences with the Steenrod algebra.

These methods and results generalize directly to odd primes: Singer shows

$$H\mathbb{F}_p^* \underline{kU}_{2k} \cong \frac{H\mathbb{F}_p^* BU}{(c_j \mid \sigma_p(j-1) < k-1)} \otimes \prod_{t=0}^{p-2} \mathrm{Op}[\beta\mathcal{P}^1 \iota_{2k-3-2t}].$$

The necessary modifications come from the structure of unstable mod-$p$ Steenrod algebra, odd-primary analogs of Wu's formulas [Sha77], and the Eilenberg–Moore spectral sequence. Specializing to $k = 3$, this again presents $H^* BU[6, \infty)$ as a quotient by $H^* BU$ by certain Chern classes whose indices satisfy a $p$-adic digital sum condition, tensored up with the Steenrod–Hopf-subalgebra of $H^* \underline{H\mathbb{Z}}_3$ generated by $\beta\mathcal{P}^1 \iota_3$. □

From the edge homomorphisms in Theorem 5.4.1, we can already see that the sequence of Hopf algebras

$$H_* \underline{H\mathbb{Z}}_3 \to H_* BU[6, \infty) \to H_* BSU$$

modeling the sequence of formal group schemes

$$\text{"}(\underline{H\mathbb{Z}}_3)_{HP}\text{"} \to BU[6, \infty)_{HP} \to BSU_{HP}$$

is neither left-exact nor right-exact. This seems bleak.

Ever the optimists, we embark on an analysis of the available algebra all the same. We begin by reusing a strategy previously employed in Lemma 3.4.12: First perform a tangent space calculation

$$T_0 C^k(\widehat{\mathbb{G}}_a; \mathbb{G}_m) \cong C^k(\widehat{\mathbb{G}}_a; \widehat{\mathbb{G}}_a),$$

then study the behavior of the different tangent directions to determine the

250  The σ-Orientation

full object $C^k(\widehat{\mathbb{G}}_a; \mathbb{G}_m)$. As a warm-up to the case $k = 3$ of interest, we will first consider the case $k = 2$. We have already performed the tangent space calculation:

**Corollary 5.4.2** (see Lemma 3.2.7)  *The collection of symmetric additive 2-cocycles of homogeneous degree n is spanned by the single element*

$$c_n(x, y) = \begin{cases} (x + y)^n - x^n - y^n & \text{if } n \neq p^j, \\ \frac{1}{p}((x + y)^n - x^n - y^n) & \text{if } n = p^j. \end{cases}$$
□

Our goal, then, is to select such an $f_+ \in C^2(\widehat{\mathbb{G}}_a; \widehat{\mathbb{G}}_a)$ and study the minimal conditions needed on a symbol $a$ to produce a point in $C^2(\widehat{\mathbb{G}}_a; \mathbb{G}_m)$ of the form $1 + af_+ + \cdots$. Since $c_n = \frac{1}{d_n}\delta(x^n)$ is itself produced by an additive formula, life would be easiest if we had access to an exponential, so that we could build

$$\text{``}\delta_{(\widehat{\mathbb{G}}_a; \mathbb{G}_m)} \exp(a_n x^n)^{1/d_n} = \exp(\delta_{(\widehat{\mathbb{G}}_a; \widehat{\mathbb{G}}_a)} a_n x^n / d_n) = \exp(a_n c_n).\text{''}$$

However, the existence of an exponential series is equivalent to requiring that $a_n$ carry a divided-power structure, which turns out not to be minimal. In fact, we can show that *no* conditions on $a_n$ are required *at all*.

**Lemma 5.4.3** ([AHS01, proposition 3.9])  *Recall from Remark 3.3.23 that the Artin–Hasse exponential is the power series*

$$E_p(t) = \exp\left(\sum_{j=0}^{\infty} \frac{t^{p^j}}{p^j}\right) \in \mathbb{Z}_{(p)}[\![t]\!].$$

*Write* $\delta_{(\widehat{\mathbb{G}}_a, \mathbb{G}_m)}: C^1(\widehat{\mathbb{G}}_a; \mathbb{G}_m) \to C^2(\widehat{\mathbb{G}}_a; \mathbb{G}_m)$ *and*

$$d_n = \begin{cases} 1 & \text{if } n = p^j, \\ 0 & \text{otherwise}. \end{cases}$$

*The class* $g_n(x, y) = \delta_{(\widehat{\mathbb{G}}_a, \mathbb{G}_m)} E_p(a_n x^n)^{1/p^{d_n}}$ *is a series in* $\mathbb{F}_p[a_n][\![x, y]\!]$ *and presents a point in* $C^2(\widehat{\mathbb{G}}_a; \mathbb{G}_m)$ *reducing to* $a_n c_n \in C^2(\widehat{\mathbb{G}}_a; \widehat{\mathbb{G}}_a)$ *on tangent spaces.*

*Proof*  Recall furthermore from Remark 3.3.23 that $E_p$ has coefficients in $\mathbb{Z}_{(p)}$, and hence it can be reduced to a series with coefficients in $\mathbb{F}_p$ if we so choose. With this in mind, we make the calculation

$$g_n(x, y) = \delta_{(\widehat{\mathbb{G}}_a, \mathbb{G}_m)} E_p(a_n x^n)^{1/p^{d_n}}$$

$$= \exp\left(\sum_{j=0}^{\infty} \frac{a_n^{p^j} \delta_{(\widehat{\mathbb{G}}_a, \widehat{\mathbb{G}}_a)} x^{np^j}}{p^{j+d_n}}\right)$$

$$= \exp\left(\sum_{j=0}^{\infty} \frac{a_n^{p^j} c_{np^j}(x,y)}{p^j}\right).$$

As claimed, the leading term is exactly $a_n c_n$, this series is symmetric, and since it is in the image of $\delta_{(\widehat{\mathbb{G}}_a, \mathbb{G}_m)}$ it is certainly a 2-cocycle. Finally, the integrality properties of $E_p$ mean that $g_n$ has coefficients in $\mathbb{Z}_{(p)}[a_n]$. □

Letting $n$ range, this culminates in the following calculation:

**Lemma 5.4.4** ([AHS01, equation 3.7]) *The map*

$$\operatorname{Spec} \mathbb{Z}_{(p)}[a_n \mid n \geq 2] \xrightarrow{\prod_{n \geq 2} g_n} C^2(\widehat{\mathbb{G}}_a; \mathbb{G}_m) \times \operatorname{Spec} \mathbb{Z}_{(p)}$$

*is an isomorphism.*[10] □

The trivariate case $k = 3$ is similar, with one important new wrinkle: Over an $\mathbb{F}_p$-algebra there is an equality $c_n^p = c_{pn}$, but this relation does not generalize to trivariate 2-cocycles. For instance, consider the following example at $p = 2$:

$$\frac{1}{2}\delta(c_6) = x^2 y^2 z^2 + x^4 yz + xy^4 z + xyz^4, \quad \left(\frac{1}{2}\delta c_3\right)^2 = x^2 y^2 z^2.$$

The following lemma states that this is the only new feature:

**Lemma 5.4.5** ([AHS01, propositions 3.20 and A.12]) *The p-primary residue of the scheme of trivariate symmetric 2-cocycles is presented by*

$$\operatorname{Spec} \mathbb{F}_p[a_d \mid d \geq 3] \times \operatorname{Spec} \mathbb{F}_p[b_d \mid d = p^j(1 + p^k)] \xrightarrow{\cong} C^3(\widehat{\mathbb{G}}_a; \widehat{\mathbb{G}}_a) \times \operatorname{Spec} \mathbb{F}_p.$$
□

Similar juggling of the Artin–Hasse exponential yields the following multiplicative classification:

**Theorem 5.4.6** ([AHS01, proposition 3.28]) *There is an isomorphism*

$$\operatorname{Spec} \mathbb{Z}_{(p)}[a_d \mid d \geq 3, d \neq 1 + p^t] \times \operatorname{Spec} \Gamma[b_{1+p^t}] \to C^3(\widehat{\mathbb{G}}_a; \mathbb{G}_m) \times \operatorname{Spec} \mathbb{Z}_{(p)}.$$

*Proof sketch* The main claim is that the Artin–Hasse exponential trick used in the case $k = 2$ works here as well, provided $d \neq 1 + p^t$ so that taking an appropriate $p$th root works out. They then show that the remaining exceptional cases extend to multiplicative cocycles only when the $p$th power of the leading coefficient vanishes. Finally, a rational calculation shows how to bind these truncated generators together into a divided-power algebra. □

---

[10] This product decomposition only happens after $p$-localization, and the decomposition is different at each prime [AHS01, section 3.3]. Compare this with results of Adams [Ada76] or of Adams and Priddy [AP76].

It is now time to clear up our confusion about the right-hand topological node by pursuing a link between $H_*\underline{H\mathbb{Z}}_3$ and the algebraic model[11]

$$\text{FormalGroups}(\widehat{\mathbb{G}}_a^{\wedge 2}, \mathbb{G}_m).$$

Analyzing the edge homomorphism from our governing Serre spectral sequence shows that the map

$$H^*BU[6, \infty) \to H^*\underline{H\mathbb{Z}}_3$$

factors through the subalgebra $A^* \subseteq H^*\underline{H\mathbb{Z}}_3$ generated by the *squares* of the polynomial generators. Accordingly, we aim to replace the right-hand node of the topological sequence with $\operatorname{Spec} A_*$ outright.

**Lemma 5.4.7** ([AHS01, lemmas 3.36 and 4.11, proposition 4.13]) *The scheme* $\operatorname{Spec} A_*$ *models* $\text{FormalGroups}(\widehat{\mathbb{G}}_a^{\wedge 2}, \mathbb{G}_m)$ *by an isomorphism $\lambda$ commuting with $e \circ \hat{\Pi}_3$.*

*Proof sketch* We can describe the $\mathbb{F}_p$-scheme $\text{FormalGroups}(\widehat{\mathbb{G}}_a^{\wedge 2}, \mathbb{G}_m)$ completely explicitly:

$$(a_{mn})_{m,n} \longmapsto \prod_{m<n} \operatorname{texp}\left(a_{mn}(x^{p^m} y^{p^n} - x^{p^n} y^{p^m})\right)$$

$$\operatorname{Spec} \mathbb{F}_p[a_{mn} \mid m < n]/(a_{mn}^p) \xrightarrow{\cong} \text{FormalGroups}(\widehat{\mathbb{G}}_a^{\wedge 2}, \mathbb{G}_m),$$

where $\operatorname{texp}(t) = \sum_{j=0}^{p-1} t^j/j!$ is the truncated exponential series. It is easy to check that this ring of functions agrees with $A^*$, and it requires hard work (although not much creativity) to check the remainder of the statement: that $e \circ \hat{\Pi}_3$ factors through $\operatorname{Spec} A_*$ and that the factorization is an isomorphism. □

We have now finally assembled our map of sequences,

$$\begin{array}{ccccccc}
\operatorname{Spec} H_*BSU & \longrightarrow & \operatorname{Spec} H_*BU[6,\infty) & \longrightarrow & \operatorname{Spec} A^* & \longrightarrow & 0 \\
\cong \downarrow \hat{\Pi}_2 & & \downarrow \hat{\Pi}_3 & & \cong \downarrow \lambda & & \\
C^2(\widehat{\mathbb{G}}_a; \mathbb{G}_m) & \xrightarrow{\delta} & C^3(\widehat{\mathbb{G}}_a; \mathbb{G}_m) & \xrightarrow{e} & \operatorname{Weil}(\widehat{\mathbb{G}}_a) & \longrightarrow & 0
\end{array}$$

which we have shown to be exact at all the indicated nodes. (The exactness of the topological sequence follows from the Serre spectral sequence analysis. The exactness of the bottom sequence follows from it receiving a map from the top exact sequence, where the left-hand vertical map is an isomorphism.) Our calculations now pay off:

---

[11] The reader might note that $\widehat{\mathbb{G}}_a$ is *not* a $p$-divisible group. Nonetheless, we can make sense of $\text{FormalGroups}(\widehat{\mathbb{G}}_a^{\wedge 2}, \mathbb{G}_m)$ as a closed subscheme of $\text{FormalSchemes}(\widehat{\mathbb{G}}_a^{\times 2}, \mathbb{G}_m)$.

## 5.4 Analysis of $BU[6, \infty)$ at Infinite Height

**Corollary 5.4.8** ([AHS01, corollary 4.14])  *The map $\hat{\Pi}_3$ is an isomorphism:*

$$\hat{\Pi}_3: \operatorname{Spec} H_* BU[6, \infty) \xrightarrow{\cong} C^3(\widehat{\mathbb{G}}_a; \mathbb{G}_m).$$

*Proof sketch*  We don't actually have to compute much about the middle map. Because the squares in the map of sequences commute and the sequences themselves are exact as indicated, we at least learn that $\hat{\Pi}_3$ is an epimorphism on group schemes, hence a monomorphism on rings of functions. But, since both source and target are affine schemes of graded finite type with equal Poincaré series in each case, this monomorphism is an isomorphism.  □

**Corollary 5.4.9** ([AHS01, theorem 2.31])  *The map $\hat{\Pi}_3$ is an isomorphism for any complex-orientable E.*

*Proof sketch*  This follows much along the lines of Corollary 4.3.7. The evenness of the topological calculation at $E = H\mathbb{F}_p$ shows that the statement holds for $H\mathbb{Z}_p^\wedge$ and $H\mathbb{Z}_{(p)}$, and since $p$ is arbitrary we conclude it for $H\mathbb{Z}$ as well. We thus learn that the statement holds for $E = MUP$ using a tangent space argument, and then an Atiyah–Hirzebruch argument gives the statement for any complex-oriented $E$.  □

*Remark 5.4.10*  This argument does *not* extend to a claim that we have an isomorphism of topological and algebraic exact sequences for any choice of complex-orientable homology theory $E$. Our trick of replacing $H_* \underline{H\mathbb{Z}}_3$ by $A_*$ has no generic analog.

Our analysis of $\operatorname{Spec} E_* BU[6, \infty)$ forms input to two related pursuits: the homology scheme $\operatorname{Spec} E_* MU[6, \infty)$ arising in the theory of Thom spectra, and the object $BU[6, \infty)_E$ predual to $\operatorname{Spec} E_0 BU[6, \infty)$. The analysis of the Thom spectrum is completely analogous to the analysis performed at the end of Lecture 5.2, and so we merely state the relevant results.

**Definition 5.4.11**  For a formal group $\widehat{\mathbb{G}}$, define maps $\mu_{ij}: \widehat{\mathbb{G}}^{\times 3} \to \widehat{\mathbb{G}}$ which multiply the $i$th and $j$th factors, discarding the remaining factor. For a line bundle $\mathcal{L}$ over $\widehat{\mathbb{G}}$, we define the scheme $C^3(\widehat{\mathbb{G}}; \mathcal{L})$ by

$$C^3(\widehat{\mathbb{G}}; \mathcal{L})(T) = \left\{ \begin{array}{c} \operatorname{triv}^{\text{ns}} \text{ of } u^* \left( \frac{e^* \mathcal{L} \otimes (\mu_{12}^* \mathcal{L} \otimes \mu_{13}^* \mathcal{L} \otimes \mu_{23}^* \mathcal{L})}{(\pi_1^* \mathcal{L} \otimes \pi_2^* \mathcal{L} \otimes \pi_3^* \mathcal{L}) \otimes \mu_{\text{all}}^* \mathcal{L}} \right) \downarrow u^* \widehat{\mathbb{G}}^{\times 3} \\ \text{which are rigid, symmetric, and } kU^0\text{-linear} \end{array} \right\}.$$

**Lemma 5.4.12** ([AHS01, theorem 2.50])  *There is a system of compatible maps*

$$\begin{array}{ccc}
\operatorname{Spec} E_0 BU[6,\infty) \times \operatorname{Spec} E_0 MU[6,\infty) & \longrightarrow & \operatorname{Spec} E_0 MU[6,\infty) \\
\| \quad \downarrow & & \downarrow \\
C^3(\mathbb{CP}_E^\infty; \mathbb{G}_m) \times C^3(\mathbb{CP}_E^\infty; \mathcal{I}(0)) & \longrightarrow & C^3(\mathbb{CP}_E^\infty; \mathcal{I}(0)),
\end{array}$$

*where the horizontal maps are the action maps defining torsors and the vertical maps are those induced by $\widehat{\Pi}_3$.* □

**Corollary 5.4.13** *Take $E$ to be even-periodic. There is then an isomorphism*
$$C^3(\mathbb{CP}_E^\infty; \mathcal{I}(0)) \cong \operatorname{Spec} E_0 MU[6,\infty).$$
*Moreover, the $E_0$-points of this scheme biject with ring spectrum maps*
$$MU[6,\infty) \to E.$$ □

**Lemma 5.4.14** *The ring map $MU[6,\infty) \to MSU$ is modeled by the map*
$$C^2(\mathbb{CP}_E^\infty; \mathcal{I}(0)) \xrightarrow{\delta} C^3(\mathbb{CP}_E^\infty; \mathcal{I}(0)),$$
$$s \in \Theta^2 \mathcal{I}(0) \mapsto \frac{\mu_{12}^* s}{\mu_1^* s \otimes \mu_2^* s} \in \Theta^1 \Theta^2 \mathcal{I}(0) \cong \Theta^3 \mathcal{I}(0).$$ □

*Remark 5.4.15* The rational conclusion of this analysis admits a very mild reformulation. Note first that there is always a natural $MU[6,\infty)$-orientation of a rational spectrum $E$ given by the composite
$$MU[6,\infty) \to MU[6,\infty) \otimes \mathbb{Q} \to H\mathbb{Q} = \mathbb{S} \otimes \mathbb{Q} \xrightarrow{\eta_E} E \otimes \mathbb{Q} = E.$$
This canonical point turns the $C^3(\mathbb{CP}_E^\infty; \mathbb{G}_m)$-torsor structure of $C^3(\mathbb{CP}_E^\infty; \mathcal{I}(0))$ into an isomorphism $C^3(\mathbb{CP}_E^\infty; \mathcal{I}(0)) \cong C^3(\mathbb{CP}_E^\infty; \mathbb{G}_m)$. In sum, an arbitrary $MU[6,\infty)$-orientation of $E$ is witnessed by a symmetric rigid trivariate power series satisfying $kU^0$-linearity, called its *characteristic series*. There are similar theories of characteristic series for rational orientations by $MSU$, $MU$, and $MUP$; in the latter two cases this recovers the *Hirzebruch series*, which associates to an orientation $\varphi \colon MUP \to \mathbb{Q} \otimes E$ the difference of trivializations $x/\exp_\varphi(x)$. Conversely, any such series gives rise to a characteristic class by the formula
$$K_\varphi(\mathcal{L}_1 \oplus \cdots \oplus \mathcal{L}_n) = \prod_{j=1}^n \frac{c_1(\mathcal{L}_j)}{\exp_\varphi(c_1(\mathcal{L}_j))},$$
where $c_1$ is the first Chern class associated to the canonical rational orientation.

Our second task is to analyze the cohomology formal scheme associated to $BU[6,\infty)$, and we begin with the choice $E = H$.

**Lemma 5.4.16** (Ando, Hopkins, and Strickland, unpublished) $DC^k(\widehat{\mathbb{G}}_a; \mathbb{G}_m)$ *are all formal varieties for $k \le 3$.*

## 5.4 Analysis of $BU[6, \infty)$ at Infinite Height

*Proof* We know that $OC^k(\widehat{\mathbb{G}}_a; \mathbb{G}_m)$ are all free $\mathbb{Z}$-modules of graded finite rank in the range $k \leq 3$, so we may write

$$O(DC^k(\widehat{\mathbb{G}}_a; \mathbb{G}_m)) \cong (OC^k(\widehat{\mathbb{G}}_a; \mathbb{G}_m))^*.$$

We must show that this Hopf algebra $O(C^k(\widehat{\mathbb{G}}_a; \mathbb{G}_m))^*$ is a power series ring.

Specialize, for the moment, to the case of $k = 2$. It will suffice to show that it is a power series ring modulo $p$ for every prime $p$. Such graded connected finite-type Hopf algebras over $\mathbb{F}_p$ were classified by Borel (and exposited by Milnor and Moore [MM65, theorem 7.11]) as either polynomial or truncated polynomial. These two cases are distinguished by the *Frobenius* operation: The Frobenius on a polynomial ring is injective, whereas the Frobenius on a truncated polynomial ring is not. It is therefore equivalent to show that the *Verschiebung* on the original ring $O(C^2(\widehat{\mathbb{G}}_a; \mathbb{G}_m)) \otimes \mathbb{F}_p$ is *surjective*. Recalling the calculation $c_n^p = c_{np}$ at the level of bivariate 2-cocycles, we compute

$$p^* a_n = a_{np}^p,$$

and since $Fa_{np} = a_{np}^p$ and $FV = p^*$, we learn

$$V(a_{np}) = a_n.$$

Essentially the same proof handles the cases $k = 1$ and $k = 0$.

The case $k = 3$ requires a small modification to cope with the two classes of trivariate 2-cocycles. On the polynomial tensor factor of $O(C^3(\widehat{\mathbb{G}}_a; \mathbb{G}_m))$ we can reuse the same Verschiebung argument to see that its dual Hopf algebra is polynomial. For the other fact, the dual of the divided-power tensor factor is, without any further argument, always a primitively generated polynomial algebra. □

**Theorem 5.4.17** (Ando, Hopkins, and Strickland, unpublished) *For $E$ an even and periodic cohomology theory, $BU[6, \infty)_E$ models the scheme $C_3 \mathbb{CP}_E^\infty$.*

*Proof sketch* Let $\widehat{\mathbb{G}}$ be an arbitrary formal group. Note first that if $C^3(\widehat{\mathbb{G}}; \mathbb{G}_m)$ is coalgebraic, then $C_3 \widehat{\mathbb{G}}$ exists and is its Cartier dual: the diagram presenting $OC^3(\widehat{\mathbb{G}}; \mathbb{G}_m)$ as a reflexive coequalizer of free Hopf algebras is also the diagram meant to present $C_3 \widehat{\mathbb{G}}$ as a coalgebraic formal scheme. So, if the coequalizing Hopf algebra has a good basis, it will follow from Theorem 5.1.8 that the resulting diagram is a colimit diagram in formal schemes, with $C_3 \widehat{\mathbb{G}}$ sitting at the cone point. It will additionally follow that the isomorphism

$$\operatorname{Spec} E_0 BU[6, \infty) \xrightarrow{\cong} C^3(\mathbb{CP}_E^\infty; \mathbb{G}_m)$$

of Corollary 5.4.9 will re-dualize to an isomorphism

$$BU[6, \infty)_E \xleftarrow{\cong} C_3 \mathbb{CP}_E^\infty.$$

By a base change argument, it suffices to take $\widehat{\mathbb{G}}$ to be the universal formal group over $O_{\mathcal{M}_{\mathrm{fgl}}}$, and we thus set about finding a good basis in this case. We hope to gain control (as in Corollary 4.3.7 or 5.4.9) of this situation using our strong knowledge of $OC^3(\widehat{\mathbb{G}}_a; \mathbb{G}_m)$. We know from Lemma 5.4.16 that $OC^3(\widehat{\mathbb{G}}_a; \mathbb{G}_m)$ is a free abelian group, and we know from Theorem 3.2.2 that $O(\mathcal{M}_{\mathrm{fgl}})$ is as well. By picking a $\mathbb{Z}$-basis $\mathbb{Z}\{\beta_j\}_j$ of $OC^3(\widehat{\mathbb{G}}_a; \mathbb{G}_m)$ and considering the specialization map from $\widehat{\mathbb{G}}$ over $\mathcal{M}_{\mathrm{fgl}}$ to $\widehat{\mathbb{G}}_a$ over $\mathrm{Spec}\,\mathbb{Z}$, we choose a map $\alpha$ of $O(\mathcal{M}_{\mathrm{fgl}})$-modules

$$\begin{array}{ccc} O(\mathcal{M}_{\mathrm{fgl}})\{\tilde{\beta}_j\}_j & \xrightarrow{\alpha} & OC^3(\widehat{\mathbb{G}}; \mathbb{G}_m) \\ \downarrow & & \downarrow \\ \mathbb{Z}\{\beta_j\}_j & \xrightarrow{\cong} & OC^3(\widehat{\mathbb{G}}_a; \mathbb{G}_m). \end{array}$$

By induction on degree, one sees that $\alpha$ is surjective, and since the source and target are abelian groups of graded finite rank *and the source is free*, we need only check that they have the same rational Poincaré series to conclude that $\alpha$ is an isomorphism. Over $\mathrm{Spec}\,\mathbb{Q}$ we can use the logarithm to construct an isomorphism

$$\mathrm{Spec}\,\mathbb{Q} \times (\mathcal{M}_{\mathrm{fgl}} \times C^k(\widehat{\mathbb{G}}; \mathbb{G}_m)) \to \mathrm{Spec}\,\mathbb{Q} \times (\mathcal{M}_{\mathrm{fgl}} \times C^k(\widehat{\mathbb{G}}_a; \mathbb{G}_m)),$$

hence the Poincaré series agree, hence $\alpha \otimes \mathbb{Q}$ is an isomorphism, and finally $\alpha$ is too.

Last, one checks that this basis gives us access to the desired collection of good subcoalgebras: These are indexed on an integer $d$, spanned by those basis vectors of degree at most $d$. □

## 5.5 Modular Forms and $MU[6, \infty)$-Manifolds

The first goal of this lecture is to give the briefest possible summary of the theory of elliptic curves that covers the topics necessary to us in the coming sections. Accordingly, we won't cover many topics that a sane introduction to elliptic curves would make a point to cover, and – perhaps worse – we will hardly prove anything. We will, however, discover a place where "$C_3\widehat{\mathbb{G}}$" appears internally to the theory of elliptic curves, and I hope nonetheless that this will give the arithmetically disinclined reader a foothold on the "elliptic" part of "elliptic cohomology."

To begin, recall that an elliptic curve in the complex setting is a torus, and it admits a presentation by selecting a lattice $\Lambda$ of full rank in $\mathbb{C}$ and forming the

## 5.5 Modular Forms and $MU[6,\infty)$-Manifolds

quotient

$$\mathbb{C} \xrightarrow{\pi_\Lambda} \mathbb{C}/\Lambda =: E_\Lambda.$$

A meromorphic function $f$ on $E_\Lambda$ pulls back to give a meromorphic function $\pi_\Lambda^* f$ on $\mathbb{C}$ which satisfies a periodicity constraint in the form of the functional equation

$$\pi_\Lambda^* f(z + \Lambda) = \pi_\Lambda^* f(z).$$

It follows immediately that there are no holomorphic such functions, save the constants – such a function would be bounded, and Liouville's theorem would apply. It is, however, possible to build the following meromorphic special function, which has poles of order 2 at the lattice points and satisfies the periodicity constraints:

$$\wp_\Lambda(z) = \frac{1}{z^2} + \sum_{\omega \in \Lambda \setminus \{0\}} \frac{1}{(z-\omega)^2} - \frac{1}{\omega^2}.$$

Its derivative is also a meromorphic function satisfying the periodicity constraint:

$$\wp'_\Lambda(z) = -2 \sum_{\omega \in \Lambda} \frac{1}{(z-\omega)^3}.$$

In fact, these two functions generate all other meromorphic functions on $E_\Lambda$, in the sense that the subsheaf spanned by the algebra generators $\wp_\Lambda$ and $\wp'_\Lambda$ is exactly $\pi_\Lambda^* \underline{\mathrm{Mer}}(E_\Lambda)$. This algebra is subject to the following relation, in the form of a differential equation:

$$\wp'_\Lambda(z)^2 = 4\wp_\Lambda(z)^3 - 60G_4(\Lambda)\wp_\Lambda(z) - 140G_6(\Lambda),$$

for some special values $G_4(\Lambda), G_6(\Lambda) \in \mathbb{C}$. Accordingly, writing $C \subseteq \mathbb{CP}^2$ for the projective curve $wy^2 = 4x^3 - G_4(\Lambda)w^2 x - G_6(\Lambda)w^3$, there is an analytic group isomorphism

$$E_\Lambda \to C,$$
$$z \pmod{\Lambda} \mapsto [1 : \wp_\Lambda(z) : \wp'_\Lambda(z)].$$

This is sometimes referred to as the *Weierstrass presentation* of $E_\Lambda$.

*Remark 5.5.1* Before proceeding, the values $G_4(\Lambda)$ and $G_6(\Lambda)$ are themselves interesting when considered as functions of the lattice. Expanding out the relation above gives an explicit formula for the $(2k)$-*th Eisenstein series*

$$G_{2k}(\Lambda) = \sum_{\ell \in \Lambda} \frac{1}{\ell^{2k}}, \quad G_{2k}(\lambda \Lambda) = \lambda^{-2k} G_{2k}(\Lambda).$$

A function on the space of lattices which satisfies such a homogeneity condition

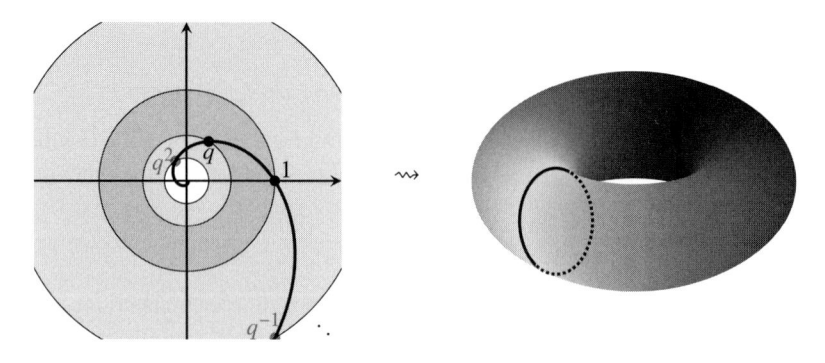

Figure 5.1 Presentation of an elliptic curve as the quotient of nested annuli.

is referred to as a *modular form* of weight $2k$, and they appear naturally as global sections over the moduli of elliptic curves on the $2k$th tensor power of the sheaf of invariant differentials. One can show that the ring of complex-analytic modular forms has the form $\mathbb{C}[G_4, G_6]$.[12]

There is a second standard embedding of a complex elliptic curve into projective space, using *θ-functions*, which are most naturally expressed with an alternative basic presentation of an elliptic curve. Select a lattice $\Lambda$ and a basis for it, and rescale the lattice so that the basis takes the form $\{1, \tau\}$ with $\tau$ in the upper half-plane. Then, the normalized exponential function $\mathbb{C} \to \mathbb{C}^\times$ given by $z \mapsto \exp(2\pi i z)$ has $1 \cdot \mathbb{Z}$ as its kernel. Setting $q = \exp(2\pi i \tau)$ to account for the missing component of the kernel of $\pi_\Lambda$, we get a second presentation of $E_\Lambda$ as $\mathbb{C}^\times / q^\mathbb{Z}$, as pictured in Figure 5.1.

**Definition 5.5.2** The basic $\theta$-function associated to $E_\Lambda$ is defined by

$$\theta_q(u) = \prod_{m \geq 1} (1 - q^m)(1 + q^{m-\frac{1}{2}} u)(1 + q^{m-\frac{1}{2}} u^{-1}) = \sum_{n \in \mathbb{Z}} u^n q^{\frac{1}{2}n^2}.$$

Given two rational numbers $0 \leq a, b \leq 1$, we can also shift the zero-set of $\theta_q$ in the 1 and $q$ directions by the fractions $a$ and $b$, giving translated $\theta$-functions:

$$\theta_q^{a,b}(u) = \left( q^{\frac{a^2}{2}} \cdot u^a \cdot \exp(2\pi i a b) \right) \cdot \theta_q(u q^a \exp(2\pi i b)).$$

The basic $\theta$-function vanishes on the set $\{\exp(2\pi i(\frac{1}{2}m + \frac{\tau}{2}n))\}$, i.e., at the

---

[12] It is possible to make sense of elliptic curves, their moduli, and modular forms in the integral context of algebraic geometry, where Deligne demonstrated an analogous theorem: There exist modular forms $c_4$, $c_6$, and $\Delta$ such that

$$H^0(\mathcal{M}_{\text{ell}}; \omega^{\otimes *}) \cong \mathbb{Z}[c_4, c_6, \Delta^{\pm}]/(c_4^3 - c_6^2 - 2^6 3^3 \Delta).$$

## 5.5 Modular Forms and $MU[6, \infty)$-Manifolds

center of the fundamental annulus. Since it has no poles, it cannot descend to give a function on $\mathbb{C}^\times/q^\mathbb{Z}$, and its failure to descend is witnessed by its imperfect periodicity relation:[13]

$$\theta_q(qu) = u^{-1} q^{\frac{-1}{2}} \theta_q(u).$$

**Lemma 5.5.3** ([Hus04, proposition 10.2.6]) *For any $N > 0$, define $V_q[N]$ to be the space of functions $f : \mathbb{C}^\times \to \mathbb{C}$ satisfying*

$$f(qu) = u^{-N} q^{-N^2/2} f(u).$$

*Then, $V_q[N]$ has $\mathbb{C}$-dimension $N^2$, and the functions $\theta_q^{a,b}$ give a basis as $a$ and $b$ range over rational numbers with denominator $N$.* □

Even though these functions do not themselves descend to $\mathbb{C}^\times/q^\mathbb{Z}$, we can collectively use them to construct a map to complex projective space, where the quasiperiodicity relations will mutually cancel in homogeneous coordinates.

**Theorem 5.5.4** ([Hus04, proposition 10.3.2]) *Consider the map*

$$\mathbb{C}/(N \cdot \Lambda) \xrightarrow{f_{(N)}} \mathbb{P}^{N^2-1}(\mathbb{C}),$$
$$z \mapsto [\cdots : \theta_q^{i/N, j/N}(z) : \cdots].$$

*For $N > 1$, this map is an embedding.* □

*Example 5.5.5* In the case of $N = 2$, the four functions involved are

$$\theta_{q^2}^{0,0}, \theta_{q^2}^{0,1/2}, \theta_{q^2}^{1/2,0}, \theta_{q^2}^{1/2,1/2},$$

and we record their zero loci in Figure 5.2. The image of $f_{(2)}$ in $\mathbb{P}^{2^2-1}(\mathbb{C})$ is cut out by the equations[14]

$$A^2 x_0^2 = B^2 x_1^2 + C^2 x_2^2, \qquad A^2 x_3^2 + B^2 x_2^2 = C^2 x_1^2,$$

where

$$x_0 = \theta_{q^2}^{0,0}(u^2), \quad x_1 = \theta_{q^2}^{0,1/2}(u^2), \quad x_2 = \theta_{q^2}^{1/2,0}(u^2), \quad x_3 = \theta_{q^2}^{1/2,1/2}(u^2)$$

and

$$A = \theta_{q^2}^{0,0}(1) = \sum_n q^{n^2}, \qquad B = \theta_{q^2}^{0,1/2}(1) = \sum_n (-1)^n q^{n^2},$$

$$C = \theta_{q^2}^{1/2,0}(1) = \sum_n q^{(n+1/2)^2},$$

---

[13] Equivalently, $\theta_q$ is properly considered as a section of a nontrivial line bundle on this quotient.
[14] This amounts to a projectivization of the embedding $S^1 \times S^1 \to \mathbb{R}^2 \times \mathbb{R}^2$.

| Function | $\theta_{q^2}^{0,0}$ | $\theta_{q^2}^{0,1/2}$ | $\theta_{q^2}^{1/2,0}$ | $\theta_{q^2}^{1/2,1/2}$ |
|---|---|---|---|---|
| Zero locus | $-q^{1+2\mathbb{Z}}$ | $q^{1+2\mathbb{Z}}$ | $-q^{2\mathbb{Z}}$ | $q^{2\mathbb{Z}}$ |

Figure 5.2 Standard $\theta$-functions and their offsets.

upon which there is the additional *Jacobi relation*

$$A^4 = B^4 + C^4.$$

*Remark 5.5.6* This embedding of $E_\Lambda$ as an intersection of quadric surfaces in $\mathbb{CP}^3$ is quite different from the Weierstrass embedding. Nonetheless, the embeddings are analytically related. Namely, there is an equality

$$\frac{d^2}{dz^2} \log \theta_q(u) = \wp_\Lambda(z).$$

Separately, Weierstrass considered a function $\sigma_\Lambda$, defined by

$$\sigma_\Lambda(z) = z \prod_{\omega \in \Lambda \setminus 0} \left(1 - \frac{z}{\omega}\right) \cdot \exp\left[\frac{z}{\omega} + \frac{1}{2}\left(\frac{z}{\omega}\right)^2\right],$$

which also has the property that its second logarithmic derivative is $\wp$ and so is "basically $\theta_q^{1/2,1/2}$." In fact, any meromorphic function on the elliptic curve $E_\Lambda$ can be written in the form

$$c \cdot \prod_{i=1}^n \frac{\sigma_\Lambda(z - a_i)}{\sigma_\Lambda(z - b_i)}.$$

Both of these embeddings enjoy algebraic analogs. The usual proof that all abelian varieties are projective is a kind of generalization of the Weierstrass embedding ([Mil08, section I.6], [Har77, remark II.7.8.2]): From a line bundle on $A$ one can extract a very ample line bundle $\mathcal{L}$, then carefully construct some generating global sections to get an embedding into $\mathbb{P}(\mathcal{L}^{\oplus n})$. In the $\theta$-function version, not only is there a generalization of the embedding but it enjoys a certain canonicality: Weil showed that quasiperiodic functions on an arbitrary complex abelian variety are characterized uniquely as an irreducible representation for a certain group, that this representation admits a separate canonical presentation and with it a canonical basis, that the functions corresponding to this basis under the isomorphism can be used to construct an embedding as above, and finally that the functions obtained in this way in the case of an elliptic curve recover the classical $\theta$-functions ([Wei64], [Car66]). Mumford showed further that there is

## 5.5 Modular Forms and $MU[6, \infty)$-Manifolds

a wholly algebraic analog of this collection of results ([Mum66], [Tat97]).[15] The full extent of this story is far beyond our scope, but a critical theorem sitting at the heart of both of these approaches is:

**Theorem 5.5.7** ("Theorem of the Cube" [Mil08, theorem I.5.1])  *Let $A$ be an abelian variety, let $p_i : A \times A \times A \to A$ be the projection onto the ith factor, and let $p_{ij} = p_i +_A p_j$, $p_{ijk} = p_i +_A p_j +_A p_k$. Then for any invertible sheaf $\mathcal{L}$ on $A$, the sheaf*

$$\Theta^3(\mathcal{L}) := \frac{p_{123}^*\mathcal{L} \otimes p_1^*\mathcal{L} \otimes p_2^*\mathcal{L} \otimes p_3^*\mathcal{L}}{p_{12}^*\mathcal{L} \otimes p_{23}^*\mathcal{L} \otimes p_{31}^*\mathcal{L} \otimes p_\emptyset^*\mathcal{L}} = \bigotimes_{I \subseteq \{1,2,3\}} (p_I^*\mathcal{L})^{(-1)^{|I|-1}}$$

*on $A \times A \times A$ is trivial. If $\mathcal{L}$ is rigid (i.e., it has a specified trivialization at the identity point of $A$), then $\Theta^3(\mathcal{L})$ is canonically trivialized by a section $s(A; \mathcal{L})$.* □

*Remark 5.5.8*  One way to read the Theorem of the Cube is that a weight zero divisor on an abelian variety is principal (i.e., it is the zeroes and poles of a meromorphic function) if and only if its nodes sum to zero. The meromorphic function you construct this way is unique up to scale, so if you impose a normalization condition at the identity point, you get a unique such function. Altogether, this gives a pairing between $(A \times A)^*$ and $A$, which in fact recovers the multiplication map $A \times A \to A$.

*Remark 5.5.9*  The section $s(A; \mathcal{L})$ satisfies three familiar properties:

- It is symmetric: Pulling back $\Theta^3 \mathcal{L}$ along a shuffle automorphism of $A^3$ yields $\Theta^3 \mathcal{L}$ again, and the pullback of the section $s(A; \mathcal{L})$ along this shuffle agrees with the original $s(A; \mathcal{L})$ across this identification.
- It is rigid: By restricting to $* \times A \times A$, the tensor factors in $\Theta^3 \mathcal{L}$ cancel out to give the trivial bundle over $A \times A$. The restriction of the section $s(A; \mathcal{L})$ to this pullback bundle agrees with the extension of the rigidifying section.
- It satisfies a 2-cocycle condition: In general, we define

$$\Theta^k \mathcal{L} := \bigotimes_{I \subseteq \{1,...,k\}} (p_I^*\mathcal{L})^{(-1)^{|I|-1}}.$$

---

[15] Mumford used this approach to say much more. The fixed representation is unchanging even as the abelian variety is varied, and hence the embeddings of the abelian varieties all target the same projective space. More than this, a given embedding is determined by the image of the identity point on the abelian variety (the "$\theta$-null point"), as the coefficients of the relevant generalizations of the Jacobi quartics can be deduced from it (see the formulas for $A$, $B$, and $C$), altogether giving an effective description of the *moduli* of abelian varieties by a generalization of the Jacobi relation ([Mum67a], [Mum67b]). Polishchuk gives an account of many of these ideas from an unusual and remarkable perspective [Pol03].

In fact, $\Theta^{k+1}\mathcal{L}$ can be written as a pullback of $\Theta^k\mathcal{L}$:

$$\Theta^{k+1}\mathcal{L} = \frac{(p_{12} \times 1)^*\mathcal{L}}{(p_1 \times 1)^*\mathcal{L} \otimes (p_2 \times 1)^*\mathcal{L}},$$

and pulling back a section $s$ along this map gives a new section

$$(\delta s)(x_0, x_1, \ldots, x_k) := \frac{s(x_0 +_A x_1, x_2, \ldots, x_k)}{s(x_0, x_2, \ldots, x_k) \cdot s(x_1, x_2, \ldots, x_k)}.$$

Performing this operation on the first and second factors yields the defining equation of a 2-cocycle.

*Remark 5.5.10* ([Bre83, section 4])  Breen presented a relative version of this part of the story that applies to arbitrary *commutative group schemes*, where the basic objects are a choice of line bundle $\mathcal{L}$ over a commutative group scheme $A$, a choice of trivialization of $\Theta^3\mathcal{L}$, and an epimorphism $\pi\colon A' \to A$ that trivializes $\mathcal{L}$.

Finally, we remark that the function

$$e\colon C^3(\widehat{\mathbb{G}}; \mathbb{G}_m) \to \underline{\mathrm{FormalGroups}}(\widehat{\mathbb{G}}^{\wedge 2}, \mathbb{G}_m)$$

considered in Lecture 5.3 also manifests in the theory of abelian varieties. Let $A$ be an abelian variety equipped with a line bundle $\mathcal{L}$. Suppose that $s$ is a symmetric, rigid section of $\Theta^3\mathcal{L}$, sometimes called a *cubical structure on $\mathcal{L}$*. Using the identification $(p_{12} - p_1 - p_2)^*\Theta^2\mathcal{L} = \Theta^3\mathcal{L}$, this induces the structure of a *symmetric biextension*[16] on $\Theta^2\mathcal{L}$, as depicted in Figure 5.3, via the multiplication maps

$$(\Theta^2\mathcal{L})_{x,y} \otimes (\Theta^2\mathcal{L})_{x',y} \to (\Theta^2\mathcal{L})_{x+x',y},$$
$$(\Theta^2\mathcal{L})_{x,y} \otimes (\Theta^2\mathcal{L})_{x,y'} \to (\Theta^2\mathcal{L})_{x,y+y'}.$$

**Definition 5.5.11**  There is a canonical piece of gluing data on this biextension, in the form of an isomorphism of pullback bundles

$$e_{p^j}\colon (p^j \times 1)^*\mathcal{L}|_{A[p^j] \times A[p^j]} \cong (1 \times p^j)^*\mathcal{L}|_{A[p^j] \times A[p^j]},$$

$$(\ell, x, y) \mapsto \left(\ell \cdot \prod_{k=1}^{p^j-1} \frac{s(x, [k]x, y)}{s(x, [k]y, y)}\right).$$

This function $e_{p^j}$ is called the *($p^j$)-th Weil pairing*.

---

[16] Biextensions themselves are a common tool in the theory of moduli of abelian varieties: A biextension on $A \times B$ is identical data to a map $A \to B^*$.

## 5.5 Modular Forms and $MU[6,\infty)$-Manifolds

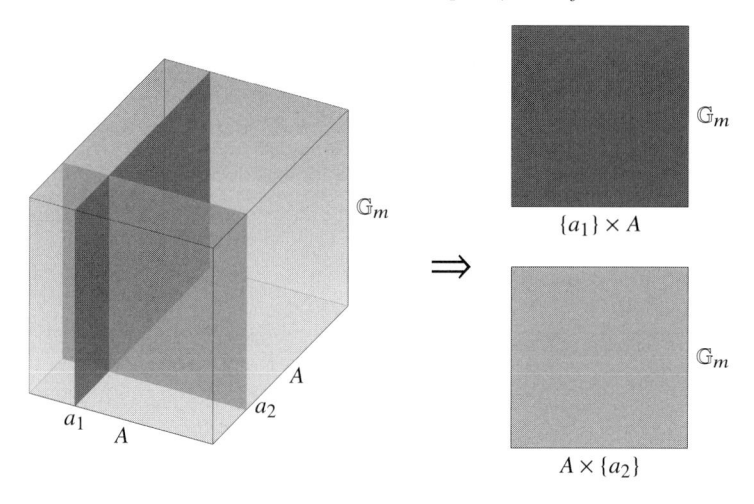

Figure 5.3 Extensions contained in a biextension.

*Remark 5.5.12* In the case that $A$ is an elliptic curve, this agrees with the usual definition of its "Weil pairing." In the case of a *complex* elliptic curve $\mathbb{C}/(1,\tau)$, this amounts to the assignment

$$\left(\frac{a}{n}, \frac{b}{n}\tau\right) \mapsto \exp\left(-2\pi i \frac{ab}{n}\right).$$

We now actually leverage this arithmetic geometry by placing ourselves in a situation where algebraic topology is directly linked to abelian varieties.

**Definition 5.5.13** An *elliptic spectrum* consists of a even–periodic ring spectrum $E$, an elliptic curve $C$ over Spec $E_0$, and a choice of isomorphism

$$\varphi\colon C_0^\wedge \xrightarrow{\cong} \mathbb{CP}_E^\infty.$$

A map of elliptic spectra consists of a map of ring spectra $f\colon E \to E'$ and an isomorphism of elliptic curves $\psi\colon f^*C \to C'$ compatible with $\varphi_E$ and $\varphi_{E'}$.[17]

*Remark 5.5.14* Our preference for *isomorphisms* of elliptic curves rather than general homomorphisms is prompted by our study of stable operations: We have seen that a morphism of complex-orientable ring spectra always gives rise to an isomorphism (after base-change) of their associated formal groups. In Appendix A we will develop a topological setting (with an attendant

---

[17] Elliptic curves and cohomology theories can be brought much closer together still, as in Lurie's framework [Lure, sections 4 and 5.3].

notion of context) which incorporates *isogenies* of elliptic curves in addition to isomorphisms.

Coupling Definition 5.5.13 to Corollary 5.4.13 and Theorem 5.5.7, we conclude the following:

**Corollary 5.5.15** *An elliptic spectrum* $(E, C, \varphi)$ *receives a canonical map of ring spectra*
$$MU[6, \infty) \to E.$$
*This map is natural in choice of elliptic spectrum: If* $(E, C, \varphi) \to (E', C', \varphi')$ *is a map of elliptic spectra, then the triangle*

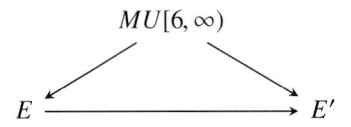

$$\begin{array}{ccc} & MU[6, \infty) & \\ \swarrow & & \searrow \\ E & \longrightarrow & E' \end{array}$$

*commutes.* □

*Example 5.5.16* Our basic example of an elliptic curve was $E_\Lambda = \mathbb{C}/\Lambda$, with $\Lambda$ a complex lattice. The projection $\mathbb{C} \to \mathbb{C}/\Lambda$ has a local inverse which defines an isomorphism of formal groups
$$\varphi \colon (E_\Lambda)_0^\wedge \xrightarrow{\cong} \widehat{\mathbb{G}}_a \times \operatorname{Spec} \mathbb{C},$$
as well as an isomorphism of cotangent spaces

$$\begin{array}{ccc} T^0(E_\Lambda)_0^\wedge & \xrightarrow{\cong} & \mathbb{C} \\ \uparrow \cong & & \| \\ T^0(\widehat{\mathbb{G}}_a \times \operatorname{Spec} \mathbb{C}) & \xrightarrow{\cong} & \mathbb{C}. \end{array}$$

Accordingly, we define an elliptic spectrum $HE_\Lambda P$ whose underlying ring spectrum is $H\mathbb{C}P$ and whose associated elliptic curve and isomorphism are $E_\Lambda$ and $\varphi$. This spectrum receives a natural map
$$MU[6, \infty) \to HE_\Lambda P,$$
which to a bordism class $M \in MU[6, \infty)_{2n}$ assigns an element
$$\Phi_\Lambda(M) \cdot u_\Lambda^n \in HE_\Lambda P_{2n},$$
where $u_\Lambda$ is the canonical section of
$$(\pi_2 HE_\Lambda P)^\sim = \omega_{\mathbb{C}P^\infty_{HE_\Lambda P}} = \omega_{(E_\Lambda)_0^\wedge}$$
and $\Phi_\Lambda(M) \in \mathbb{C}$ is some resulting complex number.

## 5.5 Modular Forms and $MU[6, \infty)$-Manifolds

*Example 5.5.17* The naturality of the $MU[6, \infty)$-orientation moves us to consider more than one elliptic spectrum at a time. If $\Lambda'$ is another lattice with $\Lambda' = \lambda \cdot \Lambda$, then the multiplication map $\lambda \colon \mathbb{C} \to \mathbb{C}$ descends to an isomorphism $E_\Lambda \to E_{\Lambda'}$ and hence a map of elliptic spectra $HE_{\Lambda'}P \to HE_\Lambda P$ acting by $u_{\Lambda'} \mapsto \lambda u_\Lambda$. The commuting triangle in Corollary 5.5.15 then begets the *modularity relation*

$$\Phi(M; \lambda \cdot \Lambda) = \lambda^{-n}\Phi(M; \Lambda).$$

*Example 5.5.18* This leads us to consider all curves $E_\Lambda$ simultaneously – or, equivalently, to consider modular forms. The lattice $\Lambda$ can be put into a standard form by picking a basis and scaling it so that one vector lies at 1 and the other vector lies in the upper half-plane, $\mathfrak{h}$. This gives a cover

$$\mathfrak{h} \to \mathcal{M}_{\text{ell}} \times \operatorname{Spec} \mathbb{C}$$

which is well-behaved (i.e., unramified) away from the special points i and $e^{2\pi i/6}$. A *complex modular form of weight n* is an analytic function $\mathfrak{h} \to \mathbb{C}$ which satisfies a certain decay condition and which is quasiperiodic for the action of $SL_2(\mathbb{Z})$, i.e.,[18]

$$f\left(M; \frac{a\tau + b}{c\tau + d}\right) = (c\tau + d)^n f(M; \tau).$$

Using these ideas, we construct a cohomology theory $HO_\mathfrak{h}P$, where $O_\mathfrak{h}$ is the ring of complex-analytic functions on the upper half-plane. The $\mathfrak{h}$-parameterized family of elliptic curves

$$\mathfrak{h} \times \mathbb{C}/(1, \tau) \to \mathfrak{h},$$

together with the logarithm, present $HO_\mathfrak{h}P$ as an elliptic spectrum $H\mathfrak{h}P$. The canonical map $\Phi \colon MU[6, \infty) \to H\mathfrak{h}P$ specializes at a point to give the functions $\Phi(-; \Lambda)$ considered above, and hence $\Phi(M) \in u^k \cdot O_\mathfrak{h}$ is itself a complex modular form of weight $k$.

In fact, this totalized map $\Phi$ is a ghost of Ochanine and Witten's modular genus from Theorem 3 of the Introduction, as a bordism class in $MU[6, \infty)_{2n}$ is, in particular, a bordism class in $MString_{2n}$. However, they know more about this function than we can presently see: For instance, they claim that it has an integral $q$-expansion. In terms of the modular form, its $q$-expansion is given by building the Taylor expansion "at $\infty$" (using that unspoken decay condition). In order to use our topological methods, it would be nice to have an elliptic spectrum embodying these $q$-expansions in the same way that $H\mathfrak{h}P$ embodied holomorphic functions, together with a comparison map that trades a modular

---
[18] That is, for the action of change of basis vectors.

form for its $q$-expansion. The main ideas leading to such a spectrum come from considering the behavior of $E_\Lambda$ as $\tau$ tends to $i \cdot \infty$.

**Definition 5.5.19** Note that as $\tau$ tends to $i \cdot \infty$, the parameter $q = \exp(2\pi i \tau)$ tends to 0. In the multiplicative model of $\mathcal{M}_{\text{ell}} \times \operatorname{Spec} \mathbb{C}$, we considered the punctured complex disk $D'$ with its associated family of elliptic curves

$$C'_{\text{an}} = \mathbb{C}^\times \times D'/(u, q) \sim (qu, q).$$

The fiber of $C'$ over a particular point $q \in D'$ is the curve $\mathbb{C}^\times/q^{\mathbb{Z}}$. The Weierstrass equations give an embedding of $C'_{\text{an}}$ into $D' \times \mathbb{C}P^2$ described by

$$y^2 w + xyw = x^3 + a_4 xw^2 + a_6 w^3$$

for certain modular forms $a_4$ and $a_6$, thought of as functions of $q$.[19] At $q = 0$, this curve collapses to the twisted cubic

$$y^2 w + xyw = x^3,$$

and over the whole open unit disk $D$ we call this extended family $C_{\text{an}}$.

Now let $A \subseteq \mathbb{Z}[\![q]\!]$ be the subring of power series which converge absolutely on the open unit disk. It turns out that the coefficients of the Weierstrass cubic lie in $A$, so it determines a generalized elliptic curve $C$ over $\operatorname{Spec} A$, and $C_{\text{an}}$ is the curve given by base-change from $A$ to the ring of holomorphic functions on $D$ [Mor89, section 5]. The *Tate curve* is the intermediate family $C_{\text{Tate}}$ over the intermediate base $D_{\text{Tate}} = \operatorname{Spec} \mathbb{Z}[\![q]\!]$, as base-changed from $A$.

The singular fiber at $q = 0$ prompts us to enlarge our notion of elliptic curve.

**Definition 5.5.20** ([AHS01, definitions B.1–B.2]) A *Weierstrass curve* is any curve of the form

$$C(a_1, a_2, a_3, a_4, a_6) := \left\{ [x : y : w] \in \mathbb{P}^2 \,\middle|\, \begin{array}{l} y^2 w + a_1 xyw + a_3 yw^2 = \\ x^3 + a_2 x^2 w + a_4 xw^2 + a_6 w^3 \end{array} \right\}.$$

A *generalized elliptic curve* over $S$ is a scheme $C$ equipped with maps

$$S \xrightarrow{0} C \xrightarrow{\pi} S$$

such that $C$ is Zariski-locally isomorphic to a system of Weierstrass curves (in a way preserving 0 and $\pi$).[20,21]

---

[19] The inquisitive reader can find expressions for $a_4$ and $a_6$ in terms of $G_4$ and $G_6$, as well as a more serious discussion of a Tate curve generally, in Silverman's text [Sil94, section VI.1].

[20] An elliptic curve in the usual sense turns out to be a generalized elliptic curve which is smooth, i.e., the discriminant of the Weierstrass equations is a unit.

[21] Unfortunately, "generalized elliptic curve" already means something in number theory, but Ando, Hopkins, and Strickland reused this phrase for this definition in their published article. In a

*Remark 5.5.21* ([AHS01, p. 670]) The singularities of a degenerate Weierstrass equation always occur outside of a formal neighborhood of the marked identity point, which in fact still carries the structure of a formal group. The formal group associated to the nodal cubic is the formal multiplicative group (indeed, the smooth locus of the nodal cubic *is the multiplicative group*), and the isomorphism making the identification extends a family of such isomorphisms $\varphi$ over the nonsingular part of the Tate curve.

**Definition 5.5.22** ([Mor89, section 5], [AHS01, section 2.7]) The generalized elliptic spectrum $K_{\text{Tate}}$, called *Tate K-theory*, has as its underlying spectrum $KU[\![q]\!]$. The associated generalized elliptic curve is $C_{\text{Tate}}$, and the isomorphism $\mathbb{CP}^\infty_{KU[\![q]\!]} \cong (C_{\text{Tate}})^\wedge_0$ is $\varphi$ from Remark 5.5.21.

The trade for the breadth of this definition is that theorems pulled from the study of abelian varieties have to be shown to extend uniquely to those generalized elliptic curves which are not smooth curves.

**Theorem 5.5.23** ([AHS01, propositions 2.57 and B.25]) *For a generalized elliptic curve C, there is a canonical[22] trivialization s of $\Theta^3 \mathcal{I}(0)$ which is compatible with change of base and with isomorphisms. If C is a smooth elliptic curve, then s agrees with that of Theorem 5.5.7.* □

**Corollary 5.5.24** *The trivializing section s associated to $C_{\text{Tate}}$ is given by $\delta^{\circ 3}\tilde{\theta}$, where $\tilde{\theta}_q$ is a slight modification of the classical $\theta$-function:*[23]

$$\tilde{\theta}_q(u) = (1-u)\prod_{n>0}(1-q^n u)(1-q^n u^{-1}), \quad \tilde{\theta}_q(qu) = -u^{-1}\tilde{\theta}_q(u).$$

*Proof* Even though $\tilde{\theta}$ is not a function on $C_{\text{Tate}}$ because of its quasiperiodicity, it does trivialize both $\pi^*\mathcal{I}(0)$ for $\pi\colon \mathbb{C}^\times \times D \to C_{\text{Tate}}$ and $\mathcal{I}(0)$ for $(C_{\text{Tate}})^\wedge_0$. Moreover, the quasiperiodicities in the factors in the formula defining $\delta^3\tilde{\theta}|_{(C_{\text{Tate}})^\wedge_0}$ cancel each other out, and the resulting function *does* descend to give a trivialization of $\Theta^3 \mathcal{I}(0)$. By the unicity and continuity clauses in Theorem 5.5.23, it must give a formula expressing $s$. □

---

number theorist's language, these are "stable curves of genus 1 with specified section in the smooth locus." No adjective other than "generalized" seems to be much better: singular, for instance, evokes the right idea but is also already taken.

[22] Canonical, with a unique continuous extension from the smooth bulk of the moduli of generalized elliptic curves, but *not* actually unique over the singular locus.

[23] Over $\mathbb{C}$, we can also identify this in terms of the Weierstrass $\sigma$-function: For any choice of $x, y \in C_\Lambda$, Abel's Theorem guarantees the existence of a rational function $f(x, y, z)$ with simple poles at $x$ and $y$ and simple roots at $-(x+y)$ and $0$. This function is unique up to scale, and in fact the fraction $\frac{f(x,y,z)}{f(x,y,0)}$ determines a cubical structure on $C_\Lambda$. Abbreviating $\sigma_\Lambda$ to just $\sigma$, Remark 5.5.6 gives a formula $f(x, y, z) = \frac{\sigma(x+y)\sigma(x+z)}{\sigma(x+y+z)\sigma(z)}$, from which it follows that $\frac{\sigma(x+y)\sigma(y+z)\sigma(z+x)\sigma(0)}{\sigma(x)\sigma(y)\sigma(z)\sigma(x+y+z)}$ gives a formula for the associated cubical structure.

**Definition 5.5.25** The induced map

$$\sigma_{\text{Tate}} : MU[6, \infty) \to K_{\text{Tate}}$$

is called the *complex $\sigma$-orientation*.

**Corollary 5.5.26** *Let $M \in \pi_{2n} MU[6, \infty)$ be a bordism class. The $q$-expansion of Witten's modular form $\Phi(M)$ has integral coefficients.*

*Proof* The span of elliptic spectra equipped with $MU[6, \infty)$-orientations

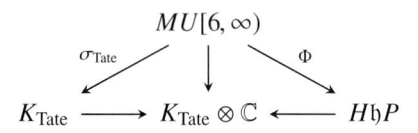

$$\begin{array}{ccccc}
 & & MU[6, \infty) & & \\
 & \overset{\sigma_{\text{Tate}}}{\swarrow} & \downarrow & \overset{\Phi}{\searrow} & \\
K_{\text{Tate}} & \longrightarrow & K_{\text{Tate}} \otimes \mathbb{C} & \longleftarrow & H\flat P
\end{array}$$

models $q$-expansion. The arrow $K_{\text{Tate}} \to K_{\text{Tate}} \otimes \mathbb{C}$ is injective on homotopy, which shows that the $q$-expansion of $\Phi(M)$ lands in the subring of integral power series. □

We can use the formula $\sigma_{\text{Tate}} = \delta^3 \tilde{\theta}$ appearing in Corollary 5.5.24 to explicitly understand the genus associated to $\sigma_{\text{Tate}}$ by passing to homotopy groups. To begin, the appearances of the map $\delta$ in Remarks 5.2.4 and 5.2.17 and Lemma 5.4.14 show that $\sigma_{\text{Tate}}$ belongs to the commutative triangle

$$MU[6, \infty) \xrightarrow{\delta} MSU \xrightarrow{\delta} MU \xrightarrow{\delta} MUP$$

$$\underset{\sigma_{\text{Tate}}}{\searrow} \quad\quad\quad\quad\quad\quad\quad\quad \downarrow \tilde{\theta}$$

$$KU[\![q]\!].$$

We will analyze this triangle by comparing $\tilde{\theta}$ to the usual $MUP$-orientation of $KU$, which selects the coordinate $f(u) = 1 - u$ on the formal completion of $\mathbb{G}_m = \operatorname{Spec} \mathbb{Z}[u^{\pm}]$. Appealing to Remark 5.2.4, the induced $MU$-orientation

$$MU \xrightarrow{\delta} MUP \xrightarrow{\text{Td}} KU$$

sends $f$ to the rigid section $\delta f$ of

$$\Theta^1 \mathcal{I}(0) = \mathcal{I}(0)_0 \otimes \mathcal{I}(0)^{-1} \cong \omega \otimes \mathcal{I}(0)$$

given by

$$\delta f = \frac{1}{1-u}\left(-\frac{du}{u}\right).$$

The difference between $\delta \operatorname{Td}$ and $\delta \tilde{\theta}$ is expressed by an element

$$\psi \in C^1(\widehat{C}_{\text{Tate}}; \mathbb{G}_m),$$

## 5.5 Modular Forms and $MU\langle 6, \infty\rangle$-Manifolds

given explicitly by the quotient formula

$$\psi = \left(\frac{\mathrm{Td}(1)}{\mathrm{Td}(u)}\right)^{-1} \cdot \frac{\tilde{\theta}_q(1)}{\tilde{\theta}_q(u)} = \prod_{n\geq 1} \frac{(1-q^n)^2}{(1-q^n u)(1-q^n u^{-1})}.$$

This gives a re-expression of $\delta\tilde{\theta}$ as the composite

$$\delta\tilde{\theta}\colon MU \xrightarrow{\eta_R} MU \wedge MU \simeq MU \wedge BU_+ \xrightarrow{\delta\,\mathrm{Td}\wedge\psi} K_{\mathrm{Tate}},$$

and hence its effect on a line bundle is determined by the evaluation of this characteristic series:

$$\psi(1-\mathcal{L}) = \prod_{n\geq 1} \frac{(1-q^n)^2}{(1-q^n\mathcal{L})(1-q^n\mathcal{L}^{-1})}.$$

Its effect on vector bundles in general is determined by the splitting principle and an exponential law, which after some computation [AHS01, section 2.7] gives the generic formula

$$\psi(\dim V \cdot 1 - V) = \bigotimes_{n\geq 1} \bigoplus_{j\geq 0} \mathrm{Sym}^j(\dim V \cdot 1 - V \otimes_{\mathbb{R}} \mathbb{C}) q^{jn} =: \bigotimes_{n\geq 1} \mathrm{Sym}_{q^n}(-\bar{V}_{\mathbb{C}}).$$

Finally, the map $(\eta_R)_*\colon MU_* \to \pi_*(MU \wedge \Sigma_+^\infty BU)$ sends a manifold $M$ with stable normal bundle $\nu$ to the pair $(M, \nu)$, so we at last compute

$$\sigma_{\mathrm{Tate}}(M \in \pi_{2n} MU\langle 6, \infty\rangle) = (\delta\,\mathrm{Td} \wedge \theta')(M, \nu)$$

$$=: \mathrm{Td}\left(M; \bigotimes_{n\geq 1} \mathrm{Sym}_{q^n}(\bar{\tau}_{\mathbb{C}})\right).$$

This is exactly Witten's formula for his genus, as applied to complex manifolds with the first two Chern classes trivialized.

*Remark 5.5.27* ([Reza, section 1.5]) Witten defines his characteristic series for *oriented* manifolds by the formula

$$K_{\mathrm{Witten}}(x) = \exp\left(\sum_{k=2}^{\infty} 2G_{2k}(\tau) \cdot \frac{x^{2k}}{(2k)!}\right)$$

$$= \frac{x/2}{\sinh(x/2)} \cdot \left(\prod_{n=1}^{\infty} \frac{(1-q^n)^2}{(1-q^n u)(1-q^n u^{-1})}\right) \cdot e^{-G_2(\tau)x^2},$$

where $G_{2k}$ is the $(2k)$-th Eisenstein series. Noting that $G_2$ is *not* a modular form, the condition that $p_1(M)/2$ vanish is precisely the condition that $G_2$ contribute nothing to the sum, so that the remainder *is* a modular form.

*Remark 5.5.28* ([AM01]) Another location where this series $\psi$ appears is in the theory of Tate $p$–divisible groups of Ando and Morava (and, in some sense, in the sense of Katz and Mazur [KM85, sections 8.7–8.8]). For a formal group $\widehat{\mathbb{G}}$ over Spf $R$ with group law $+_\varphi$, they consider the Weierstrass product

$$\Theta_\varphi(x;q) = x \cdot \prod_{\substack{k \in \mathbb{Z} \\ k \neq 0}} \frac{(x +_\varphi [k]_\varphi(q))}{[k]_\varphi(q)} \in R[\![x,q]\!],$$

which plays the role of a kind of $\theta$-function for $\widehat{\mathbb{G}}$. They also connect this series to the quotient function mapping to the group scheme $\widehat{\mathbb{G}}/q^{\mathbb{Z}}$ (in the sense of Definition A.2.22), which they further identify as the universal extension of $\widehat{\mathbb{G}}$ by a constant $p$-divisible group of dimension 1. In the presence of a global object $\mathbb{G}$ specializing to $\widehat{\mathbb{G}}$ (as in the case of $\widehat{\mathbb{G}}_a$ or $\widehat{\mathbb{G}}_m$), they show how to modify this product to an analytically convergent power series, recovering the characteristic series for the $\widehat{A}$-genus and the $L$-genus. The higher height analogs of these results seem very mysterious and very interesting.[24]

*Remark 5.5.29* ([AFG08]) Ando, French, and Ganter have given a construction that converts $MU[2k, \infty)$-orientations of a spectrum $E$ to $MU[2(k-1), \infty)$-orientations of the pro-spectrum $E^{\mathbb{CP}^\infty_{-\infty}}$. Performing this operation to the $\sigma$-orientation of $K^{\text{Tate}}$ gives the "two-variable Jacobi genus," which to $SU$-manifolds assigns certain classes in $\pi_0(K^{\text{Tate}})^{\mathbb{CP}^\infty_{-\infty}} = \mathbb{Z}[\![q]\!](\!(y)\!)$ connected to meromorphic Jacobi forms.

## 5.6 Chromatic *Spin* and *String* Orientations

We now turn to understanding $M\mathit{String}$-orientations in terms of $MU[6, \infty)$–orientations. We will approach this in successive passes, keeping the desired picture sharp the entire time but introducing generality slowly. Unfortunately, most of the original source material ([HAS99], [Stra]) for this has not been published, and hence it is difficult to give references. We have gone to some lengths to prepare the reader for the sorts of arguments that appear here: They are not so different from the arguments appearing elsewhere in this case study.

We begin at the maximally simple situation, where 2 is inverted. In this case, the Wood cofiber sequence

$$\Sigma kO \xrightarrow{\eta} kO \xrightarrow{c} kU \xrightarrow{\lambda} \Sigma^2 kO$$

---

[24] Morava is generally full of interesting ideas about genera – see, for instance, [Mor07b] and [Mor16].

## 5.6 Chromatic Spin and String Orientations

becomes split, since $\eta$ is a 2-torsion element. By considering different underlying infinite loopspaces, this gives a number of identifications of spaces as products:

$$
\begin{array}{lccc}
(-)_0: & BO \times \mathbb{Z} \longrightarrow & BU \times \mathbb{Z} \longrightarrow & kO_2, \\
(-)_2: & kO_2 \longrightarrow & BU \longrightarrow & kO_4, \\
(-)_4: & kO_4 \longrightarrow & BSU \longrightarrow & kO_6, \\
(-)_6: & kO_6 \longrightarrow & BU[6, \infty) \longrightarrow & B\mathrm{String}.
\end{array}
$$

Next, by recognizing $c \colon kO \to kU$ as the complexification map, we note that it lies in fixed points for complex-conjugation on $kU$. Having inverted 2, we avail ourselves of an idempotent selecting these fixed point subspectra:

$$P_+ = \frac{1+\xi}{2}, \qquad P_- = \frac{1-\xi}{2},$$

which reidentify the splitting $kU \simeq kO \vee \Sigma^2 kO$ in various equivalent ways:

$$kU \simeq kO \vee \Sigma^2 kO \simeq \operatorname{im} P_- \vee \operatorname{im} P_+ \simeq \ker P_+ \vee \ker P_- \simeq \left(\frac{kU}{\operatorname{im} P_+}\right) \vee \left(\frac{kU}{\operatorname{im} P_-}\right).$$

This last identification immediately gives access to the following result:

**Lemma 5.6.1** *For $E$ a complex-orientable ring spectrum with $1/2 \in \pi_0 E$, we have*

$$B\mathrm{String}_E = C_3(\mathbb{CP}_E^\infty)/([a,b,c] = -[-a,-b,-c]),$$

$$\operatorname{Spec} E_0 B\mathrm{String} = \left\{ f \in C^3(\mathbb{CP}_E^\infty; \mathbb{G}_m) \,\middle|\, f(a,b,c) = \frac{1}{f(-a,-b,-c)} \right\}$$

$$=: C_{\mathrm{is}}^3(\mathbb{CP}_E^\infty; \mathbb{G}_m),$$

*where $C_{\mathrm{is}}^3(\mathbb{CP}_E^\infty; \mathcal{I}(0))$ is the indicated subscheme of inverse-symmetric functions.*

*Proof sketch* This is a matter of translating the splittings above across Corollary 5.4.13. The complex-conjugation map $\xi$ acts on $C_k(\widehat{\mathbb{G}})$ according to

$$\xi[a_1, \ldots, a_n] = (-1)^n[-a_1, \ldots, -a_n],$$

which encodes $B\mathrm{String}_E$ as the quotient of $BU[6, \infty)_E$ by $\operatorname{im} P_-$. Finally, the claim about the homological scheme follows from the description of the cohomological formal scheme by Cartier duality. □

The repeated appearances of the terms in the above splittings suggest that the composite

$$\tau \colon kU \xrightarrow{\lambda} \Sigma^2 kO \xrightarrow{\Sigma^2 c} \Sigma^2 kU$$

of the maps in two adjacent cofiber sequences itself plays an interesting role. This is the map that encodes surjecting onto one factor in the preceding splitting, then reincluding it into the next splitting.

**Lemma 5.6.2** *At the level of formal schemes, the map $\tau$ acts by*

$$\tau \colon C_k(\widehat{\mathbb{G}}) \to C_{k+1}(\widehat{\mathbb{G}})$$
$$[a_1, \ldots, a_k] \mapsto [a_1, \ldots, a_k, -(a_1 + \cdots + a_k)].$$

*Proof* We are in pursuit of the following calculation in $kU^*(\mathbb{CP}^\infty)^{\times(k+1)}$ encoding complexification after decomplexification:

$$\beta^k x_1 \ldots x_k + \beta^k \overline{x_1} \cdots \overline{x_k} = \prod_j (1 - \mathcal{L}_j) + \prod_j (1 - \overline{\mathcal{L}}_j)$$
$$= \prod_j (1 - \mathcal{L}_j) + \prod_j \overline{\mathcal{L}}_j (\mathcal{L}_j - 1)$$
$$= \left(\prod_j (1 - \mathcal{L}_j)\right)\left(1 - (-1)^{k+1} \prod_j \overline{\mathcal{L}}_j\right)$$
$$= \beta^{k+1} x_1 \cdots x_k \cdot \xi^* \mu^*(x_1, \ldots, x_k). \qquad \square$$

Still in the case where $E$ is local away from 2, we can use this alternative description of the inclusion of the $P_-$ factor to deduce an alternative description of the formal scheme associated to $BString$:

**Corollary 5.6.3** *For $E$ a complex-orientable ring spectrum with $1/2 \in \pi_0 E$, we have*

$$BString_E = C_3(\mathbb{CP}^\infty_E)/([a, b, -a - b] = 0),$$
$$\operatorname{Spec} E_0 BString = \left\{ f \in C^3(\mathbb{CP}^\infty_E; \mathbb{G}_m) \,\middle|\, f(a, b, -a - b) = 1 \right\}. \qquad \square$$

**Definition 5.6.4** We denote this last functor by $\Sigma^3(\mathbb{CP}^\infty_E; \mathbb{G}_m)$, and more generally if $\mathcal{L}$ is a line bundle on $\mathbb{CP}^\infty_E$ then $\Sigma^3(\mathbb{CP}^\infty_E; \mathcal{L})$ will denote the subscheme of $C^3(\mathbb{CP}^\infty_E; \mathcal{L})$ of those trivializations which restrict to the rigidifying trivialization of $\Theta^3 \mathcal{L}$ on the subscheme of $(\mathbb{CP}^\infty)^{\times 3}$ specified by $[a, b, -a - b]$. Such trivializations are referred to as $\Sigma$-*structures on* $\mathcal{L}$, following Breen [Bre83, section 5].

**Corollary 5.6.5** *For $E$ a complex-orientable ring spectrum with $1/2 \in \pi_0 E$, the functor $\Sigma^3(\mathbb{CP}^\infty_E; \mathcal{I}(0))$ is isomorphic to the affine scheme* $\operatorname{Spec} E_0 MString$. *Moreover, the $E_0$-points of this scheme biject with ring spectrum maps*

$$MString \to E. \qquad \square$$

## 5.6 Chromatic Spin and String Orientations

*Remark 5.6.6* ([Hir95, section 26.1]) One of the most pleasant features of the case where 2 is inverted is that real orientations can not only be identified, but the projection idempotents mean that they can be *crafted* from complex orientations, and the idempotents have a recognizable effect on the traditional mechanisms for specifying orientations. For instance, the Conner–Floyd orientation on $K$-theory is specified by the classical logarithm, and the associated real orientation formed by projection to the positive idempotent has logarithm

$$\tanh^{-1}(x) = \frac{-\ln(1-x) + \ln(1+x)}{2},$$

i.e., the average of the logarithm and its conjugate. The resulting orientation is commonly referred to as the *Atiyah–Bott–Shapiro orientation*, whose associated genus is the $\widehat{A}$-*genus*. Similar techniques give access to connective real orientations associated to preexisting connective complex orientations.

We now turn to the *much* more complicated setting where 2 is not invertible, where we again aim to identify $MString$-orientations of a complex-orientable cohomology theory with certain $\Sigma$-structures. This is not possible in much generality, a situation which is hinted at by the analogous lower-order case: Complex-orientability is flatly *not enough* to conclude anything about whether a cohomology theory admits an $MO$-orientation, which we saw in Lemma 1.5.8 to be actually equivalent to admitting an $H\mathbb{F}_2$-algebra structure. In order to get a handle on the task in front of us, consider the following diagram of fiber sequences of infinite loopspaces:

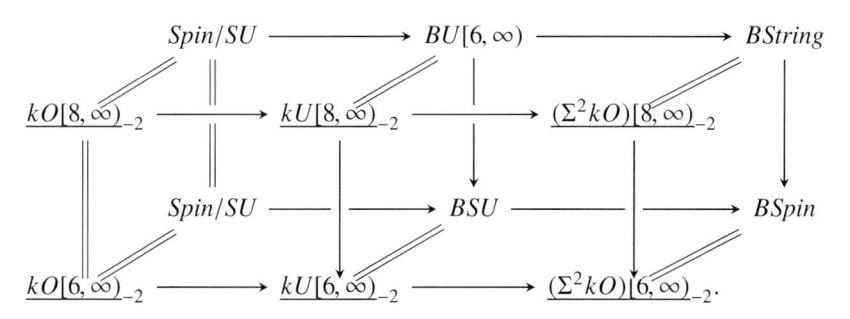

$$\begin{array}{ccccc}
& Spin/SU & \longrightarrow & BU[6,\infty) & \longrightarrow & BString \\
kO[8,\infty)_{-2} & \longrightarrow & kU[8,\infty)_{-2} & \longrightarrow & (\Sigma^2 kO)[8,\infty)_{-2} & \\
& Spin/SU & \longrightarrow & BSU & \longrightarrow & BSpin \\
kO[6,\infty)_{-2} & \longrightarrow & kU[6,\infty)_{-2} & \longrightarrow & (\Sigma^2 kO)[6,\infty)_{-2}.
\end{array}$$

Our program is to analyze the chromatic homology of $Spin/SU$ as well as the maps to $BU[6,\infty)$ and $BSU$. We hope to show that the scheme associated to $Spin/SU$ selects exactly the relations defining $\Sigma$-structures *and* that the map is flat, so that the associated bar spectral sequences

$$\operatorname{Tor}_{*,*}^{E_* Spin/SU}(E_*, E_* BSU) \Rightarrow E_* BSpin,$$
$$\operatorname{Tor}_{*,*}^{E_* Spin/SU}(E_*, E_* BU[6,\infty)) \Rightarrow E_* BString$$

collapse to give short exact sequences of Hopf algebras.

We embark on this project with an analysis of the natural bundles classified by the topological objects, so that we can guess the relevant algebraic model. We have the following complexification and decomplexification maps:

$$\begin{array}{ccc} & & kU_6 \\ & \overset{\lambda}{\nearrow} & \downarrow \delta \\ kU_4 \xrightarrow{j} & kO_6 \xrightarrow{i} & kU_4. \end{array}$$

**Lemma 5.6.7** *Let $E$ be a complex-orientable cohomology theory. The map $j$ induces an injection*

$$\operatorname{Spec} E_0(Spin/SU) \to \Sigma^2(\mathbb{CP}_E^\infty; \mathbb{G}_m).$$

*Proof, with some details omitted* Recall that our reinterpretation of Corollary 5.2.14 in Example 5.3.3 passed through the adjoint map $\widehat{\Pi}_2$ of the natural product bundle $\Pi_2 \colon (\mathbb{CP}^\infty)^{\times 2} \to BSU$. We extend the map $\Pi_2$ in two directions:

$$\mathbb{CP}^\infty \xrightarrow{(1-\mathcal{L})(1-\overline{\mathcal{L}})} \mathbb{CP}^\infty \times \mathbb{CP}^\infty \xrightarrow{\Pi_2} BSU \xrightarrow{j} Spin/SU.$$

Checking that this composite is zero on $E$-homology will give a factorization

$$\begin{array}{ccc} \operatorname{Spec} E_0 Spin/SU & \xrightarrow{E_0 j} & \operatorname{Spec} E_0 BSU \\ \vdots & & \parallel \\ \Sigma^2(\mathbb{CP}_E^\infty; \mathbb{G}_m) & \longrightarrow & C^2(\mathbb{CP}_E^\infty; \mathbb{G}_m), \end{array}$$

since $\Sigma^2(\mathbb{CP}_E^\infty; \mathbb{G}_m)$ is defined to be the closed subscheme of $C^2(\mathbb{CP}_E^\infty; \mathbb{G}_m)$ of those functions satisfying $f(x, -x) = 1$.

The ordinary homology of $Spin/SU$ is free and even, so it suffices to check that this map is null in the case $E = H\mathbb{Z}_{(2)}$, since one can then conclude the general case by the Atiyah–Hirzebruch spectral sequence. Manual calculation in a Serre spectral sequence gives the calculation

$$H\mathbb{Z}_{(2)*}(Spin/SU) \cong \Gamma_{\mathbb{Z}_{(2)}}[a_{2n+1} \mid n \geq 1]$$

as well as that the homological maps induced by $i$ and $j$ are respectively injective and surjective. In particular, we need only check that the above composite is zero on $H\mathbb{Z}_{(2)}$-homology after postcomposition with $H_*(i)$. The $SU$-bundle classified by postcomposition with $i$ is

$$(1 - \mathcal{L})(1 - \overline{\mathcal{L}}) - (1 - \overline{\mathcal{L}})(1 - \mathcal{L}),$$

which is itself trivial, hence the classifying map is null and so must be the map on homology.

## 5.6 Chromatic Spin and String Orientations

We are left with showing that the factorized map is an injection, and again appealing to Atiyah–Hirzebruch spectral sequences it suffices to show this in the case $E = H\mathbb{F}_2$. The diagram considered above extends as follows:

$$\begin{array}{ccccc} \operatorname{Spec} H\mathbb{F}_2 P_0 BSU & \xrightarrow{H\mathbb{F}_2 P_0 i} & \operatorname{Spec} H\mathbb{F}_2 P_0 Spin/SU & \xrightarrow{H\mathbb{F}_2 P_0 j} & \operatorname{Spec} H\mathbb{F}_2 P_0 BSU \\ \| & & \vdots & & \| \\ C^2(\widehat{\mathbb{G}}_a; \mathbb{G}_m) & \xrightarrow{\lambda_2} & \Sigma^2(\widehat{\mathbb{G}}_a; \mathbb{G}_m) & \longrightarrow & C^2(\widehat{\mathbb{G}}_a; \mathbb{G}_m), \end{array}$$

where

$$\lambda_2(f)\colon (x, y) \mapsto \frac{f(x, y)}{f(-_{\widehat{\mathbb{G}}} x, -_{\widehat{\mathbb{G}}} y)}.$$

Since the bottom-right map is definitionally injective and the outer rectangle commutes, it follows that the left-hand square commutes. The Serre spectral sequence calculation in integral homology indicated above shows that the top-right map is an injection, so the dotted comparison map is automatically an injection. □

**Lemma 5.6.8** *The above comparison map also belongs to a commuting diagram*

$$\begin{array}{ccccc} \operatorname{Spec} E_0 BU[6, \infty) & \longrightarrow & \operatorname{Spec} E_0 Spin/SU & \longrightarrow & \operatorname{Spec} E_0 BSU \\ \| & & \downarrow & & \downarrow \\ C^3(\mathbb{CP}_E^\infty; \mathbb{G}_m) & \xrightarrow{\lambda_3} & \Sigma^2(\mathbb{CP}_E^\infty; \mathbb{G}_m) & \longrightarrow & C^2(\mathbb{CP}_E^\infty; \mathbb{G}_m), \end{array}$$

*where*

$$\lambda_3(f)\colon (x, y) \mapsto f(x, y, -_{\widehat{\mathbb{G}}}(x +_{\widehat{\mathbb{G}}} y)).$$

*Proof* This is again a matter of checking that the outer rectangle commutes. □

This is as much as we can discern without delving into the analysis of the algebraic model. Our primary concern in that respect is to show that the algebraic map has the desired flatness property – we will actually quickly show that our comparison map is an isomorphism in our pursuit of this. Our main tool for addressing flatness is the following theorem, strongly related to the Milnor–Moore classification of Hopf algebras over a field of positive characteristic:

**Lemma 5.6.9** ([DG70, III.3.7], see also [Strb, example 12.10] and [LS17, section 2]) *Let $k$ be a field, and let $G$ and $H$ be group schemes over $\operatorname{Spec} k$. A map $f\colon G \to H$ of groups is faithfully flat if and only if for every test $k$-algebra*

$T$ and $T$-point $a \in H(T)$ there is a faithfully flat $T$-algebra $S$ and an $S$-point $b \in G(S)$ covering $a$. □

Based on this lemma, we see that if we had a sufficiently strong understanding of the possible points of $\Sigma^2(\mathbb{CP}_E^\infty; \mathbb{G}_m)$, we could manually check this condition by constructing preimages for these points. This becomes a manageable task after we note the following reductions:

1. Because of the equation $\lambda_2 = \delta \circ \lambda_3$ and because $\delta$ is surjective, checking that $\lambda_2$ is (faithfully) flat will automatically entail that $\lambda_3$ is (faithfully) flat.
2. We do not actually need *faithful* flatness to control the bar spectral sequence, but merely flatness. Accordingly, note that a map $f: G \to H$ of commutative affine group schemes is flat if and only if the map $f^\circ: G^\circ \to H^\circ$ on connected components of the identity is flat. Hence, we can reduce to checking (faithful) flatness on the identity component if necessary. In fact, because $E_0 BSU$ is polynomial, we have (Spec $E_0 BSU)^\circ$ = Spec $E_0 BSU$.
3. This last condition can be reduced further: A map of affine algebraic group schemes is flat if and only if the induced map on tangent spaces is surjective. Although $E_0 BSU$ is *infinite* polynomial, and hence not algebraic, this technicality can be overcome by means of the grading and the same condition applies.

We thus set about studying the points of $\Sigma^2(\widehat{\mathbb{G}}; \mathbb{G}_m)$ up to first order. As in Lecture 5.4, we begin with the special case $\widehat{\mathbb{G}} = \widehat{\mathbb{G}}_a$ and understand the generic case in terms of perturbation.

**Lemma 5.6.10** *The map*

$$\operatorname{Spec} \Gamma_{\mathbb{Z}_{(2)}}[a_{2n+1} \mid n \geq 1] \to \Sigma^2(\widehat{\mathbb{G}}_a; \mathbb{G}_m) \times \operatorname{Spec} \mathbb{Z}_{(2)}$$

*classifying the product*

$$\prod_{n \geq 1} \exp(a_{2n+1} c_{2n+1})$$

*is an isomorphism. In turn, there is an isomorphism*

$$\Sigma^2(\widehat{\mathbb{G}}_a; \mathbb{G}_m) \times \operatorname{Spec} \mathbb{F}_2 \cong \operatorname{Spec} \mathbb{F}_2 \left[ a_{2n+1}^{[2^j]} \bigg| n \geq 1, j \geq 0 \right] \bigg/ \left( \left( a_{2n+1}^{[2^j]} \right)^2 = 0 \right).$$

*Proof sketch*   This is a combination of three observations:

1. For $\widehat{\mathbb{G}}$ arbitrary, if $f \in \Sigma^2(\widehat{\mathbb{G}}; \mathbb{G}_m)$ expands in a coordinate to

$$1 + bc_{2^m} + \cdots,$$

then $b = 0$ as an element of $B[1/2]$ and of $B/2$.

## 5.6 Chromatic Spin and String Orientations

2. For any $f \in \Sigma^2(\widehat{\mathbb{G}}_a; \mathbb{G}_m)$ with expansion $1 + bc_n + \cdots$, $b^2 \equiv 0 \pmod 2$.
3. Modulo 2, the putative universal product series above decomposes further as

$$\prod_{n \geq 1} \left( \prod_{j \geq 0} (1 + a_{2n+1}^{[2^j]} \cdot c_{(2n+1)2^j}) \right).$$

**Corollary 5.6.11** *The injective comparison map*

$$\operatorname{Spec} E_0 \operatorname{Spin}/SU \to \Sigma^2(\mathbb{CP}_E^\infty; \mathbb{G}_m)$$

*of Lemma 5.6.7 is an isomorphism.*

*Proof* Just as in the proof of Lemma 5.6.7, a Serre spectral sequence calculation shows that the source and target of the factorization have the same Poincaré series, and we are done. □

From here, the work gets considerably more technical. In particular, we will only be able to conclude flatness in the case $\operatorname{ht} \widehat{\mathbb{G}} \leq 2$, and then only through explicit calculation. The results cleave into the following three pieces:

**Lemma 5.6.12** (Nonexistence Lemma) *Let $\widehat{\mathbb{G}}$ be a formal group of height 1 or 2 over an $\mathbb{F}_2$-algebra. For $f \in \Sigma^2(\widehat{\mathbb{G}}; \mathbb{G}_m)$ of the form $1 + ac_n + \cdots$, the following hold:*

1. *If $n = 2^m$ then $a = 0$.*
2. *If $n > 3$, $n$ is odd, and $n \neq 2^m - (2^d - 1)$, then $a = 0$.*
3. *If $d = 2$ and $n = 3$, then $a^2 - a = 0$.*

*Proof* We address the claims in turn.

1. This is a generic claim that does not rely on the height of $\widehat{\mathbb{G}}$. The 2-cocycle $c_{2^m}$ takes the form $c_{2^m}(x, y) = x^{2^{m-1}} y^{2^{m-1}}$, so that $f(x, -_{\widehat{\mathbb{G}}} x) = 1 + ax^{2^m} + \cdots$. The expected value is $f(x, -_{\widehat{\mathbb{G}}} x) = 1$, hence $a = 0$.[25]
2. This is again an explicit calculation of the leading term. Since $\widehat{\mathbb{G}}$ is of height $d$, it admits a coordinate where the negation series takes the form

$$[-1]_{\widehat{\mathbb{G}}}(x) = -x + cx^{2^d} + \cdots$$

for some $c$ not a zero divisor. In the case that $n$ is as described, one can then calculate

$$c_n(-_{\widehat{\mathbb{G}}} x, -_{\widehat{\mathbb{G}}} y) + c_n(x, y) = c_{n+(2^d-1)}(x, y) + \cdots,$$

---
[25] One can show by almost identical calculation that $a = 0$ in this case if 2 is invertible.

278  The σ-Orientation

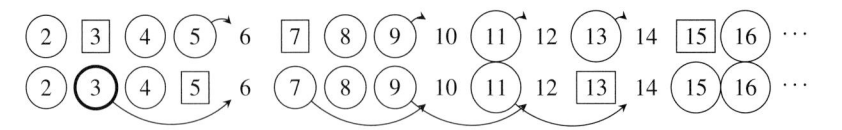

②③④⑤ 6 ☐7 ⑧⑨ 10 ⑪ 12 ⑬ 14 ☐15 ⑯ ···
②③④☐5 6 ⑦⑧⑨ 10 ⑪ 12 ☐13 14 ⑮ ⑯ ···

Figure 5.4 Application of the three lemmas in a small range. Circles denote prohibitions of $\Sigma^2$-structures by Lemma 5.6.12, arrows denote $\Sigma^2$-structures constructed by Lemma 5.6.13, and squares denote exceptional $\Sigma^2$-structures constructed by Lemma 5.6.14.

from which we make a trio of calculations:

$$\lambda_2 f = f^2, \qquad \text{(uses } f \in \Sigma^2(\widehat{\mathbb{G}}; \mathbb{G}_m)\text{)}$$
$$\lambda_2 f = 1 + ac_{n+(2^d-1)}(x,y) + \cdots, \qquad \text{(above observation)}$$
$$f^2 = 1 + a^2 c_{2n}(x,y) + \cdots. \qquad \text{(characteristic 2)}$$

For $n \geq 2^d$, this forces $a = 0$.

3. In the case $d = 2$ of the previous statement, this leaves one case open: $n = 3$. Equating the two sides then gives $a^2 = a$.  □

**Lemma 5.6.13** (Generic Existence Lemma) *Let $f = 1 + ac_n + \cdots \in C^2(\widehat{\mathbb{G}}; \mathbb{G}_m)$. If $2n+1 \geq 3$ is not of the form $2^r - (2^d - 1)$ for any $r$, then $\lambda_2 f = 1 + ac_{n+2^d-1} + \cdots$.*

*Proof sketch* This is also a consequence of the same manipulation of the negation series.  □

**Lemma 5.6.14** (Special Existence Lemma) *For $r \geq 0$ and $d \geq 1$, there exists a function $f \in C^2(\widehat{\mathbb{G}}; \mathbb{G}_m)$ with $\lambda_2 f = 1 + (a^2 + \varepsilon a + \delta)c_{2^r+(2^d-1)}$.*

*Indication of proof* This, finally, makes explicit use of height 1 and 2 formal group laws. The starting point is to set $g = \delta_{(\widehat{\mathbb{G}}_a, \mathbb{G}_m)} E_2(ax^{2^{r-1}+(2^d-1)})$ and then to carefully cancel low-order terms by multiplying in other known $\Sigma^2$-structures.  □

**Corollary 5.6.15** *For $\widehat{\mathbb{G}}$ a formal group over $\operatorname{Spec} \mathbb{F}_2$ with $\operatorname{ht} \widehat{\mathbb{G}} \leq 2$, the maps $\lambda_2$ and $\lambda_3$ are flat.*

*Proof* This is a culmination of the above calculations, summarized in Figure 5.4. As argued at the beginning of our algebraic analysis, it suffices to show just that $\lambda_2^\circ$ is faithfully flat to make the conclusion. Flatness follows by showing that the map is surjective on tangent spaces, which is exactly the content of Lemmas 5.6.12, 5.6.13, and 5.6.14.  □

### 5.6 Chromatic Spin and String Orientations

**Theorem 5.6.16** *Let $\widehat{\mathbb{G}}$ be a formal group over a perfect field of characteristic 2, and assume $\operatorname{ht}\widehat{\mathbb{G}} \leq 2$. For $E$ the associated Morava K-theory or E-theory, there are isomorphisms*

$$\operatorname{Spec} E_0 MString \xrightarrow{\cong} \Sigma^3(\mathbb{CP}_E^\infty; \mathcal{I}(0)), \quad \operatorname{Spec} E_0 MSpin \xrightarrow{\cong} C_{\text{is}}^2(\mathbb{CP}_E^\infty; \mathcal{I}(0)).$$

*The $E_0$-points of these schemes biject with ring spectrum maps $MString \to E$ and $MSpin \to E$ respectively.*[26]  □

*Proof* Our goal all along was to use the algebraic model to govern the bar spectral sequences

$$\operatorname{Tor}_{*,*}^{E_*Spin/SU}(E_*, E_*BSU) \Rightarrow E_*BSpin,$$
$$\operatorname{Tor}_{*,*}^{E_*Spin/SU}(E_*, E_*BU[6,\infty)) \Rightarrow E_*BString.$$

Our conclusion from Corollary 5.6.15 is that they are both concentrated on the 0-line and isomorphic to the respective Hopf algebra quotients

$$E_0 BSU /\!\!/ (E_0 Spin/SU) \cong E_0 BSpin,$$
$$E_0 BU[6,\infty) /\!\!/ (E_0 Spin/SU) \cong E_0 BString.$$

Our algebraic models furthermore explicitly identify these quotients on homological schemes as those subschemes on which the stated algebraic identities hold. The final claim of the theorem follows from even-ness and the universal coefficient spectral sequence.  □

**Corollary 5.6.17** *If $E$ is a finite height Morava K- or E-theory considered as an elliptic spectrum, the complex $\sigma$-orientation of Corollary 5.5.15 lifts uniquely to a ring map $MString \to E$.*

*Proof* An *immediate* consequence of the canonical cubical structure associated to an abelian variety is that it automatically satisfies the extra condition required to be a $\Sigma^3$-structure.  □

*Remark 5.6.18* ([Hop95, theorem 7.2]) More generally, the $\sigma$-orientation associated to an elliptic spectrum which either has $1/2 \in \pi_0 E$ or which has $\pi_* E$ torsion-free is supposed to lift through $MString$. Flatness in this setting is supposed to be approached via a fiber-by-fiber criterion, but the grading tricks used above are less visibly helpful in lifting the classical algebraic result to the periodic one we require to make this work. Rather than pursue this, we will give a sketch of the construction of the *String*-orientation of *tmf* in Appendix A.4, which automatically gives this much stronger result.

---

[26] This is sufficient to conclude the existence of the Atiyah–Bott–Shapiro orientation in the homotopy category.

*Remark 5.6.19* ([KLW04], [Stra]) The cohomology formal schemes of a number of other infinite loopspaces related to real $K$-theory admit reasonable descriptions, often even independent of height. A routinely useful result in this arena is due to Yagita [Yag80, lemma 2.1]: For $k_\Gamma$ the connective cover of the Morava $K$-theory $K_\Gamma$, in the Atiyah–Hirzebruch spectral sequence

$$E_2^{*,*} = Hk^*X \otimes_k k_\Gamma^* \Rightarrow k_\Gamma^* X,$$

the differentials are given by

$$d_r(x) = \begin{cases} 0 & \text{if } r \leq 2(p^d - 1), \\ \lambda Q_d x \otimes v_d & \text{if } r = 2(p^d - 1) + 1 \end{cases}$$

where $\lambda \neq 0$ and $Q_d$ is the $d$th Milnor primitive. For instance, this shows that $K_*BO$ decomposes as

$$K_*BO \cong K_*[b_2, b_4, b_{2^{d+1}-2}] \underset{K_*[b_{2j}^2 \mid j<2^d]}{\otimes} K_*[b_{2j}^2].$$

This, coupled to a theorem governing the result of the double bar spectral sequence, powers most of the results of Kitchloo, Laures, and Wilson [KLW04, section 4]. Their stronger results on the connective covers of $BO \times \mathbb{Z}$ are summarized in Figure 5.5. The remaining formal scheme $BString_K$, our prized object, is harder to access by these means: The sequence

$$(\underline{HS^1}_2)_K \to BString_K \to BSpin_K$$

is exact in the middle, but neither left- nor right-exact [KLW04, p. 234], causing significant headache. Satisfyingly, their methods also tell us that our alternative analysis fails at higher heights: The formal scheme $Spin/SU_K$ contains $\coprod_{j \geq 3}(\underline{HS^1[2]}_j)_K$ as a subscheme, and it accounts for the kernel of the map to $BSpin_K$. Once $\operatorname{ht} \Gamma \geq 3$ is satisfied, this kernel is nonzero.

*Remark 5.6.20* There are variations on this analysis that remain to be sorted out. For instance, there is a naturally occurring orientation $MU \to MSO$, whose associated genus factors through the quotient of the Lazard ring by the relation $[-1](x) = -x$. This factored map is injective and it becomes an isomorphism after inverting 2 – but, without inverting 2, $MSO_*$ itself is populated by plenty of 2-torsion. It would be nice to have available an interpretation of such real orientations in terms of formal group laws.

*Remark 5.6.21* A considerable number of related results in this area are due to Gerd Laures, ranging from splitting bordism spectra ([Lau03], see also the work of Hopkins-Hovey [HH92]), to results on the $\beta$-family in terms of modular

## 5.6 Chromatic Spin and String Orientations

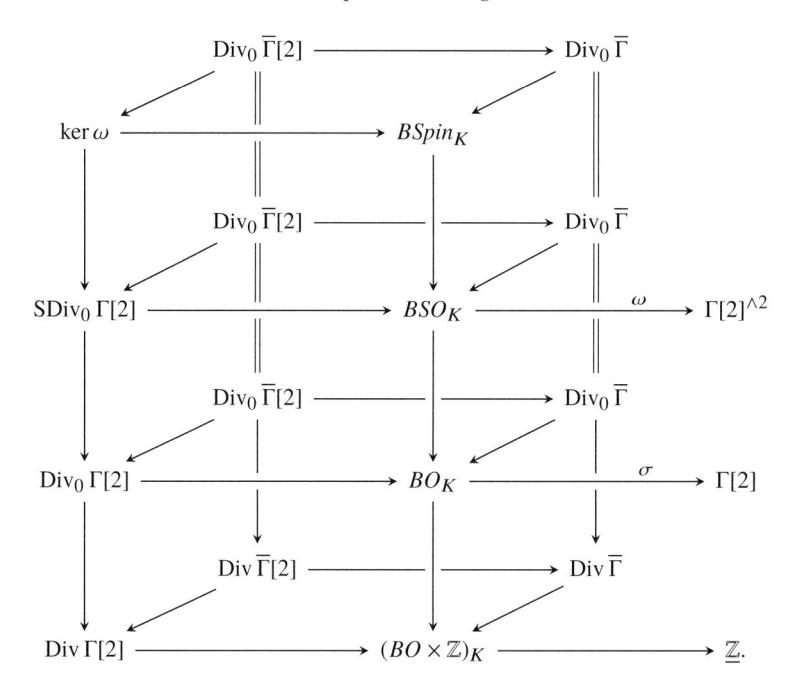

Figure 5.5 Presentations of the Morava $K$-theoretic cohomological formal schemes associated to connective covers of $BO \times \mathbb{Z}$. Each level in the prism is a bi-Cartesian square in the category of formal group schemes, and each level is the fiber of the maps off of the previous level to the Postnikov section. The formal curve $\overline{\Gamma}$ is $\mathbb{HP}_K^\infty$, which is the fiber of $\sigma \colon \mathrm{Div}_2^+ \Gamma \to \Gamma$, and $\overline{\Gamma}[2]$ is the fiber of the lift of the endo-isogeny $2 \colon \Gamma \to \Gamma$ to $\overline{\Gamma}$. The map $\sigma$ vanishes on $\mathrm{Div}_0 \overline{\Gamma}$ and acts by summation on $\mathrm{Div}_0 \Gamma[2]$. The map $\omega$ vanishes on $\mathrm{Div}_0 \overline{\Gamma}$ and acts on $\mathrm{SDiv}_0 \Gamma[2] \cong C_2 \Gamma[2]$ by $[a, b] \mapsto a \wedge b$.

forms [BL09], to $K(1)$-local versions of the results of this section [KL02], and onward.

# Appendix A

## Power Operations

**Convention: We will write $E$ for $E_\Gamma$, $\Gamma$ a fixed formal group of finite height, for the entirety of this appendix.**

Our goal in this appendix is to give a tour of the interaction of the $\sigma$-orientation with a topic of modern research, the theory of $E_\infty$-ring spectra, in a manner consistent with the rest of the topics in this book. Because the theory of $E_\infty$-ring spectra (in particular, their algebraic geometry) is still very much developing, we have no hope of stating results in their maximum strength or giving a completely clear picture – as of this writing, the maximum strength is unknown and the picture is still resolving. Although $E_\infty$-ring spectra themselves were introduced decades ago, we will even avoid giving a proper definition of them here, instead referring to the original work of May and collaborators [EKMM97] and the more recent work of Lurie [Lurc, chapter 7] for a proper treatment. In acknowledgment of this underwhelming level of rigor, we have downgraded our discussion from a case study to an appendix.[1]

As far as we are concerned, $E_\infty$-ring spectra arise in order to solve the following problem: Given two ring spectra $R$ and $S$ in the homotopy category, the set of homotopy classes of ring maps RingSpectra($R, S$) forms a subset of the set of all homotopy classes $[R, S] = \pi_0$Spectra($R, S$), selected by a homomorphism condition. There is no meaningful way to enrich this to a *space* of ring spectrum maps from $R$ to $S$, which inhibits us from understanding an obstruction theory for ring spectra, i.e., approximating $R$ by "nearby" ring spectra $R'$ in a way that relates RingSpectra($R, S$) and RingSpectra($R', S$) by a fiber sequence.

The extra data that accomplish this mapping space feat turn out to be an explicit naming of the homotopies controlling the associativity and commutativity of the

---

[1] Additionally, the contents of this appendix did not make it into the classroom version of this course, though the latter two sections did appear as contributed talks to graduate student seminars.

ring spectrum multiplication, which are subject to highly intricate compatibility conditions.[2,3] Again, rather than spell this out, it suffices for our purposes to say that there is such a notion of a structured ring spectrum that begets a mapping space between two such. We also record the following conglomerate theorem as an indication that this program overlaps with the one we have been describing already:

**Theorem A.0.1**  *The following are examples of $E_\infty$-ring spectra:*

- *([May77, section VIII.1]) The classical K-theories $KU$ and $KO$.*
- *([May77, section VIII.1]) The Eilenberg–Mac Lane spectra $HR$.*
- *([GH04, corollary 7.6–7.7]) The Morava E-theories $E_\Gamma$ and their fixed point spectra.*[4]
- *([May77, section IV.3]) The Thom spectra arising from the J-homomorphism, including $MO$, $MSO$, $MSpin$, $MString$, $MU$, $MSU$, and $MU[6,\infty)$.*
- *([Beh14], see Appendix A.3) The spectra TMF, Tmf, and tmf.* □

The forgetful map from $E_\infty$-rings down to ring spectra in the homotopy category factors through an intermediate category, that of $H_\infty$-ring spectra, which captures the extra factorizations expressing these associativity and commutativity relations. Specifically, recall the following definition from the discussion in Lecture 2.4:

**Definition A.0.2** ([BMMS86, definition I.3.1], see Lecture 2.4)  An $H_\infty$-ring spectrum is a ring spectrum $E$ equipped with factorizations $\mu_n$ as in

$$\begin{array}{ccc} E^{\wedge n} & \xrightarrow{\mu} & E \\ \downarrow & \nearrow_{\mu_n} & \\ E^{\wedge n}_{h\Sigma_n}, & & \end{array}$$

which are subject to compatibilities induced by the inclusions $\Sigma_n \times \Sigma_m \subseteq \Sigma_{n+m}$ and the inclusions $\Sigma_n \wr \Sigma_m \subseteq \Sigma_{nm}$.

**Lemma A.0.3**  *Each $E_\infty$-ring spectrum gives rise to an $H_\infty$-ring spectrum in the homotopy category.* □

---

[2] This is rather analogous to the extra data required on a *space*, beyond just a multiplication, which allows one to use the bar construction to assemble a delooping.

[3] The high degree of intricacy accomplishes this goal of constructing mapping spaces, but it interacts strangely with the classical notion of a ring spectrum in the homotopy category: There are ring spectrum maps that admit *no* enrichment to an $E_\infty$ map, and there are ring spectrum maps that admit *multiple* enrichments to $E_\infty$ maps.

[4] Notably, the Morava $K$-theories are *not* $E_\infty$-rings at finite heights, in view of Remark 2.4.19.

We care about this secondary definition because our results thus far have all concerned the cohomology of spaces, which is, at its core, a calculation at the level of *homotopy classes*. This is therefore as much of the $E_\infty$-structure as one could hope would interact with our analyses in the preceding case studies.

In Lecture 2.4 we introduced an $H_\infty$-ring structure on $MU$, and in Lectures 2.5 and 2.6 we used it to make a calculation of the coefficient ring $MU_*$. Our first goal in this appendix is to introduce an $H_\infty$-ring structure on certain chromatically interesting spectra, including Morava's theories $E_\Gamma$, and to describe the compatibility laws arising from intertwining these two $H_\infty$-structures. The culminating result is as follows:

**Theorem A.0.4** (see Corollary A.2.35) *An orientation $MU[6, \infty) \to E_\Gamma$ is $H_\infty$ if and only if the induced cubical structure is "norm-coherent" (see Definition A.2.30).* □

Before addressing this, we discuss in Appendix A.1 an important phenomenon: After deleting certain forms of torsion, the Morava $E_\Gamma$-homology of a finite spectrum can be well-approximated by its Morava $E_{\Gamma'}$-homology where $\Gamma'$ satisfies ht $\Gamma'$ < ht $\Gamma$. This is interesting in its own right, and we will see that the approximation bears directly on the study of power operations.

Second, with the homotopy category exposed, we turn to $E_\infty$-ring spectra themselves: In Appendix A.3 we introduce the spectrum *TMF*, which is the primary target of the *String*-orientation; and in Appendix A.4 indicate the construction of an $E_\infty$ orientation of *TMF*, providing full details in the simpler case of the construction of the *Spin*-orientation of $KO$.

## A.1 Rational Chromatic Phenomena (Nathaniel Stapleton)

Before embarking on the ambitious project of understanding the collection of $H_\infty$-ring maps $\varphi \colon MU[6, \infty) \to E_\Gamma$, it would be instructive to be able to answer the following: If we had such an orientation, what information could we extract from it? The main feature of such a map is that it belongs to a commutative diagram

$$\begin{array}{ccc} MU[6,\infty)^0(X) & \xrightarrow{\varphi} & E^0(X) \\ \downarrow{P^{\Sigma_n}_{MU[6,\infty)}(X)} & & \downarrow{P^{\Sigma_n}_E(X)} \\ MU[6,\infty)^0(B\Sigma_n \times X) & \xrightarrow{P^{\Sigma_n}(\varphi)} & E^0(B\Sigma_n \times X). \end{array}$$

Before understanding the maps in this commutative diagram, we must understand its nodes. There is a natural map

$$E^0(B\Sigma_n) \otimes_{E^0} E^0(X) \to E^0(B\Sigma_n \times X),$$

and whether it is an equivalence is highly dependent on the structure of $E^0(B\Sigma_n)$ itself – for instance, this would be the case if $E^0(B\Sigma_n)$ were flat as an $E^0$-module. For this reason, we are strongly motivated to feel out any foothold on this object that we are able to find, and a critical such foothold is the chromatic character theory of Hopkins, Kuhn, and Ravenel [HKR00]. Their theorems describe the Morava $E$-cohomology of a finite group with certain forms of torsion (e.g., $p$-torsion) deleted, and we give an introduction to their results in this lecture.[5]

## The $E_\Gamma$-Cohomology of Finite Abelian Groups

Let $\Gamma$ be a height $d$ formal group over $k$, a perfect field of characteristic $p$. By Theorem 3.4.5, there is a noncanonical isomorphism

$$O_{(\mathcal{M}_{\mathrm{fg}})_\Gamma^\wedge} \cong W(k)[\![u_1, \ldots, u_{d-1}]\!]$$

and this ring carries the universal deformation of $\Gamma$, which we denote as $\widehat{\mathbb{G}}$. By Definition 3.5.4, there is an associated chromatic cohomology theory

$$E = E_\Gamma$$

called height $d$ Lubin–Tate theory or Morava $E$-theory such that

$$\mathbb{CP}_E^\infty = BS_E^1 = \mathrm{Spf}(E^0(BS^1)) = \widehat{\mathbb{G}}.$$

Fixing a coordinate on $\widehat{\mathbb{G}}$ provides us with a formal group law

$$x +_{\widehat{\mathbb{G}}} y.$$

**Proposition A.1.1** ([HKR00, proposition 5.2 and lemma 5.7], see Example 5.1.12) *Let $C_m = S^1[m]$. There is an isomorphism (depending on the chosen coordinate)*

$$E^*(BC_m) \cong E^*[\![x]\!]/([m]_{\widehat{\mathbb{G}}}(x))$$

*of $E^*$-algebras. Moreover, $E^*(BC_m)$ is free as an $E^*$-module of rank $p^{kd}$ where $p^k | m$ and $p^{k+1} \nmid m$.*[6] □

---

[5] The intrepid reader who finds this material interesting should also look at: the paper by Morava [Mor12], which discusses much of the same material here in slightly different terms; the paper by Greenlees and Strickland [GS99], which discuss a "transchromatic" version of these results, where they study the object $i_d^* j_{d*}(\mathcal{M}_{\mathrm{fg}})_\Gamma^\wedge$ of Theorem 3.5.1; and the paper by Stapleton [Sta13], which streamlines and generalizes the results of Greenlees and Strickland by invoking the geometry of $p$-divisible groups.

[6] To prove this result it is most natural to set $|x| = 2$, but for the purposes of this lecture (and, indeed, this textbook) it is most natural to set $|x| = 0$. We can move back and forth between these choices by the change of coordinates $x \leftrightarrow ux$, where $u$ is a generator of $\pi_2 E$.

### A.1 Rational Chromatic Phenomena (Nathaniel Stapleton) 287

**Corollary A.1.2** ([HKR00, corollary 5.10]) *Let*

$$A \cong C_{m_1} \times \cdots \times C_{m_i}$$

*be a finite abelian group. There is then an isomorphism of $E^*$-algebras*

$$E^*(BA) \cong E^*[\![x_1, \ldots, x_i]\!]/([m_1]_{\widehat{\mathbb{G}}}(x_1), \ldots, [m_i]_{\widehat{\mathbb{G}}}(x_i)). \qquad \square$$

If $A \cong C_{p^{k_1}} \times \cdots \times C_{p^{k_i}}$ is an abelian $p$-group, it follows that $E^*(BA)$ is a free $E^*$-module of rank $p^{k_1 d + \cdots + k_i d}$. If $m$ is prime to $p$, then $[m]_{\widehat{\mathbb{G}}}(x)/x$ is a unit in $E^*[\![x]\!]$. This implies that the rank of $E^*(BA)$ only depends on the Sylow $p$-subgroup of $A$. Even more, it implies that the inclusion of the Sylow $p$-subgroup of $A$ into $A$ induces an isomorphism after applying $E$-cohomology. We will now always assume that $A$ is an abelian $p$-group.

The most computable of chromatic cohomology theories is periodic rational cohomology. For a $\mathbb{Q}$-algebra $R$, the formal group associated to $HRP$ is the (height 0) additive formal group. It is natural to want to gain information regarding $E^*(X)$ (for any chromatic cohomology theory $E$) by comparing $E$-cohomology with periodic rational cohomology.[7] For instance, this is the purpose of the classical Chern character: For $X$ a space, the Chern character is a map of commutative rings

$$KU^0(X) \to H\mathbb{Q}P^0(X)$$

induced by the map of spectra[8]

$$KU \to H\mathbb{Q} \wedge KU \simeq H\mathbb{Q}P.$$

For $X$ a finite $CW$ complex, it induces an isomorphism

$$\mathbb{Q} \otimes KU^0(X) \cong H\mathbb{Q}P^0(X).$$

When $X$ is a finite CW complex, there is an isomorphism

$$\mathbb{Q} \otimes E^*(X) \cong H(\mathbb{Q} \otimes E^0)P^*(X).$$

This follows from the fact that $\mathbb{Q} \otimes E^*$ is a flat $E^*$-algebra. When $X$ is not a finite CW complex, Proposition A.1.2 gives a non-example. It implies that

$$\mathbb{Q} \otimes E^*(BA)$$

---

[7] We have employed a similar-sounding strategy elsewhere in this text, where we have approximated an arbitrary complex-orientable theory by $H\mathbb{F}_p P$, using the fact that the infinite-height group $\widehat{\mathbb{G}}_a \otimes \mathbb{F}_p$ is a dense point in $\mathcal{M}_{\mathbf{fg}} \times \mathrm{Spec}\,\mathbb{Z}_{(p)}$. However, the height 0 point $\widehat{\mathbb{G}}_a \otimes \mathbb{Q}$ is *not* dense, and so our intention here is much more delicate.

[8] We could replace $\mathbb{Q}$ with $\mathbb{R}$ or $\mathbb{C}$ and the result would still hold.

is free as a $\mathbb{Q} \otimes E^*$-module of rank $p^{k_1 d + \ldots + k_i d}$. However,

$$H(\mathbb{Q} \otimes E^0) P^*(BA) \cong \mathbb{Q} \otimes E^*$$

is free of rank 1, a direct consequence of Machke's theorem.

To explain one of the inputs into the main theorem of this lecture, we need one more notion. For $G$ a finite group, a $G$–CW complex $X$ is a $G$-space built inductively by attaching "cells" of the form $(G/H) \times D^n$, as in the following pushout diagram:

$$\begin{array}{ccc} \coprod_{i \in I} (G/H_i \times S^{n-1}) & \longrightarrow & \coprod_{i \in I} (G/H_i \times D^n) \\ \downarrow & & \downarrow \\ X^{(n-1)} & \longrightarrow & X^{(n)}. \end{array}$$

We say that a $G$–CW complex is finite if it can be built out of finitely many cells of the form $G/H \times D^n$.

**Theorem A.1.3** (Equivariant CW Approximation [May96, theorem I.3.6])
*Every G-space is weakly equivalent to a G–CW complex.* □

Given a cohomology theory $E$, an equivariant version of the cohomology theory can be formed by setting

$$E_G^*(X) = E^*(X_{hG}),$$

where $X$ is a $G$-space and

$$X_{hG} \simeq EG \times_G X = (EG \times X)/G$$

is the homotopy orbits for the $G$-action on $X$. The resulting equivariant cohomology theory is called *Borel equivariant E-theory*.

The purpose of this case study is to explain, in the case of $E = E_\Gamma$, how to extend the above formula

$$\mathbb{Q} \otimes E^*(X) \cong H(\mathbb{Q} \otimes E^0) P^*(X),$$

originally stated for finite complexes $X$, to an analogous formula for finite $G$–CW complexes. Said another way, for each finite group, we want to approximate Borel equivariant $E$-cohomology (restricted to finite $G$–CW complexes) by some version of Borel equivariant rational cohomology. We call this "the approximation problem." In particular, this encompasses our original goal of understanding $E^*(B\Sigma_n)$, as it appears as a special case of Borel equivariant cohomology as in the formula

$$E^*(B\Sigma_n) = E_{\Sigma_n}^* ((G/G) \times *).$$

## Formal Geometry

Recall from Definition 4.6.5 that associated to a finite-height formal group is a connected $p$-divisible group built out of the $p^k$-torsion of the formal group as $k$ varies. Associated to the universal deformation $\widehat{\mathbb{G}}$ is the following $p$-divisible group of height $d$

$$\widehat{\mathbb{G}}[p] \hookrightarrow \widehat{\mathbb{G}}[p^2] \hookrightarrow \ldots,$$

where the maps are the inclusions of the $p^k$-torsion into the $p^{k+1}$-torsion. Proposition A.1.1 implies first

$$B(C_{p^k})_E \cong \widehat{\mathbb{G}}[p^k],$$

and that the $p$-divisible group associated to $\widehat{\mathbb{G}}$ can be formed using the inclusions $C_{p^k} = S^1[p^k] \hookrightarrow C_{p^{k+1}} = S^1[p^{k+1}]$. Proposition A.1.1 also implies that all of the schemes in this system are finite flat commutative groups schemes over Spf $E^0$. The $p$-divisible group is called "height $d$" because $\widehat{\mathbb{G}}[p]$ is a finite group scheme of order $p^d$ or, equivalently, since $E^0(BC_p)$ has rank $p^d$ as a free module over $E^0$.

There are $p$-divisible groups that are *not* built from formal groups, and the simplest of these plays a role in this story. We define the *constant p-divisible group of height 1* to be

$$C_{p^\infty} = S^1[p^\infty] = \text{colim}\,(C_p \hookrightarrow C_{p^2} \hookrightarrow \ldots).$$

It is called *constant* because it is made up of constant group schemes, and it is called height 1 because $|S^1[p]| = p^1$. The constant $p$-divisible group of height $d$ is then the $d$-fold product $C_{p^\infty}^{\times d} = (S^1)^{\times d}[p^\infty]$. The constant height $d$ $p$-divisible group $C_{p^\infty}^{\times d}$ is the unique height $d$ $p$-divisible group $\widehat{\mathbb{G}}[p^\infty]$ with the property that

$$O_{\widehat{\mathbb{G}}[p^k]} \cong \prod_{C_{p^k}^{\times d}} R,$$

where $R$ is the base ring.[9]

More generally, a similar description of $BA_E$ can be given using formal

---

[9] An unimportant aside: A generic $p$-divisible group contains a maximal formal subgroup, and the quotient by this subgroup is an *étale group* – and, whatever these are, they (and the attendant extension problems) form the "new" components of the theory of $p$-divisible groups. In the case of a ground field of positive characteristic, étale groups admit a pleasant characterization: They are exactly those that become isomorphic to the constant group after base-change to the algebraic closure, and so in this sense the constant $p$-divisible groups of different ranks play a role similar to the Honda formal group laws of Theorem 3.4.1.

geometry when $A$ is a finite abelian group. Let $A^*$ be the Pontryagin dual of $A$ and let

$$\text{\underline{FormalGroups}}(A^*, \widehat{\mathbb{G}})$$

be the formal scheme that associates to a complete local $E^0$-algebra $R$

$$\text{\underline{FormalGroups}}(A^*, \widehat{\mathbb{G}})(R) = \text{AbelianGroups}(A^*, \widehat{\mathbb{G}}(R)).$$

For large enough $k$, we have $A = A[p^k]$, and from this it follows that any map from $A^*$ to $\widehat{\mathbb{G}}(R)$ must land in the $p^k$-torsion $\widehat{\mathbb{G}}[p^k](R)$. There is thus an isomorphism of formal schemes

$$\text{\underline{FormalGroups}}(A^*, \widehat{\mathbb{G}}) \cong \text{\underline{FormalGroups}}(A^*, \widehat{\mathbb{G}}[p^k]).$$

**Proposition A.1.4** ([HKR00, proposition 5.12])  *There is a canonical isomorphism of formal schemes*

$$\text{Spf}(E^0(BA)) \cong \text{\underline{FormalGroups}}(A^*, \widehat{\mathbb{G}}),$$

*natural in maps of finite abelian groups.* □

The formal scheme $\text{\underline{FormalGroups}}(A^*, \widehat{\mathbb{G}})$ is finite and flat over $\text{Spf}(E^0)$. Because of this, this formal scheme can be viewed as a scheme – really a scheme, and not a formal scheme! – over $\text{Spec}(E^0)$. Without going into the technical details of formal to informal constructions, the main idea is that the topology on a complete local $E^0$-algebra is not important to $\widehat{\mathbb{G}}[p^k]$: For a continuous $E^0$-algebra $R$, there is a bijection

$$\text{Algebras}^{\text{cts}}_{E^0}(E^0(BC_p), R) \cong \text{Algebras}_{E^0}(E^0(BC_p), R).$$

The same cannot be said for $E^0(BS^1)$. In the case of our *finite* abelian group $A$, we may thus restrict our attention to the scheme

$$\text{\underline{FormalGroups}}(A^*, \widehat{\mathbb{G}}[p^k]) \colon \text{Algebras}_{E^0} \to \text{Sets}$$

sending an $E^0$-algebra $R$ to

$$\text{AbelianGroups}(A^*, \widehat{\mathbb{G}}[p^k](R)) \cong \text{GroupSchemes}(A^*, \text{Spec}(R) \underset{\text{Spec}(E^0)}{\times} \widehat{\mathbb{G}}[p^k]).$$

Since the ring map $E^0 \to \mathbb{Q} \otimes E^0$ is not continuous, this discussion is important to phrasing an algebro-geometric version of the approximation problem explained at the end of the previous section.

When $G = A$ and $X = *$, the approximation problem can now be given an algebro-geometric description: We would like to find a $(\mathbb{Q} \otimes E^0)$-algebra $R$ such that

$$\text{Spec}(R) \times \text{\underline{FormalGroups}}(A^*, \widehat{\mathbb{G}}[p^k]) \cong \text{\underline{FormalGroups}}(A^*, \text{Spec}(R) \times \widehat{\mathbb{G}}[p^k])$$

## A.1 Rational Chromatic Phenomena (Nathaniel Stapleton)

is recognizable as $\mathrm{Spec}(-)$ of the $HRP$-cohomology of a space. This would only solve the approximation problem for $E^0(BA)$ – but, astoundingly, it will turn out that this is enough.

There is a surprisingly simple answer to the approximation problem in this case. Fix[10] $\Lambda = \mathbb{Z}_p^d$ and let $\Lambda_k = \Lambda/p^k\Lambda$, so that there are noncanonical isomorphisms

$$\Lambda^* \cong (\mathbb{Q}/\mathbb{Z}_{(p)})^d \cong (S^1[p^\infty])^d$$

as well as canonical isomorphisms

$$\Lambda^* \cong (S^1[p^\infty])^d, \qquad \Lambda^*[p^k] \cong (\Lambda_k)^* \cong \mathrm{Groups}(\Lambda_k, C_{p^k}).$$

If we could find an $R$, dependent on $\widehat{\mathbb{G}}$, such that

$$\mathrm{Spec}(R) \times_{\mathrm{Spec}(E^0)} \widehat{\mathbb{G}}[p^k] \cong \Lambda_k^*$$

as abelian group schemes, then we would be done, since

$$O_{\Lambda_k^*} \cong \prod_{\Lambda_k^*} R \cong HRP^0\left(\bigsqcup_{\Lambda_k^*} *\right)$$

and

$$\mathrm{Spec}(R) \times_{\mathrm{Spec}(E^0)} \underline{\mathrm{FormalGroups}}(A^*, \widehat{\mathbb{G}}[p^k]) \cong \underline{\mathrm{FormalGroups}}(A^*, \Lambda_k^*)$$
$$\cong A^{\times d}.$$

That is, we would be able to give the simplest answer imaginable: We will have approximated $E^0(BA)$ by the $HRP$-cohomology of a collection of points. In the next section we construct a $\mathbb{Q} \otimes E^0$-algebra $C_0$ with this property.

### The Ring $C_0$

The goal of this section is to construct a $(\mathbb{Q} \otimes E^0)$-algebra $C_0$ such that

$$\mathrm{Spec}(C_0) \times_{\mathrm{Spec}(E^0)} \widehat{\mathbb{G}}[p^k] \cong \Lambda_k^*$$

for each $k > 0$. That is, there should be an isomorphism of $p$-divisible groups

$$\mathrm{Spec}(C_0) \times_{\mathrm{Spec}(E^0)} \widehat{\mathbb{G}}[p^\infty] \cong \Lambda^*.$$

---

[10] This choice of $\Lambda$ is the only choice made in this lecture, and the last subsection of this lecture addresses how natural the character map is in automorphisms of $\Lambda$. From a certain perspective, the results of Barthel and Stapleton [BS17] can be viewed as an analysis of how natural the character map is among finite index maps $\Lambda \hookrightarrow \Lambda$.

This does not uniquely specify $C_0$, but we can fix this deficiency by asking that $C_0$ carry the universal isomorphism of $p$-divisible groups

$$u \colon \Lambda^* \xrightarrow{\cong} \mathrm{Spec}(C_0) \times_{\mathrm{Spec}(E^0)} \widehat{\mathbb{G}}[p^\infty].$$

This means that, given any $E^0$-algebra $R$ and an isomorphism of $p$-divisible groups

$$f \colon \Lambda^* \xrightarrow{\cong} \mathrm{Spec}(R) \times_{\mathrm{Spec}(E^0)} \widehat{\mathbb{G}}[p^\infty],$$

there is a unique map of $E^0$-algebras

$$C_0 \xrightarrow{u_f} R$$

such that $f = \mathrm{Spec}(R) \times_{\mathrm{Spec}(C_0)} u$, where the pullback is along $u_f$.

The idea of the construction of $C_0$ is quite simple. Proposition A.1.4 implies

$$E^0(B\Lambda_k) \cong O(\underline{\mathrm{FormalGroups}}(\Lambda_k^*, \widehat{\mathbb{G}}[p^k])).$$

To construct $C_0$, it suffices to understand the open subscheme of the mapping scheme $\underline{\mathrm{FormalGroups}}(\Lambda_k^*, \widehat{\mathbb{G}}[p^k])$, which consists of the homomorphisms that are isomorphisms.

Since $E^0(B\Lambda_k)$ corepresents $\underline{\mathrm{FormalGroups}}(\Lambda_k^*, \widehat{\mathbb{G}}[p^k])$, it carries the universal homomorphism of group schemes

$$u \colon \Lambda_k^* \to \widehat{\mathbb{G}}[p^k].$$

Thus the identity map $E^0(B\Lambda_k) \xrightarrow{1} E^0(B\Lambda_k)$ corresponds to a homomorphism of abelian groups

$$\Lambda_k^* \to \widehat{\mathbb{G}}[p^k](E^0(B\Lambda_k))$$

or equivalently a map of commutative group schemes

$$u \colon \Lambda_k^* \to \mathrm{Spec}(E^0(B\Lambda_k)) \times_{\mathrm{Spec}(E^0)} \widehat{\mathbb{G}}[p^k].$$

This map is a consequence of the exponential adjunction for finite flat commutative group schemes: The identity map

$$1 \colon \underline{\mathrm{FormalGroups}}(\Lambda_k^*, \widehat{\mathbb{G}}[p^k]) \to \underline{\mathrm{FormalGroups}}(\Lambda_k^*, \widehat{\mathbb{G}}[p^k])$$

is adjoint to the evaluation map

$$\mathrm{ev} \colon \Lambda_k^* \times \underline{\mathrm{FormalGroups}}(\Lambda_k^*, \widehat{\mathbb{G}}[p^k]) \to \widehat{\mathbb{G}}[p^k].$$

Given $\alpha \in \Lambda_k^*$, the naturality of Proposition A.1.4 implies that the behavior of

$$\underline{\mathrm{FormalGroups}}(\Lambda_k^*, \widehat{\mathbb{G}}[p^k]) \to \widehat{\mathbb{G}}[p^k]$$

### A.1 Rational Chromatic Phenomena (Nathaniel Stapleton)

on rings of functions is precisely

$$E^0(B\alpha)\colon E^0(BC_{p^k}) \to E^0(B\Lambda_k).$$

Using the fixed coordinate $x$, this is equivalent to a map of $E^0(B\Lambda_k)$-algebras

$$u^*\colon E^0(B\Lambda_k)[\![x]\!]/[p^k]_{\widehat{\mathbb{G}}}(x) \to \prod_{\Lambda_k^*} E^0(B\Lambda_k).$$

Fixing a set of generators for $\Lambda_k^*$, gives an isomorphism

$$E^0(B\Lambda_k) \cong E^0[\![x_1,\ldots,x_d]\!]/([p^k]_{\widehat{\mathbb{G}}}(x_1),\ldots,[p^k]_{\widehat{\mathbb{G}}}(x_d)),$$

and the behavior of $u^*$ on the factor corresponding to $(i_1,\ldots,i_d) \in \Lambda_k^*$ is then

$$u^*\colon x \mapsto [i_1]_{\widehat{\mathbb{G}}}(x_1) +_{\widehat{\mathbb{G}}} \ldots +_{\widehat{\mathbb{G}}} [i_k]_{\widehat{\mathbb{G}}}(x_d).$$

By Corollary A.1.2, these are free $E^0(B\Lambda_k)$-modules of the same rank. To force the map to be an isomorphism, it suffices to invert the determinant. Let

$$S_k = \mathrm{im}\,(\Lambda_k^* \setminus 0 \to \widehat{\mathbb{G}}[p^k](E^0(B\Lambda_k)))$$
$$= \{[i_1]_{\widehat{\mathbb{G}}}(x_1) +_{\widehat{\mathbb{G}}} \ldots +_{\widehat{\mathbb{G}}} [i_d]_{\widehat{\mathbb{G}}}(x_d)|0 \neq (i_1,\ldots,i_d) \in \Lambda_k^*\}.$$

**Lemma A.1.5** *Inverting the determinant of $u^*$ is equivalent to inverting $S_k$.*

*Proof* For $\bar{i} = (i_1,\ldots,i_d) \in \Lambda_k^*$, set

$$[\bar{i}](\bar{x}) = [i_1]_{\widehat{\mathbb{G}}}(x_1) +_{\widehat{\mathbb{G}}} \ldots +_{\widehat{\mathbb{G}}} [i_d]_{\widehat{\mathbb{G}}}(x_d).$$

The matrix defining $u^*$ is given by the Vandermonde matrix

$$\begin{pmatrix} 1 & [\bar{j}_1](\bar{x}) & [\bar{j}_1](\bar{x})^2 & \ldots & [\bar{j}_1](\bar{x})^{p^k-1} \\ 1 & [\bar{j}_2](\bar{x}) & [\bar{j}_2](\bar{x})^2 & \ldots & [\bar{j}_2](\bar{x})^{p^k-1} \\ \vdots & \vdots & \vdots & \ddots & \vdots \\ 1 & [\bar{j}_{p^k}](\bar{x}) & [\bar{j}_{p^k}](\bar{x})^2 & \ldots & [\bar{j}_{p^k}](\bar{x})^{p^k-1} \end{pmatrix},$$

where $\{\bar{j}_1,\ldots,\bar{j}_{p^k}\} = \Lambda_k^*$. The determinant of this matrix is

$$\prod_{1 \leq s < t \leq p^k} ([\bar{j}_t](\bar{x}) - [\bar{j}_s](\bar{x})).$$

It is a basic fact of subtraction for formal group laws that

$$x -_{\widehat{\mathbb{G}}} y = v(x-y)$$

for a unit $v$. Thus the determinant is, up to multiplication by a unit,

$$\prod_{1 \le s < t \le p^k} ([\bar{j}_t](\bar{x}) -_{\widehat{\mathbb{G}}} [\bar{j}_s](\bar{x})) = \prod_{1 \le s < t \le p^k} ([\bar{j}_t - \bar{j}_s](\bar{x}))$$

$$= \prod_{0 \ne \bar{j} \in \Lambda_k^*} [\bar{j}](\bar{x})^{p^k - 1}.$$

□

**Corollary A.1.6** *Inverting $S_k$ in $E^0(B\Lambda_k)$ inverts $p$.* □

Set $C_{0,k} = S_k^{-1} E^0(B\Lambda_k)$. The set $S_k$ consists largely of zero divisors. Because of this, we should be concerned that $C_{0,k}$ is the zero ring. The following result of Hopkins, Kuhn, and Ravenel saves the day:

**Proposition A.1.7** ([HKR00, proposition 6.5]) *The ring $C_{0,k}$ is a faithfully flat $\mathbb{Q} \otimes E^0$-algebra.* □

By construction, the $E^0$-algebra $C_{0,k}$ carries the universal isomorphism

$$\Lambda_k^* \xrightarrow{\cong} \widehat{\mathbb{G}}[p^k].$$

It follows that the colimit

$$C_0 = \operatorname*{colim}_k C_{0,k}$$

carries the universal isomorphism of $p$-divisible groups

$$\Lambda^* \xrightarrow{\cong} C_0 \otimes \widehat{\mathbb{G}}.$$

This is because a map $C_0 \to R$ is a compatible system of maps $C_{0,k} \to R$ and the map $C_{0,k} \to C_{0,k+1}$ restricts the universal isomorphism

$$\Lambda_{k+1}^* \xrightarrow{\cong} \widehat{\mathbb{G}}[p^{k+1}]$$

to the $p^k$-torsion.

*Example A.1.8* Let us work out $C_0$ in the case of $p$-complete $K$-theory $E_{\widehat{\mathbb{G}}_m} = KU_p$. In this case

$$\widehat{\mathbb{G}} = \widehat{\mathbb{G}}_m,$$

the formal multiplicative group over $\mathbb{Z}_p$. Because $\widehat{\mathbb{G}}_m$ is the formal completion of the multiplicative group scheme $\mathbb{G}_m$ and the base ring is $p$-complete, we have

$$\widehat{\mathbb{G}}_m[p^k] \cong \mathbb{G}_m[p^k].$$

With the standard coordinate, this isomorphism is given by

$$\mathbb{Z}_p[x]/([p^k]_{\widehat{\mathbb{G}}_m}(x)) \cong \mathbb{Z}_p[x]/(x^{p^k} - 1)$$

$$x \mapsto x - 1,$$

## A.1 Rational Chromatic Phenomena (Nathaniel Stapleton)

from which we calculate

$$E^0(B\Lambda_k) = KU_p^0(B\mathbb{Z}/p^k) \cong \mathbb{Z}_p[x]/([p^k]_{\widehat{\mathbb{G}}_m}(x)) \cong \mathbb{Z}_p[x]/(x^{p^k} - 1).$$

We have

$$S_k = \{x, [2]_{\widehat{\mathbb{G}}_m}(x), \ldots, [p^k - 1]_{\widehat{\mathbb{G}}_m}(x)\},$$

which is

$$\{x - 1, x^2 - 1, \ldots, x^{p^k - 1} - 1\}$$

after applying the isomorphism. There is the factorization

$$x^{p^k} - 1 = \prod_{i=1,\ldots,k} \Phi_{p^i}(x),$$

where $\Phi_{p^i}(x)$ is the $p^i$th cyclotomic polynomial. But

$$x^{p^{k-1}} - 1 = \prod_{i=1,\ldots,(k-1)} \Phi_{p^i}(x)$$

and this is one of the elements in $S_k$. Clearly it is a zero divisor in $KU_p^0(B\mathbb{Z}/p^k)$, thus

$$S_k^{-1} KU_p^0(B\mathbb{Z}/p^k) \cong S_k^{-1}\mathbb{Z}_p[x]/(\Phi_{p^k}(x)).$$

We can also see that inverting the elements of $S_k$ inverts $p$. It suffices to show that the ideal generated by $p$ in $S_k^{-1}\mathbb{Z}_p[x]/(\Phi_{p^k}(x))$ is the whole ring. Recall the identity of cyclotomic polynomials

$$\Phi_{p^k}(x) = \Phi_p(x^{p^{k-1}}).$$

Thus we have

$$(S_k^{-1}\mathbb{Z}_p[x]/(\Phi_{p^k}(x)))/(p) \cong S_k^{-1}\mathbb{F}_p[x]/(\Phi_p(x^{p^{k-1}})) \cong S_k^{-1}\mathbb{F}_p[x]/(\Phi_p(x)^{p^{k-1}}),$$

but $\Phi_p(x) \in S_k$, so this is the zero ring.

Now that we have argued that inverting $S_k$ inverts $p$, we see that we have a canonical map

$$\mathbb{Q}_p \otimes_{\mathbb{Z}_p} \mathbb{Z}_p[x]/(x^{p^k} - 1) \to S_k^{-1}\mathbb{Z}_p[x]/(x^{p^k} - 1)$$

and this factors through the quotient map

$$\mathbb{Q}_p[x]/(x^{p^k} - 1) \to \mathbb{Q}_p[x]/(\Phi_{p^k}(x)).$$

However, because $\mathbb{Q}_p[x]/(\Phi_{p^k}(x))$ is a field, the canonical map

$$\mathbb{Q}_p[x]/(\Phi_{p^k}(x)) \xrightarrow{\cong} (x^{p^{k-1}} - 1)^{-1}\mathbb{Q}_p[x]/(x^{p^k} - 1)$$

is an isomorphism. Thus $C_{0,k}$ is just $\mathbb{Q}_p$ adjoin a primitive $p^k$ th root of unity and

$$C_0 = \underset{k}{\operatorname{colim}}\, C_{0,k} \cong \underset{k}{\operatorname{colim}}\, \mathbb{Q}_p(\zeta_{p^k})$$

is $\mathbb{Q}_p$ with all $p$-power roots of unity adjoined – a totally ramified extension of $\mathbb{Q}_p$ with Galois group $\mathbb{Z}_p^\times$.

## The Inertia Groupoid

Recall our goal of approximating $E^*(EG \times_G X)$ by the $HC_0P$-cohomology of some space determined by $EG \times_G X$. Stated more precisely, we'd like to find a space $F(EG \times_G X)$ and a map of cohomology theories

$$E^*(EG \times_G X) \to HC_0P^*(F(EG \times_G X))$$

such that the induced map

$$C_0 \otimes_{E^0} E^*(EG \times_G X) \to HC_0P^*(F(EG \times_G X))$$

is an isomorphism. The purpose of this section is to understand the operation $F(-)$.

One condition that we have already agreed on is that $HC_0P^0(F(BC_{p^k}))$ must be canonically isomorphic to $HC_0P^0(\Lambda_k^*) \cong \prod_{\Lambda_k^*} C_0$. Thus there is a natural first guess for what $F(-)$ might be: We could set

$$F(EG \times_G X) = \underline{\operatorname{Spaces}}(B\Lambda, EG \times_G X).$$

When $G = C_{p^k}$ and $X = *$, there is an equivalence

$$\underline{\operatorname{Spaces}}(B\Lambda, BC_{p^k}) \simeq \coprod_{\Lambda_k^*} BC_{p^k}.$$

Since $C_0$ is a rational algebra (so $|G|$ is invertible in $C_0$), we have

$$HC_0P^0(BG) \cong C_0$$

for any finite group $G$. Thus

$$HC_0P^0\left(\coprod_{\Lambda_k^*} BC_{p^k}\right) \cong \prod_{\Lambda_k^*} C_0,$$

which is what we wanted.

However, there is a problem with this construction: The functor $\underline{\operatorname{Spaces}}(B\Lambda, -)$ does not send finite complexes to finite complexes and does not preserve homotopy colimits, as can be seen by applying it to the homotopy pushout diagram

### A.1 Rational Chromatic Phenomena (Nathaniel Stapleton) 297

$$
\begin{array}{ccc}
* \sqcup * & \longrightarrow & * \\
\downarrow & & \downarrow \\
* & \longrightarrow & S^1,
\end{array}
$$

thus $HC_0 P^*(F(-))$ is not a cohomology theory. It turns out that the *inertia groupoid functor* fixes both of these problems at once, and we describe it now.

**Definition A.1.9** For a finite group $G$, let

$$G_p^d = \underline{\mathrm{TopologicalGroups}}(\Lambda, G)$$

denote the set of continuous homomorphisms from $\Lambda$ to $G$.

Choosing a basis for $\Lambda$ gives an isomorphism to the set of $d$-tuples of commuting $p$-power order elements of $G$:

$$G_p^d \cong \{(g_1, \ldots, g_d) \,|\, [g_i, g_j] = e,\ g_i^{p^k} = e \text{ for } k \text{ large enough}\}.$$

The group $G$ acts on this set by conjugation and we will write $G_p^d/G$ for the quotient by this action.

**Definition A.1.10** For a finite $G$–CW complex $X$, let

$$\mathrm{Fix}_d(X) = \coprod_{\alpha \in G_p^d} X^{\mathrm{im}\,\alpha},$$

where $X^{\mathrm{im}\,\alpha}$ is the fixed points of $X$ with respect to the image of $\alpha$.

This is a finite $G$–CW complex by intertwining the action of $G$ on $G_p^d$ by conjugation and the action of $G$ on $X$. To be precise, for $x \in X^{\mathrm{im}\,\alpha}$, we let $gx \in X^{\mathrm{im}\,g\alpha g^{-1}}$.

The $G$-space $\mathrm{Fix}_d(X)$ is also known as the $d$-fold inertia groupoid of the $G$-space $X$.[11] That is, the definition above can be made for all finite groups uniformly in the category of topological groupoids. Let $X/\!/G$ be the action topological groupoid associated to the finite $G$–CW complex $X$. This is the groupoid object in topological spaces with object space $X$ and morphism space $X \times G$. The source and target maps

$$X \times G \rightrightarrows X,$$

are the projection and action maps, respectively.

The category of topological groupoids is easy to enrich in spaces, the mapping space between two topological groupoids is just the subspace of the product

---

[11] This is really a $p$-adic form of the inertia groupoid since $\Lambda = \mathbb{Z}_p^{\times d}$ and not $\mathbb{Z}^{\times d}$.

of the space of maps between the objects and the space of maps between the morphisms that are maps of topological groupoids.

This construction can be upgraded further to an enrichment in topological groupoids so that the object space is the mapping space described above. Let $[\underline{1}] := (0 \to 1)$ be the free-standing isomorphism and let $[\underline{0}]$ be the category with one object and only the identity morphism. There are two functors

$$[\underline{0}] \to [\underline{1}]$$

picking out 0 and 1. The internal mapping topological groupoid between two action groupoids $W /\!\!/ F$ and $X /\!\!/ G$, TopologicalGroupoids$(W /\!\!/ F, X /\!\!/ G)$, can be described in the following way: The objects are the space of maps from $W /\!\!/ F \times [\underline{0}] \cong W /\!\!/ F$ to $X /\!\!/ G$ and morphisms are the space of maps from $W /\!\!/ F \times [\underline{1}]$ to $X /\!\!/ G$. The structure maps are induced by the functors $[\underline{0}] \rightrightarrows [\underline{1}]$ and the cocomposition map

$$[\underline{1}] \to [\underline{1}] \sqcup_{[\underline{0}]} [\underline{1}].$$

**Lemma A.1.11** *Let $*/\!\!/\Lambda$ be the topological groupoid with a single object and with $\Lambda$ automorphisms. There is a natural isomorphism*

$$\text{TopologicalGroupoids}(*/\!\!/\Lambda, X/\!\!/G) \cong \text{Fix}_d(X)/\!\!/G.$$

*Proof* We indicate the proof. We only need to do this for $n = 1$ as higher $n$ follow by adjunction. It also suffices to replace $*/\!\!/\mathbb{Z}_p$ by $*/\!\!/\mathbb{Z}/p^k$ for $k$ large enough. Now a functor

$$*/\!\!/\mathbb{Z}/p^k \to X/\!\!/G$$

picks out an element of $X$ and a $p$-power order element of $G$ that fixes it. Thus the collection of functors are in bijective correspondence with

$$\bigsqcup_{\alpha \in G_p^1} X^{\text{im}\,\alpha}.$$

A natural transformation between two functors is a map

$$\begin{array}{ccc} \mathbb{Z}/p^k \sqcup \mathbb{Z}/p^k \sqcup \mathbb{Z}/p^k & \longrightarrow & G \times X \\ \downarrow\downarrow & & \downarrow\downarrow \\ * \sqcup * & \longrightarrow & X. \end{array}$$

The domain here is $*/\!\!/\mathbb{Z}/p^k \times [\underline{1}]$. The two instances of $\mathbb{Z}/p^k$ on the sides come from the identity morphisms in $[\underline{1}]$. The $\mathbb{Z}/p^k$ in the middle is the important one. Given two functors picking out $x_1 \in X^{\text{im}\,\alpha_1}$ and $x_2 \in X^{\text{im}\,\alpha_2}$, the previous commuting diagram implies that a morphism from $x_1$ to $x_2$ in the inertia groupoid is an element $g \in G$ sending $x_1$ to $x_2$ in $X$. However, the composition diagram implies that $g$ must conjugate $\alpha_1$ to $\alpha_2$. □

We now compute $EG \times_G \text{Fix}_d(X)$ in several cases.

*Example A.1.12* When $X = *$,
$$EG \times_G \text{Fix}_d(*) \cong EG \times_G^{\text{conj}} G_p^d.$$
Fixing an element $\alpha$ in a conjugacy class $[\alpha]$ of $G_p^d/G$, the stabilizer of the element is precisely the centralizer of the image of $\alpha$ in $G$. Thus there is an equivalence
$$EG \times_G^{\text{conj}} G_p^d \simeq \coprod_{[\alpha] \in G_p^d/G} BC(\text{im}\,\alpha).$$

*Example A.1.13* When $G$ is a $p$-group there is an isomorphism
$$\text{TopologicalGroups}(\Lambda, G) \cong \text{TopologicalGroups}(\mathbb{Z}^{\times d}, G).$$
In this case there is an equivalence
$$EG \times_G^{\text{conj}} G_p^d \simeq L^d BG,$$
where $L$ is the free loopspace functor.

The next example shows that the inertia groupoid construction has the property that we desire when applied to abelian groups.

*Example A.1.14* When $G = C_{p^k}$,
$$EG \times_G^{\text{conj}} G_p^d \simeq \coprod_{\alpha \in \Lambda_k^*} BC_{p^k}.$$

The inertia groupoid construction has another important property:

**Proposition A.1.15** ([Kuh89, section 6]) *Let $E$ be a cohomology theory, then*
$$E^*(EG \times_G \text{Fix}_d(-))$$
*is a cohomology theory on finite $G$–CW complexes.* □

*Example A.1.16* Let $C_0^* = \pi_{-*} HC_0 P \cong C_0[u, u^{-1}]$ where $|u| = -2$. When $X = *$, Example A.1.12 gives an isomorphism
$$HC_0 P^*(EG \times_G \text{Fix}_d(*)) \cong HC_0 P^* \left( \coprod_{[\alpha] \in G_p^d/G} BC(\text{im}\,\alpha) \right)$$
$$\cong \prod_{[\alpha] \in G_p^d/G} C_0^*.$$

The last isomorphism follows from the fact that $HC_0P^*(BG) = C_0^*$. The codomain is the ring of generalized class functions on $G_p^d$ taking values in $C_0^*$:

$$\mathrm{Cl}(G_p^d, C_0^*) = \{C_0^*\text{-valued functions on } G_p^d/G\}.$$

One might want to concentrate on the degree 0 part, in which case we have an isomorphism

$$HC_0P^0(EG \times_G \mathrm{Fix}_d(*)) \cong \mathrm{Cl}(G_p^d, C_0).$$

## Complex Oriented Descent

Given a pair of cohomology theories $E$ and $F$ as well as a map of cohomology theories $g\colon E^*(-) \to F^*(-)$, the axioms of a cohomology theory (homotopy invariance, Mayer–Vietoris, ...) ensure that $g$ is an isomorphism if and only if it is an isomorphism on a point. For equivariant cohomology theories only a small change is required, a map

$$g\colon E_G^*(-) \to F_G^*(-)$$

is an isomorphism if and only if it is an isomorphism on spaces of the form $G/H$. These spaces are the generalization of the point to $G$-spaces. The purpose of complex-oriented descent is to give conditions on Borel equivariant cohomology theories $E$ and $F$ so that a map $g$ is an isomorphism if and only if it is an isomorphism on all spaces of the form $G/A$, where $A \subseteq G$ is abelian. The idea is that certain equivariant cohomology theories have the property that, for a $G$-CW complex $X$, $E_G^*(X)$ can be recovered from $E_G^*(F)$ for a certain space $F$ with the property that the cells of $F$ are all of the form $G/A \times D^n$.

In the cases that we are interested in, the space $F$ will be the bundle of complete flags associated to a vector bundle on $X$. Let $f\colon V \to X$ be an $m$-dimensional complex vector bundle. Associated to $V$ is a space $\mathrm{Flag}(V)$ over $X$ such that the fiber over $x \in X$ is the space of sequences of inclusions

$$\{0\} \subset V_1 \subset V_2 \subset \ldots \subset V_{m-1} \subset f^{-1}(x),$$

where $V_i \subset f^{-1}(x)$ is an $i$-dimensional subspace.

**Lemma A.1.17** *Let $V$ be an $m$-dimensional complex vector space, then there is an equivalence*

$$\mathrm{Flag}(V) \simeq U(m)/\mathbb{T},$$

*where $\mathbb{T} \cong (S^1)^m$ is the maximal torus in $U(m)$.*

### A.1 Rational Chromatic Phenomena (Nathaniel Stapleton)

*Proof* The idea is simply that $U(m)$ acts transitively on $\text{Flag}(V)$ and the stabilizer is $\mathbb{T}$. This can be seen by viewing a flag as a sequence of unit length normal vectors. □

**Proposition A.1.18** ([HKR00, proposition 2.4(2)]) *Let $X$ be a space and let $\text{Flag}(V) \to X$ be the bundle of complete flags of a complex vector bundle $V$ over $X$. For $E$ complex-oriented, $E^*(\text{Flag}(V))$ is a finitely generated free $E^*(X)$-module.* □

Since $f \colon V \to X$ is an $m$-dimensional complex vector bundle, it follows (see Definition 2.4.4) that $EG \times V \to EG \times X$ is an $m$-dimensional complex vector bundle (this is the external product with a zero-dimensional complex vector bundle). Since $G$ acts freely on $EG \times X$, the quotient $EG \times_G V \to EG \times_G X$ is also an $m$-dimensional complex vector bundle.

We proceed with two propositions, the proofs of which can both be found in the work of Hopkins, Kuhn, and Ravenel [HKR00, proposition 2.6]. The first proposition explains the relationship between flag bundles and the Borel construction, the second proposition gives the relationship between the inertia groupoid construction and flag bundles.

**Proposition A.1.19** *Let $X$ be a finite $G$–CW complex and let $V \to X$ be a $G$-equivariant vector bundle, then there is an equivalence*

$$EG \times_G \text{Flag}(V) \simeq \text{Flag}(EG \times_G V)$$

*of spaces over $EG \times_G X$.*

**Proposition A.1.20** *Let $X$ be a finite $G$–CW complex and let $V \to X$ be a $G$-equivariant vector bundle, then $\text{Flag}(V)^A$ is a disjoint union of fiber products of flag bundles over $X^A$ when $A \subseteq G$ is abelian.*

Recall from Remark 3.1.3 that a commutative ring map $R \to S$ gives rise to a cosimplicial commutative ring

$$\left\{ S \rightleftarrows S \otimes_R S \rightleftarrows S \otimes_R S \otimes_R S \rightleftarrows \cdots \right\}.$$

Recall also from Theorem 3.1.4 that if $S$ is faithfully flat as an $R$-algebra, then this complex is exact except at the first spot and there the homology is $R$.

This digression has important consequences for us. Since $E^*(\mathrm{Flag}(V))$ is free as an $E^*(X)$-module, there is an isomorphism

$$E^*(\mathrm{Flag}(V) \times_X \mathrm{Flag}(V)) \cong E^*(\mathrm{Flag}(V)) \otimes_{E^*(X)} E^*(\mathrm{Flag}(V))$$

and applying $E^*(-)$ to the simplicial space

$$\left\{ \mathrm{Flag}(V) \;\substack{\longrightarrow\\\longleftarrow\\\longrightarrow} \; \begin{array}{c}\mathrm{Flag}(V)\\ \times_X \\ \mathrm{Flag}(V)\end{array} \;\substack{\longrightarrow\\\longleftarrow\\\longrightarrow\\\longleftarrow} \; \begin{array}{c}\mathrm{Flag}(V)\\ \times_X \\ \mathrm{Flag}(V)\\ \times_X \\ \mathrm{Flag}(V)\end{array} \; \cdots \right\}$$

produces the descent object for the canonical map $E^*(X) \to E^*(\mathrm{Flag}(V))$. Since a finitely generated free extension is faithfully flat, Proposition A.1.18 gives a sufficient condition so that

$$E^*(X) \cong \ker\left(E^*(\mathrm{Flag}(V)) \to E^*(\mathrm{Flag}(V) \times_X \mathrm{Flag}(V))\right).$$

When this holds, we say that $E$ satisfies *complex-oriented descent*.

Since it suffices to take the trivial bundle over $X$ in these constructions, we can now recover $E^*(X)$ in a functorial way from $E^*(\mathrm{Flag}(V))$ and from $E^*(\mathrm{Flag}(V) \times_X \mathrm{Flag}(V))$.

**Proposition A.1.21** *Let $\rho\colon G \to U(n)$ be an injective map, then $U(n)/\mathbb{T}$ with the induced $G$-action is a finite $G$–CW complex with abelian stabilizers.*

*Proof* By Theorem A.1.3, since $U(n)/\mathbb{T}$ is compact it is equivalent to a finite $G$–CW complex. Let $A \subset G$ be abelian, then $\rho(A) \subset u\mathbb{T}u^{-1}$ for some $u \in U(n)$. Thus for $a \in A$, $a = utu^{-1}$ for some $t \in \mathbb{T}$, so $A$ fixes the coset $u\mathbb{T}$.

On the other hand, if $H \subset G$ is nonabelian then it cannot be contained inside a maximal torus because the representation $\rho$ is faithful. Therefore it will not fix any coset of $\mathbb{T} \subset U(n)$. □

Putting these results together, if $X$ is a finite $G$–CW complex we can take $V \cong X \times \mathbb{C}^n \to X$ to be the $G$-vector bundle where $G$ acts on $\mathbb{C}^n$ through a faithful representation $\rho\colon G \hookrightarrow U(n)$. Then

$$EG \times_G (X \times \mathbb{C}^n) \to EG \times_G X$$

is an $n$-dimensional vector bundle over $EG \times_G X$ and

$$\mathrm{Flag}(EG \times_G (X \times \mathbb{C}^n)) \simeq EG \times_G \mathrm{Flag}(X \times \mathbb{C}^n) \simeq EG \times_G (X \times U(n)/\mathbb{T}).$$

Since $E_G^*(X) = E^*(EG \times_G X)$ can be recovered from the flag bundle as above,

## A.1 Rational Chromatic Phenomena (Nathaniel Stapleton) 303

these equivalences show that $E^*(EG \times_G X)$ can be recovered from spaces with fixed points only for abelian subgroups.

**Proposition A.1.22** *Let* $g \colon E_G^*(-) \to F_G^*(-)$ *be a map of Borel equivariant cohomology theories on finite G–CW complexes satisfying complex-oriented descent, then g is an isomorphism if and only if it is an isomorphism when applied to spaces of the form* $X = G/A$ *for* $A \subset G$ *abelian.*

### The Character Map

The goal of this section is to construct a map of Borel equivariant cohomology theories

$$\chi_G \colon E^*(EG \times_G -) \to HC_0P^*(EG \times_G \operatorname{Fix}_d(-))$$

called the character map and apply Proposition A.1.22 to prove the following theorem.

**Theorem A.1.23** *For any finite G–CW complex X, the map* $\chi_G$ *induces an isomorphism*

$$C_0 \otimes_{E^0} E^*(EG \times_G X) \xrightarrow{\cong} HC_0P^*(EG \times_G \operatorname{Fix}_d(X)).$$

The map $\chi_G$ will be constructed as the composite of two maps: a "topological map" induced by a map of topological spaces and an "algebraic map" that is a consequence of the algebro-geometric description of $C_0$.

Assume that $k$ is large enough that any map $\Lambda \to G$ factors through $\Lambda_k$. Lemma A.1.11 provides us with equivalences

$$\text{TopologicalGroupoids}(*/\!/\Lambda, X/\!/G) \simeq$$
$$\text{TopologicalGroupoids}(*/\!/\Lambda_k, X/\!/G) \simeq \operatorname{Fix}_d(X)/\!/G.$$

The topological part of the character map is the map

$$E^*(EG \times_G X) \to E^*(B\Lambda_k \times EG \times_G \operatorname{Fix}_d(X))$$

that we get by applying $E^*(-)$ to the geometric realization of the evaluation map

$$\text{ev} \colon */\!/\Lambda_k \times \text{TopologicalGroupoid}(*/\!/\Lambda_k, X/\!/G) \to X/\!/G.$$

Because $E^*(B\Lambda_k)$ is finitely generated and free over $E^*$, there is a Künneth isomorphism

$$E^*(B\Lambda_k \times EG \times_G \operatorname{Fix}_d(X)) \cong E^*(B\Lambda_k) \otimes_{E^*} E^*(EG \times_G \operatorname{Fix}_d(X)).$$

Since $E^*(B\Lambda_k)$ is even-periodic, this can be further identified with

$$E^0(B\Lambda_k) \otimes_{E^0} E^*(EG \times_G \operatorname{Fix}_d(X)).$$

Recall that the definition of $C_0$ furnishes us with a canonical map

$$E^0(B\Lambda_k) \to C_0.$$

The algebraic part of the character map

$$E^0(B\Lambda_k) \otimes_{E^0} E^*(EG \times_G \mathrm{Fix}_d(X)) \to HC_0P^*(EG \times_G \mathrm{Fix}_d(X))$$

is defined to be the tensor product of the canonical ring map

$$E^0(B\Lambda_k) \to C_0$$

with the ring map induced by the map of cohomology theories

$$E \to H(\mathbb{Q} \otimes E^0)P \to HC_0P$$

given by rationalization followed by the canonical map $\mathbb{Q} \otimes E^0 \to C_0$.

The topological and algebraic maps compose to give the character map

$$\chi_G: E^*(EG \times_G X) \to HC_0P^*(EG \times_G \mathrm{Fix}_d(X)).$$

*Example A.1.24* When $X = *$, Example A.1.16 shows that the codomain of the character map is the ring of generalized class functions

$$\mathrm{Cl}(G_p^d, C_0^*) = \prod_{[\alpha] \in G_p^d/G} C_0^*.$$

Unwrapping the definition of $\chi_G$ when $X = *$ gives the following simple formula: Given $[\alpha] \in G_p^d/G$, the character map to the factor corresponding to $[\alpha]$ is just

$$E^*(BG) \xrightarrow{E^*(B\alpha)} E^*(B\Lambda_k) \xrightarrow{\mathrm{can}} C_0^*.$$

**Proposition A.1.25** *The Borel equivariant cohomology theories $E^*(EG\times_G -)$ and $HC_0P^*(EG \times_G \mathrm{Fix}_d(-))$ both satisfy complex-oriented descent.*

*Proof* For $E^*(EG\times_G -)$, this follows immediately from the previous discussion. The key ingredient is Proposition A.1.18 along with Proposition A.1.19, which say that $E^*(EG \times_G \mathrm{Flag}(V))$ is a finitely generated free $E^*(EG \times_G X)$-module.

It is a bit harder to prove that $HC_0P^*(EG \times_G \mathrm{Fix}_d(-))$ satisfies complex-oriented descent. First recall that since $G$ is finite and $C_0$ is rational, for any $G$-space $X$ there is an isomorphism

$$HC_0P^*(EG \times_G X) \cong HC_0P^*(X)^G.$$

By Proposition A.1.20, $\mathrm{Fix}_d(\mathrm{Flag}(V))$ is component-wise a disjoint union of fiber products of flag bundles over the components of $\mathrm{Fix}_d(X)$. Thus Proposition A.1.18 implies that

$$HC_0P^*(\mathrm{Fix}_d(\mathrm{Flag}(V)))$$

## A.1 Rational Chromatic Phenomena (Nathaniel Stapleton) 305

is finitely generated and free over $HC_0 P^*(\mathrm{Fix}_d(X))$. Applying fixed points $(-)^G$ to the exact sequence

$$0 \to HC_0 P^*(\mathrm{Fix}_d(X))$$
$$\to HC_0 P^*(\mathrm{Fix}_d(\mathrm{Flag}(V)))$$
$$\to HC_0 P^*(\mathrm{Fix}_d(\mathrm{Flag}(V)) \times_{\mathrm{Fix}_d(X)} \mathrm{Fix}_d(\mathrm{Flag}(V)))$$

gives an exact sequence

$$0 \to HC_0 P^*(\mathrm{Fix}_d(X))^G$$
$$\to HC_0 P^*(\mathrm{Fix}_d(\mathrm{Flag}(V)))^G$$
$$\to HC_0 P^*(\mathrm{Fix}_d(\mathrm{Flag}(V)) \times_{\mathrm{Fix}_d(X)} \mathrm{Fix}_d(\mathrm{Flag}(V)))^G$$

since fixed points is left exact. This implies that $HC_0 P^*(EG \times_G \mathrm{Fix}_d(X))$ can be recovered from $HC_0 P^*(EG \times_G -)$ of finite $G$–CW complexes with abelian stabilizers. □

*Proof of Theorem A.1.23* By Proposition A.1.22, it suffices to prove that the map is an isomorphism for $X = G/A$. Since we have an equivalence

$$(G/A)/\!\!/G \simeq */\!\!/A,$$

we are reduced to the case $G = A$ and $X = *$. Since both sides of the character map have Künneth isomorphisms for products of finite abelian groups, we are reduced to cyclic groups. We want to show that

$$C_0 \otimes_{E^0} E^*(BC_{p^k}) \xrightarrow{\cong} \mathrm{Cl}((C_{p^k})_p^d, C_0^*)$$

is an isomorphism. Since we have isomorphisms

$$C_0 \otimes_{E^0} E^*(BC_{p^k}) \cong C_0 \otimes_{E_0} E^* \otimes_{E^0} E^0(BC_{p^k}) \cong C_0^* \otimes_{C_0} C_0 \otimes_{E^0} E^0(BC_{p^k})$$

and

$$\mathrm{Cl}((C_{p^k})_p^d, C_0^*) \cong C_0^* \otimes_{C_0} \mathrm{Cl}((C_{p^k})_p^d, C_0),$$

and since $C_0^*$ is faithfully flat over $C_0$, it suffices to prove the isomorphism in degree 0. Example A.1.24 shows that this map is given by

$$C_0 \otimes_{E^0} E^0(BC_{p^k}) \xrightarrow{\prod C_0 \otimes_{E^0} E^0(B\alpha)} E^0(B\Lambda_k) \xrightarrow{\mathrm{incl}} \prod_{\alpha \in \Lambda_k^*} C_0.$$

This is the same as the map $C_0 \otimes_{E^0(B\Lambda_k)} u^*$ built previously. □

*Example A.1.26* When $n = 1$ and $X = *$ the character map produces a

$p$-complete version of the classical character map from representation theory that is due to Adams [Ada78a, section 2]). The map takes the form

$$KU_p^0(BG) \to \mathrm{Cl}(G_p^1, C_0),$$

where $C_0$ is the maximal ramified extension of $\mathbb{Q}_p$ discussed in Example A.1.8. This map is injective. Above height 1, though, this is not generally true [Kří94], but it *is* true for $G = \Sigma_n$ [HKR00, theorem 7.3], which is enough to conclude that $E^0 B\Sigma_n$ is free of finite rank.

*Remark A.1.27* It follows from the proof of Theorem A.1.23 that when working with a fixed group $G$, it suffices to take $C_{0,k}$ as the coefficients of the codomain of the character map, where $k \geq 0$ is large enough that any continuous map $\mathbb{Z}_p \to G$ factors through $\mathbb{Z}/p^k$.

### The Action of $\mathrm{GL}_d(\mathbb{Z}_p)$

There is a natural action of $\mathrm{Aut}(\Lambda) \cong \mathrm{GL}_d(\mathbb{Z}_p)$ on the isomorphism

$$C_0 \otimes \chi_G : C_0 \otimes_{E^0} E^*(EG \times_G X) \xrightarrow{\cong} HC_0 P^*(EG \times_G \mathrm{Fix}_d(X)).$$

Recall that $\mathrm{Spec}\, C_0$ carries the universal isomorphism of $p$-divisible groups

$$\Lambda^* \xrightarrow{\cong} \widehat{\mathbb{G}}[p^\infty].$$

Precomposing with the Pontryagin dual of an element in $\mathrm{Aut}(\Lambda)$ gives another isomorphism and thus induces an $E^0$-algebra automorphism of $C_0$. The quotient map

$$\mathrm{Aut}(\Lambda) \to \mathrm{Aut}(\Lambda_k)$$

gives an action of $\mathrm{Aut}(\Lambda)$ on $E^0(B\Lambda_k)$ and, by Proposition A.1.4, the inclusion $E^0(B\Lambda_k) \to C_0$ is equivariant for these actions.

There is also an action of $\mathrm{Aut}(\Lambda)$ on $\textsc{TopologicalGroupoids}(*/\!/\Lambda, X/\!/G)$ given by precomposition. These actions combine to give a conjugation action on $HC_0 P^*(EG \times_G \mathrm{Fix}_d(X))$. Explictly, for $\varphi \in \mathrm{Aut}(\Lambda)$ this action is induced by the action on

$$*/\!/\Lambda \times \textsc{TopologicalGroupoids}(*/\!/\Lambda, X/\!/G)$$

given by $\varphi \times (\varphi^{-1})^*$. There is an action of $\mathrm{Aut}(\Lambda)$ on $C_0 \otimes_{E^0} E^*(EG \times_G X)$ by just acting on $C_0$.

**Proposition A.1.28** *The character map is $\mathrm{Aut}(\Lambda)$-equivariant.*

*Proof* This follows immediately from the fact that the evaluation map

$$\text{ev}: * /\!\!/ \Lambda \times \textsc{TopologicalGroupoids}(* /\!\!/ \Lambda, X /\!\!/ G) \to X /\!\!/ G$$

is $\text{Aut}(\Lambda)$-equivariant for the trivial action on $X /\!\!/ G$ and the action described above on the domain. □

It turns out that $C_0$ is an $\text{Aut}(\Lambda)$-Galois extension of $\mathbb{Q} \otimes E^0$:

**Proposition A.1.29** ([HKR00, corollary 6.8.iii])  *There is an isomorphism*

$$C_0^{\text{Aut}(\Lambda)} \cong \mathbb{Q} \otimes E^0.$$  □

By taking $\text{Aut}(\Lambda)$-fixed points of $C_0 \otimes \chi_G$ we get an isomorphism

$$\mathbb{Q} \otimes \chi_G: \mathbb{Q} \otimes E^*(EG \times_G X) \xrightarrow{\cong} HC_0 P^*(EG \times_G \text{Fix}_d(X))^{\text{Aut}(\Lambda)}.$$

In particular, when $X = *$, we find that

$$\mathbb{Q} \otimes E^0(BG) \cong \text{Cl}(G_p^d, C_0)^{\text{Aut}(\Lambda)}.$$

That is, the rationalization of the $E$-cohomology of $BG$ is a subring of generalized class functions.

*Example A.1.30* ([HKR00, section 1.3])  In the context of Example A.1.26, the character map gives the usual ring homomorphism to class functions

$$\chi_G: R(G) \to \text{Cl}(G_p, C_0)^{\widehat{\mathbb{Z}}},$$

where $R(G)$ is the representation ring of $G$, $C_0 = \mathbb{Q}(\zeta_j : j \geq 0)$ is the maximal abelian extension of $\mathbb{Q}$, and $\widehat{\mathbb{Z}}$ is both the Galois group of the extension and $\text{Aut}(\Lambda)$. After tensoring up, this yields two isomorphisms

$$C_0 \otimes \chi_G: C_0 \otimes R(G) \to \text{Cl}(G_p, C_0),$$

$$\mathbb{Q}_p \otimes \chi_G: \mathbb{Q}_p \otimes R(G) \to \text{Cl}(G_p, C_0)^{\widehat{\mathbb{Z}}}.$$

## A.2 Orientations and Power Operations

Our introduction of $E_\infty$-rings also automatically introduces a few interesting accompanying functors:

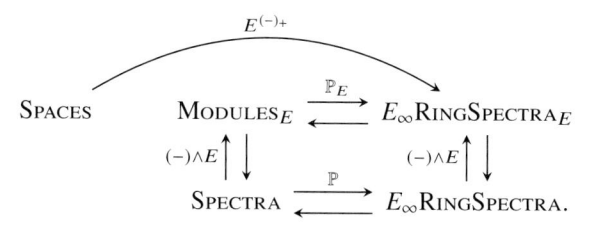

$$\begin{array}{c}
\overset{E^{(-)_+}}{\overbrace{\textsc{Spaces} \quad \textsc{Modules}_E \underset{}{\overset{\mathbb{P}_E}{\rightleftarrows}} E_\infty\textsc{RingSpectra}_E}} \\
{\scriptstyle (-)\wedge E}\uparrow\downarrow \qquad\qquad {\scriptstyle (-)\wedge E}\uparrow\downarrow \\
\textsc{Spectra} \underset{}{\overset{\mathbb{P}}{\rightleftarrows}} E_\infty\textsc{RingSpectra}.
\end{array}$$

The first functor sends a space $X$ to its spectrum of $E$-cochains $E^{\Sigma_+^\infty X}$, and the other two functors form a free/forgetful monad resolving a mapping space in $E_\infty\text{RingSpectra}_E$ by a sequence of mapping spaces in $\text{Modules}_E$. These kinds of functors are familiar to us from the discussion of contexts in Lectures 3.1 and 4.1, and the recipe applied in those situations gives an analogous story here. First, there is a natural map

$$\text{Spaces}(*, X) \to E_\infty\text{RingSpectra}_E(E^{X_+}, E),$$

which one hopes is an equivalence under (often very strong) hypotheses on $E$ and on $X$.[12] Second, the adjunction gives a mechanism for resolving $E^{X_+}$, which feeds into a spectral sequence computing this right-hand mapping space. The functors $\mathbb{P}$ and $\mathbb{P}_E$ can be given by explicit formulas:

$$\mathbb{P}(X) = \bigvee_{j=0}^\infty X_{h\Sigma_j}^{\wedge j}, \qquad \mathbb{P}_E(M) = \bigvee_{j=0}^\infty M_{h\Sigma_j}^{\wedge_E j}.$$

Finally, we expect the homotopy groups of this resolution to form a quasicoherent sheaf over a suitable $E_\infty$ *context*, which arises as the simplicial scheme associated to this resolution in the case where $X$ is a point. In this case, we can explicitly name some of the terms in this resolution: The bottom two stages take the form

$$E_\infty\text{RingSpectra}_E(E, E)$$
$$\uparrow$$
$$E_\infty\text{RingSpectra}_E(\mathbb{P}_E(E), E) \rightrightarrows E_\infty\text{RingSpectra}_E(\mathbb{P}_E^2(E), E) \substack{\longleftarrow \\ \longrightarrow \\ \longleftarrow} \cdots .$$

The available adjunctions give a more explicit presentation of these terms:

$$E_\infty\text{RingSpectra}_E(\mathbb{P}_E(E), E) \simeq \text{Modules}_E(E, E)$$
$$\simeq \text{Spectra}(\mathbb{S}, E),$$

which on homotopy groups computes the coefficient ring of $E$, and

$$E_\infty\text{RingSpectra}_E(\mathbb{P}_E^2(E), E) \simeq \text{Modules}_E(\mathbb{P}_E(E), E)$$
$$\simeq \text{Modules}_E\left(\bigvee_{j=0}^\infty E_{h\Sigma_j}^{\wedge_E j}, E\right) \simeq \prod_{j=0}^\infty \text{Spectra}\left(B\Sigma_j, E\right),$$

which on homotopy groups is made up of a product of the cohomology rings $E^*(B\Sigma_j)$. The higher terms track the compositional behavior of these summands.

---

[12] Mandell has shown this assignment for $E = H\mathbb{F}_p$ is an equivalence on connected $p$-complete nilpotent spaces of finite $p$-type [Man01]. There is an analogous unpublished theorem of Hopkins and Lurie: This assignment for $E = E_\Gamma$ is an equivalence if $X$ is a nilpotent space admitting a finite Postnikov system with at most ht $\Gamma$ stages and involving only finite groups.

*Remark A.2.1* ([Bou75], [MT92], [Bou96], [Dav95])  One of the first places these ideas appear in the literature is in the work of Mahowald and Thompson. Bousfield defined an unstable local homotopy type associated to a (simply connected) space and a homology theory. In the case of the space $S^{2n-1}$ and $p$-adic $K$-theory, Mahowald and Thompson calculated that $L_K S^{2n-1}$ appears as the homotopy fiber

$$L_K S^{2n-1} \to L_K \Omega^\infty \Sigma^\infty S^{2n-1} \to L_K \Omega^\infty \Sigma^\infty ((S^{2n-1})_{h\Sigma_p}^{\wedge p}),$$

which is an abbreviated form of the monadic resolution described above.

*Remark A.2.2* ([GH04, theorem 4.5])  In general, if $E^* B\Sigma_j$ is sufficiently nice, then the $E^2$-page of the monadic descent spectral sequence computes the derived functors of derivations, taken in a suitable category of monad-algebras for the monad specifying the behavior of power operations.

## Strickland's Theorems

We thus set out to understand the formal schemes constituting the $E_\infty$ context associated to Morava $E$-theory.[13] As described in Lectures 4.1 and 4.2, the correct language for these phenomena are schemes defined on Hopf rings, together with the adjunction between classical rings and Hopf rings consisting of the $*$-square-zero extension functor and the $*$-indecomposables functor. The rings $E^0 B\Sigma_j$ assemble into a Hopf ring using the following structure:

- The $*$-product comes from the stable transfer maps $B\Sigma_{i+j} \to B\Sigma_i \times B\Sigma_j$.
- The $\circ$-product comes from the diagonal maps $B\Sigma_j \to B\Sigma_j \times B\Sigma_j$.
- The diagonal comes from the block-inclusion maps $B\Sigma_i \times B\Sigma_j \to B\Sigma_{i+j}$.

**Definition A.2.3**  Accordingly, we set the *natural $E_\infty$ context* to be

$$E_\infty \mathcal{M}_{E_\Gamma} = \mathrm{SpH}\, E^0 B\Sigma_*.$$

The effect of this functor on classical rings follows from our work in Lecture 4.2: $E_\infty \mathcal{M}_{E_\Gamma}(T) = \mathrm{ALGEBRAS}_{E_\Gamma^0}(Q^* E^0 B\Sigma_*, T)$, where

$$Q^* E^0 B\Sigma_* = \frac{E^0 B\Sigma_*}{\bigcup_{j+k=*} \mathrm{im}\left(\mathrm{Tr}_{\Sigma_j \times \Sigma_k}^{\Sigma_{j+k}} : E^0 B\Sigma_j \times E^0 B\Sigma_k \to E^0 B\Sigma_{j+k}\right)}.$$

The ideal appearing in this equation is called the *transfer ideal*, written $I_{\mathrm{Tr}}$.

*Remark A.2.4*  In terms of the descent spectral sequence described in the previous subsection, all of the $*$-decomposables are in the image of the $d^1$-differential, so already do not contribute to the $E^2$-page of the spectral sequence.

---
[13] Much of the analysis for the case of $H\mathbb{F}_2$ can be read off from a pleasant paper of Baker [Bak15].

We dissect this Hopf ring spectrum by considering the $j$th graded piece of $E_\infty\mathcal{M}_{E_\Gamma}$ as restricted to classical rings, i.e., by understanding the formal schemes $\mathrm{Spec}(E^0 B\Sigma_j / I_{\mathrm{Tr}})$ for individual indices $j$.

*Example A.2.5* ([SS15, appendix])   To gain a foothold, it is helpful to further specialize to a particular case – say, $j = p$. In light of the results of Appendix A.1, we might begin by analyzing the (maximal) abelian subgroups of $\Sigma_p$, of which the only $p$-locally interesting one is the transitive subgroup $C_p \subseteq \Sigma_p$. In Theorem 4.6.1, we calculated $BS^1[p]_E = \mathbb{CP}^\infty_E[p]$, and we now make the further observation that the regular representation map $\rho \colon BS^1[p] \to BU(p)$ induces the following map on cohomological formal schemes:

$$\begin{array}{ccc} BS^1[p]_E & \xrightarrow{\rho_E} & BU(p)_E \\ \| & & \| \\ \underline{\mathrm{FormalGroups}}(\mathbb{Z}/p, \mathbb{CP}^\infty_E) & \longrightarrow & \mathrm{Div}^+_p\, \mathbb{CP}^\infty_E, \end{array}$$

where the bottom arrow sends such a homomorphism to its image divisor. This map belongs to a larger diagram of schemes:

$$\begin{array}{ccc} \mathrm{Spf}\, E^0 B\Sigma_p / I_{\mathrm{Tr}} \longrightarrow (B\Sigma_p)_E \longrightarrow BU(p)_E \\ \uparrow \qquad\qquad \uparrow \qquad \nearrow \\ \mathrm{Spf}\, E^0 BC_p / I_{\mathrm{Tr}} \longrightarrow (BC_p)_E. \end{array}$$

The quotient by the transfer ideal in $(BC_p)_E$ is modeled by the map

$$\begin{array}{ccc} E^0 BC_p / I_{\mathrm{tr}} & \longleftarrow & E^0 BC_p \\ \| & & \| \\ E^0[\![x]\!]/\langle p\rangle(x) & \longleftarrow & E^0[\![x]\!]/[p](x), \end{array}$$

which disallows the zero homomorphism and forces the image divisor to be a subdivisor of $\mathbb{CP}^\infty_E[p]$. Such a homomorphism is an example of a *level structure*, and the scheme of such homomorphisms is written $\mathrm{Level}(\mathbb{Z}/p, \mathbb{CP}^\infty_E)$. Finally, passing to $B\Sigma_p$ from $BC_p$ exactly destroys the choice of generator of $\mathbb{Z}/p$, i.e., it encodes passing from the homomorphism to the image divisor. This winds up giving an isomorphism

$$\mathrm{Spf}\, E^0 B\Sigma_p / I_{\mathrm{Tr}} \cong \mathrm{Sub}_p\, \mathbb{CP}^\infty_E,$$

where $\mathrm{Sub}_p$ denotes the subscheme of $\mathrm{Div}^+_p$ consisting of those effective Weil divisors of rank $p$ which are subgroup divisors.

The broad features of this example hold for a general index $j$.

## A.2 Orientations and Power Operations

**Definition A.2.6** The *abelian $E_\infty$ context* is formed by considering the inclusions

$$\bigvee_{\substack{A \leq \Sigma_j \\ A \text{ abelian}}} BA \to B\Sigma_j.$$

A consequence of Appendix A.1 is that this map is *injective* on Morava $E$-cohomology, so that we can understand the natural $E_\infty$ context in terms of this larger object. A benefit to this auxiliary context is that we can already predict its behavior in Morava $E$-theory: The cohomological formal scheme associated to an abelian group can be presented as an internal scheme of group homomorphisms, just as above. Using this auxiliary context for reference, Strickland has proven the following results:

**Theorem A.2.7** ([Str98, theorem 1.1])   *There is an isomorphism*

$$\operatorname{Spf} E^0 B\Sigma_j / I_{\mathrm{Tr}} \cong \operatorname{Sub}_j \mathbb{CP}_E^\infty,$$

*where* $\operatorname{Sub}_j$ *denotes the subscheme of* $\operatorname{Div}_j^+$ *consisting of those effective Weil divisors of rank $j$ which are subgroup divisors.*[14]

*Proof sketch, after Stapleton*   The proof of this theorem requires a very involved algebraic calculation (which can take different forms [Str97], [SS15]), so we describe only the reduction to that calculation. The object $E^0 B\Sigma_j$ is accessible via the chromatic character theory from the previous lecture, where we presented $C_0 \otimes E^0 B\Sigma_j$ as a certain ring of $C_0$-valued class functions on the symmetric group. The effect of quotienting by the transfer ideal is to delete all of the factors belonging to conjugacy classes that are not transitive. On the other side of the putative isomorphism, the formation of the scheme of subgroups commutes with base-change, and hence we have

$$\operatorname{Spec} C_0 \times \operatorname{Sub}_j \mathbb{CP}_E^\infty \cong \operatorname{Sub}_j(\operatorname{Spec} C_0 \times \mathbb{CP}_E^\infty) \cong \operatorname{Sub}_j \Lambda^*,$$

where in the second isomorphism we have used the canonical isomorphism carried by the ring $C_0$. After this character-theoretic base-change, there is a canonical bijection between the two sides: A map

$$\mathbb{Z}_p^n \xrightarrow{\text{transitive}} \Sigma_n$$

has an image factorization $\mathbb{Z}_p^n \to A \to \Sigma_n$, and the Pontryagin dual of the surjective map in the factorization gives an injective map

$$\Lambda^* \leftarrow A^*.$$

---
[14] If $j$ is not a power of $p$, this is the terminal scheme.

As described above, the permutation representation gives a natural map $B\Sigma_j \to BU(j)$, which participates in the following square:

$$\begin{array}{ccccc}
E^0 BU(j) & \longrightarrow & E^0 B\Sigma_j / I_{\mathrm{Tr}} & \xrightarrow{\mathrm{inj.}} & C_0 \otimes E^0 B\Sigma_j / I_{\mathrm{Tr}} \\
\uparrow \simeq & & \uparrow & & \uparrow \simeq \\
O(\mathrm{Div}_j^+ \, \mathbb{CP}_E^\infty) & \xrightarrow{\mathrm{surj.}} & O(\mathrm{Sub}_j \, \mathbb{CP}_E^\infty) & \xrightarrow{\mathrm{inj.}} & O(\mathrm{Sub}_j \, \Lambda^*).
\end{array}$$

The left-hand isomorphism is stated in Corollary 2.3.12; the right-hand isomorphism is the content of the previous paragraph; the bottom-left horizontal map is surjective because subgroup divisors form a closed subscheme of all divisors; the top horizontal map is an injection by a difficult theorem of Strickland [Str98, theorems 8.5–8.6]; and the bottom horizontal map is an injection by a second difficult theorem of Strickland [Str97, theorem 10.1]. In fact, the two difficult theorems say more: They show that the two terms are finite and free of equal rank. Altogether, these can be used to define the middle vertical map: Any element in the bottom can be preimaged along the surjection, translated along the isomorphism, and pushed back down along the map induced by the regular representation. One quickly checks that this gives a well-defined injective $E^0$-algebra homomorphism.

What remains to show, then, is that this last injection is an isomorphism, whether by calculating the relevant determinant [SS15, section 7.9] or by explicitly analyzing a generating element in terms of Euler classes [Str98, theorem 9.2]. □

Related methods also describe the abelian context:

**Theorem A.2.8** ([Strb, p. 45])   *For a finite abelian group A, there is a diagram*

$$\begin{array}{ccccc}
\mathrm{Spf}\, E^0 BA \Big/ \Big( \sum_{\substack{a \in A \\ a \neq 0}} \mathrm{ann}(a) \Big) & \to & \mathrm{Spf}\, E^0 BA / I_{\mathrm{Tr}} & \longrightarrow & \mathrm{Spf}\, E^0 BA \\
\| & & & & \| \\
\mathrm{Level}(A^*, \mathbb{CP}_E^\infty) & & \longrightarrow & & \underline{\mathrm{FormalGroups}}(A^*, \mathbb{CP}_E^\infty),
\end{array}$$

*where* $\mathrm{Level}(A^*, \mathbb{CP}_E^\infty)$ *denotes the subscheme of* $\underline{\mathrm{FormalGroups}}(A^*, \mathbb{CP}_E^\infty)$ *consisting of homomorphisms which are subject to the condition that* $A^*[n]$ *forms a subdivisor of* $\mathbb{CP}_E^\infty[n]$.[15]   □

*Remark A.2.9*   An important piece of intuition about the schemes of level structures $\mathrm{Level}(A^*, \mathbb{CP}_E^\infty)$ is that they form a kind of replacement for the nonexistent "scheme of monomorphisms." Specifically, the $p$-series for a

---

[15] If $A[p] \cong (\mathbb{Z}/p)^{\times k}$ has $k > \mathrm{ht}\,\Gamma$, then this is the terminal scheme.

Lubin–Tate universal deformation group is only once $x$-divisible, and hence the divisor $\widehat{\mathbb{G}}[p]$ only contains the divisor $[0]$ with multiplicity one.[16] This excludes noninjective morphisms in this case. On the other hand, the only subgroups of the formal group restricted to the special fiber are of the form $p^m \cdot [0]$. In particular, any level structure on $\widehat{\mathbb{G}}$ restricts to a morphism with this image divisor at the special fiber, and hence functoriality considerations force us to count these – which are *not* images of monomorphisms – as level structures as well.[17]

*Remark A.2.10* ([Str97])   The schemes $\mathrm{Sub}_j\,\widehat{\mathbb{G}}$ and $\mathrm{Level}(A,\widehat{\mathbb{G}})$ are known to possess many very pleasant algebraic properties: They are finite and free of predictable rank, they have Galois descent properties, the schemes $\mathrm{Level}(A,\widehat{\mathbb{G}})$ are all reduced, there are important decompositions coming from presenting a subgroup scheme as a flag of smaller subgroups, .... Indeed, these algebraic results form important ingredients to the proof of the connection with homotopy theory [Str98, section 9].

*Remark A.2.11* ([Rez12])   Rezk has shown that the $E_\infty$ descent object for Morava $E$-theory is, in a certain sense, of finite length. Specifically, there is a subobject of the descent object which consists levelwise of those flags of formal subgroups of $\widehat{\mathbb{G}}_E$ whose composition is contained in $\widehat{\mathbb{G}}[p]$, and the Koszul condition entails that this inclusion induces a weak equivalence of derived categories.

### Isogenies and the Lubin–Tate Moduli

In this subsection, we seek a comparison of the natural $E_\infty$ context and the unstable context considered in Lecture 4.1. Our model for the unstable context in Lecture 4.2 focuses on the effect of unstable operations on the cohomology of $\mathbb{CP}^\infty$, as summarized in the following result:

**Lemma A.2.12** (mild extension of Theorem 4.5.8 along the lines of Corollary 4.3.7; see also [Tho14, theorem 1.1])   *There is an isomorphism*

$$\mathrm{Spec}\, Q^* \pi_* L_\Gamma(E \wedge E) \cong \underline{\mathrm{FormalGroups}}(\mathbb{CP}^\infty_E, \mathbb{CP}^\infty_E). \quad \square$$

In order to form a comparison map between these two contexts, we will want algebraic constructions that trade a subgroup divisor (i.e., a point in the natural $E_\infty$ context) for a formal group endomorphism (i.e., a point in the unstable context). It will be useful to phrase our ideas in the language of *isogenies*.

---

[16] This kind of reasoning applies to domains of characteristic 0 generally.
[17] On the other hand, the scheme of level structures tracks *exactly* monomorphisms after tensoring up with the ring $C_0$ of Appendix A.1.

**Definition A.2.13** ([Str99b, definition 5.17]) Take $C$ and $D$ to be formal curves over $X$. A map $f \colon C \to C'$ is an *isogeny* (of degree $d$) when the induced map $C \to C \times_X C'$ exhibits $C$ as a divisor (of rank $d$) on $C \times_X C'$ as $C'$-schemes.

*Remark A.2.14* (see Remark 2.2.8) In this case, a divisor $D$ on $C'$ gives rise to a divisor $f^*D$ on $C$ by scheme-theoretic pullback:

$$\begin{array}{ccc} f^*D & \longrightarrow & D \\ \downarrow & \lrcorner & \downarrow \\ C & \xrightarrow{f} & C', \end{array}$$

altogether inducing a map

$$f^* \colon \operatorname{Div}_n^+ C' \to \operatorname{Div}_{nd}^+ C.$$

This map interacts with pushforward by $f_* f^* D = d \cdot D$, where $d$ is the degree of the isogeny.

The usual source of examples of isogenies are polynomial maps between curves. In fact, this is close to the general case, and the following result is the source of much intuition:

**Lemma A.2.15** (Weierstrass preparation [Str99b, section 5.2]) *Let $R$ be a complete local ring. Every degree $d$ isogeny $f \colon \widehat{\mathbb{A}}_R^1 \to \widehat{\mathbb{A}}_R^1$ admits a unique factorization as a coordinate change and a monic polynomial of degree $d$.* □

In the case of formal groups over a perfect field of positive characteristic, this reduces to two familiar structural results:

**Corollary A.2.16** *Every nonzero map of formal groups over a perfect field of positive characteristic can be factored as an iterate of Frobenius and a coordinate change.*[18] □

**Corollary A.2.17** *A map of formal groups over a complete local ring with a perfect positive-characteristic residue field is an isogeny if and only if the kernel subscheme of the map is a divisor.* □

This last result forms the headwaters of the connection we are seeking: Isogenies are exactly the class of formal group homomorphisms whose kernels form subgroup divisors. Again, we are seeking an assignment in the opposite direction, some special collection of endoisogenies naturally attached to prescribed kernel divisors. As a first approximation to this goal, we drop the *endo-* and aim to

---
[18] Incidentally, the Frobenius iterate appearing in the Weierstrass factorization of the multiplication-by-$p$ isogeny $p \colon \widehat{\mathbb{G}} \to \widehat{\mathbb{G}}$ is another definition of the height of $\widehat{\mathbb{G}}$.

## A.2 Orientations and Power Operations

construct just *isogenies* with this kernel property, a candidate for which is a theory of *quotient groups*.

**Definition A.2.18** For $K \subseteq \widehat{\mathbb{G}}$ a subgroup divisor, we define the quotient group $\widehat{\mathbb{G}}/K$ to be the formal scheme whose ring of functions is the equalizer

$$\mathcal{O}_{\widehat{\mathbb{G}}/K} \longrightarrow \mathcal{O}_{\widehat{\mathbb{G}}} \xrightarrow[1 \otimes \eta^*]{\mu^*} \mathcal{O}_{\widehat{\mathbb{G}}} \otimes \mathcal{O}_K.$$

**Lemma A.2.19** ([Str97, theorem 5.3.2–5.3.3])  *The functor $\widehat{\mathbb{G}}/K$ is again a one-dimensional smooth commutative formal group.* □

The inclusion of rings of functions determines an isogeny $q \colon \widehat{\mathbb{G}} \to \widehat{\mathbb{G}}/K$ of degree $|K|$. In this particular case, the induced pullback map $q^*$ of divisor schemes has an especially easy formulation:

**Lemma A.2.20**  *Pullback along the isogeny $q \colon \widehat{\mathbb{G}} \to \widehat{\mathbb{G}}/K$ is computed by*

$$q^* \colon D \mapsto D * K,$$

*where $* \colon \mathrm{Div}_n^+ \widehat{\mathbb{G}} \times \mathrm{Div}_d^+ \widehat{\mathbb{G}} \to \mathrm{Div}_{nd}^+ \widehat{\mathbb{G}}$ is the convolution product of divisors on formal groups as described in Corollary 2.3.16.* □

In the case that $K$ is specified by a level structure, this admits a further refinement:

**Corollary A.2.21**  *Let $\ell \colon A \to \widehat{\mathbb{G}}$ be a level structure parametrizing a subgroup divisor $K$. The divisor pullback map can then be computed by the expansion*

$$q^* D = \sum_{a \in A} \tau_{\ell(a)}^* D,$$

*where $\tau_g \colon \widehat{\mathbb{G}} \to \widehat{\mathbb{G}}$ is the translation by $g$ map.* □

**Definition A.2.22** This second construction can be upgraded to an assignment from *functions* on $\widehat{\mathbb{G}}$ to *functions* on $\widehat{\mathbb{G}}/K$, rather than just the ideals that they generate. Specifically, we define the *norm* of $\varphi \in \mathcal{O}_{\widehat{\mathbb{G}}}$ along a level structure $\ell \colon A \to \widehat{\mathbb{G}}$ by the formula

$$N_\ell \varphi = \prod_{a \in A} \tau_{\ell(a)}^* \varphi.$$

An often-useful property of this norm construction is that if $\varphi$ is a coordinate on $\widehat{\mathbb{G}}$, then $N_\ell \varphi$ is a coordinate on $\widehat{\mathbb{G}}/K$ [Str97, theorem 5.3.1].

*Remark A.2.23* In general, the pullback map $q^*$ admits the following description: An isogeny $q \colon C \to C'$ gives a presentation of $\mathcal{O}_C$ as a finite free

$\mathcal{O}_{C'}$-module. A function $\varphi \in \mathcal{O}_C$ therefore begets a linear endomorphism $\varphi \cdot (-) \in \mathrm{GL}_{\mathcal{O}_{C'}}(\mathcal{O}_C)$, and the determinant of this map gives an element of $q^*\varphi \in \mathcal{O}_{C'}$. Letting $\varphi$ range, this gives a multiplicative (but not typically additive) map $q^* \colon \mathcal{O}_C \to \mathcal{O}_{C'}$. If a divisor $D$ is specified as the zero-locus of a function $\varphi_D$, the divisor $q^*D$ is specified as the zero-locus of $q^*\varphi_D$.

Our last technical remark is that this definition of quotient does, indeed, have the suggested universal property:

**Lemma A.2.24** (First isomorphism theorem for formal groups [Lub67], [Str97, theorem 5.3.4])  *If $f \colon \widehat{\mathbb{G}} \to \widehat{\mathbb{H}}$ is any isogeny with kernel divisor $K$, then there is a uniquely specified commuting triangle*

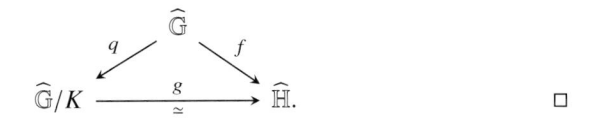

$$\begin{array}{ccc} & \widehat{\mathbb{G}} & \\ {}^{q}\swarrow & & \searrow^{f} \\ \widehat{\mathbb{G}}/K & \xrightarrow[\simeq]{g} & \widehat{\mathbb{H}}. \end{array}$$ □

We now use this lemma, along with properties of the Lubin–Tate moduli problem, to associate endo-isogenies to subgroup divisors. To begin, consider the multiplication-by-$p$ endoisogeny of the Lubin–Tate group $\widehat{\mathbb{G}}$ associated to the Honda formal group $\Gamma_d$. Since $\widehat{\mathbb{G}}$ is of finite height $d$, this map is an isogeny and the lemma gives rise to an isomorphism of formal groups

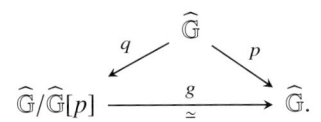

$$\begin{array}{ccc} & \widehat{\mathbb{G}} & \\ {}^{q}\swarrow & & \searrow^{p} \\ \widehat{\mathbb{G}}/\widehat{\mathbb{G}}[p] & \xrightarrow[\simeq]{g} & \widehat{\mathbb{G}}. \end{array}$$

This diagram is one of formal groups rather than of deformations – the map $p \colon \widehat{\mathbb{G}} \to \widehat{\mathbb{G}}$ does not cover the identity upon reduction to the special fiber because it disturbs coefficients. The Frobenius allows us to correct this to a diagram of deformations:

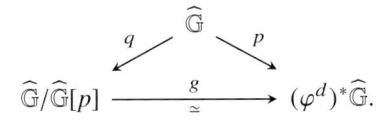

$$\begin{array}{ccc} & \widehat{\mathbb{G}} & \\ {}^{q}\swarrow & & \searrow^{p} \\ \widehat{\mathbb{G}}/\widehat{\mathbb{G}}[p] & \xrightarrow[\simeq]{g} & (\varphi^d)^*\widehat{\mathbb{G}}. \end{array}$$

This latter diagram shows that the quotient $\widehat{\mathbb{G}}/\widehat{\mathbb{G}}[p]$ is *again* a universal deformation, as witnessed by a preferred isomorphism to $\widehat{\mathbb{G}}$. For a generic formal group $\Gamma$, Lubin–Tate group $\widehat{\mathbb{G}}$, and subgroup divisor $K \leq \widehat{\mathbb{G}}$, we have access to the isogeny $q_K$ using the methods described above, but it is the magic of the Lubin–Tate moduli that furnishes us with replacements for $p$ and for $g$. Notice first that the problem simplifies dramatically for the formal group at the special

## A.2 Orientations and Power Operations

fiber: All subgroups of the special fiber formal group are of the form $p^j \cdot [0]$, and hence we can always use the Frobenius map to complete the desired triangle:

$$\begin{array}{ccc} & \Gamma & \\ {\scriptstyle q_{p^j[0]}} \swarrow & & \searrow {\scriptstyle \mathrm{Frob}^j} \\ \Gamma/(p^j \cdot [0]) & \xrightarrow[\simeq]{g_{p^j[0]}} & (\varphi^j)^*\Gamma, \end{array}$$

where $\varphi \colon k \to k$ is the Frobenius on coefficients and $\mathrm{Frob}\colon \Gamma \to \varphi^*\Gamma$ is the "geometric Frobenius," specified at the level of points by $\mathrm{Frob}(x) = x^p$ and hence at the level of formal group *laws* by

$$(x +_\Gamma y)^p = x^p +_{\varphi^*\Gamma} y^p.$$

**Lemma A.2.25** ([AHS04, section 12.3]) *Let $G$ be an infinitesimal deformation of a finite-height formal group $\Gamma$ to a complete local ring $R$. Associated to a subgroup divisor $K \leq G$, there is a commuting triangle*

$$\begin{array}{ccc} & G & \\ {\scriptstyle q_K} \swarrow & & \searrow {\scriptstyle p_K} \\ G/K & \xrightarrow[\simeq]{g_K} & \psi_K^* \widehat{\mathbb{G}} \end{array}$$

*for a map $\psi_K \colon \mathrm{Spf}\, R \to (\mathcal{M}_{\mathbf{fg}})^{\wedge}_\Gamma$ and $\widehat{\mathbb{G}}$ a universal deformation of $\Gamma$.*

*Proof* The main content of the deformation theory of finite-height formal groups, recounted in Lecture 3.4, is that there is a natural correspondence between the following two kinds of deformation data:

$$\left\{ \begin{array}{c} \Gamma \leftarrow i^*\Gamma = j^*G \to G \\ \downarrow \qquad \downarrow \qquad \downarrow \\ \mathrm{Spec}\, k \xleftarrow{i} \mathrm{Spec}\, R/\mathfrak{m} \xrightarrow{j} \mathrm{Spf}\, R \end{array} \right\} \leftrightarrow \left\{ \begin{array}{c} G = \psi^*\widehat{\mathbb{G}} \to \widehat{\mathbb{G}} \\ \downarrow \qquad \downarrow \\ \mathrm{Spf}\, R \xrightarrow{\psi} (\mathcal{M}_{\mathbf{fg}})^{\wedge}_\Gamma \end{array} \right\}.$$

Accordingly, to construct the map $\psi_K$ from the lemma statement, we need only exhibit $G/K$ as belonging to a natural diagram of the sort at left. Using the fact that finite subgroups of formal groups over a *field* are always of the form $p^j \cdot [0]$, this is exactly what the Frobenius discussion above accomplishes when coupled to Corollary A.2.16:

$$\begin{array}{ccc} \Gamma \leftarrow (\varphi^j)^*\Gamma = (j^*G)/p^j \cdot [0] \to G/K \\ \downarrow \qquad \downarrow \qquad \downarrow \\ \mathrm{Spec}\, k \xleftarrow{\varphi^j} \mathrm{Spec}\, k \xrightarrow{j} \mathrm{Spf}\, R. \end{array}$$

Transferring this to a diagram of the sort at right, this gives the map $\psi_K$ and the isomorphism $g_K$, and the map $p_K$ is constructed as their composite. □

Applying this lemma to the universal case gives the following result:

**Corollary A.2.26** *There is a unique sequence of maps*
$$P^{\Sigma_{p^k}}: \widehat{\mathbb{G}} \times \mathrm{Sub}_{p^k} \widehat{\mathbb{G}} \to \psi^*_{p^k} \widehat{\mathbb{G}}$$
*determined by the following properties:*

1. *Restricting to any point in* $\mathrm{Sub}_{p^k} \widehat{\mathbb{G}}$ *gives an isogeny with that kernel.*
2. *Deformation of Frobenius: At the special fiber, $P^{\Sigma_{p^k}}$ reduces to the kth Frobenius iterate.* □

## $H_\infty$-Orientations

Having worked through enough of the underlying algebra, we now return to our intended application: $H_\infty$ complex-orientations of Morava $E$-theory. As we have avoided endowing $E$-theory with an $H_\infty$-structure so far, there are some open questions as to what this might mean. To resolve this ambiguity, recall first from Theorem A.0.1 that there exists at least one $H_\infty$-ring structure on Morava $E$-theory. Next, notice that the set of $H_\infty$-ring maps is the subset of homotopy classes of ring maps which commute with the $\Sigma_n$-power operations on the source and target. In the case of *MUP*-orientations, this has a surprising effect: Since an ordinary ring map $MUP \to E$ corresponds to a map $\mathbb{CP}^\infty \to E$, the theory of $H_\infty$-orientations can be entirely read off from the behavior of the power operations on $\mathbb{CP}^\infty$. The behavior of power operations as restricted to this setting is wholly captured in Corollary A.2.26: An $H_\infty$ structure on Morava $E$-theory carries a $\Sigma_{p^k}$-power operation map
$$E^0 \mathbb{CP}^\infty \otimes E^0 B\Sigma_{p^k} \leftarrow E^0 \mathbb{CP}^\infty,$$
which corresponds to the map of formal schemes
$$\begin{array}{ccc} \mathrm{Spf}(E^0 \mathbb{CP}^\infty \otimes E^0 B\Sigma_{p^k}/I_{\mathrm{Tr}}) & \longrightarrow & \mathrm{Spf}\, E^0 \mathbb{CP}^\infty \\ \| & & \| \\ \mathbb{CP}^\infty_E \times \mathrm{Sub}_{p^k} \mathbb{CP}^\infty_E & \xrightarrow{P^{\Sigma_{p^k}}} & \mathbb{CP}^\infty_E. \end{array}$$
This becomes an $E^0$-algebra map (i.e., a map of schemes over $(\mathcal{M}_{\mathbf{fg}})^\wedge_\Gamma$) when the bottom map is factored as
$$\mathbb{CP}^\infty_E \times \mathrm{Sub}_{p^k} \mathbb{CP}^\infty_E \xrightarrow{P^{\Sigma_{p^k}}} \psi^*_{\Sigma_{p^k}} \mathbb{CP}^\infty_E \to \mathbb{CP}^\infty_E.$$
This map has exactly the prescribed kernel, and by its very nature as a *power operation* it reduces to the Frobenius on the special fiber – and thus the unicity

## A.2 Orientations and Power Operations

of Corollary A.2.26 applies.[19] Since this analysis applies to *any* putative theory of $E$-theoretic power operations, we need not specify what $H_\infty$-structure we take on $E$-theory for the purpose of studying orientations.

Beyond these observations, there are two reductions, both due to McClure, that lighten our workload from infinitely many conditions to check to two:

**Theorem A.2.27** ([BMMS86, proposition VIII.7.2]) *Let $E$ and $F$ be $H_\infty$-ring spectra with $F$ $p$-local, and let $f\colon E \to F$ be a map of ring spectra in the homotopy category. Then $f$ is furthermore an $H_\infty$-ring map if and only if the following equation is satisfied:*

$$f \circ P_E^{\Sigma_p} = P_F^{\Sigma_p} \circ f_{h\Sigma_p}^{\wedge p}.$$
□

**Theorem A.2.28** ([HL16, proposition 33]) *Let $E$ be an $E_\infty$-ring spectrum and let $X$ be a space such that $E^{\Sigma_+^\infty X}$ is a wedge of copies of $E$. The map*

$$\iota^* \otimes \Delta^*\colon \widetilde{E}^* X_{hG}^{\wedge G} \to \widetilde{E}^* X^{\wedge G} \oplus \widetilde{E}^*(X \wedge BG_+)$$

*is then injective.*[20]  □

**Corollary A.2.29** ([AHS04, p. 271], [BMMS86, proposition VII.7.2]) *A ring map $x\colon MUP \to E_\Gamma$ is $H_\infty$ if and only if the internal power operations commute:*

$$x \circ \mu_{MUP}^{C_p} \circ \Delta = \mu_{E_\Gamma}^{C_p} \circ x_{hC_p}^{\wedge p} \circ \Delta \in E_\Gamma^0 T(\mathcal{L} \otimes \rho \downarrow \mathbb{C}P^\infty \times BC_p).$$
□

We now expand the algebraic condition that this last corollary encodes. The power operation for *MUP* was described in Lemma 2.5.6, where we found that it applies the norm construction to $f$ for the universal $\mathbb{Z}/p$-level structure on $\widehat{\mathbb{G}}$. The cyclic power operation on Morava $E$-theory was determined in the previous section to act by pullback along the deformation of Frobenius isogeny associated to the same universal level structure. This condition is important enough to warrant a name:

**Definition A.2.30** A coordinate $\varphi\colon \widehat{\mathbb{G}} \to \widehat{\mathbb{A}}^1$ on a Lubin–Tate group $\widehat{\mathbb{G}}$ is said to be *norm-coherent* when for all subgroups $K \subseteq \widehat{\mathbb{G}}$, we have that $N_K \varphi$ and $\psi_K^* \varphi$ are related by the isomorphism $g_K$.

---

[19] This identification is also compatible with the map sending a power operation to its constituent sum of unstable operations, i.e., the map from the natural $E_\infty$ context to the unstable context.

[20] The actual statement of McClure's result [BMMS86, proposition VIII.7.3] has several additional hypotheses: $\pi_* E$ is taken to be even-concentrated and free over $\mathbb{Z}_{(p)}$, $X$ has homology free abelian in even dimensions and zero in odd dimensions, and $X$ and $E$ are both taken to be finite type. Although this is the theorem cited in the source material ([And95, section 4], [AHS04, proof of proposition 6.1]), the version that has the weak hypotheses that we require only appeared in print much later.

**Corollary A.2.31** *The subset of those orientations which are $H_\infty$ correspond exactly to those coordinates which are norm-coherent, and it suffices to check the norm-coherence condition just for the universal $\mathbb{Z}/p$-level structure.* □

This characterization is already somewhat interesting, but it will only be truly interesting once we have found examples of such coordinates. Our actual main result is that such coordinates are remarkably (and perhaps unintuitively) common. In order to set up the statement and proof of this result, we consider the following somewhat more general situation:

**Definition A.2.32** A line bundle $\mathcal{L}$ on $\widehat{\mathbb{G}}$ together with a level structure $\ell \colon A \to \widehat{\mathbb{G}}$ induces a line bundle $N_\ell \mathcal{L}$ on $N_\ell \widehat{\mathbb{G}}$ according to the formula

$$N_\ell \mathcal{L} = \bigotimes_{a \in A} \tau^*_{\ell(a)} \mathcal{L}.$$

This line bundle interacts with the norm-coherence triangles well: $N_\ell \mathcal{L} = g^*_\ell \psi^*_\ell \mathcal{L}$. However, the operations on individual sections can be different: A section $s \in \Gamma(\mathcal{L} \downarrow \widehat{\mathbb{G}})$ is said to be *norm-coherent* when for any choice of level structure we have $N_\ell s = g^*_\ell \psi^*_\ell s$.

*Example A.2.33* In particular, functions $\varphi \colon \widehat{\mathbb{G}} \to \widehat{\mathbb{A}}^1$ can be thought of as sections of the trivial line bundle $O_{\widehat{\mathbb{G}}}$, and coordinates can be thought of as trivializations of the same trivial line bundle. We also remarked in Definition A.2.22 that the property of being a coordinate is stable under the operations on both sides of the norm-coherence equation.[21]

**Theorem A.2.34** ([And95, theorem 2.5.7], [PS], [Zhu17]) *Let $\mathcal{L}$ be a line bundle on a Lubin–Tate formal group $\widehat{\mathbb{G}}$, and let $s_0$ be a section of $\mathcal{L}_0$, the restriction of $\mathcal{L}$ to the formal group at the special fiber. Then there exists exactly one norm-coherent section $s$ of $\mathcal{L}$ itself which restricts to $s_0$.*

*Proof sketch* The first observation is that the norm-coherence condition can be made sense of after reducing modulo any power of the maximal ideal in the Lubin–Tate ring, and that the resulting condition is trivially satisfied in the case where the entire maximal ideal is killed. We then address the problem inductively: Given a norm-coherent section modulo $\mathfrak{m}^j$, we seek a norm-coherent section modulo $\mathfrak{m}^{j+1}$ extending this. By picking *any* lift extending this, we can test the norm-coherence condition and produce an error term that measures its failure to hold. In the specific case of the canonical subgroup $\widehat{\mathbb{G}}[p] \leq \widehat{\mathbb{G}}$, this error term is contained in the Lubin–Tate ring, and so we can modify our lift by subtracting

---

[21] This is reflected in topology as the assertion that the two power operations give two Thom classes for the regular representation, which must therefore differ by a unit.

off this error term. This perturbation has no effect on the "norm" side of the norm-coherence condition, and it has a linear effect on the "$\psi_{[p]}$" side, so it cancels with the error term to give a section which is norm-coherent *only when tested against the canonical subgroup*.

One can already show that the unicity clause applies: There is only one section $s$ of $\mathcal{L}$ which reduces to $s_0$ and which is norm-coherent for the canonical subgroup $\widehat{G}[p]$. In order to conclude that $s$ is truly norm-coherent, one shows that $N_\ell s$ satisfies its own $[p]$-norm-coherence condition (for the new canonical subgroup $(\widehat{G}/\ell)[p] \leq \widehat{G}/\ell$), forcing it to agree with $\psi_\ell^* s$. □

**Corollary A.2.35** (see [And95, theorem 5]) *For $\Gamma$ a finite-height formal group over a perfect base field and for any choice of $H_\infty$-ring structure on $E_\Gamma$, every orientation $s_0 \colon MUP \to K_\Gamma$ extends uniquely to a diagram*

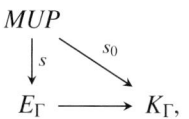

$$\begin{array}{ccc} & MUP & \\ {\scriptstyle s}\swarrow & \downarrow {\scriptstyle s_0} & \\ E_\Gamma & \longrightarrow & K_\Gamma, \end{array}$$

*where $s$ is an $H_\infty$-ring map.* □

*Example A.2.36* ([And95, section 2.7], see Remark 3.3.22) The usual coordinate on $\widehat{\mathbb{G}}_m$ with $x +_{\widehat{\mathbb{G}}_m} y = x + y - xy$ satisfies the norm-coherence condition. We compute directly in the case of the canonical subgroup:

$$N_{[2]}(x) = x(x +_{\widehat{\mathbb{G}}_m} 2) \qquad (p = 2)$$
$$= x(x + 2 - 2x) = 2x - x^2 = 2^*(x),$$

$$N_{[p]}(x) = \prod_{j=0}^{p-1}(x +_{\widehat{\mathbb{G}}_m} (1 - \zeta^j)) \qquad (p > 2)$$

$$= \prod_{j=0}^{p-1}(1 - \zeta^j(1-x)) = (1 - (1-x)^p) = p^*(x).$$

This is exactly the computation that $g_{[p]}^* \psi_{[p]}^*(x) = p^*$ and $N_{[p]}(x)$ agree for this choice of $x \in O_{\widehat{\mathbb{G}}_m}$.

The generality with which we approached the proof of Theorem A.2.34 not only clarifies which operations apply to the objects under consideration,[22] but also applies naturally to the other kinds of orientations discussed in Case Study 5.

---

[22] This is meant in contrast to Ando's original proofs [And95, section 2.6], where he deals only with coordinates and uses uncomfortable composition operations.

**Theorem A.2.37** *Orientations $MU[6, \infty) \to E_\Gamma$ which are $H_\infty$ correspond to norm-coherent cubical structures. If $\operatorname{ht} \Gamma \leq 2$, orientations $M\mathit{String} \to E_\Gamma$ which are $H_\infty$ correspond to norm-coherent $\Sigma$-structures.* □

*Example A.2.38* ([AHS04, sections 15–16]) Let $C_0$ be an elliptic curve over a perfect field of positive characteristic. The Serre–Tate theorem says that the infinitesimal deformation theory of $C_0$ is naturally isomorphic to the infinitesimal deformation theory of its $p$-divisible group $C_0[p^\infty]$, and we let $C$ be the universally deformed elliptic curve. Our discussion of norm-coherence for Lubin–Tate groups can be repeated almost verbatim for elliptic curves, and we note further that level structures on $\widehat{C}$ inject into level structures on $C$.

We can apply these observations to $E_{\widehat{C}}$, the Morava $E$-theory for the formal group $\widehat{C}$, considered as an elliptic spectrum. The natural orientation

$$MU[6, \infty) \to E_{\widehat{C}}$$

from Corollary 5.5.15 is determined by the natural cubical structure on $\mathcal{I}(0)$, which by Theorem 5.5.7 is *uniquely* specified by the elliptic curve. Our main new observation, then, is that the $N_\ell$ and $\psi_\ell^*$ constructions both convert this to a cubical structure on $C/\ell$, and hence are forced to agree by unicity. In turn, it follows that the $\sigma$-orientation is a map of $H_\infty$-rings.

*Remark A.2.39* In the next lecture we will show that Tate $K$-theory (see Definition 5.5.22) not only arises as an elliptic spectrum, but that this endows it with a natural $E_\infty$ structure. From the perspective of this section, the induced $H_\infty$ structure is controlled by quotients by finite subgroups of the Tate curve, which has been studied from various perspectives by Huan [Hua16] and Ganter [Gan07, G+13]. This distinguishes it from the $E_\infty$-ring spectrum $KU^{\mathbb{CP}_+^\infty}$, which otherwise has the same underlying ring spectrum in the homotopy category. This distinction also arises in classical homotopy theory: The total exterior power operation in classical $K$-theory can be shown to give an $E_\infty$ map to $K^{\text{Tate}}$ with its $E_\infty$-ring structure but *not* to $KU^{\mathbb{CP}_+^\infty}$.

## A.3 The Spectrum of Modular Forms

The introduction of the geometry of $E_\infty$-ring spectra has borne out a second version of the $\sigma$-orientation, summarized as a map of $E_\infty$-ring spectra

$$\sigma \colon M\mathit{String} \to \mathit{tmf}.$$

This map has *extremely* good properties, not only owing to it being a map of structured ring spectra but to the object *tmf* itself, not heretofore discussed.

## A.3 The Spectrum of Modular Forms

This is one variant of the spectrum of *topological modular forms*, which comes about from the following daydream: According to Corollary 5.5.15 and Corollary 5.6.17, every elliptic spectrum is naturally oriented, and the system of orientations should give rise to a *String*-orientation of the homotopy limit over all elliptic spectra, itself a kind of "universal" elliptic spectrum. There are several delicate points to this: the limit of a diagram of rings need not be a complex-orientable ring spectrum (see the $C_2$-equivariant spectrum $KU$); this diagram is very large; and the diagram exists only in the homotopy category[23] and contains loops, so "homotopy limit" is not automatically defined without finding a lift to a more structured context.

The goal of this lecture is to sketch out both the ingredients and the recipe for constructing this object. We will not prove the main theorem in full detail, as the details are so thick that they do not admit a more reasonable presentation than what Behrens has already given [Beh14]. The idea is to make use of the obstruction theory that $E_\infty$-rings garner us, and to "work locally" on a particular class of examples where the obstruction theory is well-behaved, using these to carefully exhaust the problem. We begin with a precise definition of what "locally" means in this setting:

**Definition A.3.1** ([Lur09, definition 6.2.2.6, section 6.5])    A D-*valued sheaf* on a site C is a functor $F \colon C \to D$ that converts Čech diagrams of covers in C to homotopy limit diagrams in D.

**Definition A.3.2** ([Goe10, remark 4.1])    Let $f \colon \mathcal{N} \to \mathcal{M}_\mathbf{fg}$ be a flat, representable morphism of stacks. A *topological enrichment* of $\mathcal{N}$ is a sheaf of $E_\infty$-ring spectra $O$ on $\mathcal{N}$ such that[24]

$$(\pi_n \circ O)(U \subseteq \mathcal{N}, \text{ affine and open}) \cong \begin{cases} (f^*\omega^{\otimes k})|_U & \text{if } n = 2k \text{ is even,} \\ 0 & \text{if } n \text{ is odd.} \end{cases}$$

For a fixed map $f \colon \mathcal{N} \to \mathcal{M}_\mathbf{fg}$, this has its own associated moduli problem of topological enrichments of $f$.

**Theorem A.3.3** (Goerss, Hopkins, and Miller [HM14, GH04]; Behrens [Beh14, Theorem 1.1]; Lurie [Lure])    *The moduli of topological enrichments of* $\mathcal{M}_\mathbf{ell}$, *the moduli of elliptic curves equipped with the étale topology, is connected.*

We now outline the weaker result that the moduli of enrichments is *nonempty*.

---

[23] Many of our elliptic spectra we produced by coupling Theorem 3.5.1 to Brown representability, which has poor functoriality properties.
[24] In particular, $\pi_0 O$ recovers the structure sheaf of $\mathcal{N}$.

## The Moduli of Elliptic Curves

In order to give a coherent strategy for proving Theorem A.3.3, we need to know something about the moduli of elliptic curves itself. Recalling from Lecture 5.5 the idea of a Weierstrass presentation of an elliptic curve, we define a general Weierstrass curve to be a projective curve specified by an equation of the form

$$C_a := \{y^2 + a_1 xy + a_3 y = x^3 + a_2 x^2 + a_4 x + a_6\},$$

the universal one of which is defined over

$$\mathcal{M}_{\text{Weier}} = \operatorname{Spec} \mathbb{Z}[a_1, a_2, a_3, a_4, a_6][\Delta^{-1}],$$

where the invertibility of the function $\Delta$ guarantees that these curves are nonsingular. The point at infinity in the set of projective solutions gives the curve a canonical marked point and hence the structure of an abelian variety. The fraction $y/x$ gives a coordinate in a neighborhood of the point at infinity, and hence Taylor expansion in this coordinate describes a map

$$\mathcal{M}_{\text{Weier}} \to \mathcal{M}_{\text{fgl}}.$$

Just as several formal group laws give the same formal group, several Weierstrass curves present the same elliptic curve, which are related by transformations of the form

$$f_{\lambda,s,r,t} \colon C_a \to C_{a'},$$
$$x \mapsto \lambda^2 x + r,$$
$$y \mapsto \lambda^3 y + sx + t,$$

the universal one of which is defined over

$$\mathcal{M}_{\text{Weier.trans.}} = \mathcal{M}_{\text{Weier}} \times \operatorname{Spec} \mathbb{Z}[\lambda^{\pm}, r, s, t].$$

The evident structure map belongs to a groupoid scheme structure on $\mathcal{M}_{\text{ell}} = (\mathcal{M}_{\text{Weier}}, \mathcal{M}_{\text{Weier.trans.}})$, for which there is a descended map

$$\mathcal{M}_{\text{ell}} \to \mathcal{M}_{\text{fg}}.$$

With the moduli of elliptic curves now specified, we will construct a topological enrichment of $\mathcal{M}_{\text{ell}}$ by doing so locally, then gluing the resulting local definitions together along common subsets. We first divide $\mathcal{M}_{\text{ell}}$ up over primes by passing to the $p$-completion, and then we further divide the $p$-complete moduli into two regions via the following result:

**Lemma A.3.4** *Over a base field of characteristic $p$, the $p$-divisible group of an elliptic curve is either formal of height 2 (called the supersingular case)*

## A.3 The Spectrum of Modular Forms

*or an extension of an étale p-divisible group of height 1 by a formal group of height 1 (called the ordinary case).*

*Proof* The category of Dieudonné modules with quasi-isogenies inverted becomes semisimple, and the simple components all take the form

$$M_{m,n} = \mathrm{Cart}_{\overline{\mathbb{F}}_p}/(V^m = F^n),$$

for $m$ and $n$ coprime [Man63]. An abelian variety is always isogenous to its dual [Mil08, section 7], and hence in this semisimple category the Dieudonné module associated to an abelian variety decomposes into a Cartier self-symmetric sum of generators.[25] An abelian variety of dimension $d$ has $p$-divisible group of height $2d$, and Cartier duality on these simple components obeys the formula $DM_{m,n} = M_{n,m}$, from which it follows that the only possibilities for the quasi-isogenous components of a $p$-divisible group associated to an elliptic curve are $M_{1,1}$ and $M_{1,0} \oplus M_{0,1}$. □

The names supersingular and ordinary are partially explained by the following result, which says that ordinary curves form the generic case and that supersingular curves are comparatively very rare:

**Lemma A.3.5** ([Sil86, theorem V.4.1.c], [Beh14, corollary 4.3]) *The supersingular locus of $\mathcal{M}_{\mathrm{ell}}$ is zero-dimensional.*[26] □

We write $i\colon \mathcal{M}_{\mathrm{ell}}^{\mathrm{ord}} \to (\mathcal{M}_{\mathrm{ell}})_p^\wedge$ for the open inclusion of the ordinary locus. We then plan to recover a topological enrichment $(\mathcal{O}_{\mathrm{top}})_p^\wedge$ of $\mathcal{M}_{\mathrm{ell}}$ by constructing the pieces of the following pullback:

$$\begin{array}{ccc} (\mathcal{O}_{\mathrm{top}})_p^\wedge & \longrightarrow & (\mathcal{O}_{\mathrm{top}})_{\mathcal{M}_{\mathrm{ell}}^{\mathrm{ss}}}^\wedge = \mathcal{O}^{\mathrm{ss}} \\ \downarrow & & \downarrow \\ \mathcal{O}^{\mathrm{ord}} = i_*i^*\mathcal{O}_{\mathrm{top}} & \longrightarrow & i_*i^*\left((\mathcal{O}_{\mathrm{top}})_{\mathcal{M}_{\mathrm{ell}}^{\mathrm{ss}}}^\wedge\right). \end{array}$$

This decomposition is compatible with the perspective on homotopy theory taken up in the rest of this textbook, as it is an instantiation of the chromatic fracture square. The top-right node forms the $\widehat{L}_2$-local component, the bottom-left forms the $\widehat{L}_1$-local component, and the bottom-right is the gluing data: the $\widehat{L}_1$-localization of the $\widehat{L}_2$-local component.

---

[25] This is called the *Riemann–Manin symmetry condition*.
[26] In fact, its image in the coarse moduli is the zero-locus of a polynomial, the $p$th Hasse invariant, which is of degree $(p-1)/2$. Many of these roots are repeated: There are actually at most $\lfloor (p-1)/12 \rfloor + 2$ distinct isomorphism classes over $\overline{\mathbb{F}}_p$.

## The Supersingular Locus

Our task in this section is to define $O^{ss}$ on $\widehat{\mathcal{M}^{ss}_{ell}}$, the infinitesimal neighborhood of the supersingular part of the topological enrichment in the larger moduli, and it suffices to specify its behavior on formal étale affines. Since the moduli is itself zero-dimensional, these are exactly the affine covers of the deformation spaces of the individual supersingular curves in the larger moduli $\mathcal{M}_{ell}$. The following arithmetic result gives us a crucial reduction:

**Theorem A.3.6** (Serre–Tate)  *The map $\mathcal{M}_{ell} \to \mathcal{M}_{pdiv}(2)$ is formally étale, where $\mathcal{M}_{pdiv}(2)$ is the moduli of p-divisible groups of height $2$.*[27]  □

**Lemma A.3.7** ([BL10, theorem 7.1.3])  *The deformation theory of a connected p-divisible group of height $d$ is isomorphic to the deformation theory of the associated formal group of height $d$.*  □

This reduces us to finding a topological enrichment for $(\mathcal{M}_{fg})^\wedge_C$, i.e., a version of Morava $E$-theory. A *very* extravagant application of the resolution tools for $E_\infty$-ring spectra yields the following theorem, essentially owing to the very nice (i.e., formally smooth) deformation space and very nice (i.e., formally smooth) space of operations:

**Theorem A.3.8** (Goerss, Hopkins, and Miller [GH04, corollary 7.6–7.7])  *Let $\Gamma$ be a finite-height formal group over a perfect field. The moduli of topological enrichments of $(\mathcal{M}_{fg})^\wedge_\Gamma$ is homotopy equivalent to $B \operatorname{Aut} \Gamma$. An element $\gamma$ of $\pi_1$ of this moduli based at a specific realization $E_\Gamma$ gives a cohomology operation $\psi^\gamma \colon E_\Gamma \to E_\Gamma$ whose behavior on $\mathbb{CP}^\infty_{E_\Gamma}$ is to induce the automorphism $\gamma$.*  □

Now we use the reduction above to extract from this a topological enrichment of $\widehat{\mathcal{M}^{ss}_{ell}}$: The enrichment sheaf arises as the pullback of the Goerss–Hopkins–Miller sheaf along the Serre–Tate map

$$\widehat{\mathcal{M}^{ss}_{ell}} = \coprod_{\text{s.s. } C} (\mathcal{M}_{ell})^\wedge_C \xrightarrow{\text{f.é.}} \coprod_{\text{s.s. } C} (\mathcal{M}_{pdiv}(2))^\wedge_{C[p^\infty]} \xleftarrow{\cong} \coprod_{\text{s.s. } C} (\mathcal{M}_{fg})^\wedge_C.$$

*Remark A.3.9* ([Beh14, section 8, step 1])  This buys more than just a bouquet of Morava $E$-theories, or even the global sections

$$O^{ss}\left(\widehat{\mathcal{M}^{ss}_{ell}}\right) = \prod_{\text{supersingular } C} E_{\widehat{C}}^{h \operatorname{Aut} C}.$$

For instance, the moduli $\mathcal{M}^{ss}_{ell}(N)$ of supersingular elliptic curves $C$ equipped with a level-$N$ structure[28] forms an étale cover of $\mathcal{M}^{ss}_{ell}$ whenever $p \nmid N$, and

---

[27] In general, $\mathcal{M}^d_{ab} \to \mathcal{M}_{pdiv}(2d)$ is formally étale [Del81, appendix 1].
[28] A *level-N structure* is a specified isomorphism $C[N] \cong (\mathbb{Z}/N)^{\times 2}$, i.e., a choice of basis for the $N$-torsion.

A.3 The Spectrum of Modular Forms    327

hence this sheaf produces a spectrum $TMF(N)^{ss} = O^{ss}(\mathcal{M}_{ell}^{ss}(N))$ satisfying $(TMF(N)^{ss})^{hGL_2(\mathbb{Z}/N)} \simeq TMF^{ss}$.

## The Ordinary Locus

We now turn to the ordinary locus, which constitutes the bulk of the problem: Remember that the supersingular locus was essentially discrete, and we are setting out to construct a sheaf, which means that we will be manufacturing a *lot* of spectra. The main tool for analyzing this situation is a specialization of the obstruction theory of Goerss, Hopkins, and Miller. First, note that completed $p$-adic $K$-homology (i.e., continuous Morava $E$-theory for $\widehat{\mathbb{G}}_m$) carries an action by $\operatorname{Aut} \widehat{\mathbb{G}}_m$, and using the results of Appendix A.2 this extends to an action by $\operatorname{End} \widehat{\mathbb{G}}_m$ using the $p$th power operation. In turn, the completed $p$-adic $K$-homology of an $E_\infty$-ring spectrum carries an action by $\operatorname{End} \widehat{\mathbb{G}}_m$, which is sometimes referred to as the structure of a $\theta$-*algebra*. To be more explicit:

**Theorem A.3.10** ([McC83])   *The homotopy of a $K(1)$-local commutative $K_p$-algebra spectrum $R$, such as $\widehat{L}_1(K_p \wedge E)$, carries an extra family of ring operations $\psi^k$ indexed on $k \in \mathbb{Z}_p$, as well as a ring map $\theta$,[29] such that*

$$\psi^1(x) = x, \qquad \psi^k(\psi^{k'} x) = \psi^{kk'}(x),$$
$$\psi^p(x) = x^p + p\theta(x), \qquad \psi^k(x) \cdot \psi^{k'}(x) = \psi^{k+k'}(x). \qquad \square$$

Goerss, Hopkins, and Miller obstruction theory reverses this information flow by seeking answers to the questions:

- We say that an $E_\infty$-ring *realizes* to a $\theta$-algebra $A$ when $\widehat{L}_1 E \wedge K_p \cong A$. Given a $\theta$-algebra $A$, what is the moduli of $E_\infty$-rings which realize to $A$?[30]
- Given a map $f : A \to B$ of $\theta$-algebras, as well as specified realizations $R$ and $S$ of $A$ and $B$ respectively, what is the moduli of maps $R \to S$ of $E_\infty$-rings which realize to $f$?

**Theorem A.3.11** (Goerss and Hopkins, $K(1)$-locally; [Beh14, theorem 7.1]) *Given a map of $\theta$-algebras $f : A_* \to B_*$, the following André–Quillen cohomology groups (internal to $\theta$-algebras) measure various obstructions:*

| Moduli problem | Existence | Uniqueness |
|---|---|---|
| A model $E$ for $A$ | $H_\theta^{s \geq 3}(A_*, \Omega^{s-2} A_*)$ | $H_\theta^{s \geq 2}(A_*, \Omega^{s-1} A_*)$ |
| A map $E \to F$ of models | $H_\theta^{s \geq 2}(A_*, \Omega^{s-1} B_*)$ | $H_\theta^{s \geq 1}(A_*, \Omega^s B_*)$ |

---

[29] If $E_*$ is torsion-free, then $\theta$ is redundant data, but that it is a ring map is not a redundant condition.

[30] Note that the homotopy of $E$ itself can be recovered from that of $R = \widehat{L}_1(K_p \wedge E)$ by taking fixed points for the $\mathbb{Z}_p^\times$-action, i.e., by an Adams spectral sequence.

*Finally, given such a map $f$, there is a spectral sequence computing the homotopy groups of the $E_\infty$ mapping space:*

$$E_2^{s,t} = H_\theta^s(A_*, \Omega^{-t}B_*) \Rightarrow \pi_{-s-t}(E_\infty \text{R\scriptsize INGSPECTRA}(E, F), f).\qquad\square$$

*Remark A.3.12* ([Beh14, lemmas 7.5–7.6]) These enhanced André–Quillen cohomology groups can be computed using a Grothendieck-type spectral sequence, intertwining classical André–Quillen cohomology groups for commutative rings with the extra task of checking compatibility with the $\theta$-algebra structure. In practice, this means that if the underlying ring of a $\theta$-algebra is especially nice, it is immediately guaranteed that the relevant obstruction groups vanish.

In order to apply this theorem, we need a guess as to what $\theta$-algebra should correspond to the completed $p$-adic $K$-theory of $TMF$. The discussion in Remarks 3.5.3 and 3.5.8 and Definition 3.5.4 provide the foothold we need. We expect the $\theta$-algebra to appear as the corner in the right-most of the following two pullback squares:

$$\begin{array}{ccccccc}
\cdots & \longrightarrow & \operatorname{Spf} W_1 & \longrightarrow & \operatorname{Spf} V_\infty^\wedge & \longrightarrow & \operatorname{Spf} \mathbb{Z}_p \\
& & \downarrow & \lrcorner & \downarrow & \lrcorner & \downarrow{\scriptstyle \text{f.é.}} \\
\cdots & \longrightarrow & \mathcal{M}_{\text{ell}}^{\text{ord}}(p^1) & \longrightarrow & \mathcal{M}_{\text{ell}}^{\text{ord}} & \longrightarrow & \mathcal{M}_{\text{fg}}.
\end{array}$$

Defined via these moduli, $V_\infty^\wedge$ has a natural structure as a solution to a moduli problem itself: It parameterizes pairs $(C, \eta: \widehat{\mathbb{G}}_m \xrightarrow{\cong} \widehat{C})$ of ordinary elliptic curves and markings of their associated formal groups. It carries a natural interpretation as a $\theta$-algebra [Beh14, equation 5.3], where the interesting operation $\psi^p$ acts by

$$\psi^p: \left(C, \eta: \widehat{\mathbb{G}}_m \xrightarrow{\cong} \widehat{C}\right) \mapsto \left(C^{(p)}, \eta^{(p)}: \widehat{\mathbb{G}}_m \to \widehat{C^{(p)}}\right),$$

where $\eta^{(p)}$ is the factorization

$$\begin{array}{ccccc}
\widehat{\mathbb{G}}_m[p] & \longrightarrow & \widehat{\mathbb{G}}_m & \xrightarrow{p} & \widehat{\mathbb{G}}_m \\
\downarrow & & \downarrow{\scriptstyle \eta} & & \downarrow{\scriptstyle \eta^{(p)}} \\
C[p] & \longrightarrow & C & \xrightarrow{p} & C^{(p)}.
\end{array}$$

Unfortunately, this $\theta$-algebra is not nice enough to apply the Goerss–Hopkins–Miller theorem. In order to fix this, it becomes convenient to work at $p \geq 5$ for simplicity, and then pass to a slightly more rigid moduli: We introduce a formal $(\mathbb{Z}/p)$-level structure, i.e., an isomorphism $\mathbb{Z}/p \cong (\widehat{C})[p]$.[31] This

---

[31] This partial rigidification of the formal component of the elliptic curve is related to the

## A.3 The Spectrum of Modular Forms

étale $(\mathbb{Z}/p)^{\times}$-cover of $\mathcal{M}_{\text{ell}}^{\text{ord}}$, known as the first Igusa cover, has the following exceptional property:

**Lemma A.3.13** ([Hid04, theorem 2.9.4], [Beh14, lemma 5.2]) *For $p \geq 5$, the moduli $\mathcal{M}_{\text{ell}}^{\text{ord}}(p)$ is affine.* □

**Corollary A.3.14** ([Beh14, theorem 7.7]) *Write $W_1$ for the associated $\theta$-algebra. It has vanishing Goerss–Hopkins–Miller obstruction groups, hence realizes uniquely to an ordinary $E_\infty$-ring spectrum $TMF(p)^{\text{ord}}$, and the action of $(\mathbb{Z}/p)^{\times}$ on the level structure enhances to a coherent $(\mathbb{Z}/p)^{\times}$-action on $TMF(p)^{\text{ord}}$.* □

We define $TMF^{\text{ord}}$, our candidate for the global sections of $O^{\text{ord}}$, to be the $(\mathbb{Z}/p)^{\times}$-fixed points of $TMF(p)^{\text{ord}}$, and indeed its $p$-adic $K$-theory is $V_\infty^{\wedge}$ [Beh14, lemma 7.9]. More than this, it turns out that the $\theta$-algebra associated to any formal étale affine open of $\mathcal{M}_{\text{ell}}^{\text{ord}}$ has a unique realization *as an algebra under $TMF(p)^{\text{ord}}$*, and maps between such also lift uniquely [Beh14, section 7, step 2]. Altogether, this gives us the desired sheaf $O^{\text{ord}}$ – and it shows that the potential complexity introduced by working with sheaves in an $\infty$-category does not arise in this case.

*Remark A.3.15* This is a common strategy: First find a topological enrichment of an affine cover of your stack of interest, then descend it to the stack itself.

### Gluing Data

We must also manufacture a map of sheaves

$$i_*i^* O_{\text{top}} \to i_*i^* \left((O_{\text{top}})^{\wedge}_{\mathcal{M}_{\text{ell}}^{\text{ss}}}\right)$$

to complete our putative pullback square. This is rather similar to the construction of $O^{\text{ord}}$ itself: We construct a candidate map

$$TMF^{\text{ord}} \to (TMF^{\text{ss}})^{\text{ord}} =: \widehat{L}_1(TMF^{\text{ss}})$$

of global sections, and then we use this to control the map of sheaves using relative Goerss–Hopkins obstruction theory. The main results that marry algebra to topology are the following two facts about $(TMF^{\text{ss}})^{\text{ord}}$. The first is that $(TMF^{\text{ss}})^{\text{ord}}$ counts as an elliptic spectrum:

**Lemma A.3.16** ([Beh14, lemma 6.8]) *There is an elliptic curve $C^{\text{alg}}$ over an affine $\text{Spf}((V_\infty^{\wedge})^{\text{ss}})$ such that $(TMF^{\text{ss}})^{\text{ord}}$ is an elliptic spectrum for this curve.*

observations of Hopkins on the connection between homology and stacky pullbacks mentioned in Remark 3.5.8 ([Hop14a, section 3.1], [Rezc, section 12]).

*Remarks on proof* This comes down to *algebraization*: In certain cases involving formal schemes, one can guarantee the existence of extensions of the following form:

$$\begin{array}{ccc} \mathrm{Spf}\, A & \longrightarrow & X \\ \downarrow & & \downarrow \\ \mathrm{Spec}\, A & \stackrel{\exists}{\dashrightarrow} & Y. \end{array}$$

Such a theorem appears here when studying the homotopy ring of $\widehat{L}_1 E_{\widehat{C}}$, which can be calculated to be

$$\pi_0 \widehat{L}_1 E_{\widehat{C}} = (\pi_0 E_{\widehat{C}})[u_1^{-1}]_p^\wedge,$$

which is no longer easily viewed as the ring of functions on a formal scheme. However, if the classifying map $\mathrm{Spec}\, \pi_0 E_{\widehat{C}} \to \mathcal{M}_{\mathrm{ell}}$ is first algebraized, these operations of localization and completion can be performed on ordinary affine schemes. This has its own wrinkle: Algebraization is hard to understand, for one, but we are also briefly obligated to replace $X = \mathcal{M}_{\mathrm{ell}}$ by a certain compactified moduli $Y = \overline{\mathcal{M}_{\mathrm{ell}}}$ of cubic curves with nodal singularities allowed. □

This specification of a map $\mathrm{Spf}((V_\infty^\wedge)^{\mathrm{ss}}) \to \mathcal{M}_{\mathrm{ell}}$ gives two candidates for a $\theta$-algebra structure on the $p$-adic $K$-theory of $TMF^{\mathrm{ss}}$: there is one coming from transfer of structure along the map of schemes, and there is one coming from the sheer fact that $(TMF^{\mathrm{ss}})^{\mathrm{ord}}$ is an $E_\infty$-ring spectrum, and hence topology simply imbues it with such algebraic structure.

**Theorem A.3.17** ([Beh14, theorem 6.10]) *The natural $\theta$-algebra structure on $\mathrm{Spf}((V_\infty^\wedge)^{\mathrm{ss}})$ induced by the map $\mathrm{Spf}((V_\infty^\wedge)^{\mathrm{ss}}) \to \mathrm{Spf}\, V_\infty^\wedge$ agrees with the Goerss–Hopkins–Miller $\theta$-algebra structure on $K_p(TMF^{\mathrm{ss}})$.* □

This is to be read as a recognition theorem for the $\theta$-algebra structure on the topological object $(TMF^{\mathrm{ss}})^{\mathrm{ord}}$: it matches the algebraic model. Once this is established, the Goerss–Hopkins–Miller obstructions to *existence* can be shown to vanish after introducing a suitable level structure; it follows that the map lifts to a $(\mathbb{Z}/p)^\times$-equivariant map of the $E_\infty$-rings of the global sections over the moduli with level structure; this descends to a map of global sections over the original module after taking $(\mathbb{Z}/p)^\times$-fixed points; and one finally produces the map of sheaves by further applications of relative obstruction theory [Beh14, section 8, step 2].

*Remark A.3.18* ([Beh14, section 9]) Arithmetic fracture is dealt with similarly, but it is *far* simpler. Because $\mathbb{Q} \otimes TMF$ has a smooth $\mathbb{Q}$-algebra as its homotopy, the Goerss–Hopkins–Miller obstructions governing the existence of maps of commutative $H\mathbb{Q}$-algebras vanish, letting us lift algebraic results into homotopy theory wholesale.

## Variations on these Results

*Remark A.3.19* ([Beh14, lemma 5.2, case 2])   At the prime 3, the proof of Igusa's theorem needs amplification, but the statement remains the same and the rest of the story goes through smoothly.

*Remark A.3.20* ([Beh14, lemma 5.2, case 3; section 8, step 1])   At the prime 2, two further things go wrong: one must pass to the Igusa cover $\mathcal{M}_{\text{ell}}^{\text{ord}}(4)$ before it becomes affine, but then the Galois group of this cover is $C_2$, which has infinite cohomological dimension at 2. Appealing to the equivalence $KO = KU^{hC_2}$, one works with 2-adic *real* $K$-theory instead, which pre-computes the Galois action.

*Remark A.3.21* ([Hop14b], [Lau04])   There is another way to construct $TMF^{\text{ord}}$ at low primes, given by a complex consisting of two $E_\infty$ cells attached to $\mathbb{S}$. The way this is done, essentially, is by constructing a complex whose $p$-adic $K$-theory matches the expected value: First it must have the right dimension, and then the action of $\theta$ must be corrected.

*Remark A.3.22* ([Law09, section 7])   There is an analogous (and much easier) picture for the moduli of forms of the multiplicative group: Any unordered pair of puncture points in $\mathbb{A}^1$ can be used to give $\mathbb{P}^1$ the unique structure of a group with identity at $\infty$, and the associated formal group is classified by a map $\mathcal{M}_{\mathbb{G}_m} \to \mathcal{M}_{\text{fg}}$; there is an equivalence $\mathcal{M}_{\mathbb{G}_m} \simeq BC_2$; and $KU$ forms the global sections of a topological enhancement of $\text{Spec}\,\mathbb{Z} \to \mathcal{M}_{\text{fg}}$ which descends using the complex-conjugation action to $BC_2 \to \mathcal{M}_{\text{fg}}$.

*Remark A.3.23*   With some effort, the construction of $\mathcal{O}_{\text{top}}$ outlined here extends to the compactified moduli $\overline{\mathcal{M}_{\text{ell}}}$ where Weierstrass curves with nodal singularities are allowed, i.e., where $\Delta$ is *not* inverted (as in $y^2 + xy = x^3$). The resulting global sections yields a spectrum $Tmf$, a close cousin to $TMF$. The connective truncation of that spectrum is denoted $tmf$, and it arises as the global sections of a topological enrichment of a stack of generalized cubics, i.e., where cuspidal singularities are also allowed (as in $y^2 = x^3$).

*Remark A.3.24* ([Sto12])   A topological enrichment can be thought of as an enhancement of a classical algebro-geometric object to a *spectral* (or *derived*) one. This opens the door for exploring all sorts of phenomena: For instance, there is a very interesting manifestation of Serre duality on $\mathcal{M}_{\text{ell}}$ in this enhanced setting, whose exploration is due to Stojanoska.

### Descent on Homotopy

One of the main upsides of producing a topological enrichment is that it is naturally equipped with a spectral sequence computing the homotopy of its global sections, coming from recovering $O(N)$ as the homotopy limit of finer and finer covers of $N$.

**Lemma A.3.25** *For $O$ a topological enrichment of a map $N \to M_{\text{fg}}$, there is a Bousfield–Kan spectral sequence*

$$E_2^{s,t} = H^s(N; \pi_t \circ O) \Rightarrow \pi_{t-s}O(N).$$  □

**Lemma A.3.26** *This spectral sequence is isomorphic to the $MU$-Adams spectral sequence for $O(N)$.*

*Main observation in the case of TMF*  Consider the Čech complex associated to the affine cover

$$M_{\text{Weier}} \to M_{\text{ell}}.$$

We claim that the complex making up the $E_1$-term of the descent spectral sequence is isomorphic to the complex making up the $E_1$-term of the $MU$-Adams spectral sequence. To illustrate, we compute the first two terms of each and compare them.

1. Consider the pullback diagram of stacks

$$\begin{array}{ccc} M_{\text{Weier}} & \longrightarrow & M_{\text{fgl}} \\ \downarrow & \lrcorner & \downarrow \\ M_{\text{ell}} & \longrightarrow & M_{\text{fg}}. \end{array}$$

   This is also the pullback diagram computing $\operatorname{Spec} MU_*TMF$.
2. Now consider the pair of cubes of iterated pullbacks pictured in Figure A.1. These compute the pullback of the cube in two different ways, producing an isomorphism $M_{\text{Weier.trans}} \cong (M_{\text{fg}} \times M_{\text{ps}}^{\text{gpd}}) \times_{M_{\text{fgl}}} M_{\text{Weier}}$.
$n$. The general case is similar, but requires stomaching iterated pullbacks in $n$-cubes.  □

*Example A.3.27*  We now appeal to basic results about $M_{\text{ell}} \times \operatorname{Spec} \mathbb{Z}[1/6]$ to compute $\pi_* TMF[1/6]$ using these methods.[32] After inverting 2 and 3, we can use scaling and translation transformations to complete both the cube and the

---

[32] This is admittedly a rather elaborate way of recovering the homotopy of the *complex-orientable* ring spectrum $TMF[1/6]$.

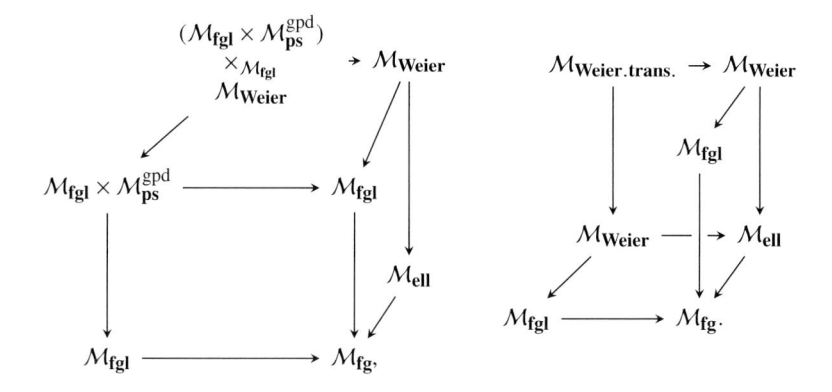

Figure A.1 Two expressions of the same cubical pullback.

square, replacing an arbitrary Weierstrass curve with a *unique* one of the form $y^2 = x^3 + c_4 x + c_6$. This exhausts the morphisms in the groupoid $\mathcal{M}_{ell}$: The map

$$\operatorname{Spec} \mathbb{Z}[c_4, c_6, \Delta^{-1}][1/6] \to \mathcal{M}_{ell} \times \operatorname{Spec} \mathbb{Z}[1/6]$$

is an equivalence of stacks (see Remark 5.5.1). Since the quasicoherent sheaf cohomology of affines is always amplitude 0, this spectral sequence is concentrated on the 0-line, and we recover

$$\pi_* TMF[1/6] \cong \mathbb{Z}[c_4, c_6, \Delta^{-1}][1/6].$$

*Remark A.3.28* The calculation of $\pi_* TMF$ has been completed integrally by Hopkins and Mahowald; see Bauer for detailed exposition of these results [Bau03]. One of the central features of the calculation is that a power of the periodicity generator from degree 2 survives, so that $\pi_* TMF$ becomes 576-periodic. By contrast, neither *Tmf* nor *tmf* are periodic spectra.

## A.4 Orientations by $E_\infty$ Maps

We now recount a more modern take on the story of the $\sigma$-orientation which passes directly through the algebra of $E_\infty$-ring spectra. Though technically intensive, our reward for grappling with this will be the modularity of the *String*-orientation, enriching Corollary 5.5.26 to the real setting. Luckily, most of the basic ideas are classically familiar, centering on a particular functor

$$gl_1 \colon E_\infty \text{R{\scriptsize ING}S{\scriptsize PECTRA}} \to \text{S{\scriptsize PECTRA}}.$$

This functor derives its name from two compatible sources: First, its underlying infinite loopspace is the construction $GL_1$ described in Lecture 1.1; and second, it participates in an adjunction

$$\text{ConnectiveSpectra} \xrightleftharpoons[gl_1]{\Sigma^\infty_+ \Omega^\infty} E_\infty\text{RingSpectra}$$

analogous to the adjunction between the group of units and the group-ring constructions in classical algebra. Its relevance to us is its participation in the theory of highly structured Thom spectra. Let $j \colon g \to gl_1 \mathbb{S}$ be a map of connective spectra, begetting a map $J \colon G \to GL_1 \mathbb{S}$ of infinite loopspaces, where we have written $G = \Omega^\infty g$.

**Lemma A.4.1** ([ABG+14, section 4], [May77, section IV.2]) *The Thom spectrum of the map $BJ$ is presented by the pushout of $E_\infty$-rings*[33]

$$\begin{array}{ccc} \Sigma^\infty_+ GL_1 \mathbb{S} & \xrightarrow{\Omega^\infty \Sigma j} & \Sigma^\infty_+ \Omega^\infty gl_1 \mathbb{S}/g \\ \downarrow & \ulcorner & \downarrow \\ \mathbb{S} & \longrightarrow & MG. \end{array}$$

□

**Corollary A.4.2** ([ABG+14, section 4], [May77, section IV.3]) *There is a natural equivalence between the space of null-homotopies of the composite*

$$g \xrightarrow{j} gl_1 \mathbb{S} \xrightarrow{gl_1 \eta_R} gl_1 R$$

*and the space of $E_\infty$-ring maps $MG \to R$, where $MG$ is the Thom spectrum of the stable spherical bundle classified by $J$.*[34]

*Proof* Applying the mapping space functor $E_\infty\text{RingSpectra}(-, R)$ to the pushout diagram in the lemma, we have a pullback diagram of mapping spaces:

$$\begin{array}{ccc} E_\infty\text{RingSpectra}(\Sigma^\infty_+ GL_1 \mathbb{S}, R) & \longleftarrow & E_\infty\text{RingSpectra}(\Sigma^\infty_+ \Omega^\infty gl_1 \mathbb{S}/g, R) \\ \uparrow & \ulcorner & \uparrow \\ E_\infty\text{RingSpectra}(\mathbb{S}, R) & \longleftarrow & E_\infty\text{RingSpectra}(MG, R). \end{array}$$

---

[33] This is a kind of "twisted group-ring" construction.
[34] The underlying infinite loopspace $GL_1 R$ also plays a role in the theory of Thom isomorphisms at the level of individual bundles: $R$-orientations of a spherical bundle $X \to GL_1 \mathbb{S}$ biject with null-homotopies of the composite $X \to GL_1 \mathbb{S} \to GL_1 R$. As an example consequence, first note that the sequence of maps

$$BO \xrightarrow{j} GL_1 \mathbb{S} \to GL_1 KO$$

is a 2-equivalence. It thus follows that if the tangent bundle of a manifold $\tau \colon M \to BO$ admits a Thom isomorphism in $KO$-homology, then there exists a lift $\tilde{\tau} \colon M \to BO[3, \infty) = BSpin$.

### A.4 Orientations by $E_\infty$ Maps

We can reidentify each of the three terms to get

$$\begin{array}{ccc} \text{SPECTRA}(gl_1\mathbb{S}, gl_1 R) & \longleftarrow & \text{SPECTRA}(gl_1\mathbb{S}/g, gl_1 R) \\ \uparrow & \ulcorner & \uparrow \\ \{gl_1\eta_R\} & \longleftarrow & E_\infty\text{RINGSPECTRA}(MG, R), \end{array}$$

hence $E_\infty\text{RINGSPECTRA}(MG, R)$ appears as the fiber at $gl_1\eta_R$ of the restriction map, which coincides with the space of null-homotopies as claimed. □

**Corollary A.4.3** ([AHRb, section 2.3])  *There exist $E_\infty$ maps $Mj \to R$ if and only if $gl_1\eta_R \circ j$ is null-homotopic. If this is the case, then the set*

$$E_\infty\text{RINGSPECTRA}(Mj, R)$$

*is a torsor for* $[\Sigma g, gl_1 R]$. □

Ando, Hopkins, and Rezk [AHRb] have used this presentation to understand the mapping space $E_\infty\text{RINGSPECTRA}(MString, tmf)$. In this lecture, we will use this same technology to understand the mapping space

$$E_\infty\text{RINGSPECTRA}(MSpin, KO_{(p)}),$$

which proceeds along entirely similar lines but is a *considerably* simpler computation.[35] The approach to this computation is to mix the presentation above with chromatic fracture applied to the target:

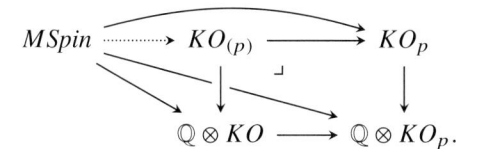

$$\begin{array}{ccc} MSpin \dashrightarrow KO_{(p)} & \longrightarrow & KO_p \\ & \downarrow & \downarrow \\ & \mathbb{Q} \otimes KO & \longrightarrow & \mathbb{Q} \otimes KO_p. \end{array}$$

So, we seek a pair of $E_\infty$-ring maps into the rationalization and the $p$-completion of $KO$ which agree on the $p$-local adèles, which involves understanding not just the mapping spaces but also the pushforward maps between them.

### Rational Orientations

We begin with the two rational nodes in the pullback diagram. As a first approximation to our goal, consider the problem of giving a complex-orientation $MU \to \mathbb{Q} \otimes R$ of a rational ring spectrum $\mathbb{Q} \otimes R$. There is an automatic such orientation granted by

---

[35] Ando, Hopkins, and Rezk also do $E_\infty\text{RINGSPECTRA}(MSpin, KO)$ as a warm-up computation [AHRb, section 7], and we are further $p$-localizing that result so as not to have to think about arithmetic fracture. Working arithmetically globally should be an easy exercise for the reader.

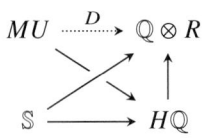

$$MU \xdashrightarrow{D} \mathbb{Q} \otimes R$$
$$\mathbb{S} \longrightarrow H\mathbb{Q}$$

constructed out of the unit map $\mathbb{S} \to MU$, the unit map $\mathbb{S} \to \mathbb{Q} \otimes R$, the rationalization map $\mathbb{S} \to \mathbb{Q} \otimes \mathbb{S} \cong H\mathbb{Q}$, and the standard additive orientation $MU \to H\mathbb{Q}$ of an Eilenberg–Mac Lane spectrum. When $E_\infty \text{RingSpectra}(MU, T)$ is nonempty, it is a torsor for $[bu, gl_1 T]$, and since we have a preferred orientation $D$ we thus have isomorphisms

$$\pi_0 E_\infty \text{RingSpectra}(MU, \mathbb{Q} \otimes R) \xleftarrow{\cong} [bu, gl_1 \mathbb{Q} \otimes R]$$
$$\xleftarrow{\cong} [bu, \mathbb{Q} \otimes gl_1 R]$$
$$\xrightarrow{\cong} [\mathbb{Q} \otimes bu, \mathbb{Q} \otimes gl_1 R],$$

the last of which is specified by a sequence $(t_{2k})_{k \geq 1} \in \pi_*(\mathbb{Q} \otimes gl_1 R)$. The role played by the sequence $(t_{2k})$ is to perturb the Thom class.

**Lemma A.4.4** ([AHRb, proposition 3.12]) *Write $x$ for the Thom class of $\mathcal{L}$ on $\mathbb{C}P^\infty$ in $(\mathbb{Q} \otimes R)$-cohomology as furnished by the automatic orientation $D$. The Thom class associated to some other orientation of $\mathbb{Q} \otimes R$ is tracked by a difference series $x / \exp_F(x)$, and the sequence $(t_k)$ is expressed by*

$$x / \exp_F(x) = \exp\left(\sum_k t_k / k! \cdot x^k\right).$$

*Proof sketch* Let $v^k \colon S^{2k} \to BU$ be the $k$th power of the class $\mathcal{L}$, so that it comes from a restriction

$$S^{2k} \to (\mathbb{C}P^\infty)^{\wedge k} \xrightarrow{\mathcal{L}^{\boxtimes k}} BU.$$

The Thom class for this bundle comes from the top Chern class, which is the top coefficient in the product of total Chern classes applied to the individual bundles. Following the usual formulas shows the map $v^k$ to behave on homotopy by multiplication by $(-1)^k t_k$. □

Now we move away from $MU$. There are three directions for generalization: connective orientations, real orientations, and non-complex targets.

1. The rational analysis of Ando, Hopkins, and Strickland identifies the set

$$[BU[2k, \infty), \mathbb{Q} \otimes R]$$

with $k$-variate symmetric multiplicative 2-cocycles over $R$, every one of which arises as $\delta^1$ repeatedly applied to a univariate series. In homotopy

theoretic terms, this means that every $MU[2k, \infty)$-orientation of a rational spectrum factors through an $MU$-orientation.
2. The cofiber sequence $kO \to kU \to \Sigma^2 kO$ splits rationally, using the idempotents $\frac{1 \pm \chi}{2}$ on $kU$. Accordingly, $MU$-orientations of rational spectra that factor through $MSO$-orientations have an invariance property under $\chi$: $-[-1](x) = x$, corresponding to the idempotent factor +. This pattern continues for the characteristic series of connective orientations.
3. This same cofiber sequence and idempotent splitting also tells us that rational $KU$-cohomology classes in the image of $KO$-cohomology are $\chi$–invariant, i.e., they belong to the − factor.

Our main example is the usual orientation $MU \to KU$ that selects the formal group law $x + y - xy$. This is associated to the difference Thom class $x/(e^x - 1) = x/\exp_{\widehat{\mathbb{G}}_m}(x)$. To make this difference $[-1]$-invariant (and hence give a complex-orientation of $\mathbb{Q} \otimes KO$), we use the averaged exponential class $(e^{x/2} - 1) - (e^{-x/2} - 1)$.[36] In turn, we use the lemma to calculate the behavior on homotopy of the associated orientation:[37]

$$\frac{x}{e^{x/2} - e^{-x/2}} = \exp\left(-\sum_{k=2}^{\infty} \frac{B_k}{k} \cdot \frac{x^k}{k!}\right).$$

Finally, we calculate the effect of the orientation on the second half of the factorization

$$MSU \to MSpin \to KO,$$

again using the relevant idempotent, which has the effect of halving the coefficients in the characteristic series: $-\frac{B_k}{2k}$.[38]

This accounts for both of the mapping sets

$E_\infty\text{RingSpectra}(MSpin, \mathbb{Q} \otimes KO), \quad E_\infty\text{RingSpectra}(MSpin, \mathbb{Q} \otimes KO_p).$

The set of rational characteristic series includes into the set of adèlic characteristic series as the subset with rational coefficients.

**Finite Place Orientations**

We want to understand $E_\infty\text{RingSpectra}(MSpin, KO_p)$ and the map

$E_\infty\text{RingSpectra}(MSpin, KO_p) \to E_\infty\text{RingSpectra}(MSpin, \mathbb{Q} \otimes KO_p).$

Here is the initial set-up:

---
[36] Incidentally, this is equal to $2\sinh(x/2)$.
[37] This comes out of applying d log to the fraction.
[38] While we're here, you might want to observe that elements in $[bu, gl_1 R]$ push forward to elements in $[bu, gl_1 \mathbb{Q} \otimes R]$ which do not disturb the denominators of the elements $t_k$. (On the other hand, the "Miller invariant" associated to a rational ring spectrum is *zero*, because arbitrary elements in $[bu, gl_1 \mathbb{Q} \otimes R]$ can completely destroy the denominators.)

$$\text{spin} \xrightarrow{j} gl_1\mathbb{S} \xrightarrow{\phantom{gl_1\eta_{KO_p}}} Cj$$
$$\phantom{\text{spin} \xrightarrow{j} gl_1\mathbb{S}} {}_{gl_1\eta_{KO_p}}\searrow \quad \downarrow A$$
$$\phantom{\text{spin} \xrightarrow{j} gl_1\mathbb{S} \xrightarrow{gl_1\eta_{KO_p}}} gl_1KO_p.$$

We are looking to understand the space of filler diagrams $A$ (i.e., vertical maps with choice of homotopy of the precomposite to $gl_1\eta_{KO_p}$). Notice first that there is a natural cofiber sequence to be placed on the bottom row:

$$\text{spin} \xrightarrow{j} gl_1\mathbb{S} \longrightarrow Cj$$
$$\searrow \quad \downarrow {}_{gl_1\eta_{KO_p}} \quad \downarrow A$$
$$\Sigma^{-1}\mathbb{Q}/\mathbb{Z} \otimes gl_1KO_p \to gl_1KO_p \to \mathbb{Q} \otimes gl_1KO_p \to \mathbb{Q}/\mathbb{Z} \otimes gl_1KO_p.$$

There is a canonical bolded vertical lift of $gl_1\eta_{KO_p}$ since $gl_1\mathbb{S}$ is a torsion spectrum, and this precomposes with $j$ to give the diagonal map. Notice now that selecting a filler triangle $A$ gives a commuting square with choice of homotopy and that $[gl_1\mathbb{S}, \mathbb{Q} \otimes gl_1KO_p] = 0$, and hence we would get a natural map (and natural homotopy) off of the homotopy cofibers:

$$\text{spin} \xrightarrow{j} gl_1\mathbb{S} \longrightarrow Cj \longrightarrow \text{bspin} \longrightarrow bgl_1\mathbb{S}$$
$$\searrow \quad \downarrow {}_{gl_1\eta_{KO_p}} \quad \downarrow A \; B \quad \downarrow C \quad \searrow \quad \downarrow$$
$$\Sigma^{-1}\mathbb{Q}/\mathbb{Z} \otimes gl_1KO_p \to gl_1KO_p \to \mathbb{Q} \otimes gl_1KO_p \to \mathbb{Q}/\mathbb{Z} \otimes gl_1KO_p,$$

where $C$ is a map making the triangle it belongs to commute. This all gives a function assigning $A$ to $B$ and $A$ to $C$ (and, in fact, the latter assignment factors through the former).

In order to show nonconstructively that the set of $A$s is nonempty, we might try to discern that $gl_1\eta_{KO_p} \circ j \in [\text{spin}, gl_1KO_p]$ is zero by investigating the mapping set $[\text{spin}, gl_1KO_p]$ itself. We proceed by a sequence of quite improbable steps, beginning with the following theorem original to Ando, Hopkins, and Rezk:

**Theorem A.4.5** ([AHRb, theorem 4.11])  *Let $R$ be a $E(d)$-local $E_\infty$-ring spectrum, and set $F$ to be the fiber*

$$F \to gl_1R \to L_d gl_1R,$$

*known as the discrepancy spectrum. Then $\pi_*F$ is torsion and $F$ satisfies the coconnectivity condition $F \simeq F(-\infty, d]$.* □

It follows that $gl_1KO_p \to L_1 gl_1KO_p$ is a 1-connected map, and hence

$$[\text{spin}, gl_1KO_p] = [\text{spin}, L_1 gl_1KO_p].$$

A.4 Orientations by $E_\infty$ Maps 339

In fact, we can even pass to the $K(1)$-localization, if we digress for a moment to introduce Rezk's logarithmic cohomology operation.

**Lemma A.4.6** ([Kuh08, theorem 1.1]) *For each $d \geq 1$, the Bousfield–Kuhn functor is a functor $\Phi_d$: $\text{Spaces}_{*/} \to \text{Spectra}$ which commutes with finite limits, is insensitive to upward truncation, and which evaluates on infinite loopspaces to give $\Phi_d(\Omega^\infty X) = \widehat{L_d} X$.*[39,40]   □

**Definition A.4.7** ([Rez06, section 3])   The natural equivalence

$$(GL_1 R)[1, \infty) \to (\Omega^\infty R)[1, \infty)$$

gives rise to a map $\ell_d$ as in the diagram

$$
\begin{array}{ccc}
\Phi_d(GL_1 R)[1, \infty) & \xrightarrow{\simeq} & \Phi_d(\Omega^\infty R)[1, \infty) \\
\| & & \| \\
gl_1 R \longrightarrow & \widehat{L_d} gl_1 R \xrightarrow{\simeq} & \widehat{L_d} R.
\end{array}
$$

$$\underbrace{\qquad\qquad\qquad}_{\ell_d}$$

*Remark A.4.8* Applying the logarithm to the corners in the height 1 chromatic fracture square yields the following identification:

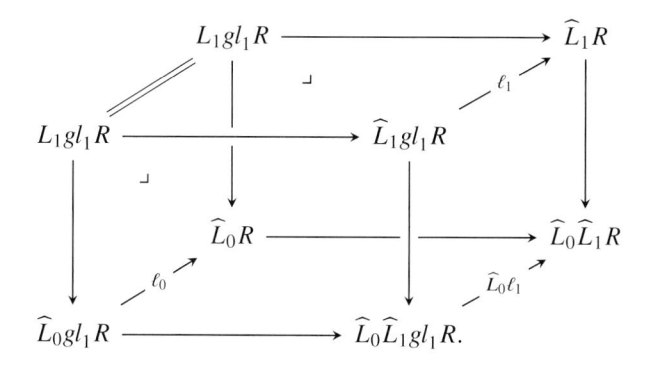

[diagram: cube with vertices $L_1 gl_1 R$, $\widehat{L_1} R$, $L_1 gl_1 R$, $\widehat{L_1} gl_1 R$, $\widehat{L_0} R$, $\widehat{L_0}\widehat{L_1} R$, $\widehat{L_0} gl_1 R$, $\widehat{L_0}\widehat{L_1} gl_1 R$, with maps $\ell_0$, $\ell_1$, $\widehat{L_0}\ell_1$]

The front and back faces are connected by logarithms of *different* heights – or, equivalently, the bottom horizontal arrow of the back face is *twisted* from the usual chromatic fracture presentation of $L_1 R$. The identification of this map is the usual sticking point in this approach.

**Theorem A.4.9** ([Rez06, theorem 1.9])   *For $R$ a $K(1)$-local $E_\infty$-ring with $\pi_0 R$*

---

[39] Importantly, $\Phi_d$ does *not* care about the actual infinite loopspace structure on $\Omega^\infty X$, just that it has *some* lift to a spectrum $X$.
[40] There is also a version of this theorem for $d = 0$, but since rational localization has no periodic behavior the results as not nearly as striking.

torsion-free, the map $\pi_0\ell_1 \colon \pi_0 R^\times \to \pi_0 R$ is given by the formula[41]

$$\ell_1(x) = \frac{1}{p}\log\left(\frac{x^p}{\psi^p x}\right) = \sum_{k=1}^\infty \frac{p^{k-1}}{k}\left(\frac{\theta(x)}{x^p}\right)^k. \qquad \square$$

**Corollary A.4.10** *The natural map $L_1 gl_1 KO_p \to \widehat{L}_1 gl_1 KO_p$ is a connective equivalence.*

*Proof* We specialize the above square to $R = KO_p$:

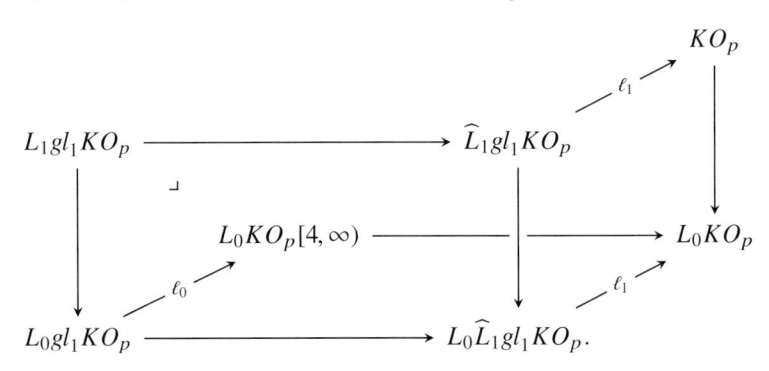

The behavior of the back horizontal map is determined by Rezk's formula for the logarithm. It acts by some nonzero number in every positive degree, hence the fiber has the form $\prod_{k=-\infty}^0 \Sigma^{4k-1} H\mathbb{Q}$. Since the front face is a fiber square, this is also a calculation of the fiber of the map in the lemma statement.[42] $\square$

As a consequence, we have identifications

$$gl_1 \eta_{KO_p} \circ j \in [spin, gl_1 KO_p] \cong [spin, L_1 gl_1 KO_p] \cong [spin, \widehat{L}_1 gl_1 KO_p].$$

A direct application of the Rezk logarithm replaces $\widehat{L}_1 gl_1 KO_p$ with $KO_p$, and the $K(1)$-localization of *spin* recovers $\Sigma^{-1} KO_p$. Altogether, this identifies

$$spin \xrightarrow{j} gl_1 \mathbb{S} \xrightarrow{gl_1 \eta_{KO_p}} gl_1 KO_p$$

with a point in the mapping set $[\Sigma^{-1} KO_p, KO_p]$ – and we mark this as a point where we would like to understand the space of $KO_p$-operations.

We claim also that the kernel of the assignment $A \mapsto C$ is easy to understand: Two fillers $A$ are related by an element of $[bspin, gl_1 KO_p]$, and their corresponding $C$s are related by the corresponding element of $[bspin, \mathbb{Q} \otimes gl_1 KO_p]$. This set

---

[41] The analog of this formula for $E_\Gamma$ (but not an arbitrary $K(d)$-local $E_\infty$-ring spectrum) is given in [Rez06, section 1.10].

[42] As a corollary of this same method, the Rezk logarithm for $R = KU_p^\wedge$ gives an equivalence $gl_1 KU_p^\wedge[3, \infty) \to KU_p^\wedge[3, \infty)$. This was previously known by nonconstructive methods to Adams and Priddy [AP76, corollary 1.4].

## A.4 Orientations by $E_\infty$ Maps

is rational, hence factors through the rationalization of $[bspin, gl_1 KO_p]$ where it must already be null, and hence it is a torsion element of $[bspin, gl_1 KO_p]$. Meanwhile, the same argument as above identifies

$$[bspin, gl_1 KO_p] = [KO_p, KO_p],$$

which we mark as a second point where we would like to understand $KO_p$-operations. If we were to find the group of degree-preserving $KO_p$-operations to be torsion-free, then the assignment $A \mapsto C$ would be *injective*.

We would like to understand the behavior of $C$ on homotopy based on some data about $A$. This serves two purposes: There is the necessary condition that the triangle formed by $C$ and the canonical map $bspin \to \mathbb{Q}/\mathbb{Z} \otimes gl_1 KO_p$ commute, and then also the composite

$$bspin \xrightarrow{C} \mathbb{Q} \otimes gl_1 KO_p \to (gl_1(\mathbb{Q} \otimes KO_p))[1, \infty)$$

describes the map into the adèlic component. In order to gain access to $C$, first notice that we can postcompose $B$ with the localization map off of $gl_1 KO_p$ as in Figure A.2.[43] This gives a new map $B' \colon KO_p \to KO_p$ – a third reason to understand $KO_p$-operations.

We are now in a position to compute the action of $C$ on a homotopy class in $\pi_* bspin$ by chasing through the following steps:

1. We push such a class forward to $\widehat{L}_1 bspin \simeq KO_p$ along the localization map.
2. We then pull it back to $\widehat{L}_1 Cj \simeq KO_p$ along $KO_p \xrightarrow{1-\psi^c} KO_p$, which acts by multiplication by $(1 - c^k)$ on $\pi_{4k}$.
3. We push it down along $B'$ to $\widehat{L}_1 gl_1 KO_p \simeq KO_p$, which acts by an unknown factor.
4. We include it into the rational component of $\mathbb{Q} \otimes \widehat{L}_1 gl_1 KO_p$, using the fact that $\pi_* \widehat{L}_1 gl_1 KO_p$ is torsion-free.
5. Finally, we pull it back to $\mathbb{Q} \otimes gl_1 KO_p$ along the logarithm $\ell_1$, which acts by multiplication by $(1 - p^{k-1})$ using Rezk's $K(1)$-local formula.[44]

The effect of this sequence of steps is

$$t_{4k} = (1 - c^k)^{-1} b_{4k} (1 - p^{k-1})^{-1},$$

where $t_{4k}$ and $b_{4k}$ are the effects on $\pi_{4k}$ of the maps $C$ and $B'$ respectively. In

---

[43] Importantly, and differently from what every source says, this isn't a map of cofiber sequences and so there is no commuting map with signature $\widehat{L}_1 bspin \to \mathbb{Q} \otimes \widehat{L}_1 gl_1 KO_p$.

[44] The formula for the logarithm in nonzero degrees comes from thinking of the logarithm as a *natural transformation* and applying it to the mapping set $\ell \colon gl_1 KO^0(S^{2n}) \to KO^0(S^{2n})$.

Figure A.2 A diagram showing the interconnections among the main components of the $p$-primary part of the Ando–Hopkins–Rezk argument.

## A.4 Orientations by $E_\infty$ Maps

the course of this proof, we are using the fact that division in the ring $\mathbb{Z}_p$ is unique when it is possible – the more responsible-looking equation to write is

$$b_{4k} = (1 - c^k)t_{4k}(1 - p^{k-1}).$$

Now, finally, the diagonal map $bspin \to \mathbb{Q}/\mathbb{Z} \otimes gl_1 KO_p$ becomes relevant. To check the commutativity of the triangle with $C$, we need only compare the results of the composite on homotopy since the map $C$ targets a rational spectrum and hence is determined by its effect on homotopy. The following invariance property makes this map accessible:

**Theorem A.4.11** ([AHRb, proposition 3.15, corollary 3.16])  *For any $A_\infty$-orientation $\varphi\colon MU \to R$ of an $A_\infty$-ring spectrum $R$, the denominators of the characteristic series associated to $\mathbb{Q} \otimes \varphi$ compute the behavior of the map $\pi_* BU \to \mathbb{Q}/\mathbb{Z} \otimes GL_1 R$.*  □

**Corollary A.4.12**  *The numbers $t_{4k}$ describing the effect of $C$ satisfy the congruences*

$$t_{4k} \equiv -\frac{B_k}{2k} \pmod{\mathbb{Z}}.$$

*Proof sketch*  The Todd orientation $MU \to KU$ is known to be $A_\infty$ [EKMM97, theorem V.4.1], and the characteristic series of the Todd orientation has coefficients $B_k$. The extra division by 2 is picked up by studying the map $\pi_* BSU \to \pi_* BSpin$ and the map $\pi_* KO \to \pi_* KU$.  □

We have thus identified the legal fillers $C$ as those sequences of rational numbers $t_{4k}$ satisfying conditions:

1. $t_{4k}$ has the correct denominators: for $k \geq 1$, $t_{4k} \equiv -B_k/(2k) \pmod{\mathbb{Z}}$.
2. $b_{4k}$ is the effect on homotopy of some map $B'\colon KO_p \to KO_p$.

### Stable $KO$-Operations

We have identified three points where we want to understand the collection of stable $KO_p$-operations (which we will abbreviate to $KO$, since in this subsection we are concerned only with the $p$-adic case). Although much of the main text of this book has been concerned with this sort of subject, this does not appear to be so immediately accessible: We want operations rather than cooperations, and $KO$ is *not* a complex-orientable ring spectrum. It is close to one, though, and we gain access to it through familiar approximation.

The easy initial calculation is the continuous $p$-adic $KU$-homology $KU^\vee KU$ can be computed to be $KU^\vee KU = \text{Spaces}(\mathbb{Z}_p^\times, \mathbb{Z}_p)$, the ring of $\mathbb{Z}_p$-valued functions[45] on $\mathbb{Z}_p^\times$ which are continuous for the adic topologies on the domain and the target. This comes out of the stable cooperations of Landweber flat homology theories discussed in Definition 3.5.4, where we showed that $E_\Gamma$ has cooperations given by the ring of functions on the pro-étale group scheme $\operatorname{Aut}\Gamma$. For $\Gamma = \widehat{\mathbb{G}}_m$, this group scheme $\operatorname{Aut}\widehat{\mathbb{G}}_m$ is constant at $\mathbb{Z}_p^\times$, so that $KU^\vee KU$ is the ring of $\mathbb{Z}_p$-valued functions on $\mathbb{Z}_p^\times$. Turning to cohomology, it follows by the universal coefficient spectral sequence that $KU^0 KU = \operatorname{Groups}(\operatorname{Spaces}(\mathbb{Z}_p^\times, \mathbb{Z}_p), \mathbb{Z}_p)$ and that $KU^1 KU = 0$. These correspondences behave as follows [AHRa]:

1. The Kronecker pairing

$$\mathbb{S}^0 \xrightarrow{c} KU \wedge KU \xrightarrow{1 \wedge f} KU \wedge KU \xrightarrow{\mu} KU$$

   is computed by the evaluation pairing

$$(c \in KU^\vee KU, f \in K^0 KU) \mapsto f(c).$$

2. The stable operation $\psi^\lambda$ attached to $[\lambda] \in \operatorname{Aut}\widehat{\mathbb{G}}_m$ operates by evaluation at $\lambda$.
3. The stable cooperation $v^{-k} \wedge v^k \in \pi_0 KU \wedge KU$ corresponds to the polynomial function $x \mapsto x^k$, as justified by the computation

$$\operatorname{ev}_\lambda(v^{-k} \wedge v^k) = \frac{\psi^\lambda v^k}{v^k} = \frac{\lambda^k v^k}{v^k} = \lambda^k.$$

These last two facts mean that the behavior of a stable operation on homotopy is identical information to the values of a functional $f$ on the standard polynomial functions $x^k$. We record this algebraic model as follows:

**Lemma A.4.13** *For any $N \geq 0$, the assignment*

$$\operatorname{Groups}(\operatorname{Spaces}(\mathbb{Z}_p^\times, \mathbb{Z}_p), \mathbb{Z}_p) \xrightarrow{(f(x \mapsto x^k))_k} \prod_{k \geq N} \mathbb{Z}_p$$

*is injective. A sequence $(x_k)$ is said to be a Kümmer sequence when it lies in this image.*[46]   □

*Remark A.4.14*   An interesting feature of the lemma is the auxiliary index $N$, which is *not* part of the property of being Kümmer. In $p$-adic geometry, this is reflected by the $p$-adic convergence of the sequence

$$d + (p-1)p^r \xrightarrow{r \to \infty} d,$$

---

[45] *Functions*, not homomorphisms!
[46] A bit more explicitly: $(x_k)$ is Kümmer when for all $h(x) = \sum_{k=N}^n a_k x^k \in \mathbb{Q}[x]$ we have $\sum_{k=N}^m a_k x_k \in \mathbb{Z}_p$.

### A.4 Orientations by $E_\infty$ Maps

and hence the continuous reconstruction property

$$x_d = \lim_{r \to \infty} x_{d+(p-1)p^r}.$$

In homotopy theory, this is reflected by the $p$-adic reconstruction property

$$KU \wedge KU[2k, \infty) \simeq KU \wedge KU.$$

*Remark A.4.15* With this computation in hand, the spectrum of $p$-local cooperations $KU_{(p)} \wedge KU_{(p)}$ can be recovered from arithmetic fracture, as can the global operations $KU \wedge KU$. The answer is quite similar: $\pi_0 KU \wedge KU$ is populated by rational polynomials which evaluate to integers on all integer inputs, called *numerical polynomials*.

To pass from $KU$ to $KO$, we begin with the Tate trick ([CM17], [GS96], [HS96]):

$$\begin{aligned}KU \wedge KO &\simeq KU \wedge (K^{hC_2}) & (KO \text{ is a homotopy fixed-point spectrum}) \\ &\simeq KU \wedge (KU_{hC_2}) & (\text{Tate objects vanish } K(1)\text{-locally}) \\ &\simeq (KU \wedge KU)_{hC_2} & (\text{homotopy colimits pull past smash products}) \\ &\simeq (KU \wedge KU)^{hC_2}, & (\text{Tate objects vanish } K(1)\text{-locally})\end{aligned}$$

so that $\pi_0 KU \wedge KO = \text{Spaces}(\mathbb{Z}_p^\times/C_2, \mathbb{Z}_p)$. Taking fixed points again, we then also have $\pi_* KO \wedge KO = \text{Spaces}(\mathbb{Z}_p^\times/C_2, KO_*)$, and $KO^*KO$ is the $KO_*$-linear dual. From this we gain the computations

$$[\Sigma^{-1}KO, KO] = 0, \quad [KO, KO] \cong \text{Groups}(\text{Spaces}(\mathbb{Z}_p^\times/C_2, \mathbb{Z}_p), \mathbb{Z}_p),$$

where the isomorphism is given by sending an operation $f : KO \to KO$ to the Kümmer sequence $(f(v^k)/v^k)_k \in \pi_0 KO = \mathbb{Z}_p$. We additionally see that $[KO, KO]$ torsion-free, which accounts for all of our outstanding claims.

### Mazur's Construction of Kubota–Leopoldt $p$-adic $L$-functions

Having learned enough about $KO$-operations to justify the program enacted in the previous subsections, we now need to show that there exist sequences of $p$-adic integers satisfying those criteria.

**Theorem A.4.16** ([MSD74]) *For any auxiliary $c \in \mathbb{Z}_p^\times$, there is a functional*

$f_c$ satisfying[47,48]

$$f_c(x^{k\geq 1}) = \frac{-B_k}{k}(1-p^{k-1})(1-c^k).$$

□

This theorem is stated in exactly the generality it was originally proven, and so you might wonder why Mazur had already proven *exactly* what we needed. To understand his program, recall these two facts about $\zeta$:

1. Except for a real Euler factor, $\zeta$ is basically the Mellin transform of the measure $\frac{dx}{e^x-1}$ (i.e., its sequence of moments):

$$\zeta(s) = \frac{1}{\Gamma(s)} \int_0^\infty x^{s-1} \frac{dx}{e^x-1}.$$

2. For any $k \in \mathbb{Z}_{>0}$, $\zeta(1-k) = -B_k/k$, where $\frac{t}{e^t-1} = \sum_{k=0}^\infty B_k \frac{t^k}{k!}$.

Mazur's idea was to build a $p$-adic $\zeta$-function by investigating similar $p$-adic integrals, beginning with certain finitary approximations to this one. To begin, a Bernoulli polynomial for $k \in \mathbb{Z}_{>0}$ is

$$\sum_{k=0}^\infty B_k(x) \frac{t^k}{k!} = \frac{te^{tx}}{e^t-1}.$$

These polynomials beget Bernoulli distributions according to the rule

$$\mathbb{Z}/p^n\mathbb{Z} \xrightarrow{E_k} \mathbb{Q} \subseteq \mathbb{Q}_p$$
$$x \in [0, p^n) \mapsto k^{-1} p^{n(k-1)} B_k(xp^{-n}).$$

A distribution in general is a function on $\mathbb{Z}_p$ such that its value at any node in the $p$-adic tree is equal to the sum of the values of its immediate children, and the $p$-adic integral of a locally constant function with respect to such a distribution is defined by their convolution. For example, the constant function 1 factors through $\mathbb{Z}/p$, hence

$$\int_{\mathbb{Z}_p} dE_k = \frac{1}{k} \sum_{a=0}^{p-1} B_k\left(\frac{a}{p}\right) = \overbrace{\frac{B_k(0)}{k}}^{\text{non-obvious}} = \frac{B_k}{k}.$$

---

[47] Explicitly,

$$f_c(h) = \int_{\mathbb{Z}_p^\times} h(x) d\mu_c = \lim_{r \to \infty} \frac{1}{p^r} \sum_{\substack{0 \leq i < p^r \\ p \nmid i}} \int_i^{ci} \frac{h(t)}{t} dt, \quad \int_{\mathbb{Z}_p^\times} d\mu_c = \frac{1}{p} \log(c^{p-1}).$$

[48] With considerable effort, this output can be halved [AHRb, section 10.3].

## A.4 Orientations by $E_\infty$ Maps

However, this distribution is not a *measure*, in the sense that it is not bounded and hence does not extend to a functional on all continuous functions (rather than just locally constant ones). The standard fix for this is called *regularization*: Pick $c \in \mathbb{Z}$ with $p \nmid c$, and set $E_{k,c}(x) = E_k(x) - c^k E_k(c^{-1}x)$. This is a measure, and for $k \geq 1$ it has total volume given by

$$\int_{\mathbb{Z}_p} dE_{k,c} = \int_{\mathbb{Z}_p} dE_k - c^k \int_{\mathbb{Z}_p} dE_k(c^{-1}x) = \frac{B_k}{k}(1 - c^k).$$

These measures interrelate: $E_{k,c} = x^{k-1} E_{1,c}$, and hence the single measure $E_{1,c}$ has all of these values as moments. We would like to perform $p$-adic interpolation in $k$ to remove the restriction $k \geq 1$, but this is not naively possible: If $k = 0$, say, then we naively have $E_{0,c} = x^{-1} E_{1,c}$, which will not make sense whenever $x \in p\mathbb{Z}_p$. This is most easily solved by restricting $x$ to lie in $\mathbb{Z}_p^\times$, which has a predictable effect for $k \in \mathbb{Z}_{>0}$:

$$\int_{\mathbb{Z}_p^\times} x^{k-1} dE_{1,c} = \int_{\mathbb{Z}_p} x^{k-1} dE_{1,c} - \int_{p\mathbb{Z}_p} x^{k-1} dE_{1,c}$$
$$= \int_{\mathbb{Z}_p} x^{k-1} dE_{1,c} - p^{k-1} \int_{\mathbb{Z}_p} x^{k-1} dE_{1,c}$$
$$= \frac{B_k}{k}(1 - c^k)(1 - p^{k-1}).$$

Hence, the Mellin transform of the measure $dE_{1,c}$ on $\mathbb{Z}_p^\times$ gives a sort of $p$-adic interpolation of the $\zeta$-function.

It also has *exactly* the properties we need to guarantee the existence of an $E_\infty$-orientation $MSpin \to KO$. It is remarkable that the three factors in

$$\int_{\mathbb{Z}_p^\times} x^{k-1} dE_{1,c} = \frac{B_k}{k}(1 - c^k)(1 - p^{k-1})$$

have discernable provenances in the two fields. In stable homotopy theory these arise respectively in the characteristic series of the orientation $MU \to KU$, in the finite Adams resolution for the $K(1)$-local sphere, and in the Rezk logarithm. In $p$-adic analytic number theory, they arise as the special values of the $\zeta$-function, the regularization to make it a measure, and the restriction to perform $p$-adic interpolation. It is completely mysterious how or if these operations correspond.[49]

---

[49] Walker's Ph.D. thesis gives a kind of generalization of this correspondence: For a height 1 formal group $\Gamma$ over a field of characteristic $p > 2$, he associates to an $H_\infty$-orientation $MU \to E_\Gamma$ a sequence of generalized Bernoulli numbers and demonstrates that they appear as the moments of a $p$-adic measure on $\mathbb{Z}_p^\times$ [Wal09].

## Notes on the *tmf* Case

The case of $MString \to tmf$ has all of the same trappings, but while many of the steps remain the same, many details become much more intricate:

1. Begin with a rational orientation, which is basically the Witten genus valued in holomorphic expansions of modular forms.
2. Analyze the homotopy type of $\widehat{L}_1 tmf$ and compare it to that of $KO$. This lets us use another universal coefficient theorem to lift our description of $KO^*KO$ as $KO^*$-valued measures to $(\widehat{L}_1 tmf)^* KO$ as $\widehat{L}_1 tmf^*$-valued measures.
3. The homotopy type of $\widehat{L}_2 tmf$ is "naively irrelevant" in the chromatic fracture square: maps $bspin \to \widehat{L}_2 tmf$ factor through $\widehat{L}_2 bspin = \widehat{L}_2 KO_p = 0$.
4. However, the logarithm's presence $\widehat{L}_1 tmf \to \widehat{L}_1 \widehat{L}_2 tmf$ has a real effect that must be understood. This is not easy: The higher height logarithm is not so accessible, and in general it is best cast in the language of $p$-divisible groups.[50] In the specific case of *tmf*, it requires a real understanding of its theory of power operations, which must be assembled out of the discussion of power operations for $E$-theories given above.
5. You also have to calculate the Miller invariant associated to *tmf*. In the case of $E_\infty \text{R{\scriptsize ING}S{\scriptsize PECTRA}}(MSpin, KO)$, one uses the $A_\infty$-orientation $MU \to KU$, as well as a rational understanding of the maps $\pi_* BU \to \pi_* BO$ and $\pi_* KO \to \pi_* KU$, and one must replace each of these ingredients with *tmf* analogs. The main point is that there are very few modular forms whose constant term is a given Bernoulli number.[51]
6. Finally, you have to ramp up the algebraic part of the calculation by identifying the analogs of the Mazur moments in $\pi_* \widehat{L}_1 tmf$. These turn out to be normalized Eisenstein series.

---

[50] Behrens has claimed, but has not yet put into print, that the transchromatic logarithms are described by an analog of Rezk's formulas for the logarithm at height $n$ [Rez06, sections 1.10 and 1.12], where the collection of subgroups appearing in the formula are replaced by those subgroups of a certain $p$-divisible group which have no interaction with the étale component.

[51] There is an alternative approach, along the lines of Miller's original calculation [Mil82], that takes as input a calculation of a certain $S^1$-transfer. There is, supposedly, a second alternative approach that applies the Bousfield–Kuhn functor to an $H_\infty$-orientation to determine the Miller invariant, but I am unable to reconstruct this argument.

# Appendix B

## Loose Ends

Thus ends our technical discussion of the interaction of algebraic geometry and algebraic topology. Before closing the volume entirely, we include a brief account of the historical context of these results, in the senses of adjacent mathematics, mathematical history, and future directions as yet unexplored.

### B.1 Historical Retrospective (Michael Hopkins)

I've been asked to describe what I remember about the genesis of the $\sigma$-orientation and related matters. I'll do my best, with the caveat that my collaborators may have different memories, and if they do, theirs are just as right as mine.

The story starts around the fall of 1989. I had spent a good deal of time in the late 1980s thinking about elliptic cohomology and not really getting anywhere, and I wanted to just think about homotopy theory for a while. At the time Haynes Miller and I, inspired by work of Andy Baker and Jim McClure, had shown that the space of $A_\infty$ automorphisms of the Morava $E$-theory spectrum associated with the universal deformation of a height $d$ formal group law over $\overline{\mathbb{F}}_p$ was homotopy discrete and equivalent to the semidirect product of the Morava stabilizer group and the Galois group of $\overline{\mathbb{F}}_p/\mathbb{F}_p$. In fact, Haynes and I worked out this proof at a reception for new faculty, in the very first conversation we had when I arrived at MIT in September 1989. The reason for wanting this action was to construct the spectrum $EO_d = E_d^{hH}$ associated to a finite subgroup $H$ of the Morava stabilizer. At the time that just seemed like a cool thing to do, but in retrospect the possibility of doing so must have been inspired by Ravenel's paper [Rav78a]. This led to the project of computing as many of the homotopy groups of $EO_d$ as one could. The first interesting examples were for $d = (p-1)$, which is the smallest value of $d$ for which there is an element of order $p$ in the stabilizer group. This was the case Ravenel pointed to in [Rav78a] and it had immediate applications to the non-existence of Smith–Toda complexes. The next case after that was the binary tetrahedral group in the height 2 stabilizer group at $p = 2$.

Back in those days I drove an Alfa Romeo Spider, and at some point it needed a new head gasket. I had dropped the car off at a small local shop, and when I went to pick it up the next day the mechanic told me he hadn't been able to get to it. I asked how long it would be, and he told me four hours. Conveniently, I was in that wonderful state of being obsessed with a math problem, so I just shrugged my shoulders and said that was cool, I'd wait. (Also, conveniently, this was before cell phones and the distractions of the internet.) I pulled out a pencil and paper, and I managed in that time to get a formula for the action of the group on the ring of functions and compute the cohomology. I also had a method for getting the first differential, and after writing it down I realized it formally implied all the other differentials. When I got home I wrote – by hand – a postscript file for a picture of the whole spectral sequence.

It happened that Mahowald made a visit to MIT shortly after that and I showed him the computation. He immediately recognized it and told me he had published a paper with Don Davis proving that such a spectrum could not exist. He also said he had never fully believed the proof but never could find anything wrong with it. I still don't know how he recognized the computation. What I had drawn was what we now think of as the Adams–Novikov spectral sequence for the $K(2)$-localization of $tmf$, and what Mahowald had related it to was a spectrum whose cohomology is $\mathcal{A}^*/\!\!/\mathcal{A}(2)^*$. Technically there wasn't quite a contradiction. However, we soon convinced ourselves this spectrum $EO_2$ probably did imply the existence of a spectrum whose cohomology is $\mathcal{A}^*/\!\!/\mathcal{A}(2)^*$, and that the Davis–Mahowald argument probably applied to $EO_2$ as well. We both worked pretty hard trying to find the resolution. I was worried about something foundational in the theory of $A_\infty$-ring spectra and devoted a lot of time to that, and Mahowald perused his argument with Davis. Mark and I went through the Davis–Mahowald argument very, very carefully. It involved a long series of incredibly dexterous moves, and I think I learned more about how homotopy groups work in checking that argument than from any other experience – but we couldn't find a mistake. On April Fool's Day Mark found the error: His paper with Davis was completely fine, and the error was in the accepted computations of the homotopy groups of spheres. Though I don't recall if this was 1990 or 1991, the day has stayed with me. Adams hadn't been gone long and it was his tradition to give a lecture every April Fool's Day proving two contradictory statements, and challenge the audience to find a mistake. Mark and I felt he had given us one last private April Fool's lecture.

## B.1 Historical Retrospective (Michael Hopkins)

Davis and Mahowald had really wanted to have a spectrum whose cohomology is $\mathcal{A}^*/\!/\mathcal{A}(2)^*$, and without it they had made do with the Thom spectrum $MO[8, \infty)$ associated to the 7-connected cover $BO[8, \infty)$. Bahri and Mahowald [BM80] had shown that there was an isomorphism of $\mathcal{A}$-modules

$$H\mathbb{F}_2^*(MO[8, \infty)) \approx \mathcal{A}^*/\!/\mathcal{A}(2)^* \oplus M$$

in which $M$ is 15-connected. This situation was meant to be an analog of the situation with *Spin* cobordism (with cohomology $\mathcal{A}^*/\!/\mathcal{A}(1)^* \oplus N$ for some 7-connected $N$) and connected $K$-theory $kO$ (with cohomology $\mathcal{A}^*/\!/\mathcal{A}(1)^*$). This made it natural to construct a non-periodic version $eo_2$ of $EO_2$ with cohomology $\mathcal{A}^*/\!/\mathcal{A}(2)^*$. Mahowald and I succeeded in doing so at the Mittag-Leffler Institute in the fall of 1993. It also made it natural to look for an "orientation"

$$MO[8, \infty) \to eo_2$$

analogous to the Atiyah–Bott–Shapiro orientation $MSpin \to kO$. At the time there was no known method of construction of the Atiyah–Bott–Shapiro orientation that did not rely on the interpretation of $KO$-theory in terms of vector bundles, so this seemed to be a tricky problem. I conceived of a two-stage program for doing this: The first step was to produce a map $MO[8, \infty) \to E_2$ invariant up to homotopy under the action of the binary tetrahedral group, and the second step was to rigidify everything in sight by requiring all of the maps to be $A_\infty$ or $E_\infty$.

When I got back to MIT in the winter of 1994, Matt Ando and Neil Strickland were around and we got to thinking pretty hard about trying to understand the $E_\infty$ or $A_\infty$ maps from $MO[8, \infty)$ to $E_2$. We weren't getting anywhere when Mark Hovey told us that computations he and Ravenel had done seemed to indicate that there couldn't be a map of spectra $MO[8, \infty) \to E_2$ invariant up to homotopy under the action of the binary tetrahedral group. This didn't look good for the first step of the program, so Matt and Neil and I started thinking about how one might understand the cohomology of $MO[8, \infty)$ in terms of formal groups. We found an answer in terms of cubical structures on formal groups and realized that there was a canonical map from $MO[8, \infty)$ to any complex-oriented cohomology theory $E$ whose formal group was the formal completion of an elliptic curve [AHS01]. This led to the picture conjectured in my 1994 ICM talk [Hop95] of the *tmf* sheaf and the relationship between the conjectured $MO[8, \infty)$ orientation and the Witten genus. It took a while, but eventually the *tmf* sheaf was constructed, we found the right way to think about $E_\infty$ orientations of Thom spectra, and with Charles Rezk were able to produce the $\sigma$-orientation as well as the Atiyah–Bott–Shapiro orientation using homotopy theory. I announced those results at the 2002 ICM [Hop02].

In the end the homotopy fixed point spectra $E_d^{hH}$ turned out to have many applications, from the non-existence of Smith–Toda complexes, computations in chromatic homotopy theory at low primes, and even to the Kervaire invariant problem. However, generalizing the whole package with the orientation (which, after all, was the problem that led to *tmf*) is still quite a mystery. In the early 1990s Hovey found an ingenious argument showing that for height $d > 2$ there can't be an orientation $MO[d, \infty) \to E_d^{hH}$ if $H$ contains a nontrivial element of order $p$. What should play the role of an orientation in those cases is pretty much up for grabs.

## B.2 The Road Ahead

In this section, we discuss various topics (sometimes very sketchily, sometimes in more detail) which are currently poorly understood but which also are now coming over the horizon. With any luck, a number of these will be resolved by the time a second printing of this book becomes a possibility, though many others will probably take the better part of a century to fully resolve. Accordingly, this is obviously a rather eclectic collection of loose ends; we have made only the barest effort to be comprehensive, and we have certainly missed huge swaths of relevant active research subjects out of preference for the author's own interests. While the inclusion of this section may "date" the book as the problems pass from unresolved to resolved, or as conjectures are confronted by counterevidence, for pedagogical reasons alone it seems important to name the kinds of questions that one might pursue using the tools we have built up.

These are sorted very roughly by their order of appearance in the main thread of the book.

### Splitting Bordism Spectra

Our first major project in this text, which culminated in Lemma 1.5.8, was to split $MO$ into a wedge of Eilenberg–Mac Lane spectra. This is actually the first of several splittings of bordism spectra, all proven by similar[1] methods:

1. First, one calculates the homology $H\mathbb{F}_{2*}MX$ of the bordism spectrum as a comodule over the dual Steenrod algebra.
2. Then, one shows that the associated sheaf $(H\mathbb{F}_{2*}MX)^{\sim}$ is the pushforward of a simpler sheaf for some smaller structure group $G \le \underline{\mathrm{Aut}}_1(\widehat{\mathbb{G}}_a)$. (This

---
[1] Similar, at least, at some macroscopic scale.

## B.2 The Road Ahead

amounts to understanding the Steenrod subcomodule generated by the unit class.)

3. Finally, one demonstrates some algebraic structure theorem for sheaves arising through such pushforwards. These are referred to generally as "Milnor–Moore-type theorems."
4. These results then assemble to yield a splitting at the level of 2-complete spectra through Adams spectral sequence methods.[2]

For instance, by studying the homology of $MSO$, one finds that the unit class in $H\mathbb{F}_{2*}MSO$ generates the submodule [Swi02, lemma 20.38]

$$1 \cdot \mathcal{A}_*/\!\!/(\mathrm{Sq}^1) \subseteq H\mathbb{F}_{2*}MSO,$$

and an appropriate generalization of Milnor–Moore then gives a splitting

$$MSO \simeq \left(\bigvee_j \Sigma^{n_j} H\mathbb{F}_2\right) \vee \left(\bigvee_k \Sigma^{m_k} H\mathbb{Z}\right).$$

Similarly, the homology of $MSpin$ reveals an inclusion ([ABS64], [ABP67], [GP84])

$$1 \cdot \mathcal{A}_*/\!\!/(\mathrm{Sq}^1, \mathrm{Sq}^2) \subseteq H\mathbb{F}_{2*}MSpin,$$

and a further generalization of Milnor–Moore ultimately produces a splitting[3]

$$MSpin \simeq \left(\bigvee_j kO[n_j, \infty)\right) \vee \left(\bigvee_k \Sigma^{m_k} H\mathbb{F}_2\right).$$

When studying $MString$, one finds that there is an inclusion

$$1 \cdot \mathcal{A}_*/\!\!/(\mathrm{Sq}^1, \mathrm{Sq}^2, \mathrm{Sq}^4) \subseteq H\mathbb{F}_{2*}MString,$$

but at this point the Adams spectral sequence becomes too intricate to analyze effectively and no analogous splitting of $MString$ has yet been produced.[4,5]

In addition to these splittings, there is also a collection of Landweber-type results that fall along similar lines:

---

[2] Replacing 2 by $p$ everywhere, one can also produce odd-primary results.
[3] However, this time the splitting is not one of ring spectra.
[4] See, however, recent work of Laures and Schuster [LS17, section 2].
[5] There are also $p$-local complex analogs of these results:

$$MU \simeq \bigvee_j \Sigma^{n_j} BP, \quad MSU \simeq \left(\bigvee_j \Sigma^{n_j} BP\right) \vee \left(\bigvee_k \Sigma^{m_k} BoP\right),$$

where $BoP$ is a certain amalgam of $BP$ and $kO$ described in the work of Pengelley [Pen82].

- Conner and Floyd [CF66] demonstrate the following:
$$MU_*(X) \otimes_{MU_*} KU_* \xrightarrow{\cong} KU_*(X),$$
$$MSp_*(X) \otimes_{MSp_*} KO_* \xrightarrow{\cong} KO_*(X).$$

- Hopkins and Hovey [HH92, theorem 1] demonstrate the following:
$$MSpin_*(X) \otimes_{MSpin_*} KO_* \xrightarrow{\cong} KO_*(X),$$
$$MSpin_*^{\mathbb{C}}(X) \otimes_{MSpin_*^{\mathbb{C}}} KU_* \xrightarrow{\cong} KU_*(X),$$
where $Spin^{\mathbb{C}} := O \times_{O[0,2]} U[0,2]$ belongs to a family of amalgams of the orthogonal and unitary groups.

- Ochanine [Och87a] demonstrates the following non-result:
$$MSU_*(X) \otimes_{MSU_*} KO_* \xrightarrow{\not\cong} KO_*(X).$$
In particular, this highlights the importance of the Hopkins–Hovey result: Although $KO$ also receives an $MSU$-orientation, this information is not enough to recover $KO$ by Landweber-type techniques.

- Landweber, Ravenel, and Stong ([Lan88], [LRS95]) originally constructed a variant of elliptic cohomology according to the following formula:
$$MSO_*(X) \otimes_{MSO_*} \mathbb{Z}\left[\frac{1}{6}, \delta, \varepsilon, \Delta, \Delta^{-1}\right] \Big/ \left(2^6 \varepsilon (\delta^2 - \varepsilon)^2 - \Delta\right) =: Ell^*(X).$$

These splittings and these flatness isomorphisms – and the absence of a known splitting in the *String* case! – are of great interest to a homotopy theorist. In the more obvious direction, they reveal a great deal about bordism spectra in terms of well-understood spectra, but less obviously they also promise to let information flow in the opposite direction. As a thought experiment, imagine an alternative history where we had never encountered real $K$-theory, but we had nonetheless embarked on a study of bordism theories. An intensive study of *Spin*-bordism would have led us inexorably toward $kO$, and once we had isolated $kO$ from the rest of *MSpin* we could then attempt to rig a geometric model for this dramatically simpler spectrum, ultimately leading us to the highly interesting and rewarding theory of vector bundles [Hov97, p. 338]. Similarly, one might hope that a splitting of *MString* would furnish us with spectra that themselves stand a high chance of admitting interesting geometric models, perhaps via as-yet undiscovered geometric constructions. Many of the original spectra under *MString* that were considered, like $KO^{\text{Tate}}$ and $EO_2$, have both nonconnected and completed homotopy, and so are unlikely to have direct geometric interpretation – but these complaints do not apply to *tmf*. Indeed, to a large extent, the start of this program (though not the promised geometric model) has been realized by the existence of *tmf* and of the $\sigma$-orientation: It is known that the $\sigma$-orientation produces a split inclusion

$$\mathcal{A}^*/\!/\mathcal{A}^*(2) \cong H\mathbb{F}_2^* tmf \to H\mathbb{F}_2^* MString$$

which is supported on the unit class. Since $\mathcal{A}^*/\!/\mathcal{A}^*(2)$ is an indecomposable Steenrod module, any splitting of $H\mathbb{F}_{2*}MString$ into indecomposable components will attach the unit class to a version of *tmf*.[6] The idea, then, is to take this as evidence that *tmf* wants to admit a geometric model, and we need only uncover what geometry does the job. In the more restrictive case of *Ell*, there actually *is* a geometric model, as uncovered by Kreck and Stolz [KS93]. They produced a geometrically defined integral cohomology theory using symplectic vector 2-bundles that agrees with the Landweber–Ravenel–Stong functor after inverting 6. Even ignoring the specific case of *tmf*, these generic methods may continue to point the way for geometry associated to the bordism spectra $MO[k,\infty)$ for still larger values of $k$. There is a great deal of ongoing work surrounding this problem, especially in the research programs of Hovey and Laures.

Even if $KO$ and $\widehat{L}_2TMF$ appear un-geometric, they appear to belong to the beginning of a recognizable pattern: $KO_p^\wedge = E_1^{hC_2}$ and $\widehat{L}_2 tmf = E_2^{hM}$ both arise as fixed point spectra for maximal finite subgroups of their respective stabilizer groups. One might therefore analogously define *higher real K-theories* by the formula $EO_d = E_d^{hM}$ for a maximal finite subgroup $M \leq \mathrm{Aut}\,\Gamma_d$. These spectra are interesting in their own right, but in light of the discussion above one might furthermore search for a connection between them and higher bordism spectra. An extremely interesting result of Hovey [Hov97, proposition 2.3.4] says that this likely cannot be made to work out as stated: For $p > 3$ and $k$ arbitrary, there is no map of ring spectra

$$MO[k,\infty) \to EO_{p-1}.$$

The meat of this theorem comes from two competing forces:

- The homotopy of $EO_{p-1}$ is known by computations of Hopkins and Miller to contain $\alpha_1$. Since the homotopy of $MO[k,\infty)$ agrees with that of the sphere below $k$ and destroys the $\alpha$ elements above $k$, this forces the value of $k$ to be large enough to accommodate the elements present in $\pi_* EO_{p-1}$.
- Because $EO_{p-1}$ is local, any orientation of it factors through $\widehat{L}_{p-1}MO[k,\infty)$. The Ravenel–Wilson acyclicity results for Eilenberg–Mac Lane spaces show that the natural map $\widehat{L}_{p-1}MO[k,\infty) \to \widehat{L}_{p-1}MO[p,\infty)$ is an equivalence

---

[6] However, as in the case of the Atiyah–Bott–Shapiro orientation $MSpin \to kO$, even this piece cannot participate in a splitting of ring spectra, as shown by McTague [McT13].

for any $k \geq p$. It follows that such an $MO[k, \infty)$-orientation of $EO_{p-1}$ is equivalent data to an $MO[p, \infty)$-orientation.

For $p > 3$, these bounds pass through each other and there is no satisfactory value of $k$. At the same time, because of the tight connection between stable homotopy theory and the Morava $E$-theories, this might be read as a failure of the connective bordism spectra $MO[k, \infty)$ more than a failure of higher real $K$-theories. It is interesting to try to imagine a continuation of the sequence $MO$, $MSO$, $MSpin$, $MString$ which does not suffer from Hovey's negative result. There is a slightly different continuation of the sequence $MU$, $MSU$, $MU[6, \infty)$ that has recently borne fruit. In the notation of Remark 4.3.10, the underlying spaces $BU$, $BSU$, and $BU[6, \infty)$ can be identified with the Wilson spaces $Y_2$, $Y_4$, and $Y_6$ at any prime. Although $MU[k, \infty)$ also suffers from Hovey's theorem, Hood Chatham has observed that by fixing a prime $p$ and appealing to the Adams splitting $Y_{2p} = \underline{BP\langle 1 \rangle}_{2p} \leq \underline{kU}_{2p}$, the resulting Thom spectrum is instead free from Hovey's theorem, and indeed $EO_{p-1}$ is $MY_{2p}$-oriented. The relation of this result to the program outlined above remains unexplored.[7]

## Blueshift and Redshift

This procedure of passing to (higher-order) vector bundles, as in the Kreck–Stolz construction, is embodied by passing from a ring spectrum to its algebraic $K$-theory spectrum. There are ongoing research programs with the goal of demonstrating that the algebraic $K$-theory of a ring spectrum of "chromatic complexity $d$" is itself of chromatic complexity one larger, as in the transition from $kO$ to $tmf$. This is one instance of a family of operations that (conjecturally) modulate chromatic height; such operations that raise height are called *redshifting*, and operations that lower height are called *blueshifting*, in reference to the periodicity perspective of Theorem 3.5.21.

As an example of a blueshifting construction, an important observation of Ando, Morava, and Sadofsky [AMS98] is that the $C_p$–Tate construction on the ring spectrum $E(d)$ is a variant of $E(d-1)$.[8] In fact, Hovey, Sadofsky, and Kuhn show that something similar happens quite generally: If $R$ is an $E(d)$-local ring spectrum and $G$ is a finite $p$-group, then the Tate object $R^{tG}$ is an $E(d-1)$-local

---

[7] Some footholds for the *algebraic* geometry of the large $k$ case extending Theorem 5.4.6 have been worked out [HLP13]. There is mild evidence that there is an isomorphism $O_{C^k(\hat{\mathbb{G}}_a;\hat{\mathbb{G}}_m)} \otimes \mathbb{Z}_{(2)} \to H\mathbb{Z}_{(2)*}BU[2k, \infty)$, corresponding to some of the input required by Corollary 5.4.9. However, this algebra contains 2-torsion (whose base-change to $\mathbb{F}_2$ appears to account for at least *some* of the complaints raised by Hugh, Lau, and Peterson [HLP13, remark 8.1]), which stymies the rest of that proof and hence blocks the analysis for a general complex-orientable cohomology theory. Additionally, the known results require $p = 2$, which means that any tether between this approach and Wilson spaces will involve higher-order truncated Brown–Peterson spectra, which directly connects neither with Chatham's result nor with $BU[2k, \infty)$. The ultimate fate of the schematic interpretation of connective complex-orientations is presently unknown.

[8] Somewhat more specifically, it gives $E(d-1)$ with homotopy tensored up by a quite *large* ring.

ring spectrum. For a striking example of this phenomenon, we consider the case of the 2-adic sphere spectrum, which is $K(\infty)$-local. We specialize to the case $G = C_2$ in order to perform an explicit analysis of $\mathbb{S}^{tC_2}$, which belongs to a stable fiber sequence

$$\mathbb{S}^{hC_2} \to \mathbb{S}^{tC_2} \to \Sigma \mathbb{S}_{hC_2}.$$

The orbit spectrum can be identified as

$$\mathbb{S}_{hC_2} \simeq \Sigma_+^\infty EC_2/C_2 = \Sigma_+^\infty BC_2 \simeq \Sigma_+^\infty \mathbb{RP}^\infty,$$

and similarly we can identify the fixed point spectrum as

$$\mathbb{S}^{hC_2} \simeq F_{C_2}(\Sigma_+^\infty EC_2, \mathbb{S}) \simeq D\Sigma_+^\infty \mathbb{RP}^\infty.$$

A remarkable theorem of Atiyah [Ati61] presents the Spanier–Whitehead dual of projective space. First, the calculation $T(\mathcal{L} \downarrow \mathbb{RP}^n) = \mathbb{RP}^{n+1}$ of Example 1.1.3 extends by the formula $T(m\mathcal{L} \downarrow \mathbb{RP}^n) = \Sigma_+^\infty \mathbb{RP}^{n+m}/\mathbb{RP}^{m-1}$ to positive values of $m$, and the same formula can be used to *define* complexes $\mathbb{RP}^n_m$ for negative values of $m$. Then, Atiyah demonstrates the formula

$$D\mathbb{RP}^{n-1}_m \simeq \Sigma \mathbb{RP}^{m-1}_n.$$

In our case, we thus calculate

$$\begin{array}{ccccc} \mathbb{S}^{hC_2} & \longrightarrow & \mathbb{S}^{tC_2} & \longrightarrow & \Sigma \mathbb{S}_{hC_2} \\ \| & & \| & & \| \\ \Sigma \mathbb{RP}^{-1}_{-\infty} & \longrightarrow & \Sigma \mathbb{RP}^{\infty}_{-\infty} & \longrightarrow & \Sigma \mathbb{RP}^\infty. \end{array}$$

By shifting this presentation slightly, we highlight an interesting comparison map. Note that the bottom cell of $\mathbb{RP}^n_0$ is unattached, and hence there is a natural wrong-way map

$$\mathbb{RP}^n_0 \to \mathbb{S}^0.$$

Taking Spanier–Whitehead duals and applying Atiyah's formula, we produce a map

$$\mathbb{S}^0 \to D\mathbb{RP}^{(n+1)-1}_{-(0)} = \Sigma \mathbb{RP}^{(0)-1}_{-(n+1)} \to \Sigma \mathbb{RP}^\infty_{-(n+1)}.$$

Taking the inverse limit gives a map

$$\mathbb{S}^0 \to \Sigma \mathbb{RP}^\infty_{-\infty},$$

and Lin showed this map to be a 2-adic equivalence [Lin80].[9] From the perspective of blueshift, the qualitative form of this is not entirely unexpected: The sphere has infinite chromatic complexity, and so blueshifting it by one stage yields another spectrum of infinite chromatic complexity. The precise form, however, is quite surprising: We produced exactly the sphere spectrum again!

---
[9] Carlsson proved a generalization of this to all other groups, known as the Segal conjecture.

This alternative presentation of the sphere spectrum as a Tate spectrum yields a filtration spectral sequence

$$\pi_* \mathbb{S}^* \Rightarrow \pi_*(\mathbb{S}^{-1})_2^\wedge.$$

In terms of this spectral sequence, any stable stem is represented by a (coset of) element(s) on the $E_1$-page, and this reverse reading of the spectral sequence is known as the *Mahowald root invariant* [MS88]. The root invariant appears to possess deeply interesting redshifting properties – not wholly surprising, since we are reading a blueshifting construction in reverse. For instance, the root of $p$ lies in the $\alpha$-family, and the root of an $\alpha$-family element lies in the $\beta$-family [Beh06b]. The root invariant has many striking connections to other areas of topology [MR93].

In addition to the chromatic properties waiting to be explored, the Tate construction itself has some truly puzzling features. Mahowald's original perspective on the Tate construction was through the definition

$$R^{tC_2} := \lim_{n \to \infty} \Sigma R \wedge \mathrm{P}^\infty_{-n},$$

whereas the general construction of Greenlees–May [GM95][10] uses the formula

$$R^{tG} = (F(EG_+, R) \wedge \widetilde{EG})^G,$$

whose pieces in the case at hand consist of $EG_+ = S(\infty \mathcal{L})$ and $\widetilde{EG} = S^{\infty \mathcal{L}}$, which rearranges to give

$$R^{tC_2} := \operatorname*{colim}_{m \to \infty} F(\mathrm{P}^\infty_{-m}, R).$$

Something quite spooky has happened: A colimit and a limit were inexplicably interchanged in the formulas

$$\mathbb{RP}^\infty_{-\infty} = \operatorname*{colim}_n \lim_m \mathbb{RP}^n_{-m} = \lim_m \operatorname*{colim}_n \mathbb{RP}^n_{-m}.$$

Understanding why these extremely different constructions give the same answer will likely yield some very important background theory.[11]

---

[10] Stroilova's Ph.D. thesis [Str12] gives a variation of this construction that converts an $E(n)$-local ring spectrum to an $E(n-k)$-local ring spectrum.

[11] The Telescope Conjecture is a famous problem which is at least superficially related. Namely, there are various "finitary" flavors of chromatic localization, which are typically less categorically robust but more computable. They assemble into a diagram:

$$\begin{array}{ccccc} E & \to & L_d^{\mathrm{fin}} E & \to & L_d E \\ \downarrow & & \downarrow & & \downarrow \\ L_{X(d)} E & \to & \widehat{L}_d^{\mathrm{fin}} E & \to & \widehat{L}_d E, \end{array}$$

where $X(d)$ is a finite complex of type exactly $d$, $v$ is a $v_d$-self-map of $X(d)$, $T(d) = X(d)[v^{-1}]$ is the localizing telescope, $\widehat{L}_d^{\mathrm{fin}}$ is Bousfield localization with respect to $T(d)$ (which can be shown to be independent of choice of $X(d)$ and of $v$), and $L_d^{\mathrm{fin}}$ denotes localization with respect to the class of *finite* $E(d)$-acyclics. Much is known about these functors: for instance, $L_{X(d)} L_d = \widehat{L}_d$,

## B.2 The Road Ahead

The blueshifting behavior of the Tate construction begs for insertion into the framework for understanding chromatic homotopy theory described in the main thread of this text. For instance, the formal-geometric perspective on the isomorphism in Lin's theorem is that the residue map $\mathbb{F}_2((x)) \to \mathbb{F}_2\{x^{-1}\}$ is a quasi-isomorphism of continuous Steenrod comodules [Str99b, remark 8.34].[12] The framework of $p$-divisible groups appearing around the edges of this book seems to show a lot of promise for serving as a general organizing principle for these "transchromatic" results, but this largely has not been worked out.[13,14]

### Why Formal Groups?

A question we have cheerfully left unresolved is: What is so special about $MU$ that makes it such an effective tool in studying stable homotopy theory? There are many approaches to this question that appear to lead in many different directions; we address several of them in turn.

First, one can ask this question just considering $MU$ as a ring spectrum. From this perspective, we summarized the most important results about $MU$ in Corollary 3.6.4: The context functor perfectly detects the Balmer spectrum of the global stable category. Given a general ring spectrum $R$, we can ask two questions analogous to this result:

1. Can one find an $R$-algebra $S$ whose context functor induces a homeomorphism of Balmer spectra $\mathrm{Spec}(\mathrm{Modules}_R^{\mathrm{perf}}) \to \mathrm{Spec}(\mathrm{Coh}(\mathcal{M}_{S/R}))$?[15]
2. Given a thick $\otimes$-ideal $\alpha \subseteq \mathrm{Modules}_R$, is there a complementary localizer

$$L_\alpha \colon \mathrm{Modules}_R \to \mathrm{Modules}_{R,(\alpha)}?$$

---

there is a chromatic fracture square relating $L_d^{\mathrm{fin}}$ to $\widehat{L}_{\leq d}^{\mathrm{fin}}$, and $L_d^{\mathrm{fin}} E \simeq L_d E$ if and only if $\widehat{L}_{\leq d}^{\mathrm{fin}} E \simeq \widehat{L}_{\leq d} E$. One major question about these functors remains open, corresponding to the last unsettled nilpotence and periodicity conjecture of Ravenel [Rav84, conjecture 10.5]: Is the map $\widehat{L}_d^{\mathrm{fin}} E \to \widehat{L}_d E$ an equivalence? Multiple proofs and disproofs have been offered ([Rav93], [MS95], [Kra00], [MRS01], ...) but the literature (and, indeed, the conjecture) remains unsettled. Our interest in this problem here is in the formula

$$\widehat{L}_d^{\mathrm{fin}} E = \mathrm{colim}\left( \cdots \to X(d) \wedge E \xrightarrow{v} X(d) \wedge E \to \cdots \right).$$

This formula uses a colimit, whereas the formula for the right-hand side given in Lemma 3.6.8 uses a limit.

[12] Reader beware: It is, of course, illegal to commute ordinary homology past the inverse limit.
[13] For instance, contact between $p$-divisible groups and the Tate construction is visible in work of Greenlees and Strickland ([GS99], [Strb, p. 10]), and $p$-divisible groups have played a central role in the height-modulating phenomena of Hopkins–Kuhn–Ravenel character theory (see Appendix A.1, [HKR00]) and in its extensions by Stapleton (see, for example, [Sta13], [Sta15]).
[14] It is formally obvious that $L_\Gamma$ is blueshifting, but in most cases this operation is computationally opaque. A theme in this area of homotopy theory is to recognize one form of height modulation in terms of another, so that their respective technical benefits can be simultaneously employed – for instance, this is the meat of chromatic character theory.
[15] Note that the right-hand side is a completely algebraic construction: these are simplicial sheaves of modules.

Can these localizers be presented via Bousfield's framework as homological localizations for auxiliary $S$-algebra spectra $S_\alpha$ (see Theorem 3.6.5)? Do the contexts $\mathcal{M}_{S_\alpha/R}$ admit compatible localizers with $\mathcal{M}_{S/R}$?

For $R = \mathbb{S}$, this is precisely the role that the $R$-algebra $S = MU$ and the $S$-algebras $S_d = E(d)$ play.[16]

There are some obvious restrictions on the $R$-algebra $S$: For instance, because we intend to use $S$ to form a context, we require that $\pi_* S$ (and $\pi_* S^{\wedge_R (j)}$ generally) be even. There are, in fact, two results in the literature that shed light on exactly this requirement. First, Priddy showed that by iteratively attaching cells to the $p$-local sphere in order to ensure that it has only even-dimensional homotopy, one arrives at $BP$ [Pri80]. More recently, Beardsley showed that $MU_{(p)}$ arises similarly by inductively attaching $A_\infty$-algebra cells to kill the odd homotopy on the $p$-local sphere [Bea17].[17] It would be satisfying to understand whether some construction like this holds in any sort of generality.

For a second perspective, using the adjunction

$$\text{RingSpectra}(MU, E) \cong \text{Spectra}_{\mathbb{S}/}(\Sigma^{-2}\Sigma^\infty \mathbb{CP}^\infty, E),$$

we might ask: What's so special about the space $\mathbb{CP}^\infty$? Much of the content of this book supports the following perspective: Given a space $X$, we consider the $E_\infty$-ring pro-spectrum $DX_+ = \{F(X_\alpha, \mathbb{S})\}_\alpha$. Because each $X_\alpha$ is a compact object, base-change along the unit map $\eta \colon \mathbb{S} \to E$ is computed by the following formula:

$$\eta^* DX_+ = E \wedge \{F(X_\alpha, \mathbb{S})\}_\alpha = \{E \wedge F(X_\alpha, \mathbb{S})\}_\alpha = \{F(X_\alpha, E)\}_\alpha.$$

Applying the functor $\text{Spf} \circ \pi_0$ to this pro-system yields the formal scheme $X_E$ considered in Definition 2.1.13. One of the overarching themes of this book has been to think of the objects $X$ and $DX_+$ as spectral incarnations of some algebro-geometric recipe, which base-change along $\eta$ to form classical algebro-geometric constructions which have been "bound" to $E$. The situation, as pointed out around Remark 3.1.21, is somewhat analogous to that of Lubin and Tate's explicit local class field theory, where a certain recipe associates to a local number field a governing formal group in terms of which much of the structure of the number field can be cast.[18] In the setting of algebraic topology,

---

[16] A potentially useful observation is that this does not appear to be a question in the domain of highly structured ring spectra. After all, the ring spectra $E(d)$ are not known to be $E_\infty$.

[17] He also identifies certain intermediate spectra in this process as the spectra $X(n)$ that arise for Devinatz, Hopkins, and Smith in their proof of the nilpotence conjectures [DHS88]. He *also* shows that this procedure works integrally, but there is lingering unexplained geometric information in how the cells are attached.

[18] Jack Morava is very insistent on parameterizing his $K$-theories not by a formal group but by a local number field $L$ and its Lubin–Tate $p$-divisible group, a trend not picked up on by most

such a "recipe" is embodied by $\mathbb{C}P^\infty$ itself. Understanding in what sense $\mathbb{C}P^\infty$ is encoding anything (or any of the other spaces discussed in this text) is an important challenge for homotopy theorists in years to come.[19]

Lastly, one can ask why these important structure theorems for stable homotopy theory, which have such a glittering link to complicated algebraic constructions, have *manifold geometry* ultimately underpinning them. Here I have nothing to offer but surprise and confusion.

## Adams Filtration Asymptotics

Various large-scale behaviors of the $MU$-Adams spectral sequence are poorly understood and would yield interesting information about stable homotopy theory as a whole. For instance, the following question is pulled from Mike Hopkins [Hop08, section 10]: Let $g(n)$ denote the largest $MU$-Adams filtration degree of an element of $\pi_n \mathbb{S}$ (i.e., $g$ traces the vanishing curve on the $E_\infty$ page of the $MU$-Adams spectral sequence). The nilpotence conjectures are all equivalent to the statement

$$\lim_{n\to\infty} \frac{g(n)}{n} = 0,$$

but little is known about the asymptotics of $g$ beyond this statement. For instance, one might ask: For what $\varepsilon > 0$ does the asymptotic formula $g(n) = O(n^\varepsilon)$ hold, and what is the infimum $\varepsilon_{\inf}$ over such values? Various values of the infimum have various consequences: Sufficiently small values entail the Telescope Conjecture (see earlier footnotes), and sufficiently large values entail its failure. Hopkins and Smith claim a plausibility argument that $\varepsilon_{\inf} = 1/2$, which is a Goldilocks value [HD90] where no consequence for the Telescope Conjecture can be deduced.[20]

---

other algebraic topologists. In connection with the "topological Langlands program" hinted at below, he is also highly interested in the Weil–Shafarevich theorem [Wei74, appendix III], which shows that all such number fields arise as maximal tori in certain (fixed) division algebras (dependent on the degree of the extension), and their Galois groups can be understood through this embedding. This forms a thread through his entire body of literature, but some of his most recent thoughts on this can be found in [Mor16].

[19] An extremely interesting and recent preprint of Lurie [Lurb] gives a real foothold on "spectral formal geometry," including non-complex-orientable examples of spectral formal groups, a characterization of Morava $E$-theory as an $E_\infty$-ring spectrum in this framework, and an analysis of what is "special" about the (spectral) formal group associated to $\mathbb{C}P^\infty$. This portends a more delicate description of the spectral role of $MU$: Whereas $MU_*$ classifies formal groups in the large, the behavior of $MU$-orientations must be somehow tightly bound to the particular spectral formal group associated to $\mathbb{C}P^\infty$, as in Remark 3.1.21.

[20] Variations on this question with other homology theories are also interesting. For instance, a consequence of the Hopkins–Smith periodicity theorems is that a finite complex is type $d$ if and only if $\lim_{n\to\infty} g(n)/n = (2(p^d - 1))^{-1}$ holds for $g$ formed from the $H\mathbb{Z}/p$-based Adams spectral sequence [Hop95, section 3.5].

Another important observation is that $\eta$ is not nilpotent in the $E_2$-term of the $MU$-Adams spectral sequence, and so generally the nilpotence theorems – the most impactful theorems known about the global stable category – do not hold in the algebraic model QCoH($\mathcal{M}_{\mathrm{fg}}$). Rather, the nilpotency of $\eta$ is enforced by a differential further in the spectral sequence. This differential is actually also algebraic, but of a different nature: It can be deduced from the $C_2$-fixed point spectral sequence for $KU^{hC_2} \simeq KO$. Understanding what bouquet of extra algebraic techniques account for the general nilpotency of stable homotopy elements would be very interesting.

Yet another interesting observation about this same homotopy fixed point spectral sequence is that it has a horizontal vanishing line on the $E_4$ page. This phenomenon is quite generic [MM15], and in particular it is also true of the descent spectral sequence for $TMF$. This is quite intriguing: The moduli stack of elliptic curves is Artinian (as are most "well-behaved" moduli stacks considered in arithmetic geometry), meaning that it has finite stabilizer groups. Such stacks are especially amenable to geometric study. However, this same finiteness is the source of the infinite cohomological dimension of the moduli of elliptic curves: After all, every nontrivial finite group has infinite group cohomology with coefficients in the trivial integral representation. In this sense, the *derived* moduli of elliptic curves enjoys both of these benefits: It has finite stabilizer groups, and it simultaneously is, in a certain sense, of *finite* cohomological dimension. Surely this is useful for something: There must be some facts that arithmetic geometers wish were true, but which are stymied by the infinite cohomological dimension of $\mathcal{M}_{\mathrm{ell}}$. Finding a use for this may well allow information to flow out of homotopy theory and into arithmetic geometry, itself an exciting prospect.

Finally, a precise understanding of vanishing curves in other Adams spectral sequences – for instance González's results for the $BP\langle 1 \rangle$-Adams spectral sequence [Gon00] – also gives rise to sparsity results in the $MU$-Adams spectral sequence, and hence control of the overall behavior.

## Local Analysis on $\mathcal{M}_{\mathrm{fg}}$

The stabilizer action $\mathrm{Aut}\,\Gamma_d \circlearrowright E_d^*$ is enormously, inhumanly complicated, as even a passing look at the computations originally pursued by Miller, Ravenel, and Wilson [MRW77] will make clear, never mind their many extensions by other authors over the decades since.[21] However quantitatively inaccessible, this action may yet be amenable to qualitative analysis, and there are a great many outstanding conjectures about its behavior in this sense. The largest one is

---

[21] *Unstable* chromatic homotopy theory is even worse off; it is so complicated that it has scarcely begun to be explored. Heuts [Heu18] and Wang [Wan15] are among the few modern results.

## B.2 The Road Ahead

the chromatic splitting conjecture [Hov95, conjecture 4.2], which asserts the following claims, listed in ascending order of severity:

1. $H^*(\mathrm{Aut}\,\Gamma_d; \mathbb{W}(\mathbb{F}_{p^d})[u^\pm])$ contains an exterior algebra $\Lambda[x_1, \ldots, x_d]$, where $x_j$ has cohomological degree 1 and transforms in the $(1 - 2j)$-th character of $\mathbb{G}_m$.[22]

2. Each nonzero class $x_{i_1} \cdots x_{i_j}$ in the exterior algebra pushes forward to a nonzero class in $H^*(\mathrm{Aut}\,\Gamma_d; (\mathcal{M}_{\mathbf{fg}})^{\wedge}_{\Gamma_d}[u^\pm])$, and it survives the $E_d$-Adams spectral sequence to give a nonzero homotopy class

$$x_{i_1} \cdots x_{i_j} : \mathbb{S}_p^{j-2i_+} \to \widehat{L}_d \mathbb{S}.$$

3. The composite

$$\mathbb{S}_p^{j-2i_+} \xrightarrow{x_{i_1} \ldots x_{i_j}} \widehat{L}_d \mathbb{S}^0 \to \Sigma F(L_{d-1}\mathbb{S}^0, L_d \mathbb{S}_p^0)$$

factors through

$$L_{d-\max i_k} \mathbb{S}_p^{j-2i_+}.$$

4. The maps above split $F(L_{d-1}\mathbb{S}^0, L_d\mathbb{S}_p^0)$ into $2^d - 1$ summands.

5. The cofiber sequence

$$F(L_{d-1}\mathbb{S}^0, L_d\mathbb{S}_p^0) \to L_{d-1}\mathbb{S}_p^0 \to L_{d-1}\widehat{L}_d\mathbb{S}^0$$

splits, so that

$$L_{d-1}\widehat{L}_d\mathbb{S}^0 \simeq L_{d-1}\mathbb{S}_p^0 \vee \Sigma F(L_{d-1}\mathbb{S}^0, L_d\mathbb{S}_p^0).$$

By exhaustive computation, this conjecture has been verified in the case $d = 1$ and in the case $d = 2, p \geq 5$. Very recently, Beaudry has shown that the final claim of the conjecture is *false* in the case $d = 2$ and $p = 2$ [Bea15], but leaves open the other statements in the full conjecture. Essentially everything else is unknown.

A subtle point in the above statement of the splitting conjecture is that the classes $x_j$ lie in the cohomology of $\mathbb{W}(\mathbb{F}_{p^d})$, which is *not* the Lubin–Tate ring. Remarkably, the natural map

$$H^*(\mathrm{Aut}\,\Gamma_d; \mathbb{W}(\mathbb{F}_{p^d})) \to H^*(\mathrm{Aut}\,\Gamma_d; (\mathcal{M}_{\mathbf{fg}})^{\wedge}_{\Gamma_d})$$

has turned out to be an isomorphism in the cases where the conjecture has been verified, and this has turned out to be a linchpin in the rest of the computation – and we do not have a conceptual reason for either of these facts. Specializing to

---

[22] One of these elements has a concrete description: $x_1$ is described by the determinant homomorphism, which sends a stabilizer element in $\mathrm{Aut}\,\Gamma_d$ to the determinant of its matrix representation as described in Theorem 3.6.16.

the case $d = 2$, Goerss has observed that this statement is equivalent to several others, any of which could be a hint toward a conceptual explanation:

- The natural map
$$H^*(\text{Aut }\Gamma_2; \mathbb{F}_{p^2}) \to H^*(\text{Aut }\Gamma_2; \mathbb{F}_{p^2}[\![u_1]\!])$$
is an isomorphism.
- The Frobenius map
$$\text{Frob}: H^*(\text{Aut }\Gamma_2; \mathbb{F}_{p^2}[\![u_1]\!]) \to H^*(\text{Aut }\Gamma_2; \mathbb{F}_{p^2}[\![u_1]\!])$$
is an isomorphism.
- The multiplication-by-$v_1^k$ map
$$v_1^k: H^*(\text{Aut }\Gamma_2; \mathbb{F}_{p^2}\{v_1^k\}) \to H^*(\text{Aut }\Gamma_2; \mathbb{F}_{p^2})$$
is zero.

In a different direction, one could hope to check that the natural map is an isomorphism just on the relevant torsion-free parts, by finding methods by which to study the maps

$$H^*(\text{Aut }\Gamma_d; \mathbb{W}(\mathbb{F}_{p^d})) \otimes \mathbb{Q} \to H^*(\text{Aut }\Gamma_d; (\mathcal{M}_{\text{fg}})^{\wedge}_{\Gamma_d}) \otimes \mathbb{Q}$$
$$\to H^*(\text{Aut }\Gamma_d; (\mathcal{M}_{\text{fg}})^{\wedge}_{\Gamma_d} \otimes \mathbb{Q}).$$

As announced by Morava [Mor85, remark 2.2.5], methods from the theory of $p$-adic analytic Lie groups show that the source of these maps has exactly the desired form, but comparing the rationalized cohomology with the cohomology of the rational representation is a delicate affair, since the group cohomology of Aut $\Gamma_d$ is being taken in a "profinite" sense.

In any event, this only addresses the "easiest" part of the conjecture, and the other parts take considerable effort even to parse properly. The last part is the most interesting: It is a statement about $L_{d-1}\widehat{L_d}\mathbb{S}$, and hence about the interplay between two different chromatic heights. Torii has spent much of his career analyzing algebraic models of this phenomenon – in particular, the action of Aut $\Gamma_{d-1}$ on the "punctured Lubin–Tate ring" $\mathbb{F}_{p^d}(\!(u_d)\!)[u^{\pm}]$ ([Tor03], [Tor07], [Tor10a], [Tor10b], [Tor11]). Recent activity on this front indicates that there is much to mine from this vein, and that Torii's program deserves more attention than it has so far garnered.

A related point of interest is the action of Aut $\Gamma_d$ on $\mathbb{F}_{p^d}[\![u_d]\!][u^{\pm}]$. The action of Aut $\Gamma_d$ on the special fiber of the Lubin–Tate ring is somewhat well-understood [Rav77]; for instance, in the case $d = 2$ we have a calculation

$$H^*(\text{Aut }\Gamma_2; \mathbb{F}_{p^2}) \cong (\mathbb{F}_{p^2}[\zeta][u^{\pm(p^2-1)}])\{1, h_0, h_1, g_0, g_1, t\} \bigg/ \left( \begin{array}{l} h_0 g_1 = t, \\ h_1 g_0 = t \end{array} \right),$$

with all unlabeled products of the Roman elements equal to zero. This forms the input to a Bockstein spectral sequence

$$H^*(\mathrm{Aut}\,\Gamma_2; \mathfrak{m}^j/\mathfrak{m}^{j+1}) \Rightarrow H^*(\mathrm{Aut}\,\Gamma_2; \mathbb{F}_{p^2}[\![u_1]\!][\![u^{\pm}]\!]),$$

where $\mathfrak{m} = (u_1)$ is the maximal ideal in this local graded ring, and hence the representation $\mathfrak{m}^j/\mathfrak{m}^{j+1}$ is (a twist of) the representation $\mathbb{F}_{p^2}$. This Bockstein spectral sequence displays a number of intriguing phenomena[23] – for instance, it has "periodic" differentials [Sad93]. One such periodic family is specified by

$$d_1 u^{1-p^2} = (u_1 u^{1-p}) \cdot h_1,$$
$$d_p u^{p(1-p^2)} = (u_1 u^{(1-p)})^p \cdot u^{(p-1)(1-p^2)} \cdot h_0,$$
$$d_{p^n+p^{n-1}-1} u^{p^n(1-p^2)} = 2(u_1 u^{(1-p)})^{p^n+p^{n-1}-1} \cdot u^{(p^n-p^{n-1})(1-p^2)} \cdot h_0.$$

Even more interestingly, there are explicit lifts of these cohomology classes to cochains in the cobar complex that witness these differentials, given as follows:

$$v_2^{(n)} = (u^{1-p^2})^{p^n} \prod_{j=0}^{n-2}\left(1 - u_1^{(p+1)(p^{n-1}-p^j)}\right) \quad (\mathrm{mod}\ u_1^{p^n+p^{n-1}}).$$

As $n$ grows large, this formula looks curiously like an analytic Weierstrass product. It would be interesting to know what function it names, which could perhaps help illuminate the behavior of the spectral sequence itself.

### *p*-adic Interpolation

In projective geometry, one studies a projective variety by calculating its global sections against different line bundles:

$$X \mapsto [(\mathcal{L} \in \mathrm{Pic}(X)) \mapsto H^0(X; \mathcal{L})].$$

Done correctly, this can be used to recover a graded ring $R$ with a natural map $\mathrm{Proj}(R) \to X$ which is definitionally an isomorphism in the case that $X$ is affine – that is, this construction captures the isomorphism type of $X$.

This is somewhat analogous to Whitehead's theorem in homotopy theory: The Picard group of the stable category (i.e., the group of isomorphism classes of spectra $Y$ such that there exists a $Y^{-1}$ with $Y \wedge Y^{-1} \simeq \mathbb{S}^0$) consists of precisely

---

[23] For $d > 2$ there are a succession of similar Bockstein spectral sequences, where the Lubin–Tate generators are reintroduced to the Lubin–Tate ring one at a time. The first of these spectral sequences always displays these same intriguing phenomena, but the later ones are much more poorly understood.

the stable spheres. However, in the $\widehat{\mathbb{G}}_m$-local category there are many more invertible objects – in fact, the Picard group of this category has the form

$$\operatorname{Pic}(\operatorname{Spectra}_{\widehat{\mathbb{G}}_m}) \cong \mathbb{Z}_p \times \mathbb{Z}/(2p-2),$$

where the right-hand factor exactly tracks the degree of a generator of the $K(1)$-homology of an invertible spectrum. Although Whitehead's theorem only requires testing against the "standard spheres," now a mere subgroup of the *much* larger Picard group, the homotopy groups of $K(1)$-local spectra as graded over the larger group display considerable extra structure. For instance, the homotopy groups of the $\widehat{\mathbb{G}}_m$-local sphere graded over this larger group are described by the same formula given in Example 3.6.19:

$$\pi_t \widehat{L}_1 \mathbb{S}^0 = \begin{cases} \mathbb{Z}_p & \text{when } t = 0, \\ \mathbb{Z}_p/(pk) & \text{when } t = k|v_1| - 1, \\ 0 & \text{otherwise,} \end{cases}$$

where the symbol $k$ can now be taken to be any $p$-adic integer. This exactly enforces a kind of $p$-adic continuity of this family of groups.

The idea is that these kinds of observations can be used to bundle the behavior of the homotopy of the $\Gamma_d$-local sphere into a digestible format. The computation of the homotopy of the $\Gamma_2$-local sphere has been fully executed by Shimomura and collaborators ([Shi86], [SY94], [SY95], [Beh12]), but it is *exceedingly* complicated. The role of continuity properties is to reduce the seemingly erratic behavior of arbitrary functions down to a specification on a dense set. Further properties – analyticity, say, or more seriously a direct relationship to $p$-adic analytic number theory – could reduce the statement to something genuinely tractable. This idea is meant to be analogous to the relative ease of studying number theoretic $L$-functions over trying to understand them through painstaking computation of their special values. Such a program was initiated by Hopkins [Str92], from which a partial collection of results have emerged.[24] Most strikingly, Mitchell ([Mit05], [HM07]) has fully elucidated this program in the $\widehat{\mathbb{G}}_m$-local category, Hovey and Strickland have shown a restricted continuity result in the general setting [HS99, section 14],[25] and the original work of Hopkins, Mahowald, and Sadofsky [HMS94] shows that the following pair of squares are both pullbacks:

---

[24] Behrens has also pursued a program encoding this problem in terms of modular forms ([Beh09], [Beh06a], [Beh07]).

[25] Also interesting are some negative results, such as: The number of Picard-places where the homotopy of $\widehat{L}_2 M(p)$ has infinite order has positive Haar measure [HS99, section 15.2].

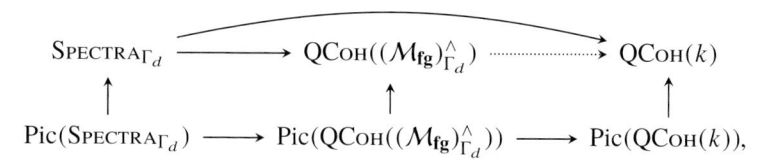

$$\text{SPECTRA}_{\Gamma_d} \longrightarrow \text{QCoH}((\mathcal{M}_{\mathbf{fg}})^{\wedge}_{\Gamma_d}) \dashrightarrow \text{QCoH}(k)$$
$$\uparrow \qquad\qquad \uparrow \qquad\qquad \uparrow$$
$$\text{Pic}(\text{SPECTRA}_{\Gamma_d}) \longrightarrow \text{Pic}(\text{QCoH}((\mathcal{M}_{\mathbf{fg}})^{\wedge}_{\Gamma_d})) \longrightarrow \text{Pic}(\text{QCoH}(k)),$$

allowing for the easy detection of invertible $\Gamma_d$-local spectra.[26]

There is further hope that the analogy to special values of $L$-functions can be strengthened to a precise connection: The orders of the $\widehat{\mathbb{G}}_m$-local homotopy groups of the sphere (or, equivalently, the orders of the stable image of $j$ elements) are controlled by Bernoulli denominators, which also appear as the negative special values of the Riemann $\zeta$-function.[27,28] This portends to be more than coincidence: The conjectured $p$-adic local Langlands correspondence promises a comparison between certain "nice" representations of the groups $GL_d(\mathbb{Q}_p)$, $\text{Gal}(\overline{\mathbb{Q}_p}/\mathbb{Q}_p)$, and $\text{Aut}\,\Gamma_d$ in such a way that the $L$-functions naturally associated to each representation are equal across the correspondence. At $d = 1$, the relevant spaces of representations are identically equal (and, in particular, the groups $\text{Aut}\,\Gamma_1$, $GL_1(\mathbb{Q}_p)$, and $\text{Gal}(\overline{\mathbb{Q}_p}/\mathbb{Q}_p)^{ab}$ are themselves all equal), and the $L$-function associated to the $\text{Aut}\,\Gamma_1$-representation $E_{\widehat{\mathbb{G}}_m}(\mathbb{S}^0)$ is, indeed, the Riemann $\zeta$-function. It is rather a lot to hope that this correspondence continues at higher heights, but with the current rate of progress on the $p$-adic Langlands correspondence (see, e.g., [Kni16]) this promises to soon be testable, if not provable.[29,30]

---

[26] There are interesting basic open questions about the behavior of the horizontal map $\text{Pic}(\text{SPECTRA}_{\Gamma_d}) \to \text{Pic}(\text{QCoH}((\mathcal{M}_{\mathbf{fg}})^{\wedge}_{\Gamma_d}))$. Is it injective? Is it surjective? It can be shown to be injective in the case of $p \gg \text{ht}\,\Gamma$, and it is known to fail to be injective when $p$ is small – but then the precise degree to which it fails to be injective becomes of interest. Picard elements in the kernel of this map are called *exotic*, and they are responsible for rotating differentials in certain Adams spectral sequences with cyclic symmetries.

[27] Generalizations of these phenomena (to larger complexes, to the prime 2, and to higher heights) are the subject of work in progress by Salch.

[28] It would be truly amazing to have an analog of the Beilinson conjectures in this topological Langlands program.

[29] An interesting feature of the local Langlands correspondence is that it has *two* geometric instantiations, stemming from the Lubin–Tate moduli of infinite level and from the *Drinfeld moduli* of infinite height. Drinfeld modules, the constituent points of the Drinfeld moduli, have the many desirable properties, but among their stranger properties is that they are naturally positive objects – a situation that a modern homotopy theorist essentially never enters. Any kind of introduction of equicharacteristic algebraic geometry into homotopy theory would be a welcome invitation and possible point of homotopical contact between these two kinds of halves of the local Langlands program. (One such place this is coming into view is in the "ultrachromatic program" of Barthel, Behrens, Schlank, and Stapleton [BSS17].)

[30] In-progress work of Stapleton and collaborators shows that a version of character theory can be used to convert the $E$-cohomology of a space into a sheaf over the Lubin–Tate moduli of infinite level with an intricate action of a certain large matrix group relevant to the Jacquet–Langlands program.

The previous program is not the only way that homotopy theory appears to admit $p$-adic interpolation, as the following striking result of Yanovski demonstrates. A *generalized homotopy cardinality function* is a function $\chi$, defined only on spaces generated under finite colimits by $\pi$-finite spaces and valued in rational numbers, which satisfies

1. Homotopy invariance: if $A \simeq B$, then $\chi(A) = \chi(B)$.
2. Normalization: $\chi(*) = 1$.
3. Additivity: $\chi(A \cup_C B) = \chi(A) + \chi(B) - \chi(C)$.
4. Multiplicativity: $\chi(A \times B) = \chi(A) \cdot \chi(B)$.

There is a natural family of examples of such functions: At any prime $p$, set

$$\chi_{d,p}(X) = \dim K(d)^0(X) - \dim K(d)^1(X).$$

The special cases $\chi_{0,p}$ and $\chi_{\infty,p}$ recover the Euler characteristic in the cases where they converge, but this quickly gets fussy. For instance, $B\mathbb{Z}/p$ is a $\pi$-finite space but is cohomologically infinite dimensional – but its Euler characteristic is *regularized* to give closed sums like

$$\chi_{\infty,p}(B\mathbb{Z}/p) = 1 - (p-1) + (p-1)^2 - \cdots = \frac{1}{p}.$$

The finite-height versions are much better behaved: Using the results of Lecture 4.6, we can quickly compute

$$\chi_{d,p}(B\mathbb{Z}/p) = p^d, \qquad \chi_{d,p}(H\mathbb{Z}/p_m) = p^{\binom{d}{m}},$$

without any handwaving about summation.

In another direction, the function

$$|X| = \sum_{x_0 \in \pi_0 X} \prod_{n=1}^{\infty} |\pi_n(X, x_0)|^{(-1)^n} \in \mathbb{Q}_{\geq 0}$$

was shown to be a homotopy cardinality function (on $p$-local $\pi$-finite spaces) by Baez and Dolan [Bae03]. In the same toy setting as above, we compute

$$|B\mathbb{Z}/p| = \frac{1}{p}.$$

It is quite a curiosity to get the same value as the regularized sum, and Baez and Dolan conjectured that whenever these formulas could be simultaneously made sense of, they would agree. Yanovsky demonstrated this conjecture to be true: Taking $p > 2$, he showed that the function $n \mapsto \chi_{n,p}(X)$ extends uniquely to a dyadic analytic function $\widehat{L}_{X,p} \colon \mathbb{Z}_2^\wedge \to \mathbb{Z}_2^\wedge$ and that the Baez–Dolan function appears as $\widehat{L}_{X,p}(-1)$. This justifies $p$-adic analytic continuation formulas relating the two invariants, such as

$$\widehat{L}_{\underline{H\mathbb{Z}/p_m},p}(-1) = p^{\binom{-1}{m}} = p^{(-1)^m} = |\underline{H\mathbb{Z}/p_m}|.$$

## Interaction with Dieudonné Theory

In Lecture 4.4, we gave a pair of definitions of a Dieudonné module, one contravariant and one covariant, and in Theorem 4.4.18 we announced that the resulting modules were linearly dual to one another. This is a remarkably difficult theorem to prove, it is hard to find an accessible proof in the literature,[31] and a *geometric* proof appears not to exist. This doesn't have the same grandeur as the preceding open problems, but the rabbit hole still seems to be quite a lot deeper than one might first expect.

To set the stage, we summarize the proof of Mazur and Messing [MM74, section II.15]. First, the functor of curves is both representable and corepresentable in the category of formal group schemes by the *formal (co)Witt scheme* ([Zin84, chapter 3], [Laz75, section III.4]):

$$\text{FormalSchemes}(\widehat{\mathbb{A}}^1, \widehat{\mathbb{G}})^{p\text{-typ}} \cong \text{FormalGroups}(\widehat{\mathbb{W}}_p, \widehat{\mathbb{G}})$$
$$\cong \text{FormalGroups}(\widehat{\mathbb{G}}^*, \widehat{C\mathbb{W}}_p).$$

There is a canonical short exact sequence

$$0 \to \widehat{\mathbb{G}}_a \to \widehat{C\mathbb{W}}_p \xrightarrow{V} \widehat{C\mathbb{W}}_p \to 0,$$

and a co-curve $\gamma^*\colon \widehat{\mathbb{G}} \to \widehat{C\mathbb{W}}_p$ gives a pullback sequence

$$\begin{array}{ccccccccc}
0 & \longrightarrow & \widehat{\mathbb{G}}_a & \longrightarrow & \widehat{\mathbb{W}}_p & \xrightarrow{V} & \widehat{\mathbb{W}}_p & \longrightarrow & 0 \\
& & \| & & \uparrow & & \uparrow \gamma^* & & \\
0 & \longrightarrow & \widehat{\mathbb{G}}_a & \longrightarrow & E & \longrightarrow & \widehat{\mathbb{G}} & \longrightarrow & 0.
\end{array}$$

This latter sequence is a *rigidified extension* of $\widehat{\mathbb{G}}$ by $\widehat{\mathbb{G}}_a$, as in Remark 4.4.14. The conclusion of Mazur and Messing is that this assignment is an isomorphism.

We suspect that the two Dieudonné functors can be connected via a *residue pairing* between forms on $\widehat{\mathbb{G}}$ over $\mathbb{W}(k)$ and curves on $\widehat{\mathbb{G}}_0$ over $k$. For inspiration, the residue pairing for an equicharacteristic local field

$$\langle -, - \rangle \colon k((z)) \times k((z))^* \to k$$

---

[31] In fact, Grothendieck introduced the de Rham–Dieudonné functor and crystalline Dieudonné theory in the same landmark paper, but elected not to provide a proof of an equivalence of his methods with past ones.

is given by the formula

$$\langle g, f \rangle = \operatorname{Res}_{z=0}\left(g \cdot \frac{\mathrm{d}f}{f}\right) = \operatorname{Res}_{z=0}(g \cdot \mathrm{d}\log f).$$

There are several ingredients in its construction for which we must find formal group analogs. First, we must deal with the different ground objects $\mathbb{W}_p(k)$ and $k$ for $\widehat{\mathbb{G}}$ and $\widehat{\mathbb{G}}_0$:

**Lemma B.2.1** ([Laz75, lemma VII.7.5]) *There is a $(\mathbb{W}, F)$-linear section $\sigma$ of the base-change map $D_*\widehat{\mathbb{G}} \to D_*\widehat{\mathbb{G}}_0$.*

*Construction* Cartier constructs an auxiliary Dieudonné module $M$ by picking a presentation[32]

$$D_*\widehat{\mathbb{G}}_0 = \mathbb{W}_p(k)\{\gamma, V\gamma, \ldots, V^{d-1}\gamma\} \Big/ \left(FV^i\gamma = \sum_{j=0}^{d-1} c_{i,j}V^j\gamma\right)$$

and forming the module

$$M := \mathbb{W}_p(\mathbb{W}_p(k))\langle\!\langle V \rangle\!\rangle \{\widetilde{\gamma}_0, \ldots, \widetilde{\gamma}_{d-1}\} \Big/ \left(F\widetilde{\gamma}_i = \sum_{j=0}^{d-1} \Delta(c_{i,j})\widetilde{\gamma}_j\right).$$

There is a $(\mathbb{W}, F)$-linear map $\tau\colon D_*\widehat{\mathbb{G}}_0 \to M$ given by base-change along $\Delta$, and Cartier shows that this map has a universal property [Laz75, VII.2.9]: $(\mathbb{W}, F)$-linear objects under $D_*\widehat{\mathbb{G}}_0$ agree with Dieudonné modules under $M$ by restriction along $\tau$. Reapplying the change-of-rings functor $\pi_*M$ gives the Cartier–Dieudonné module for the universal additive extension of $\widehat{\mathbb{G}}_0$ [Laz75, V.6.22 and VII.2.28], and hence $D_*\widehat{\mathbb{G}}$ is a Dieudonné module under $\pi_*M$, and hence under $M$. This gives the desired map

$$\sigma\colon D_*\widehat{\mathbb{G}}_0 \to D_*\widehat{\mathbb{G}}. \qquad \square$$

Second, we must produce an analog of logarithmic differentiation, which is responsible for the bilinearity on the side of the differential form. This arises naturally when trying to find analogs of differentiation internal to formal groups:

1. The fiber sequence

$$T_0V \to T_*V \to V$$

described around Lemma 2.1.4 does not naturally split: An arbitrary formal variety has no intrinsic notion of "constant vector field" [Laz75, V.11.12].

---

[32] The formal group so constructed does not actually depend upon the presentation.

## B.2 The Road Ahead

However, in the presence of a formal group structure $\widehat{\mathbb{H}}$ on $V$, the tangent space acquires a natural splitting analogous to the splitting in classical Lie theory:[33]

$$\widehat{\mathbb{H}} \times (\widehat{\mathbb{G}}_a \otimes \operatorname{Lie} \widehat{\mathbb{H}}) \xrightarrow{\cong} T_* \widehat{\mathbb{H}},$$

$$(x, \xi) \mapsto x +_{\widehat{\mathbb{H}}} \varepsilon \cdot \xi.$$

This gives rise to an invariant notion of differentiation: Given two formal groups $\widehat{\mathbb{G}}, \widehat{\mathbb{H}}$ as well as any pointed map $f \colon \widehat{\mathbb{H}} \to \widehat{\mathbb{G}}$ of formal varieties, there is a function

$$D_{\widehat{\mathbb{H}}}^{\widehat{\mathbb{G}}} f \colon \widehat{\mathbb{H}} \to \widehat{\mathbb{G}}_a \otimes (\operatorname{Lie} \widehat{\mathbb{H}})^* \otimes \operatorname{Lie} \widehat{\mathbb{G}}$$

characterized by

$$T_* f(x +_{\widehat{\mathbb{H}}} \varepsilon \cdot \xi) = f(x) +_{\widehat{\mathbb{H}}} \varepsilon \cdot (D_{\widehat{\mathbb{G}}}^{\widehat{\mathbb{H}}} f)(x) \xi.$$

2. By taking $\widehat{\mathbb{G}}_a$ as a model for $\widehat{\mathbb{A}}^1$, any curve $\gamma \colon \widehat{\mathbb{A}}^1 \to \widehat{\mathbb{G}}$ can be interpreted as a map of formal varieties $\gamma \colon \widehat{\mathbb{G}}_a \to \widehat{\mathbb{G}}$. Applying the recipe above gives rise to a function $\widehat{\mathbb{A}}^1 \to \widehat{\mathbb{G}}_a \otimes \operatorname{Lie} \widehat{\mathbb{G}}$, i.e., a series with coefficients in $\operatorname{Lie} \widehat{\mathbb{G}}$. This series can actually be given explicitly [Laz75, V.7.3]:

$$\gamma \mapsto \sum_{n=1}^{\infty} t^{n-1} \operatorname{Lie}(F_n \gamma).$$

In particular, for $\widehat{\mathbb{G}} = \widehat{\mathbb{G}}_a$, this computes the classical derivative of $\gamma$ expanded in the canonical coordinate [Laz75, V.7.13].

3. This assignment is compatible with the Dieudonné module structures on curves on $\widehat{\mathbb{G}}$ and curves on $\widehat{\mathbb{G}}_a \otimes \operatorname{Lie} \widehat{\mathbb{G}}$. The inverse Dieudonné functor then gives rise to a map

$$\check{D}_{\widehat{\mathbb{G}}} \colon \widehat{\mathbb{G}} \to \widehat{\mathbb{G}}_a \otimes \operatorname{Lie} \widehat{\mathbb{G}},$$

called the *reduced derivative*. This construction is natural in $\widehat{\mathbb{G}}$, so if $\widehat{\mathbb{G}}$ had a logarithm, there would be a commuting square

$$\begin{array}{ccc}
\widehat{\mathbb{G}} & \xrightarrow{\log_{\widehat{\mathbb{G}}}} & \widehat{\mathbb{G}}_a \otimes \operatorname{Lie} G \\
\downarrow{\check{D}_{\widehat{\mathbb{G}}}} & & \downarrow{\check{D}_{\widehat{\mathbb{G}}_a \otimes \operatorname{Lie} G}} \\
\widehat{\mathbb{G}}_a \otimes \operatorname{Lie} \widehat{\mathbb{G}} & = & \widehat{\mathbb{G}}_a \otimes \operatorname{Lie} G.
\end{array}$$

---

[33] This topic also arises when trying to understand what is special about curves in the image of Cartier's section $\sigma$: they are *horizontal* in a related sense.

The bottom logarithm is the identity, since the source is already additive. If we evaluate these maps on a curve $\gamma \in C_*\widehat{\mathbb{G}}$, we arrive at the equation [Laz75, V.8.5]

$$\check{D}_{\widehat{\mathbb{G}}} \circ \gamma = \check{D}_{\widehat{\mathbb{G}}_a \otimes_{\operatorname{Lie} G} G} \circ \log_{\widehat{\mathbb{G}}} \circ \gamma = \frac{d}{dt}\left((\log_{\widehat{\mathbb{G}}} \circ \gamma)(t)\right).$$

Hence, $\check{D}_{\widehat{\mathbb{G}}}$ plays the role of a logarithmic derivative.

The conjecture, then, is that these pieces form the foundations of a local residue pairing internal to the theory of formal groups, which, appropriately interpreted, gives the duality paring between the covariant and contravariant geometric Dieudonné modules. But there are still several details left to the enterprising reader, such as: Should the de Rham complex be replaced? Relatedly, can $\check{D}_{\widehat{\mathbb{G}}}$ or $D_{\widehat{\mathbb{H}}}^{\widehat{\mathbb{G}}}$ be used to give a coordinate-free description of the deformation complex of Definition 3.4.9? Where do *primitives* in de Rham cohomology enter play? How do the formulas for $\sigma$ in terms of $\check{D}_{\widehat{\mathbb{G}}}$ [Laz75, VII.6.14] enter this story – presumably in demonstrating perfection?

In a more topological direction, one wonders to what extent it is critical to use $H\mathbb{F}_p$ in the machinery built up in Lecture 4.5. In particular, could $H\mathbb{F}_p$ be replaced by another field spectrum to build Morava $K$-theoretic analogs of Brown–Gitler spectra? There is some existing work on this in the height 1 case, due to Bousfield [Bou96], but the matter remains unsettled.

## Structured Ring Spectra

As indicated in the introduction to Appendix A, the algebraic geometry of $E_\infty$-ring spectra is still very much in flux, and accordingly there are a lot of interesting open questions about their interaction with the story presented here.

One of the most obvious ones, almost directly cribbed out of Appendix A.3, is whether $\mathcal{M}_{fg}$ itself admits a topological enrichment [Goe09]. This question is quite flexible: Varying the Grothendieck topology chosen on $\mathcal{M}_{fg}$ will almost certainly affect the positivity of the answer, and there are versions of the question that apply to $E_\infty$-rings, $A_\infty$-rings, or just spectra. Still, even admitting the existence of a huge family of implicit questions, very little is known. On the looser end, we do not have an example of a formal group that cannot arise as the formal group associated to a complex-orientable cohomology theory. If we had such an example of a prohibited formal group, then if the selecting map $\operatorname{Spec} R_0 \to \mathcal{M}_{fg}$ were furthermore – for example – flat, then we could conclude that there cannot exist a topological enrichment of the flat site. We do have one rather extreme example: One cannot adjoin $p$th roots to $E_\infty$-rings ([SVW99], [Dev17]), from which it follows that certain forms of $K$-theory do not have $E_\infty$ structures, and hence the fpqc site of $\mathcal{M}_{\mathrm{fg}}$ does not admit a topological enrichment with a sheaf of $E_\infty$-rings. A recent result of Lawson also shows that $BP$ does *not* admit an $E_\infty$ structure – at least at $p = 2$, and an extension of his techniques to odd primes yield similar results in that setting ([Law17], [Sen17]). This does not yield any information over what the above example of $KU[\zeta_p]$ shows, but it is worth pointing out that May asked the question of whether $BP$ admits an $E_\infty$ structure over four decades prior [May74] (and it has had a long and storied intervening history), and the delay in its resolution is some indication of the difficulty of this problem. Some other richer stacks than $\mathcal{M}_{\mathrm{fg}}$ are known not to admit enrichment: For instance, for this same reason it follows that there cannot exist a topological enrichment of the moduli of (almost any piece of) formal groups equipped with level-$p$ structures,[34] and Lawson has shown that there cannot exist spectra associated to certain stacks of formal $A$-modules [Law07]. Meanwhile, other nearby stacks do admit enrichments: There are spectra $TAF$, an abbreviation for "topological automorphic forms," which generalize $TMF$ [BL10].

Meanwhile, even though we know that the spectrum $E_\Gamma$ admits an $E_\infty$-ring structure, and despite our analysis of Appendix A.2, it is not known whether there is an $E_\infty$-orientation. This question has received plenty of attention: Most recently, Hopkins and Lawson have produced a spectral sequence computing the space of such maps, the first differential of which encodes the norm coherence condition of Appendix A.2 [HL16]. Their work organizes a process that is manually understandable at low heights: The natural map $\mathbb{P}(\Sigma^{-2}\mathbb{CP}^\infty) \to MU$ is a rational isomorphism of $E_\infty$-rings, but at height 1 it is not, essentially because the collection of power operations for $\widehat{\mathbb{G}}_m$-local $E_\infty$-rings acts freely on the source but not on the target. This can be corrected for with a single rationally acyclic term, altogether begetting a two-term free resolution of $MU$ as an $L_1$-local $E_\infty$-ring spectrum. Again, this resolution is insufficient when localized at $\Gamma_2$, and the process continues ad infinitum as in the Hopkins–Lawson paper. Understanding this spectral sequence even in the height 2 case, as specialized to $E_{\Gamma_2}$, would be an extremely interesting exercise, as it would illuminate the single extra condition at height 2 needed to enrich a norm coherent orientation to an $E_\infty$-orientation.

Let us turn our attention away from the (rather burning) question of realizability. As we observed in the main proof of Appendix A.4, the $E_\infty$-ring structure on a spectrum can be mined for interesting arithmetic information in rather unexpected ways: Computation of the Miller invariant for $KO$ yields the Bernoulli numbers, and the same for $TMF$ yields the normalized Eisenstein series. The original

---

[34] In particular, this appears to inhibit us from producing a spectrum embodying the Lubin–Tate tower at infinite level, which is a bummer from the perspective of the local Langlands program discussed above.

computation of the Miller invariant for $MU$ was performed by studying the $S^1$-transfer map, and it was concluded that the results are essentially the same for any complex-orientable cohomology theory: There is a set of "universal Bernoulli numbers" lying over $\mathcal{M}_{\mathbf{fgl}}$ [Mil82], and the Miller invariant for other complex-orientable theories is always computed by the pushforward of these universal values along the classifying map. Baker, Carlisle, Gray, Hilditch, Richter, and Wood [BCG+88] have shown that one can produce other interesting number theoretic phenomena by instead using the iterated $S^1$-transfer (or, equivalently, the transfer for a torus). Since Morava's theories $E_\Gamma$ are known to be $E_\infty$-rings, one wonders what functions on Lubin–Tate space these iterated transfers select.

Hopkins has also highlighted the potential connections between $gl_1 tmf$, number theory, and manifold geometry [Hop14c], in analogy to the connections between $gl_1 kO$ (through the image of $j$ spectrum), number theory (through Bernoulli numbers), and manifold geometry (through the Hopf invariant one problem). One place the analysis of $gl_1 KO$ could be strengthened (and which would, presumably, also strengthen the analysis of $gl_1 tmf$) would be to *uniquely* characterize the Bernoulli sequences appearing in Appendix A.4. More than one sequence satisfies those simultaneous congruences [SN14], but the natural one appearing in the Bernoulli numbers is in some sense the "smallest" one. It would be interesting to have a non-tautological encoding of this statement, in such a way that certain orientations in other contexts might also be singled out as preferable. It would also be interesting if this could be encoded as a "real place" condition, which had a topological incarnation in terms of a smooth cohomology theory like differential real $K$-theory.

An unpublished theorem of Hopkins and Lurie describes the $E$-theoretic *discrepancy spectrum*, defined as in by the fiber sequence

$$F_\Gamma \to gl_1 E_\Gamma \to L_d gl_1 E_\Gamma.$$

Over an algebraically closed field $k$, the theorem states that $F_\Gamma$ satisfies

$$\Omega^\infty F_\Gamma \simeq \Omega^\infty (\Sigma^d I_{\mathbb{Q}_p/\mathbb{Z}_p} \vee H(k^*)^{\text{tors}}),$$

where $(k^*)^{\text{tors}}$ is the group of roots of unity of $k$ and $I_{\mathbb{Q}_p/\mathbb{Z}_p}$ is the $p$–local Brown–Comenetz dualizing spectrum, itself defined so as to satisfy the relation

$$\pi_0 F(E, I_{\mathbb{Q}_p/\mathbb{Z}_p}) = \text{AbelianGroups}(\pi_0 E, \mathbb{Q}_p/\mathbb{Z}_p).$$

By setting $E = \mathbb{S}_{(p)}$, it follows that the homotopy groups of $I_{\mathbb{Q}_p/\mathbb{Z}_p}$ mostly match those of $\mathbb{S}_{(p)}$ itself (or, rather, their Pontryagin duals), so computing them is equivalent to computing the homotopy groups of spheres. The Hopkins–Lurie theorem gives an interesting approach to performing this computation: The

## B.2 The Road Ahead

chromatic fracture square for $L_d gl_1 E_\Gamma$ couples to the logarithm to rewrite this square in terms of $\widehat{L}_j$-localizations of $E_\Gamma$, intertwined by very complicated power operation formulas. It would be an extremely interesting exercise to see this play out even at low heights.

Finally there is a substantially different approach to associating algebro-geometric objects to the elliptic cohomology of spaces, as pioneered in work of Grojnowski [Gro07] and of Ginzburg, Kapranov, and Vasserot [GKV95], which has been used to enormous effect by Lurie in his wide-ranging program to realize various pieces of elliptic cohomology (or, more broadly, algebraic geometry) completely internally to the algebric geometry of $E_\infty$-ring spectra [Lure].[35,36] Any serious student of elliptic cohomology should also become familiar with this framework. It has one feature that is especially curious and worth emphasizing here: The study of power operations is incidental to the Lurie-style approach of working with $E_\infty$-ring spectra, but his framework uses in a critical way the geometry of $p$-divisible groups. It must therefore be the case that the structure of a $p$-divisible group – and, more presumptuously and more specifically, its subgroup lattice as in Appendix A.2 – must entirely encode its theory of power operations. It is very much worthwhile to understand this connection in more generality than just in the setting of Morava $E$-theory.

### Equivariance

Since the resolution of the Kervaire invariant one question by Hill, Hopkins, and Ravenel [HHR16], the field of equivariant homotopy theory has emerged from slumber to enjoy a serious revitalization. There is a lot to study here, and most of it is better elucidated elsewhere, but one of the central points of the Hill–Hopkins–Ravenel program is exactly the employ of formal geometry. Namely, one of their main tools is the moduli of formal $\mathbb{Z}[\sqrt[8]{i}]$-modules, which are formal groups equipped with a certain $C_8$-action effecting multiplication by the associated powers of $\sqrt[8]{i}$ on the tangent space. They take as a model for this a certain bouquet of copies of $MU$, which selects four coordinates on a given formal group, intertwined by this action of $\sqrt[8]{i}$. This approach to this problem has historical precedent: The Kervaire invariant one problem is the

---

[35] The $\sigma$-orientation arises essentially from considering the objects associated to $BU(n)$ and $BO(n)$ using these machines (see [Lure, section 5.1] especially) and identifying certain naturally occurring local systems (of spectra).

[36] A curious feature of the current understanding of spectral algebraic geometry is its assumption that the ring spectra involved are *connective*. There are some results in the nonconnective (and specifically 2-periodic) direction, but we are a long ways from giving truly satisfactory structural descriptions of the phenomena observed there. In this setting, chromatic homotopy theory is meant to be give guiding principles: Its theorems are meant to shed light on the behavior of the nonconnective spectral algebraic geometry of the sphere spectrum.

$p = 2$ case of a family of prime-indexed problems, and the versions satisfying $p \geq 5$ were simultaneously resolved by Ravenel by similar methods with formal $A$-modules [Rav78a]. This leaves one case open: The Kervaire problem at $p = 3$ remains unresolved. From the perspective of formal $A$-modules, this is somewhat predictable: At large primes, a certain governing map is $p$-adically continuous, but it fails to be continuous at small primes.[37]

In a different direction, the moduli of formal $A$-modules has been used to great effect by Salch to make computations deep into the chromatic layers [Sal16]. In essence, the theory of formal $A$-modules grants access to multiplicative "height amplification" theorems: Deep knowledge of the homotopy groups of spheres at height 2 can yield extensive information at height $2 \cdot 2$, for instance. Salch's methods also promise to yield connections to the $L$-functions perspective described above, fitting into a larger program that describes the homotopy groups of higher-height localizations of spheres in much the same way that polyzeta functions are associated to arithmetic schemes of cohomological dimension larger than 1. All this is quite frustrating, then, in the face of Lawson's result that the moduli of formal $A$-modules does not admit a topological enrichment [Law07].

In a yet different direction, Lurie's approach to the reinterpretation of the $\sigma$-orientation through spectral algebraic geometry relies on a concept of "2-equivariance" – spectra equipped with a notion of equivariance not just against a pointed homotopy 1-type (i.e., a $BG$ for some $G$) but against general homotopy 2-types. Relatively little about this has been written down, but one can find an overview in his survey [Lure, section 5.3].

## Index Theorems

In closing the book, we finally address the lurking physical inspiration for the $\sigma$-orientation. One of the most remarkable developments of modern physics, from the perspective of a mathematician, is their ability to "guess" the Atiyah–Singer index theorem [Tak08, section 8.6]: There is a path integral formulation for the supertrace of the Dirac operator on a *Spin* manifold $M$ which expresses it as the time evolution of a massive supersymmetric particle evolving through $M$, and standard expansions for this physical system lead directly to the discovery of the $\widehat{A}$-genus and, ultimately, the index formula

$$\text{ind } \partial_+ = \text{Tr}_s \, e^{-H} = (2\pi i)^{-\frac{n}{2}} \int_M \widehat{A}(M).$$

---

[37] Behrens gave a talk at the MSRI Hot Topics session on the Kervaire problem in 2010. He addresses this point at the timecode 1h04m20s in the recorded video.

## B.2 The Road Ahead

The $\sigma$-orientation was uncovered by Witten when studying the free loopspace $\mathcal{L}M$ of the manifold $M$ – indeed, a natural space to consider when studying string theory on $M$. The condition that $M$ be *String* is exactly the precondition that $\mathcal{L}M$ be *Spin*, so that an analogous Dirac operator exists. The $\sigma$-orientation then arises when trying to express a formula for the index of the Dirac operator on the free loopspace in terms of invariants of $M$ and using analogous path integral methods [Seg88].

Because of the predictive power of the physical apparatus, one might hope to uncover other types of analytic data from it – for instance, a geometric theory of elliptic cohomology.[38] In addition to the variant of bordism theory described far above, the most long-standing program is due to Stolz and Teichner ([ST04], [ST11]), though it is not alone even among physically inspired theories (for instance, see [DH11] as well as many other papers by the same authors).

One of the remarkable features of loop groups is that this process does not appear to be iterable: The theory of compact Lie groups is exceedingly nice (for instance, they have surjective exponential maps), the theory of loop groups formed from compact Lie groups is only slightly less nice (for instance, they have dense exponential maps), and any extension of this seems to immediately go to pot.[39] Finding the "next step" after loop groups is an extremely difficult problem that requires such inventive thinking that it could only open enormous avenues for generalization.

---

[38] In a different direction, there is also a less-discussed physical system that has produced models for the infinite loopspaces associated to topologically familiar spectra, such as $KU$ and $KO$ [Kit09]. Variations on the same idea are also conjectured to produce models for many more, such as the Anderson duals of various bordism spectra, including (in the case of stably framed bordism) the Anderson dualizing object itself – largely considered to be a resolutely un-geometric object!

[39] For instance, little positive is known about the group of free "tori," i.e., the loop group of a loop group.

# References

[AA66]    J. F. Adams and M. F. Atiyah. K-theory and the Hopf invariant. *Q. J. Math.*, 17(1):31–38, 1966. [163]

[ABG+14]  M. Ando, A, J. Blumberg, D. Gepner, M. J. Hopkins, and C. Rezk. An ∞-categorical approach to $R$-line bundles, $R$-module Thom spectra, and twisted $R$-homology. *J. Topol.*, 7(3):869–893, 2014. [334]

[ABP67]   D. W. Anderson, E. H. Brown, Jr., and F. P. Peterson. The structure of the Spin cobordism ring. *Ann. Math. (2)*, 86(2):271–298, 1967. [353]

[ABS64]   M. F. Atiyah, R. Bott, and A. Shapiro. Clifford modules. *Topology*, 3(suppl. 1):3–38, 1964. [353]

[Ada62]   J. F. Adams. Vector fields on spheres. *Topology*, 1(1):63–65, 1962. [39]

[Ada66]   J. F. Adams. On the groups $J(X)$: IV. *Topology*, 5:21–71, 1966. [148, 161, 163]

[Ada76]   J. F. Adams. Primitive elements in the $K$-theory of $BSU$. *Q. J. Math.*, 27(2):253–262, 1976. [251]

[Ada78a]  J. Frank Adams. Maps between classifying spaces: II. *Invent. Math.*, 49(1):1–65, 1978. [306]

[Ada78b]  J. F, Adams. *Infinite loop*. Princeton University Press, Princeton, NJ; University of Tokyo Press, Tokyo, 1978. [v, 78]

[Ada95]   J. F. Adams. *Stable homotopy and generalised homology*. University of Chicago Press, Chicago, IL, 1995. Reprint of the 1974 original. [x, 43, 85, 94, 95]

[Ade52]   J. Adem. The iteration of the Steenrod squares in algebraic topology. *PNAS*, 38(8):720–726, 1952. [83]

| | |
|---|---|
| [AFG08] | M. Ando, C. P. French, and N. Ganter. The Jacobi orientation and the two-variable elliptic genus. *Algebr. Geom. Topol.*, 8(1):493–539, 2008. [270] |
| [AHRa] | M. Ando, M. J. Hopkins, and C. Rezk. Moments of measures on $\mathbb{Z}_p^\times$, with applications to maps between $p$-adic $K$-theory spectra. Unpublished. [344] |
| [AHRb] | M. Ando, M. J. Hopkins, and C. Rezk. Multiplicative orientations of $ko$-theory and the spectrum of topological modular forms. www.math.illinois.edu/~mando/papers/koandtmf.pdf, accessed: November 16, 2016. [335, 336, 338, 343, 346] |
| [AHS01] | M. Ando, M. J. Hopkins, and N. P. Strickland. Elliptic spectra, the Witten genus and the theorem of the cube. *Invent. Math.*, 146(3):595–687, 2001. [4, 226, 227, 228, 232, 235, 237, 241, 250, 251, 252, 253, 266, 267, 269, 351] |
| [AHS04] | M. Ando, M. J. Hopkins, and N. P. Strickland. The sigma orientation is an $H_\infty$ map. *Amer. J. Math.*, 126(2):247–334, 2004. [66, 78, 317, 319, 322] |
| [AL17] | V. Angeltveit and J. A. Lind. Uniqueness of $BP\langle n\rangle$. *J. Homotopy Relat. Struct.*, 12(1):17–30, 2017. [194] |
| [AM01] | M. Ando and J. Morava. A renormalized Riemann–Roch formula and the Thom isomorphism for the free loop space. In *Topology, geometry, and algebra: interactions and new directions (Stanford, CA, 1999)*, pages 11–36. American Mathematical Society, Providence, RI, 2001. [270] |
| [Ami59] | S. A. Amitsur. Simple algebras and cohomology groups of arbitrary fields. *Trans. Amer. Math. Soc.*, 90:73–112, 1959. [100] |
| [AMS98] | M. Ando, J. Morava, and H. Sadofsky. Completions of $\mathbb{Z}/(p)$-Tate cohomology of periodic spectra. *Geom. Topol.*, 2(1):145–174, 1998. [356] |
| [And95] | M. Ando. Isogenies of formal group laws and power operations in the cohomology theories $E_n$. *Duke Math. J.*, 79(2):423–485, 1995. [319, 320, 321] |
| [AP76] | J. F. Adams and S. B. Priddy. Uniqueness of $BSO$. *Math. Proc. Cambridge Philos. Soc.*, 80(3):475–509, 1976. [251, 340] |
| [Ara73] | S. Araki. *Typical formal groups in complex cobordism and K-theory*. Kinokuniya Book-Store Co., Ltd., Tokyo, 1973. [112, 125] |
| [AS01] | M. Ando and N. P. Strickland. Weil pairings and Morava $K$-theory. *Topology*, 40(1):127–156, 2001. [227, 232, 236, 241, 242, 243, 244, 246, 247] |

| | |
|---|---|
| [Ati61] | M. F. Atiyah. Thom complexes. *Proc. Lond. Math. Soc.*, 3(1):291–310, 1961. [357] |
| [Bae03] | J. Baez. Euler characteristic versus homotopy cardinality. Lecture at the Program on Applied Homotopy Theory, Fields Institute, Toronto, September, 2003. [368] |
| [Bak87] | A. Baker. Combinatorial and arithmetic identities based on formal group laws. In *Algebraic Topology Barcelona 1986*, pages 17–34. New York: Springer, 1987. [112] |
| [Bak15] | A. Baker. Power operations and coactions in highly commutative homology theories. *Publ. Res. Inst. Math. Sci.*, 51(2):237–272, 2015. [309] |
| [Bal10] | P. Balmer. Spectra, spectra, spectra: tensor triangular spectra versus Zariski spectra of endomorphism rings. *Algebr. Geom. Topol.*, 10(3):1521–1563, 2010. [154, 155] |
| [Bar] | T. Barthel. Chromatic completion. http://people.mpim-bonn .mpg.de/tbarthel/CC.pdf, accessed: October 26, 2016. [158] |
| [Bau03] | T. Bauer. Computation of the homotopy of the spectrum *tmf*. *ArXiv Mathematics e-prints*, November 2003. [333] |
| [Bau14] | T. Bauer. Formal plethories. *Adv. Math.*, 254:497–569, 2014. [178] |
| [BCG+88] | A. Baker, D. Carlisle, B. Gray, S. Hilditch, N. Ray, and R. Wood. On the iterated complex transfer. *Math. Z.*, 199(2):191–207, 1988. [374] |
| [BCM78] | M. Bendersky, E. B. Curtis, and H. R. Miller. The unstable Adams spectral sequence for generalized homology. *Topology*, 17(3):229–248, 1978. [169, 171, 177, 178] |
| [BD] | A. A. Beilinson and V. Drinfeld. *Quantization of Hitchin's integrable system and Hecke eigensheaves*. www.math.uchicago .edu/~mitya/langlands/hitchin/BD-hitchin.pdf, accessed July 20, 2018. [54] |
| [Bea15] | A. Beaudry. The chromatic splitting conjecture at $n = p = 2$. *ArXiv e-prints*, February 2015. [363] |
| [Bea17] | J. Beardsley. Constructions of MU by attaching $E_k$-cells in odd degrees. *ArXiv e-prints*, August 2017. [360] |
| [Beh06a] | M. Behrens. A modular description of the $K(2)$-local sphere at the prime 3. *Topology*, 45(2):343–402, 2006. [366] |
| [Beh06b] | M. Behrens. Root invariants in the Adams spectral sequence. *Trans. Amer. Math. Soc.*, 358(10):4279–4341, 2006. [358] |
| [Beh07] | M. Behrens. Buildings, elliptic curves, and the $K(2)$-local sphere. *Amer. J. Math.*, 129(6):1513–1563, 2007. [366] |

# References

[Beh09] M. Behrens. Congruences between modular forms given by the divided $\beta$ family in homotopy theory. *Geom. Topol.*, 13(1):319–357, 2009. [366]

[Beh12] M. Behrens. The homotopy groups of $S_{E(2)}$ at $p \geq 5$ revisited. *Adv. Math.*, 230(2):458–492, 2012. [366]

[Beh14] M. Behrens. The construction of *tmf*. In *Topological modular forms*, pages 149–206. American Mathematical Society, Providence, RI, 2014. [284, 323, 325, 326, 327, 328, 329, 330, 331]

[Bei80] A.A. Beilinson. Residues and adeles. *Functional Analysis and Its Applications*, 14(1):34–35, 1980. [164]

[BF15] T. Barthel and M. Frankland. Completed power operations for Morava E-theory. *Algebr. Geom. Topol.*, 15(4):2065–2131, 2015. [103]

[BJW95] J. M. Boardman, D.C. Johnson, and W. S. Wilson. Unstable operations in generalized cohomology. In *Handbook of algebraic topology*, pages 687–828. North-Holland, Amsterdam, 1995. [170, 176, 187]

[BL07] V. Buchstaber and A. Lazarev. Dieudonné modules and $p$-divisible groups associated with Morava $K$-theory of Eilenberg–Mac Lane spaces. *Algebr. Geom. Topol.*, 7:529–564, 2007. [207]

[BL09] M. Behrens and G. Laures. $\beta$-family congruences and the $f$-invariant. *Geom. Topol. Monographs*, 16:9–29, 2009. [281]

[BL10] M. Behrens and T. Lawson. Topological automorphic forms. *Mem. Amer. Math. Soc.*, 204(958):xxiv–141, 2010. [326, 373]

[BM73] N.A. Baas and I. Madsen. On the realization of certain modules over the Steenrod algebra. *Mathematica Scandinavica*, 31(1):220–224, 1973. [159]

[BM80] A. P. Bahri and M. E. Mahowald. A direct summand in $H^*(MO\langle 8\rangle, Z_2)$. *Proc. Amer. Math. Soc.*, 78(2):295–298, 1980. [351]

[BMMS86] R. R. Bruner, J. P. May, J. E. McClure, and M. Steinberger. $H_\infty$ *ring spectra and their applications*. Springer-Verlag, Berlin, 1986. [75, 76, 77, 78, 83, 148, 164, 284, 319]

[Boa82] J. M. Boardman. The eightfold way to BP-operations or $E_*E$ and all that. In *Current trends in algebraic topology, Part 1 (London, Ont., 1981)*, pages 187–226. American Mathematical Society, Providence, RI, 1982. [105]

[Boa99] J. M. Boardman. Conditionally convergent spectral sequences. In *Homotopy invariant algebraic structures (Baltimore, MD, 1998)*,

| | |
|---|---|
| | pages 49–84. American Mathematical Society, Providence, RI, 1999. [104] |
| [Bou75] | A. K. Bousfield. The localization of spaces with respect to homology. *Topology*, 14:133–150, 1975. [309] |
| [Bou79] | A. K. Bousfield. The localization of spectra with respect to homology. *Topology*, 18(4):257–281, 1979. [104, 110, 155] |
| [Bou96] | A. K. Bousfield. On $\lambda$-rings and the $K$-theory of infinite loop spaces. *K-Theory*, 10(1):1–30, 1996. [309, 372] |
| [BPS92] | D. Barnes, D. Poduska, and P. Shick. Unstable Adams spectral sequence charts. In *Adams memorial symposium on algebraic topology: Manchester 1990*, volume 2, page 87. Cambridge University Press, Cambridge, 1992. [168] |
| [Bre83] | L. Breen. *Fonctions thêta et théorème du cube*. Springer-Verlag, Berlin, 1983. [262, 272] |
| [Bro69] | W. Browder. The Kervaire invariant of framed manifolds and its generalization. *Ann. Math. (2)*, 90(2):157–186, 1969. [164] |
| [BS17] | T. Barthel and N. Stapleton. The character of the total power operation. *Geom. Topol.*, 21(1):385–440, 2017. [291] |
| [BSS17] | T. Barthel, T. Schlank, and N. Stapleton. Chromatic homotopy theory is asymptotically algebraic. *ArXiv e-prints*, November 2017. [367] |
| [BT00] | J.Y. Butowiez and P. Turner. Unstable multiplicative cohomology operations. *Q. J. Math.*, 51(4):437–449, 2000. [165] |
| [Car66] | P. Cartier. Quantum mechanical commutation relations and theta functions. In *Proc. Sympos. Pure Math*, 9: 361–383, 1966. [260] |
| [CF66] | P. E. Conner and E. E. Floyd. *The relation of cobordism to K-theories*. Springer-Verlag, Berlin, 1966. [146, 354] |
| [Cha82] | K. Chan. A simple proof that the unstable (co)-homology of the Brown–Peterson spectrum is torsion free. *J. Pure Appl. Algebra*, 26(2):155–157, 1982. [193] |
| [CM17] | D. Clausen and A. Mathew. A short proof of telescopic Tate vanishing. *Proc. Amer. Math. Soc.*, 145(12):5413–5417, 2017. [345] |
| [Coh81] | R. L. Cohen. Odd primary infinite families in stable homotopy theory. *Mem. Amer. Math. Soc.*, 30(242):viii–92, 1981. [211] |
| [Dav95] | D. M. Davis. Computing $v_1$-periodic homotopy groups of spheres and some compact Lie groups. In *Handbook of algebraic topology*, pages 993–1048. North-Holland, Amsterdam, 1995. [309] |

[Del81] P. Deligne. Cristaux ordinaires et coordonnées canoniques. In *Algebraic surfaces (Orsay, 1976–78)*, pages 80–137. Springer, Berlin, 1981. [326]

[Dem86] M. Demazure. *Lectures on p-divisible groups*. Springer-Verlag, Berlin, 1986. Reprint of the 1972 original. [197, 221, 226]

[Dev17] S. K. Devalapurkar. Roots of unity in $K(n)$-local $E_\infty$-rings. *arXiv preprint arXiv:1707.09957*, 2017. [372]

[DG70] M. Demazure and P. Gabriel. *Groupes algébriques. Tome I: Géométrie algébrique, généralités, groupes commutatifs*. Masson & Cie, Éditeur, Paris; North-Holland Publishing Co., Amsterdam, 1970. [275]

[DH95] E. S. Devinatz and M. J. Hopkins. The action of the Morava stabilizer group on the Lubin–Tate moduli space of lifts. *Amer. J. Math.*, 117(3):669–710, 1995. [160, 202]

[DH11] C. L. Douglas and A. G. Henriques. Topological modular forms and conformal nets. In *Mathematical foundations of quantum field theory and perturbative string theory*, pages 341–354. American Mathematical Society, Providence, RI, 2011. [377]

[DHS88] E. S. Devinatz, M. J. Hopkins, and J. H. Smith. Nilpotence and stable homotopy theory: I. *Ann. Math. (2)*, 128(2):207–241, 1988. [148, 360]

[DL14] D. G. Davis and T. Lawson. Commutative ring objects in pro-categories and generalized Moore spectra. *Geom. Topol.*, 18(1):103–140, 2014. [156]

[Dri74] V. G. Drinfeld. Elliptic modules. *Mat. Sb. (N.S.)*, 94(136):594–627, 656, 1974. [136]

[EKMM97] A. D. Elmendorf, I. Kriz, M. A. Mandell, and J. P. May. *Rings, modules, and algebras in stable homotopy theory*. American Mathematical Society, Providence, RI, 1997. [74, 102, 283, 343]

[Fox93] T. F. Fox. The construction of cofree coalgebras. *J. Pure Appl. Algebra*, 84(2):191–198, 1993. [186]

[G+13] N. Ganter. Power operations in orbifold Tate $K$-theory. *Homol. Homotopy Appl*, 15(1):313–342, 2013. [322]

[Gan07] N. Ganter. Stringy power operations in Tate K-theory. *ArXiv Mathematics e-prints*, January 2007. [322]

[GH04] P. G. Goerss and M. J. Hopkins. Moduli spaces of commutative ring spectra. In *Structured ring spectra*, pages 151–200. Cambridge University Press, Cambridge, 2004. [284, 309, 323, 326]

References 385

[GKV95] V. Ginzburg, M. Kapranov, and E. Vasserot. Elliptic algebras and equivariant elliptic cohomology. *arXiv preprint q-alg/9505012*, 1995. [375]

[GLM93] P. Goerss, J. Lannes, and F. Morel. Hopf algebras, Witt vectors, and Brown–Gitler spectra. In *Algebraic topology (Oaxtepec, 1991)*, pages 111–128. American Mathematical Society, Providence, RI, 1993. [204, 205, 206, 209, 210, 211]

[GM95] J. P. C. Greenlees and J. P. May. *Generalized Tate cohomology*. American Mathematical Society, Providence, RI, 1995. [358]

[Goe99] P. G. Goerss. Hopf rings, Dieudonné modules, and $E_*\Omega^2 S^3$. In *Homotopy invariant algebraic structures (Baltimore, MD, 1998)*, pages 115–174. American Mathematical Society, Providence, RI, 1999. [174, 182, 185, 207, 208, 209]

[Goe08] P. G. Goerss. Quasi-coherent sheaves on the moduli stack of formal groups. Unpublished monograph, 2008. [27, 128]

[Goe09] P. G. Goerss. Realizing families of Landweber exact homology theories. *New topological contexts for Galois theory and algebraic geometry (BIRS 2008)*, 16:49–78, 2009. [372]

[Goe10] P. G. Goerss. Topological modular forms [after Hopkins, Miller and Lurie]. *Astérisque*, (332):Exp. No. 1005, viii, 221–255, 2010. Séminaire Bourbaki. Volume 2008/2009. Exposés 997–1011. [323]

[Gon00] J. González. A vanishing line in the $BP\langle 1\rangle$-adams spectral sequence. *Topology*, 39(6):1137–1153, 2000. [362]

[GP84] V. Giambalvo and D. J. Pengelley. The homology of *MSpin*. *Math. Proc. Cambs. Philos. Soc.*, 95:427436, 1984. [353]

[Gro71] A. Grothendieck. Revêtement étales et groupe fondamental (sga1). *Lecture Notes in Math.*, 288, 1971. [56, 102]

[Gro74] A. Grothendieck. *Groupes de Barsotti-Tate et cristaux de Dieudonné*. Les Presses de l'Université de Montréal, Montreal, Que., 1974. [196, 197, 200, 201, 214]

[Gro86] B. H. Gross. On canonical and quasi-canonical liftings. *Invent. Math.*, 84(2):321–326, 1986. [202]

[Gro07] I. Grojnowski. Delocalised equivariant elliptic cohomology. In *Elliptic cohomology*, pages 114–121. Cambridge University Press, Cambridge, 2007. [375]

[GS96] J. P. C. Greenlees and H. Sadofsky. The Tate spectrum of $v_n$-periodic complex oriented theories. *Mathematische Zeitschrift*, 222(3):391–405, 1996. [345]

[GS99] J. P. C. Greenlees and N. P. Strickland. Varieties and local cohomology for chromatic group cohomology rings. *Topology*, 38(5):1093–1139, 1999. [286, 359]

[Har77] R. Hartshorne. *Algebraic geometry*. Springer-Verlag, New York, 1977. [54, 64, 65, 260]

[Har80] B. Harris. Bott periodicity via simplicial spaces. *J. Algebra*, 62(2):450–454, 1980. [38]

[HAS99] M. Hopkins, M. Ando, and N. Strickland. Elliptic cohomology of $BO\langle 8\rangle$ in positive characteristic. 1999. Unpublished. [270]

[Haz12] M. Hazewinkel. *Formal groups and applications*. AMS Chelsea Publishing, Providence, RI, 2012. Corrected reprint of the 1978 original. [57, 122, 131]

[HD90] R. H. Dicke. Dirac's cosmology and Mach's principle. *Nature* 192:440–441, 1990. [361]

[Hed14] S. M. H. Hedayatzadeh. Exterior powers of $\pi$-divisible modules over fields. *J. Number Theory*, 138:119–174, 2014. [218]

[Hed15] S. M. H. Hedayatzadeh. Exterior powers of Lubin-Tate groups. *J. Théor. Nombres Bordeaux*, 27(1):77–148, 2015. [218]

[Heu18] G. Heuts. Lie algebras and $v_n$-periodic spaces. *ArXiv e-prints*, March 2018. [362]

[HG94a] M. J. Hopkins and B. H. Gross. Equivariant vector bundles on the Lubin–Tate moduli space. In *Topology and representation theory (Evanston, IL, 1992)*, pages 23–88. American Mathematical Society, Providence, RI, 1994. [200, 202]

[HG94b] M. J. Hopkins and B. H. Gross. The rigid analytic period mapping, Lubin–Tate space, and stable homotopy theory. *Bull. Amer. Math. Soc. (N.S.)*, 30(1):76–86, 1994. [202]

[HH18] M. A. Hill and M. J. Hopkins. Real Wilson Spaces I. *ArXiv e-prints*, June 2018. []

[HH92] M. J. Hopkins and M. A. Hovey. Spin cobordism determines real $K$-theory. *Mathematische Zeitschrift*, 210(1):181–196, 1992. [280, 354]

[HH95] M. J. Hopkins and J. R. Hunton. On the structure of spaces representing a Landweber exact cohomology theory. *Topology*, 34(1):29–36, 1995. [187, 209]

[HHR16] M. A. Hill, M. J. Hopkins, and D. C. Ravenel. On the nonexistence of elements of Kervaire invariant one. *Ann. Math.*, 184(1):1–262, 2016. [164, 187, 375]

[Hid04] H. Hida. *p-adic automorphic forms on Shimura varieties*. Springer-Verlag, New York, 2004. [329]

| | |
|---|---|
| [Hir95] | F. Hirzebruch. *Topological methods in algebraic geometry*. Springer-Verlag, Berlin, 1995. Reprint of the 1978 edition. [273] |
| [HKR00] | M. J. Hopkins, N. J. Kuhn, and D. C. Ravenel. Generalized group characters and complex oriented cohomology theories. *J. Amer. Math. Soc.*, 13(3):553–594, 2000. [78, 286, 287, 290, 294, 301, 306, 307, 359] |
| [HL] | M. J. Hopkins and J. Lurie. Ambidexterity in $K(n)$-local stable homotopy theory. www.math.harvard.edu/~lurie/papers/Ambidexterity.pdf, accessed: January 9, 2015. [211, 212, 215, 216, 217, 218, 221] |
| [HL16] | M. J. Hopkins and T. Lawson. Strictly commutative complex orientation theory. *arXiv preprint arXiv:1603.00047*, 2016. [319, 373] |
| [HLP13] | A. Hughes, J. M. Lau, and E. Peterson. Multiplicative 2-cocycles at the prime 2. *J. Pure Appl. Algebra*, 217(3):393–408, 2013. [356] |
| [HM07] | R. Hahn and S. Mitchell. Iwasawa theory for $K(1)$-local spectra. *Trans. Amer. Math. Soc.*, 359(11):5207–5238, 2007. [366] |
| [HM14] | M. J. Hopkins and H. R. Miller. Elliptic curves and stable homotopy theory I. In *Topological modular forms*, pages 224–275. American Mathematical Society, Providence, RI, 2014. [323] |
| [HMS94] | M. J. Hopkins, M. Mahowald, and H. Sadofsky. Constructions of elements in Picard groups. In *Topology and representation theory (Evanston, IL, 1992)*, pages 89–126. American Mathematical Society, Providence, RI, 1994. [160, 366] |
| [Hop] | M. J. Hopkins. Complex oriented cohomology theories and the language of stacks. Unpublished. [43, 106, 115, 127, 140, 144] |
| [Hop95] | M. J. Hopkins. Topological modular forms, the Witten genus, and the theorem of the cube. In *Proceedings of the International Congress of Mathematicians, Vol. 1, 2 (Zürich, 1994)*, pages 554–565. Birkhäuser, Basel, 1995. [279, 351, 361] |
| [Hop02] | M. J. Hopkins. Algebraic topology and modular forms. In *Proceedings of the International Congress of Mathematicians, Vol. I (Beijing, 2002)*, pages 291–317. Higher Ed. Press, Beijing, 2002. [351] |
| [Hop08] | M. J. Hopkins. The mathematical work of Douglas C. Ravenel. *Homol Homotopy Appl.*, 10(3):1–13, 2008. [xii, 361] |
| [Hop14a] | M. J. Hopkins. From spectra to stacks. In *Topological modular forms*, pages 118–126. American Mathematical Society, Providence, RI, 2014. [110, 147, 329] |

[Hop14b] M. J. Hopkins. $K(1)$-local $E_\infty$ ring spectra. In *Topological modular forms*, pages 301–216. American Mathematical Society, Providence, RI, 2014. [331]

[Hop14c] M. J. Hopkins. The string orientation. In *Topological modular forms*, pages 127–142. American Mathematical Society, Providence, RI, 2014. [374]

[Hov95] M. Hovey. Bousfield localization functors and Hopkins' chromatic splitting conjecture. In *The Čech centennial (Boston, MA, 1993)*, pages 225–250. American Mathematical Society, Providence, RI, 1995. [156, 363]

[Hov97] M. Hovey. $v_n$-elements in ring spectra and applications to bordism theory. *Duke Math. J.*, 88(2):327–356, 1997. [354, 355]

[Hov02] M. Hovey. Morita theory for Hopf algebroids and presheaves of groupoids. *Amer. J. Math.*, 124(6):1289–1318, 2002. [34, 101, 106]

[Hov04] M. Hovey. Homotopy theory of comodules over a Hopf algebroid. In *Homotopy theory: relations with algebraic geometry, group cohomology, and algebraic K-theory*, pages 261–304. American Mathematical Society, Providence, RI, 2004. [107]

[HS96] M. Hovey and H. Sadofsky. Tate cohomology lowers chromatic Bousfield classes. *Proc. Amer. Math. Soc.*, 124(11):3579–3585, 1996. [345]

[HS98] M. J. Hopkins and J. H. Smith. Nilpotence and stable homotopy theory: II. *Ann. Math. (2)*, 148(1):1–49, 1998. [148, 149, 150, 151, 152, 153, 159]

[HS99] M. Hovey and N. P. Strickland. Morava $K$-theories and localisation. *Mem. Amer. Math. Soc.*, 139(666):viii–100, 1999. [103, 144, 366]

[HSS00] M. Hovey, B. Shipley, and J. Smith. Symmetric spectra. *J. Amer. Math. Soc.*, 13(1):149–208, 2000. [74]

[HT98] J. R. Hunton and P. R. Turner. Coalgebraic algebra. *J. Pure Appl. Algebra*, 129(3):297–313, 1998. [207]

[Hua16] Z. Huan. Quasi-elliptic cohomology and its power operations. *arXiv preprint arXiv:1612.00930*, 2016. [322]

[Hub91] A. Huber. On the Parshin–Beilinson adeles for schemes. In *Abhandlungen aus dem Mathematischen Seminar der Universität Hamburg*, volume 61, page 249. Springer, New York, 1991. [164]

[Hus04] D. Husemöller. *Elliptic curves*. Springer-Verlag, New York, second edition, 2004. [259]

[JN10] N. Johnson and J. Noel. For complex orientations preserving power operations, p-typicality is atypical. *Topology Appl.*, 157(14):2271–2288, 2010. [77, 131]

[JW73] D. C. Johnson and W. S. Wilson. Projective dimension and Brown–Peterson homology. *Topology*, 12:327–353, 1973. [193]

[JW75] D. C. Johnson and W. S. Wilson. $BP$ operations and Morava's extraordinary $K$-theories. *Math. Z.*, 144(1):55–75, 1975. [149]

[Kat81] N. M. Katz. Crystalline cohomology, Dieudonné modules, and Jacobi sums. In *Automorphic forms, representation theory and arithmetic (Bombay, 1979)*, pages 165–246. Tata Institute of Fundamental Research, Bombay, 1981. [195, 196, 197, 200]

[Kit09] A. Kitaev. Periodic table for topological insulators and superconductors. In *AIP conference proceedings*, volume 1134, pages 22–30. American Institute of Physics, College Park, MD, 2009. [377]

[KL02] N. Kitchloo and G. Laures. Real structures and Morava $K$-theories. *K-Theory*, 25(3):201–214, 2002. [281]

[KLW04] N. Kitchloo, G. Laures, and W. S. Wilson. The Morava $K$-theory of spaces related to $BO$. *Adv. Math.*, 189(1):192–236, 2004. [147, 280]

[KM85] N. M. Katz and B. Mazur. *Arithmetic moduli of elliptic curves*. Princeton University Press, Princeton, NJ, 1985. [270]

[Kni16] E. Knight. *A p-adic Jacquet–Langlands correspondence*. PhD thesis, Harvard University Cambridge, MA, 2016. [367]

[Koc78] S. O. Kochman. A chain functor for bordism. *Trans. Amer. Math. Soc.*, 239:167–196, 1978. [1]

[Kra00] H. Krause. Smashing subcategories and the telescope conjecture: an algebraic approach. *Invent. Math.*, 139(1):99–133, 2000. [359]

[Kří94] I. Kříž. All complex Thom spectra are harmonic. In *Topology and representation theory (Evanston, IL, 1992)*, pages 127–134. American Mathematical Society, Providence, RI, 1994. [158, 306]

[KS93] M. Kreck and S. Stolz. $\mathbb{H}^2$-bundles and elliptic homology. *Acta Mathematica*, 171(2):231–261, 1993. [355]

[Kuh89] N. J. Kuhn. Character rings in algebraic topology. In *Advances in homotopy theory (Cortona, 1988)*, pages 111–126. Cambridge University Press, Cambridge, 1989. [299]

[Kuh08] N. J. Kuhn. A guide to telescopic functors. *Homol Homotopy Appl.*, 10(3):291–319, 2008. [339]

[Lan75] P. S. Landweber. Invariant regular ideals in Brown–Peterson homology. *Duke Math. J.*, 42(3):499–505, 1975. [130]

| | |
|---|---|
| [Lan76] | P. S. Landweber. Homological properties of comodules over $MU_*(MU)$ and $BP_*(BP)$. *Amer. J. Math.*, 98(3):591–610, 1976. [140] |
| [Lan88] | P. S. Landweber. Elliptic cohomology and modular forms. In *Elliptic curves and modular forms in algebraic topology*, pages 55–68. Springer, New York, 1988. [354] |
| [Lau03] | G. Laures. An $E_\infty$ splitting of spin bordism. *Amer. J. Math.*, 125(5):977–1027, 2003. [280] |
| [Lau04] | G. Laures. $K(1)$-local topological modular forms. *Invent. Math.*, 157(2):371–403, 2004. [331] |
| [Law07] | T. Lawson. Realizability of the Adams–Novikov spectral sequence for formal $A$-modules. *Proc. Amer. Math. Soc.*, 135(3):883–890, 2007. [373, 376] |
| [Law09] | T. Lawson. An overview of abelian varieties in homotopy theory. In *New topological contexts for Galois theory and algebraic geometry (BIRS 2008)*, pages 179–214. Geometry and Topology Publishers, Coventry, 2009. [331] |
| [Law10] | T. Lawson. Structured ring spectra and displays. *Geom. Topol.*, 14(2):1111–1127, 2010. [203] |
| [Law17] | T. Lawson. Secondary power operations and the Brown–Peterson spectrum at the prime 2. *arXiv preprint arXiv:1703.00935*, 2017. [131, 373] |
| [Laz55] | M. Lazard. Sur les groupes de Lie formels à un paramètre. *Bull. Soc. Math. France*, 83:251–274, 1955. [111, 115, 132] |
| [Laz75] | M. Lazard. *Commutative formal groups*. Springer-Verlag, Berlin, 1975. [55, 131, 204, 369, 370, 371, 372] |
| [Laz97] | A. Lazarev. Deformations of formal groups and stable homotopy theory. *Topology*, 36(6):1317–1331, 1997. [136, 137, 216] |
| [Lin80] | W. H. Lin. On conjectures of Mahowald, Segal and Sullivan. *Math. Proc. Cambridge Philos. Soc.*, 87(3):449–458, 1980. [357] |
| [LN12] | T. Lawson and N. Naumann. Commutativity conditions for truncated Brown–Peterson spectra of height 2. *J. Topol.*, 5(1):137–168, 2012. [131, 194] |
| [LRS95] | P. S. Landweber, D. C. Ravenel, and R. E. Stong. Periodic cohomology theories defined by elliptic curves. *Contemp. Math.*, 181:317–317, 1995. [4, 354] |
| [LS17] | G. Laures and B. Schuster. Towards a splitting of the $K(2)$-local string bordism spectrum. *ArXiv e-prints*, October 2017. [275, 353] |
| [LT65] | J. Lubin and J. Tate. Formal complex multiplication in local fields. *Ann. Math. (2)*, 81(2): 380–387, 1965. [202] |

[LT66] J. Lubin and J. Tate. Formal moduli for one-parameter formal Lie groups. *Bull. Soc. Math. France*, 94:49–59, 1966. [134, 136, 138]

[Lub67] J. Lubin. Finite subgroups and isogenies of one-parameter formal lie groups. *Ann. Math.*, 85(2):296–302, 1967. [316]

[Lura] J. Lurie. Chromatic homotopy. www.math.harvard.edu/~lurie/252x.html, accessed: September 22, 2016. [26, 73, 119, 124, 130, 138, 140, 157]

[Lurb] J. Lurie. Elliptic cohomology II: orientations. http://math.harvard.edu/~lurie/papers/Elliptic-II.pdf, accessed: February 4, 2018. [361]

[Lurc] J. Lurie. Higher algebra. www.math.harvard.edu/~lurie/papers/HA.pdf, accessed: September 22, 2016. [74, 77, 79, 103, 110, 167, 283]

[Lurd] J. Lurie. Spectral algebraic geometry. www.math.harvard.edu/~lurie/papers/SAG-rootfile.pdf, accessed: September 22, 2016. [80, 104]

[Lure] J. Lurie. A survey of elliptic cohomology. www.math.harvard.edu/~lurie/papers/survey.pdf, accessed: April 4, 2017. [263, 323, 375, 376]

[Lur09] J. Lurie. *Higher topos theory*. Princeton University Press, Princeton, NJ, 2009. [323]

[Mah79] M. Mahowald. Ring spectra which are Thom complexes. *Duke Math. J.*, 46(3):549–559, 1979. [13]

[Man63] Yu I Manin. The theory of commutative formal groups over fields of finite characteristic. *Russ. Math. Surv.*, 18(6):1, 1963. [325]

[Man01] M. A. Mandell. $E_\infty$ algebras and $p$-adic homotopy theory. *Topology*, 40(1):43–94, 2001. [308]

[Mar83] H. R. Margolis. *Spectra and the Steenrod algebra*. North-Holland Publishing Co., Amsterdam, 1983. [155]

[Mat89] H. Matsumura. *Commutative ring theory*. Cambridge University Press, Cambridge, second edition, 1989. Translated from the Japanese by M. Reid. [145]

[Mat16] A. Mathew. The Galois group of a stable homotopy theory. *Adv. Math.*, 291:403–541, 2016. [104]

[May74] J. P. May. Problems in infinite loop space theory. In *Notas de Matemfiticas y Simposio, Ntimero*, volume 1, pages 111–125, 1974. [373]

[May75] J. P. May. Classifying spaces and fibrations. *Mem. Amer. Math. Soc.*, 1(155):xiii–98, 1975. [11]

[May77]  J. P. May. $E_\infty$ ring spaces and $E_\infty$ ring spectra. Springer-Verlag, Berlin, 1977. [16, 284, 334]

[May96]  J. P. May. *Equivariant homotopy and cohomology theory*. American Mathematical Society, Providence, RI, 1996. [77, 79, 288]

[May99]  J. P. May. *A concise course in algebraic topology*. University of Chicago Press, Chicago, IL, 1999. [248]

[McC83]  J. E. McClure. Dyer–Lashof operations in $K$-theory. *Bull. Amer. Math. Soc. (N.S.)*, 8(1):67–72, 1983. [327]

[McT13]  C. McTague. *tmf* is not a ring spectrum quotient of *String* bordism. *arXiv preprint arXiv:1312.2440*, 2013. [355]

[Mes72]  W. Messing. *The crystals associated to Barsotti–Tate groups: with applications to abelian schemes*. Springer-Verlag, Berlin, 1972. [196]

[Mic03]  W. Michaelis. Coassociative coalgebras. In *Handbook of algebra*, volume 3, pages 587–788. North-Holland Publishing Co., Amsterdam, 2003. [221, 222]

[Mila]  H. Miller. Notes on Cobordism. www-math.mit.edu/~hrm/papers/cobordism.pdf, accessed: October 4, 2016. [127]

[Milb]  H. Miller. Sheaves, gradings, and the exact functor theorem. Unpublished. [102]

[Mil58]  J. Milnor. The Steenrod algebra and its dual. *Ann. Math. (2)*, 67:150–171, 1958. [26]

[Mil81]  H. R. Miller. On relations between Adams spectral sequences, with an application to the stable homotopy of a Moore space. *J. Pure Appl. Algebra*, 20(3):287–312, 1981. [110]

[Mil82]  H. R. Miller. Universal Bernoulli numbers and the $S^1$-transfer. In *Current trends in algebraic topology, Part 2 (London, Ont., 1981)*, pages 437–449. American Mathematical Society, Providence, RI, 1982. [348, 374]

[Mil08]  J. S. Milne. Abelian varieties. 2008. www.jmilne.org/math/CourseNotes/av.html, accessed July 20, 2018. [260, 261, 325]

[Mit83]  S. A. Mitchell. Power series methods in unoriented cobordism. In *Proceedings of the Northwestern Homotopy Theory Conference (Evanston, Ill., 1982)*, pages 247–254. American Mathematical Society, Providence, RI, 1983. [45]

[Mit05]  S. A. Mitchell. $K(1)$-local homotopy theory, Iwasawa theory, and algebraic $K$-theory. In *Handbook of K-theory*. Springer Science & Business Media, New York, 2005. [157, 366]

[MM65]  J. W. Milnor and J. C. Moore. On the structure of Hopf algebras. *Ann. Math. (2)*, 81:211–264, 1965. [38, 255]

[MM74] B. Mazur and W. Messing. *Universal extensions and one dimensional crystalline cohomology.* Springer-Verlag, Berlin, 1974. [200, 201, 369]

[MM15] A. Mathew and L. Meier. Affineness and chromatic homotopy theory. *J. Topol.*, 8(2):476–528, 2015. [362]

[MNN15] A. Mathew, N. Naumann, and J. Noel. On a nilpotence conjecture of j. p. may J. P. May. *J. Topol.*, 8(4):917–932, 2015. [156]

[Mor85] J. Morava. Noetherian localisations of categories of cobordism comodules. *Ann. Math. (2)*, 121(1):1–39, 1985. [201, 202, 364]

[Mor89] J. Morava. Forms of $K$-theory. *Math. Z.*, 201(3):401–428, 1989. [ix, 4, 198, 266, 267]

[Mor07a] J. Morava. Complex cobordism in algebraic topology. *arXiv*, 2007. [155]

[Mor07b] J. Morava. The motivic Thom isomorphism. In *Elliptic cohomology*, pages 265–285. Cambridge University Press, Cambridge, 2007. [270]

[Mor11] J. Morava. Abelian varieties and the Kervaire invariant. *arXiv preprint arXiv:1105.0539*, 2011. [164]

[Mor12] J. Morava. Local fields and extraordinary $K$-theory. *arXiv preprint arXiv:1207.4011*, 2012. [286]

[Mor16] J. Morava. Complex orientations for $THH$ of some perfectoid fields. arXiv, 2016. [270, 361]

[MR93] M. E. Mahowald and D. C. Ravenel. The root invariant in homotopy theory. *Topology*, 32(4):865–898, 1993. [358]

[MRS01] M. Mahowald, D. Ravenel, and P. Shick. The triple loop space approach to the telescope conjecture. In *Homotopy methods in algebraic topology (Boulder, CO, 1999)*, pages 217–284. American Mathematical Society, Providence, RI, 2001. [359]

[MRW77] H. R. Miller, D. C. Ravenel, and W. S. Wilson. Periodic phenomena in the Adams–Novikov spectral sequence. *Ann. Math. (2)*, 106(3):469–516, 1977. [108, 163, 164, 362]

[MS88] M. Mahowald and P. Shick. Root invariants and periodicity in stable homotopy theory. *Bull. Lond. Math. Soc.*, 20(3):262–266, 1988. [358]

[MS95] M. Mahowald and H. Sadofsky. $v_n$ telescopes and the Adams spectral sequence. *Duke Math. J.*, 78(1):101–129, 1995. [359]

[MSD74] B. Mazur and P. Swinnerton-Dyer. Arithmetic of Weil curves. *Invent. Math.*, 25(1):1–61, 1974. [345]

[MT68] R. E. Mosher and M. C. Tangora. *Cohomology operations and applications in homotopy theory*. Harper & Row, New York, 1968. [26, 32]

[MT92] M. Mahowald and R. D. Thompson. The $K$-theory localization of an unstable sphere. *Topology*, 31(1):133–141, 1992. [309]

[Mum66] D. Mumford. On the equations defining abelian varieties: I. *Invent. Math.*, 1:287–354, 1966. [261]

[Mum67a] D. Mumford. On the equations defining abelian varieties: II. *Invent. Math.*, 3:75–135, 1967. [261]

[Mum67b] D. Mumford. On the equations defining abelian varieties: III. *Invent. Math.*, 3:215–244, 1967. [261]

[Nau07] N. Naumann. The stack of formal groups in stable homotopy theory. *Adv. Math.*, 215(2):569–600, 2007. [107]

[Nav10] L. S. Nave. The Smith–Toda complex $V((p+1)/2)$ does not exist. *Ann. Math. (2)*, 171(1):491–509, 2010. [148]

[Nis73] G. Nishida. The nilpotency of elements of the stable homotopy groups of spheres. *J. Math. Soc. Japan*, 25:707–732, 1973. [148]

[Och87a] S. Ochanine. Modules de $SU$-bordisme: applications. *Bull. Soc. Math. Fr*, 115:257–289, 1987. [354]

[Och87b] S. Ochanine. Sur les genres multiplicatifs définis par des intégrales elliptiques. *Topology*, 26(2):143–151, 1987. [3]

[Och91] S. Ochanine. Elliptic genera, modular forms over $KO_*$, and the Brown–Kervaire invariant. *Math. Z.*, 206(1):277–291, 1991. [3]

[Pen82] D. J. Pengelley. The homotopy type of $MSU$. *Amer. J. Math.*, 104(5):1101–1123, 1982. [353]

[Pol03] A. Polishchuk. *Abelian varieties, theta functions and the Fourier transform*. Cambridge University Press, Cambridge, 2003. [261]

[Pri80] S. Priddy. A cellular construction of $BP$ and other irreducible spectra. *Math. Z.*, 173(1):29–34, 1980. [360]

[PS] E. Peterson and N. Stapleton. Isogenies of formal groups laws and power operations in the cohomology theories $E_n$, revisited. In preparation. [320]

[Qui69] D. Quillen. On the formal group laws of unoriented and complex cobordism theory. *Bull. Amer. Math. Soc.*, 75:1293–1298, 1969. [73]

[Qui71] D. Quillen. Elementary proofs of some results of cobordism theory using Steenrod operations. *Adv. Math.*, 7:29–56, 1971. [82, 83, 90, 91, 92, 94]

[Rav77] D. C. Ravenel. The cohomology of the Morava stabilizer algebras. *Math. Z.*, 152(3):287–297, 1977. [364]

# References

[Rav78a]  D. C. Ravenel. The non-existence of odd primary Arf invariant elements in stable homotopy. *Math. Proc. Camb. Phil. Soc.* 83:429–443, 1978. [164, 349, 350, 376]

[Rav78b]  D. C. Ravenel. A novice's guide to the Adams–Novikov spectral sequence. In *Geometric applications of homotopy theory (Proc. Conf., Evanston, Ill., 1977), II*, pages 404–475. Springer, Berlin, 1978. [41, 108]

[Rav84]  D. C. Ravenel. Localization with respect to certain periodic homology theories. *Amer. J. Math.*, 106(2):351–414, 1984. [104, 147, 148, 149, 151, 359]

[Rav86]  D. C. Ravenel. *Complex cobordism and stable homotopy groups of spheres*. Academic Press, Orlando, FL, 1986. [x, 32, 33, 36, 106, 109, 127, 163]

[Rav92]  D. C. Ravenel. *Nilpotence and periodicity in stable homotopy theory*. Princeton University Press, Princeton, NJ, 1992. [156, 158]

[Rav93]  D. C. Ravenel. Life after the telescope conjecture. In *Algebraic K-theory and algebraic topology (Lake Louise, AB, 1991)*, pages 205–222. Kluwer Academic Publishers, Dordrecht, 1993. [359]

[Reza]  C. Rezk. Elliptic cohomology and elliptic curves (Felix Klein lectures, Bonn 2015). www.math.uiuc.edu/~rezk/felix-klein-lectures-notes.pdf, accessed: October 7, 2016. [133, 269]

[Rezb]  C. Rezk. Notes on the Hopkins–Miller theorem. www.math.uiuc.edu/~rezk/hopkins-miller-thm.pdf, accessed: October 10, 2016. [134]

[Rezc]  C. Rezk. Supplementary notes for Math 512. www.math.uiuc.edu/~rezk/512-spr2001-notes.pdf, accessed: October 28, 2016. [161, 162, 329]

[Rez06]  C. Rezk. The units of a ring spectrum and a logarithmic cohomology operation. *J. Amer. Math. Soc.*, 19(4):969–1014, 2006. [339, 340, 348]

[Rez12]  C. Rezk. Rings of power operations for Morava E-theories are Koszul. *ArXiv e-prints*, April 2012. [313]

[Rud98]  Y. B. Rudyak. *On Thom spectra, orientability, and cobordism*. Springer-Verlag, Berlin, 1998. [12, 76, 89, 90]

[RW80]  D. C. Ravenel and W. S. Wilson. The Morava $K$-theories of Eilenberg–Mac Lane spaces and the Conner–Floyd conjecture. *Amer. J. Math.*, 102(4):691–748, 1980. [172, 174, 180, 211, 212, 213, 216, 218]

| | |
|---|---|
| [RW77] | D. C. Ravenel and W. S. Wilson. The Hopf ring for complex cobordism. *J. Pure Appl. Algebra*, 9(3):241–280, 1976/1977. [181, 183, 188, 189, 191, 192, 194] |
| [RWY98] | D. C. Ravenel, W. S. Wilson, and N. Yagita. Brown–Peterson cohomology from Morava $K$-theory. *K-Theory*, 15(2):147–199, 1998. [218] |
| [Sad93] | H. Sadofsky. Hopkins' and Mahowald's picture of Shimomura's $v_1$-Bockstein spectral sequence calculation. In *Algebraic topology (Oaxtepec, 1991)*, pages 407–418. American Mathematical Society, Providence, RI, 1993. [365] |
| [Sal16] | A. Salch. The cohomology of the height four Morava stabilizer group at large primes. *arXiv preprint arXiv:1607.01108*, 2016. [376] |
| [Sch70] | C. Schoeller. Étude de la catégorie des algèbres de Hopf commutatives connexes sur un corps. *Manuscripta Math.*, 3:133–155, 1970. [204, 205, 206] |
| [Seg70] | G. Segal. Cohomology of topological groups. In *Symposia Mathematica, Vol. IV (INDAM, Rome, 1968/69)*, pages 377–387. Academic Press, London, 1970. [172] |
| [Seg88] | G. Segal. Elliptic cohomology [after Landweberstong, Ochanine, Witten, and others]. *Astérisque*, 161–162:187–201, 1988. [377] |
| [Sen17] | A. Senger. The Brown–Peterson spectrum is not $E_{2(p^2+2)}$ at odd primes. *ArXiv e-prints*, October 2017. [373] |
| [Sha77] | P. Brian Shay. mod $p$ Wu formulas for the Steenrod algebra and the Dyer–Lashof algebra. *Proc. Amer. Math. Soc.*, 63(2):339–347, 1977. [249] |
| [Shi86] | K. Shimomura. On the Adams–Novikov spectral sequence and products of $\beta$-elements. *Hiroshima Math. J.*, 16(1):209–224, 1986. [366] |
| [Sil86] | J. H. Silverman. *The arithmetic of elliptic curves*. Springer-Verlag, New York, 1986. [325] |
| [Sil94] | J. H. Silverman. *Advanced topics in the arithmetic of elliptic curves*. Springer-Verlag, New York, 1994. [266] |
| [Sin68] | W. M. Singer. Connective fiberings over BU and U. *Topology*, 7:271–303, 1968. [4, 249] |
| [SN14] | J. Sprang and N. Naumann. $p$-adic interpolation and multiplicative orientations of KO and *tmf* (with an appendix by Niko Naumann). *ArXiv e-prints*, September 2014. [374] |

References 397

[SS15]   T. M. Schlank and N. Stapleton. A transchromatic proof of Strickland's theorem. *Adv. Math.*, 285:1415–1447, 2015. [310, 311, 312]

[ST04]   S. Stolz and P. Teichner. What is an elliptic object? *London Math. Soc. Lecture Note Ser.*, 308:247, 2004. [377]

[ST11]   S. Stolz and P. Teichner. Supersymmetric field theories and generalized cohomology. In *Mathematical foundations of quantum field theory and perturbative string theory*, pages 279–340. American Mathematical Society, Providence, RI, 2011. [377]

[Sta13]  N. Stapleton. Transchromatic generalized character maps. *Algebr. Geom. Topol.*, 13(1):171–203, 2013. [286, 359]

[Sta14]  The Stacks Project Authors. Stacks Project. http://stacks.math.columbia.edu, accessed July 20, 2018, 2014. [101]

[Sta15]  N. Stapleton. Transchromatic twisted character maps. *J. Homotopy Relat Struct*, 10(1):29–61, 2015. [359]

[Ste53a] N. E. Steenrod. Cyclic reduced powers of cohomology classes. *PNAS*, 39(3):217–223, 1953. [83]

[Ste53b] N. E. Steenrod. Homology groups of symmetric groups and reduced power operations. *PNAS*, 39(3):213–217, 1953. [83]

[Sto63]  R. E. Stong. Determination of $H^*(BO(k, \cdots, \infty), \mathbb{Z}_2)$ and $H^*(BU(k, \cdots, \infty), \mathbb{Z}_2)$. *Trans. Amer. Math. Soc.*, 107:526–544, 1963. [4, 249]

[Sto12]  V. Stojanoska. Duality for topological modular forms. *Doc. Math.*, 17:271–311, 2012. [331]

[Stra]   N. P. Strickland. Formal schemes for $K$-theory spaces. Unpublished. [270, 280]

[Strb]   N. P. Strickland. Functorial philosophy for formal phenomena. http://hopf.math.purdue.edu/Strickland/fpfp.pdf, accessed: January 9, 2014. [6, 23, 51, 65, 72, 160, 198, 202, 204, 275, 312, 359]

[Str92]  N. P. Strickland. On the $p$-adic interpolation of stable homotopy groups. In *Adams Memorial Symposium on Algebraic Topology, 2 (Manchester, 1990)*, pages 45–54. Cambridge University Press, Cambridge, 1992. [366]

[Str97]  N. P. Strickland. Finite subgroups of formal groups. *J. Pure Appl. Algebra*, 121(2):161–208, 1997. [133, 311, 312, 313, 315, 316]

[Str98]  N. P. Strickland. Morava $E$-theory of symmetric groups. *Topology*, 37(4):757–779, 1998. [311, 312, 313]

[Str99a] N. P. Strickland. Products on MU-modules. *Trans. Amer. Math. Soc.*, 351(7):2569–2606, 1999. [131, 147, 194]

[Str99b]  N. P. Strickland. Formal schemes and formal groups. In *Homotopy invariant algebraic structures (Baltimore, MD, 1998)*, pages 263–352. American Mathematical Society, Providence, RI, 1999. [x, 20, 23, 34, 52, 54, 56, 61, 65, 70, 72, 181, 187, 222, 223, 224, 225, 314, 359]

[Str00]  N. P. Strickland. Gross–Hopkins duality. *Topology*, 39(5):1021–1033, 2000. [159]

[Str06]  N. P. Strickland. Formal groups. https://neil-strickland.staff.shef.ac.uk/courses/formalgroups/fg.pdf, accessed July 20, 2018, 2006. [44, 45, 130, 132, 133]

[Str12]  O. Stroilova. *The generalized Tate construction*. PhD thesis, Massachusetts Institute of Technology, 2012. [358]

[SVW99]  R. Schwänzl, R. M. Vogt, and F. Waldhausen. Adjoining roots of unity to $E_\infty$ ring spectra in good cases: a remark. *Contemp. Math*, 239:245–249, 1999. [372]

[Swi02]  R. M. Switzer. *Algebraic topology: homotopy and homology*. Springer-Verlag, Berlin, 2002. Reprint of the 1975 original [Springer, New York]. [12, 42, 69, 71, 176, 353]

[SY94]  K. Shimomura and A. Yabe. On the chromatic $E_1$-term $H^*M_0^2$. In *Topology and representation theory (Evanston, IL, 1992)*, pages 217–228. American Mathematical Society, Providence, RI, 1994. [366]

[SY95]  K. Shimomura and A. Yabe. The homotopy groups $\pi_*(L_2 S^0)$. *Topology*, 34(2):261–289, 1995. [366]

[Tak08]  L. A. Takhtajan. *Quantum mechanics for mathematicians*. American Mathematical Society, Providence, RI, 2008. [376]

[Tat97]  J. Tate. The work of David Mumford. In *Fields medallists' lectures*, pages 219–223. World Scientific, Singapore, 1997. [261]

[Tho14]  R. Thompson. An unstable change of rings for Morava $E$-theory. *ArXiv e-prints*, September 2014. [313]

[Tor03]  T. Torii. On degeneration of one-dimensional formal group laws and applications to stable homotopy theory. *Amer. J. Math.*, 125(5):1037–1077, 2003. [364]

[Tor07]  T. Torii. Milnor operations and the generalized Chern character. In *Proceedings of the Nishida Fest (Kinosaki 2003)*, pages 383–421. Geometry and Topology Publishers, Coventry, 2007. [364]

[Tor10a]  T. Torii. Comparison of Morava $E$-theories. *Math. Z.*, 266(4):933–951, 2010. [364]

| | |
|---|---|
| [Tor10b] | T. Torii. HKR characters, $p$-divisible groups and the generalized Chern character. *Trans. Amer. Math. Soc.*, 362(11):6159–6181, 2010. [364] |
| [Tor11] | T. Torii. $K(n)$-localization of the $K(n+1)$-local $E_{n+1}$-Adams spectral sequences. *Pacific J. Math.*, 250(2):439–471, 2011. [364] |
| [TW80] | R. W. Thomason and W. S. Wilson. Hopf rings in the bar spectral sequence. *Quart. J. Math. Oxford Ser. (2)*, 31(124):507–511, 1980. [172] |
| [Vak15] | R. Vakil. *The rising sea: foundations of algebraic geometry*. 2015. Preprint. http://math.stanford.edu/~vakil/216blog, accessed January 15, 2013. [64] |
| [Vis05] | A. Vistoli. Grothendieck topologies, fibered categories and descent theory. In *Fundamental algebraic geometry*, pages 1–104. American Mathematical Society, Providence, RI, 2005. [102] |
| [Wal09] | B. J. Walker. Orientations and p-adic analysis. *ArXiv e-prints*, April 2009. [347] |
| [Wan15] | G. Wang. *Unstable chromatic homotopy theory*. PhD thesis, Massachusetts Institute of Technology, 2015. [362] |
| [Wei64] | A. Weil. Sur certains groupes d'opérateurs unitaires. *Acta Mathematica*, 111(1):143–211, 1964. [260] |
| [Wei74] | A. Weil. *Basic number theory*. Springer-Verlag, New York, third edition, 1974. [361] |
| [Wei11] | J. Weinstein. The geometry of Lubin–Tate spaces. http://math.bu.edu/people/jsweinst/FRGLecture.pdf, accessed July 20, 2018, 2011. [195] |
| [Wil73] | W. S. Wilson. The $\Omega$-spectrum for Brown–Peterson cohomology: I. *Comment. Math. Helv.*, 48:45–55, 1973; corrigendum, ibid. 194. [194] |
| [Wil75] | W. S. Wilson. The $\Omega$-spectrum for Brown–Peterson cohomology: II. *Amer. J. Math.*, 97:101–123, 1975. [194] |
| [Wil82] | W. S. Wilson. *Brown–Peterson homology: an introduction and sampler*. Conference Board of the Mathematical Sciences, Washington, DC, 1982. [129, 130, 164, 174, 175, 183, 189, 193] |
| [Wit87] | E. Witten. Elliptic genera and quantum field theory. *Comm. Math. Phys.*, 109(4):525–536, 1987. [3] |
| [Wit88] | E. Witten. The index of the Dirac operator in loop space. In *Elliptic curves and modular forms in algebraic topology (Princeton, NJ, 1986)*, pages 161–181. Springer, Berlin, 1988. [3] |
| [Yag80] | N. Yagita. On the Steenrod algebra of Morava $K$-theory. *J. London Math. Soc. (2)*, 22(3):423–438, 1980. [280] |

[Yu95] J.-Kang Yu. On the moduli of quasi-canonical liftings. *Compositio Math.*, 96(3):293–321, 1995. [202]

[Zhu17] Y. Zhu. Norm coherence for descent of level structures on formal deformations. *ArXiv e-prints*, June 2017. [320]

[Zin84] T. Zink. *Cartiertheorie kommutativer formaler Gruppen.* BSB B. G. Teubner Verlagsgesellschaft, Leipzig, 1984. [135, 201, 369]

[Zin02] T. Zink. The display of a formal $p$-divisible group. *Astérisque*, (278):127–248, 2002. [203]

# Index

abelian variety, 260, 325
Adams spectral sequence, 31, 104, 332
   unstable, 168, 171
adjoint map, 239
adèles, 158
algebraization, 330
André–Quillen cohomology, 327
arithmetic fracture, 330
Artin–Hasse exponential, 131, 250
bar spectral sequence, 172, 191, 211, 249, 273
Barsotti–Tate group, *see* $p$-divisible group
Bernoulli
  distribution, 346
  measure, 347
  number, 337, 348
  polynomial, 346
Bockstein, 74, 147, 159
  2-adic, 37
bordism
  complex, $MU$, 49
  geometric chains, 1
  unoriented, $MO$, 12
Bousfield localization, 155
Bousfield–Kan spectral sequence, 328
Bousfield–Kuhn functor, 339
Brown–Peterson homology, 145

Brown–Comenetz dual, 202
Brown–Gitler spectrum, 206, 211
canonical bundle, 66, 86, 226
Cartan formula, 24, 85
Cartier dual, 225
Čech complex, 101, 332
characteristic series, 254, 269, 336
  Hirzebruch series, 254
Chern class, 56, 69, 227, 229
  Chern polynomial, 69
  Conner–Floyd, 84
  total symmetric, 85
chromatic localization
  completeness, 158
  convergence, 158
  fracture, 325, 335
  harmonic, 158
chromatic spectral sequence, 164
coalgebra
  good basis, 222
Cohen ring, 135
commutativity hypothesis, **CH**, 104
comodule, 33, 101, 105, 140
  coalgebra, 171, 221
  cofree, 33
  cotensor product, 36
completion, 103
  unstable, 168

context, 104
  abelian $E_\infty$, 311
  natural $E_\infty$, 309
  unstable, 168, 313
coordinate, 16
  system, 54
cotangent space, 51
crystal, 196
  crystalline site, 201
cubical structure, 4, 262, 272, 322
  symmetric biextension, 262
  theorem of the cube, 261

de Rham complex, 195
deformation complex, 136
derived fixed points, 37
descent
  data, 100, 167
  effective, 99
  faithfully flat, 99, 102
  monadic, 167, 308
  object, 103
    unstable, 168
detects nilpotence, 148
Dieudonné module, 197, 200, 215, 226, 325
  algebra, 207
  contravariant functor, 197
  display, 203
  graded, 205
  tensor product, 207
discrepency spectrum, 338
division algebra, 160
divisor, 61, 69
  Cartier, 64
  convolution, 72, 315
  divisorial section, 64
  flag of subgroups, 313
  pullback, 314
  pushforward, 64

special, 233
subgroup, 311
$E_\infty$-ring spectrum, 283
Eisenstein series, 257, 348
elliptic curve, 5, 256
  generalized, 266
  moduli, 323
  ordinary, 325
  $p$-divisible group, 324
  supersingular, 324
  Tate curve, 266
  Weierstrass curve, 266, 324
elliptic spectrum, 4, 263, 322
  generalized, 267
equivariant spectrum, 74
  homotopy fixed points, 79
  homotopy orbits, 74
  representation sphere, 75

field spectrum, 150
flatness hypothesis, **FH**, 104, 169
  unstable, **UFH**, 169
flatness hypothesis, **FH**
  unstable generation, **UGH**, 177
formal group, 20, 57
  additive, $\widehat{\mathbb{G}}_a$, 20
  additive homomorphism, 28
  bud, 114
  Cartier dual, 255, 271
  coarse moduli, 133
  cohomology, 116, 137, 215, 234
  exponential, 120
  height, 121, 129
  homomorphism, 29, 177, 187
  Honda group, 156, 199
    law, 132
  $\ell$-series, 200, 212
    divided, 91
    law, 58
  multiplicative, $\widehat{\mathbb{G}}_m$, 21

quotient, 270, 315
⋆-isomorphism, 134
strict isomorphism, 37, 96, 120
universal constant extension, 270
formal plethories, 178
formal scheme, 20, 54
    affine $n$-space, 54
    closed subscheme, 61
    coalgebraic, 255
    formal neighborhood, 54, 133, 147
    formal variety, 54
    from a space, 21, 28, 56, 226
    smooth, 54
    tangent space, 249
Frobenius, 123, 197, 201, 255

grading, *see* $\mathbb{G}_m$ action
Greek letter family, 163
group scheme, 17
    additive group, $\mathbb{G}_a$, 18
    multiplicative group, $\mathbb{G}_m$, 18
        action, 23, 178

$H_\infty$-ring spectrum, 75, 284
$H_\infty^2$-structure on $MU$, 76
homology suspension, 89, 175, 187, 193, 208, 213
homothety, 123, 200
Hopf algebra, 17, 105, 217, 224
    alternating, 217
    bipolynomial, 193
    group-ring, 35
    tensor product, 207
Hopf algebroid, 105
Hopf invariant 1 elements, 39
Hopf ring, 170, 206
    algebraic, 183
    enriched, 171
    free, 175, 181
    Kronecker, 182
    mixed unstable cooperations, 180
    ring-ring, 180
    spectrum, SpH, 185
    ∗-indecomposables, 175
    topological, 180
Hurewicz homomorphism, 33, 106

Igusa tower, 329
inverse function theorem, 55
inverse symmetric function, 271
isogeny, 64, 264, 313, 314

$J$-homomorphism, 12, 49
Johnson–Wilson homology, 145

Kähler differentials, 53, 58
    cohomologically invariant, 195
    invariant, 58, 67
Kronecker pairing, 75, 182, 344
Kudo–Araki map, 127
Kümmer sequence, 344

Landweber flat, 207, 344
level structure, 310, 312, 326
    formal, 328
line bundle, 60
    tensor product, 72
logarithm, 58, 95, 256
Miščenko, 112
Lubin–Tate stack, 133, 201
    automorphisms, 139
    canonical lift, 160
        quasicanonical, 202
    stabilizer, 160, 199

Mayer–Vietoris sequence, 158
Mellin transform, 346
Milnor primitive, 147, 280
Milnor–Moore, 38, 192, 255, 275
modular form, 4, 258, 265
    modularity relation, 265
    $q$-expansion, 4, 265
    topological, *tmf*, 323
Moore spectrum, 153
Morava $E$-theory, 145, 211, 238, 326

Morava $K$-theory, 146, 211, 279
Morava stabilizer group, *see* Lubin–Tate stack
norm, 309, 315
  binomial formula, 78
  coherence, 285, 319
  equivariant, 78
numerical polynomials, 345
obstruction theory, 283
$\theta$-algebra, 327
operations
  Adams, 162
  Landweber–Novikov, 84
  mixed, 179, 211
  unstable, 175, 211
orientation, 56, 73
  $\widehat{A}$-genus, 270, 273
  Adams's condition, 43
  Atiyah–Bott–Shapiro, 273
  complex, 49, 226
  genus, 2
  $L$-genus, 270
  $MSU$, 232
  $MU[6, \infty)$, 254
  real, 13
  *String*, 272
  Witten genus, 3
$p$-adic interpolation, 347
$p$-divisible group, 214, 242
  alternating power, 218, 241, 252
  connected, 214
  elliptic curve, 324
  Tate, 270
$p$-typical, 188
  curve, 201
  logarithm, 122
parameter, 16
  system, 54
period map, 201

periodified spectrum, 24
Poincaré Lemma, 196
Poincaré series, 38, 253, 256, 277
Pontryagin–Thom construction, 2
Postnikov tower, 3, 8, 232
power operation, 74
projectivization, 67
push–pull, 78
Ravenel–Wilson relation, 184, 189
regularization, 347
restriction, 77
Rezk logarithm, 339
Riemann–Manin symmetry, 325
ring of functions, 51
scheme
  affine, 16, 51
  affine $n$-space, 17
  closed, 51
  coalgebraic, 52, 222
  Sch, 221
  nilpotent thickening, 56
  simplicial, 101
Segal condition, 101, 169
Serre duality, 331
Serre–Tate theorem, 322, 326
sheaf
  flat, 140
  ideal, 60, 72
  $\infty$-sheaf, 323
  quasicoherent, 34, 140
  simplicial, 101
  surjection, 63
$\sigma$-orientation, 4, 268
$\Sigma$-structure, 272, 322
spectrum of units, $gl_1$, 11, 333
spherical bundle, 11
splitting principle, 69, 227, 269
  for $BSU$, 230

stack
  associated, 106
  closed, 128
  cover, 111
  equivalence, 128
  invariant function, 128
  invariant ideal, 128
  open, 130
  representable, 110
  topological enrichment, 323
Steenrod algebra, 24
  odd primes, 74
  periodified, dual, 26
  unstable module, 210
Stiefel–Whitney class, 42
symmetric 2-cocycle, 114, 232
  lemma, 114, 250
symmetric function
  monomial, 84
  Newton's theorem, 62
tangent space, 52
tangent spectral sequence, 138, 249, 276
Tate construction, 79
Tate curve, 4
Tate $K$-theory, 267
Tate vanishing, 345
tautological bundle, see canonical bundle

$\theta$-function, 258
  Jacobi relation, 260
$\Theta$-structure, see cubical structure
thick subcategory, 151
  spectrum, 154
  $\otimes$-ideal, 154
Thom isomorphism, 12, 49, 226
  cohomological, 15
Thom sheaf, 66, 226
Thom spectrum, 12
  Thom diagonal, 228
transfer, see norm, equivariant
transpotence, 213
$v_d$-self-map, 148, 151, 163
Verschiebung, 123, 197, 201, 255
Weierstrass preparation, 314
Weierstrass presentation, 257
Weil pairing, 246, 262
Wilson space, 194
Witt vectors, 135
  Hopf algebra, 204
Wood cofiber sequence, 38
Wu formulas, 248
$X(n)$, 147
$\zeta$-function
  Euler factor, 346
  $p$-adic, 346